Food crops of the lowland tropics

Food crops of the lowland tropics

EDITED BY

C. L. A. LEAKEY

AND

J. B. WILLS

OXFORD UNIVERSITY PRESS

1977

Oxford University Press, Walton Street, Oxford OX2 6DP

OXFORD LONDON GLASGOW NEW YORK
TORONTO MELBOURNE WELLINGTON CAPE TOWN
IBADAN NAIROBI DAR ES SALAAM LUSAKA ADDIS ABABA
KUALA LUMPUR SINGAPORE HONG KONG TOKYO
DELHI BOMBAY CALCUTTA MADRAS KARACHI

ISBN 0 19 854517 7

© OXFORD UNIVERSITY PRESS 1977

PRINTED IN GREAT BRITAIN
AT THE UNIVERSITY PRESS, OXFORD
BY VIVIAN RIDLER
PRINTER TO THE UNIVERSITY

Foreword

D. RHIND, C.M.G., O.B.E., B.Sc., F.L.S., F.I.Biol.

Formerly Agricultural Research Adviser, Ministry of Overseas Development

---

This account of the diverse agricultural industry of West Africa and the approaches being made towards its progress contains a mass of information which it is hoped will be valuable to those who have the task of improving agricultural production. I know of no other book which treats the subject so fully.

There have already been large changes in land-use in West Africa, though it must be said that much of the farming industry is still at the slash-and-burn, swidden stage. Change to a settled system such as predominates in Asia is only likely to occur by 'evolution' rather than 'revolution', but much can be done to quicken the pace. Only with close attention to detail can changes be adopted, because large changes entail a number of concurrent alterations to the farming system—social, economic, land tenure, and all the multifarious factors which together affect the farmer. An easy improvement, such as a better cultivar of crop, may produce some benefit, but almost always it requires changes in crop husbandry to achieve its full potential.

Haswell has described 'Six Hands' of farming (Tropical Farming Economics, 1973). Africa conforms to only the second 'Hand' of the series, and so has a long way to go. But those who must assist the essential evolution should not feel frustrated if progress seems slow; the farmer has his problems which the sheltered research worker must learn to understand and with which he must have patience.

The most outstanding difference between agriculture in the tropics of Asia and Africa is that Asia uses draft animals nearly everywhere whereas Africa is often still in the 'hoe' stage, even in areas where trypanosomaiasis is no problem. It is now, at last, becoming generally recognized that a sudden leap from hoes to tractors often brings more problems than it solves, especially with high fuel costs and scarce servicing facilities. Rapidly rising costs of inputs for the farmer in Africa calls for a careful examination of what should be recommended to him, bearing in mind the low value of much of his produce. This book is concerned with all aspects of such examination and with the ways of formulating useful practical advice. Evolution there must and will be, but it can only be achieved if each advance is profitable to the farmer.

London, 1975

D. RHIND

Preface

This book was first suggested by Haldore Hanson of the Ford Foundation in September 1971, after the completion of twelve seminars sponsored jointly by the Ford Foundation and the Institut de Recherches Agronomiques Tropicales et des Cultures Vivrières and held at the then recently founded International Institute of Tropical Agriculture in Nigeria.

The intention then was, and still is, to make available to researchers, agricultural educators, and agricultural administrators a body of knowledge about food crops and food-crop research in tropical Africa that would serve as a library tool and desk reference for professionals in this field. The level of the book is intended to be such that it is clear enough to be read and understood by administrators, research workers, and agricultural extension personnel.

Since it is based to a large extent on subject-matter presented at the IITA Seminars, to which many distinguished scientists from developing countries in the lowland tropics around the world were invited, we have used the title *Food crops of the lowland tropics* and have included, where appropriate, comparative subject-matter from outside West Africa. However, anyone scanning the pages of this book will quickly realize that by far the greatest proportion of the subject-matter is concerned with West African food-crop agriculture.

There have inevitably been great differences in approach between the authors of different chapters. In the interests of the reader the editors have made some of these differences less noticeable. However, it was arranged that several chapters should be under the joint authorship of pairs of scientists working in institutes remote from each other and having very different backgrounds and approaches to the subject. In doing this the intention was to build a bridge across the language barrier separating agricultural development, philosophy, and practice in francophone and anglophone West Africa. In some cases the authors have been able to make opportunities to work closely together in preparing a joint chapter; in other cases the work has been the responsibility of a single author; and in yet others, the editors have reconciled, into a single chapter, two distinct, independently prepared texts on the same subject.

It had originally been hoped that the book could be prepared and published by the latter part of 1972, but even in 1971 it was recognized that there might be a hiatus while the contributors wrote their chapters. Authors have been given the opportunity of updating their chapters, and most of the material now included covers publications and information that became available up to about mid 1974. The book does not set out to be totally comprehensive. The balance of contents of each chapter inevitably reflects in some degree the importance attached by its author or authors to different aspects. The editors regret the omission of accounts of maize and of the soils in the lowland tropics, but these are the subjects of other recent or forthcoming books.†

The intention is that staff at the International Institute of Tropical Agriculture will be nominated to collect and collate new data and comments relevant to each chapter, so that subsequent editions may be prepared from time to time, bringing up to date a book which will provide, on a continuing basis, a common background and understanding to support the development of all West African food-crop agriculture. The book is also to be published in a French edition.

C. L. A. L.
J. B. W.

Cambridge
October 1975

---

† National Academy of Sciences (1972). *Soils of the Humid Tropics.* Publication 1948, National Academy of Sciences, Washington, D.C. 219 pp.

Acknowledgements

The editors would like to thank all those who have had a part in bringing this book into being. Haldore Hanson of the Ford Foundation was active in promoting the seminar series held at research centres in West Africa in order to provide opportunities to bring together existing knowledge from which he hoped a green revolution would emerge. Dr. G. Vallaeys on behalf of IRAT and Dr. H. R. Albrecht on behalf of IITA merit special thanks for jointly sponsoring the seminars with the Ford Foundation and subsequently for ensuring that the co-operation could be obtained of agricultural scientists familiar with the work and publications of both anglophone and francophone Africa. Many of the authors who have prepared chapters for this book are members of staff of those two organizations. Dr. Walter Rockwood, as Head of the Communications and Information Department at IITA, and his successor Dr. Ray Woodis have played important administrative roles in planning for publication.

The following have assisted in the translation of texts: Mme de Venoge, M. Carlander, and Mrs. Susan Jellis.

Permission is gratefully acknowledged for the following: FAO for allowing us to use the map on p. 131, which has been redrawn from *The grass cover of Africa* (1960) by J. M. Rattray; to Oxford University Press (Eastern Africa) for permission to base the map on p. 128 on a map in *The climate of Africa* by B. W. Thompson; to R. W. J. Keay and the Clarendon Press, Oxford for the map on p. 129 reproduced from *Vegetation map of Africa south of the Sahara*.

The editors gratefully acknowledge professional advice from the following: Dr. A. O. Abifarin, Dr. P. Ahn, Dr. D. J. Andrews, Dr. J. Cock, Dr. H. Doggett, Dr. R. G. Fennah, and staff of the Commonwealth Institute of Entomology, Mr. M. Grehan, Mr. G. Hill, Professor Margaret Keay, Dr. E. J. H. Khan, Mr. A. Johnston, and staff of the Commonwealth Mycological Institute, Dr. Robert Lornay, Dr. D. Rose Innes, Dr. Thurston Shaw, Dr. Merrick Posnansky, Dr. E. Tidbury, and Professor Orrin Webster.

Special thanks are due to Mrs. Susan Leakey, who has typed the edited copy of the book, except for two chapters which were prepared by Miss Janet Baker.

List of Contributors

R. O. ADEGBOYE
University of Ife, Nigeria

H. R. ALBRECHT
Recently Director, The International Institute of
Tropical Agriculture, Ibadan

R. H. BOOTH
The Tropical Products Institute, Ministry of
Overseas Development, London

C. DES BOUVRIE
Food and Agriculture Organization of the United
Nations

O. J. BURDEN
The Tropical Products Institute, Ministry of
Overseas Development, London

R. CHABROLIN
Institute for Tropical Agronomic Research and
Food Crops (IRAT)

H. R. CHHEDA
University of Ibadan, Nigeria

D. G. COURSEY
The Tropical Products Institute, Ministry of
Overseas Development, London

L. V. CROWDER
Cornell University, Ithaca, New York

M. DELASSUS
Institute for Tropical Agronomic Research and
Food Crops (IRAT)

C. ÉTASSE
Institute for Tropical Agronomic Research and
Food Crops (IRAT)

C. GAURY
Précy-sur-Vrin, France. (Recently of the Centre
for Experimental Studies on the Mechanization
of Tropical Agriculture (CEEMAT) France)

D. J. GREENLAND
The International Institute of Tropical Agriculture, Ibadan

D. W. HALL
Ministry of Overseas Development, London

J. C. MOOMAW
The International Institute of Tropical Agriculture, Ibadan

K. O. RACHIE
The International Institute of Tropical Agriculture (IITA), Ibadan

J. R. RYDZEWSKI
Southampton University, England

P. SILVESTRE
Institute for Tropical Agronomic Research and
Food Crops (IRAT)

H. D. TINDALL
The National College of Agricultural Engineering,
Silsoe, Bedfordshire

R. TOURTE
Institute for Tropical Agronomic Research and
Food Crops (IRAT)

K. J. TREHARNE
The International Institute of Tropical Agriculture, Ibadan

W. K. WHITNEY
American Cynamid Company, Princeton, U.S.A.
(Recently of the International Institute of Tropical
Agriculture, Ibadan)

G. VALLAEYS
Director, Institute for Tropical Agronomic Research and Food Crops (IRAT)

Contents

1. Research strategies

H. R. ALBRECHT AND G. VALLAEYS

1.1. Background

The head of the agricultural economics programme of the US Department of Agriculture, Dr. Paalberg, recently stated: 'Over the period 1954–1972 there was an upward trend in the *per capita* production of food throughout the world. Africa is the exception.' He went on to say that, while *per capita* food production rose about $1\frac{1}{2}$ per cent annually in the developed regions, the rise was less than $\frac{1}{2}$ per cent in the less-developed regions. The losing battle may be less a function of food production per adult engaged in agriculture than of population increase, since those of young age who comprise the expanding base of the population pyramid must be expected to be non-producers. In some of these countries population growth has been at a rate of about 3 per cent per annum. This illustrates the urgent need for a concentrated research effort with the food crops of West Africa.

Commercial crops such as cocoa, rubber, and palm oil, have all been given considerable attention by researchers over the years, mainly, perhaps, because these kinds of crops have been the principal source of foreign exchange for many sub-Saharan African nations.

A series of seminars during the period 1970–2 was sponsored jointly by the Ford Foundation (FF), the Institut de Recherches Agronomiques Tropicales et des Cultures Vivrières (IRAT), and the International Institute of Tropical Agriculture (IITA). The seminar programme was designed to survey the present status of agricultural research, especially with respect to food production, in order to determine where the gaps may be, and to establish more widely and effectively the professional and working relationships among agricultural scientists of the whole region. It was recognized that the rapid expansion of population, increasing urbanization, and rising consumer incomes in West Africa would inevitably result in increasing demands for food. And it was considered desirable to examine the role of agri-cultural research in improving the economic and social welfare of the farming public and, especially, in helping to make high-quality foodstuffs available in ample quantities to meet the expected future needs. People can forgo or postpone many wants that money can buy, but supplies of daily food are a basic necessity.

In the opinion of most thinking people, the principal job facing agricultural research workers in the tropics, therefore, should be the conquest of famine, or at least of hunger, and the improvement of levels of nutrition—for malnutrition is always present or incipient, sapping vitality and endangering health. Research workers must be joined in this task by the farmers of the tropics. The latter must be the ultimate producers of more, better, and cheaper food for both rural and urban people as well as the producers of the increasing quantities of agricultural raw materials needed in a developing indigenous industrial complex, both for use inside the country and for export.

1.2. The research strategies seminars of 1970–2

The word 'strategy' conjures up ideas of the art of war; and in this case we can say that we are planning the war on hunger. Such strategic planning inevitably involves deciding the geographical and chronological order of priorities for guiding combined efforts.

It has to be admitted that the deliberations of the first FF/IITA/IRAT (1970) seminar did not in fact succeed in defining a limited number of specific objectives for concentrated priority effort. One of the difficulties which faced participants in assessing priorities was the very wide range of food crops that form a characteristic feature of the region as a whole. Such variation in food crops is not found to the same extent in any other tropical regions, e.g. people in South and South-East Asia live mainly on rice, and levels of food production in Latin America can be enhanced considerably simply by improving the maize and wheat crops.

The 1970 seminar reached the conclusion that agricultural research in Africa must strive, at least in the immediate future, to cover a wide range of activities. No attempt was made at this first seminar to place crops according to the order of their importance, but participants endeavoured to define the main objectives of research and to assess how much work is required to develop effective research in the region and permit the results of this research to be applied. The deliberations at the seminar revealed the immensity of the work to be undertaken, and it was realized that this could be successful only with constant close collaboration between all the research workers involved in the advancement of West African agriculture.

1.3. The present situation of food production in West Africa

1.3.1. *Limitations to food production in the past*

In the recent past, West African countries, with the exception of Liberia, have been under foreign jurisdiction. Where the colonial administrations considered that the indigenous people were not short of food they endeavoured to develop crops for export: groundnuts and cotton in the savanna zones; coffee, cocoa, oil palm, and rubber in the forest zones. Although foodstuffs were generally available in sufficient amounts considerable social and economic difficulties were always encountered when attempts were made to raise production. The following characteristics provide a background to any proposed changes: populations are sparse (some states have less than 10 inhabitants per km², and the whole of the region has an average population density of about 20 inhabitants per km²); traditional agriculture is based on systems whose main feature was collective land tenure which inhibits any innovation by individuals; there are generally insufficient staff to carry out extension in the scattered rural population; agricultural production tends to depend on rainfall, which is unreliable from season to season. Hence many African populations are constantly threatened with famine.

Before trying to define a new strategy for agricultural research in West Africa, the present situation should be outlined briefly.

1.3.2. *Demography*

The total population in West Africa at present amounts to about 120 million, but it is increasing rapidly, e.g. 1·4 per cent per annum in Sierra Leone

and 3 per cent per annum in the Ivory Coast. The Food and Agriculture Organization (FAO) have estimated the mean population growth rate at 2·2 per cent.

The population density is low on average, but high densities occur in some countries such as Nigeria (75 inhabitants per km²) and pockets of high-density population occur in some countries which are apparently sparsely populated (Mauritania, 1 inhabitant per km²). For historical reasons, farms are often concentrated in a few areas, and in such areas there may be a local shortage of land.

The proportion of the population engaged in agriculture is always high, e.g. 91 per cent in Niger and Mali, 67 per cent in Nigeria, and 52 per cent in Dahomey. Town populations are now expanding much faster than the rural populations. Thus African agriculture, which in the past had to meet only the domestic needs of those in the areas surrounding the farms, will, in future, have to provide increasingly greater surpluses for transport to urban areas, in order to feed the growing proportion of the population who live in towns.

It is clearly important to try to establish how African agriculture is facing up to the twofold challenge of a larger total population and a decreasing proportion of people engaged in agriculture.

1.3.3. *The main crops in West Africa*

The main cereals grown in West Africa are sorghum and bulrush or pearl millet (*Pennisetum typhoideum*) (6·4 million t†), maize (2·4 million t), and rice (1·9 million t). Sorghum and bulrush millet are the most widespread cereals, and these are adapted in the Sudanian zone. These generally have quite high *per capita* production per annum, e.g. Niger, 235 kg; Mali, 230 kg; Senegal, 225 kg. The populations in the humid regions are, however, relatively short of grain (Nigeria, 67 kg; Ghana, 68 kg; Ivory Coast, 120 kg).

Root and tuber crops such as yams, sweet potatoes, and cassava play an important part in the West African diet, especially in the humid zone. In comparison with a total cereal production of approximately 11 million t, root and tuber production exceeds 30 million t fresh weight. Allowing for the adjustment from wet to dry weights this implies that roots and tubers supply approximately the same quantity of food calories as cereals within the whole region. Roots and tubers available *per caput* per

† A metric ton is now called a tonne, symbol t.

annum vary from country to country in more-or-less inverse proportion to the amounts of cereals. They are high in the southern part of West Africa (e.g. Dahomey, 494 kg; Ivory Coast, 417 kg; Ghana, 363 kg; Nigeria, 306 kg) and often low in the north (e.g. Senegal, 29 kg; Upper Volta, 8 kg; Mauritania, 3 kg; Gambia, 2 kg).

Grain legumes represent a relatively modest part of agricultural output: 1·15 million t, including 975 000 t of cowpea (*Vigna unguiculata*). The amounts available *per caput* vary considerably from country to country, e.g. Niger, 37 kg; Upper Volta, 31 kg; Nigeria, 13 kg; Ivory Coast, 2·2 kg; Ghana, 1 kg.

1.3.4. Food supply in terms of energy and protein

In terms of energy value, the food supply appears on average to be just adequate in West Africa as a whole at 22 000 cal *per caput* per day.† This is, however, 1980 cal *per caput* per day in Dahomey but 2430 cal *per caput* per day in the Ivory Coast.

The protein supply also seems adequate in the region as a whole, amounting to about 60 g per day and ranging from 43 g in Ghana to 77 g in Niger. There is mostly a shortage of animal protein, and thus of certain essential amino acids under-represented in plant-protein sources, but animal protein is relatively plentiful in Mauritania (37 g per day) and Senegal (21 g per day).

1.3.5. Food imports

There have been increasingly large food imports in West Africa. Whereas in 1965 the region imported approximately 500 000 t rice (in paddy equivalent) and 300 000 t wheat and flour (in wheat equivalent), by 1970 these figures had risen to 610 000 t and 570 000 t respectively.

1.3.6. Yields of food crops

Yields per unit area or *per caput* have hardly increased and are very low in comparison with those known from research to be possible.

For maize, the yields vary, according to country, from 11·1 q/ha (Nigeria) to 4·5 q/ha (Togo); for bulrush millet and sorghum from 16·9 q/ha (Sierra Leone) to 4·3 q/ha (Togo); and for rice from 25 q/ha (Niger) to 8·8 q/ha (Mali).‡ Root crops generally yield only from 3 t/ha to 16 t/ha.

† *Per caput* = per head.
‡ q = quintal = 100 kg. This measure is approximately equal to a hundredweight in Imperial measure.

1.4. Previous food-crop research

There has for some time been a need for an abundant and diversified research effort concerned with food crops in West Africa. Even though modern technology does not in itself guarantee increased yields, total production of the region could be improved both in respect of quality and quantity. All the governments of West Africa have inaugurated well-organized food-crop research programmes, and some of them, such as those at the Centre de Bambey in Senegal, Rokupr in Sierra Leone, Bouaké in the Ivory Coast, and Moor Plantation in Nigeria were well established long before the countries in which they are located gained their independence. Nevertheless, serious gaps do exist in the agricultural research effort in West Africa and a need for the following may be identified:

(1) a freer and more abundant interchange of research information and materials;
(2) an increased study of successful research carried out elsewhere, which may usefully be applied locally;
(3) a greater emphasis on an interdisciplinary approach to problem-solving.

1.4.1. Past neglect of the economic and social aspects of food production

Some of the disciplines needed for ensuring agricultural development are often not included in research programmes in individual countries or, if present, are seriously under-manned. For example, it has become especially apparent that there is a low level of involvement of the economic and social sciences, not only are these disciplines ignored in such studies as marketing distribution and farm management but they are also neglected when studying the methodology for transferring new technology to the farms.

1.4.2. A wide view of agriculture

Attempts to bring about the improvement of agriculture in general in West Africa must take account of the predominance of smallholdings, shortages of capital, complicated land-tenure systems, minimal opportunities for the adoption of improved tools and practices, inadequate distribution and marketing systems, and penalizing environmental influences (heat, drought, pests, diseases, and the erosion, physical deterioration, and excessive leaching of soils). Detailed consideration was given to these wider aspects of agriculture in the 1970-2 FF/IRAT/ IITA seminars, and the topics will be discussed in various chapters of this book.

1.5. Conclusions drawn from the 1970-2 seminars

Until now, African farmers have practised unintensive farming on ever-increasing areas. This was an easy and obvious way to maintain *per capita* food production when an unlimited area of cultivable land was available. However, as is generally recognized, it has been possible to maintain the productivity of African soils at a constant and low level only through periodical use of long fallows. Extending the areas under crops as new land begins to become scarce leads to a reduction in the length of the fallow, which then only partially restores soil fertility. To compensate for reduced productivity per unit area of cropped land even greater areas are then cultivated, and the fallows made progressively shorter. Some areas in West Africa are already familiar with this vicious circle which, if not corrected, becomes a downward spiral.

To prevent this loss of the capital asset of the land, priority must be given, in agricultural research work, to techniques for maintaining or raising soil fertility as economically as possible. This will allow safe transformation from shifting to settled agriculture.

1.6. The role of crop improvement

Plant breeding and selection for developing high-yielding and adapted cultivars involves a great deal of work, for which all the necessary resources should be provided without delay. It is now quite generally accepted that the use of high-yielding cultivars is a prerequisite condition for having any other improved or intensifying production techniques applied by the farmer. Only in this way for example can the costs incurred for fertilizer and plant protection be justified.

1.7. Financial, social, and political aspects of schemes to increase food production

Because of the consequent advantages (which have too frequently not been realized), research is sorely needed on the costs, feasibility, and social and political implications of mechanization, large-scale land clearance, and irrigation and drainage schemes.

1.7.1. Farming-systems research as an alternative to single-crop studies

Research on farming (production) systems, rather than research on single crops or single aspects of

single crops is considered to be of very high strategic priority. To obtain a basis for recommending either 'improvement' or 'transformation' approaches we must study in detail a gamut of systems of increasing intensity from shifting cultivation through continuous cropping to multiple cropping. All specialists have a place in the farming-systems research; they can all provide a contribution to an essentially multidisciplinary exercise.

1.7.2. Overcoming the language barrier

In a region historically divided by language barriers, stress must be placed on the advantages to be gained from co-operative research efforts, especially within ecological zones, regardless of national and language boundaries. Some commendable examples of proven, workable co-operation in the region, apart from the FF/IRAT/IITA 1970-2 seminar series (of which this book is one result), are the joint research efforts developed between IRAT and IITA and with national agencies. The co-ordinating role being played in the region by the West African Rice Development Association (see p. 23) provides another good example.

1.8. Other requirements for food development

Other strategic requirements which have become clear involve the need to increase the interaction between agricultural and nutritional policy and the need to identify, evaluate, and maintain the genetic germplasm† of many food crops of importance in West Africa, and the need for seed multiplication and distribution schemes. So far, there has been far too little research which has been designed to improve the quality of food crops in Africa from the nutritional standpoint. Malnutrition due to food imbalance is much more common than hunger. The development of quick methods for the screening of plant materials for total protein and essential-amino-acid content is required. Such methods would be especially useful to plant breeders, who have at present to investigate thousands of lines to satisfy themselves that the new cultivars of food crops they create are superior to the

† The term 'germplasm' is widely used and understood as the sum of gene assemblies carried by the widest possible range of actual genotypes. Gene assemblies of high adaptive value are likely to occur in primitive landraces of crops (according to the views of Vavilov). The conservation and study of collections of wild and primitive material of crop plants has become an important aspect of modern crop study. For a fuller account of such work see Frankel, O. H. and Bennett, E. (1970). *Genetic resources in plants*, Blackwell, Oxford and Edinburgh.

old not only in their capacity to yield but in quality as well. The wide ranges in protein quantity and quality within many crop species has added a new dimension to plant-breeding research; undoubtedly cultural and storage practices also influence quality. Few African research centres are yet equipped to provide adequate biochemical services in support of food and nutrition programmes.

There are few facilities in West Africa for the maintenance of germplasm of tropical food crops. The establishment of such facilities on a regional or sub-regional basis is needed to expedite the efforts of plant breeders. Coupled with this is the need to arrange for safe but more expeditious passage from nation to nation of breeding stocks and cultivars, so that nowhere is anyone—research workers, farmers, processors, or consumers—denied the use and benefits that improved food crops can offer.

In Africa there is almost a complete absence, both in the public and the private sectors, of facilities and procedures for the multiplication and distribution of seed (or other planting materials) of new cultivars of crop plants.

1.9. Green Revolution or Green Evolution?

In opening the first seminar of the 1970–2 Research Strategies series, Haldore Hanson challenged agricultural scientists in Africa to match the Green Revolution of Asia, and expressed optimism that this could be achieved. Without questioning in detail the reasons for his optimism, we may wonder whether the whole of West Africa is ready for a Green Revolution in the near future. The conditions required for a swift, radical change may not occur in large parts of the region, either because there are no possibilities for irrigation, because of the great distances involved and associated transport difficulties, or because the existing social structures prove far more conservative than anticipated.

This suggests that two different lines of research should be followed simultaneously: selecting highly productive cultivars to be used in areas where the conditions for change are relatively favourable; and, for the areas where a break with the past is hardly possible in view of the existing obstacles, gradually improving less sophisticated techniques while using, if necessary, hardier cultivars of lower potential. We thus see two possible broad conceptions for agricultural development strategy:

(1) a gradual and carefully planned and phased evolution;

(2) a complete package of new techniques in the manner of the Green Revolution in Asia.

The Green Revolution in Asia can indeed be cited as a good, modern example of the effectiveness of co-operation in problem-solving in research, both among nations and between specialist disciplines. Too often in Africa, agricultural research has been unilateral rather than interdisciplinary in its nature, and often conducted at isolated locations operating independently of one another. Isolated research, even if technically excellent, for the most part remains isolated in that its application seldom becomes widespread.

It is recognized that in Asia the Green Revolution has succeeded generally in increasing production, but it has not always succeeded in improving the lot of the farmers, especially the small-scale farmers. Short-comings of the high-yielding wheat and rice cultivars themselves, lack of capital resources to promote optimum growing conditions over wide areas, and the fear that increased production would soon exceed demand and thereby cause prices to farmers to fall have all combined to lead those sceptical of the movement to anticipate far-reaching and usually unfavourable social, political, and economic consequences arising from this major transformation of agriculture. Nevertheless, Dr. Guy Hunter has pointed out that 'it seems unlikely—and quite probably undesirable in terms of human welfare—that violent, apocalyptic change will be precipitated by the Green Revolution . . . before the scientists brought the new seeds to Asia, there was not even a winning chance'.

Admitting then that there are lessons and warnings to be learned from the Asian experience, perhaps different principles should be devised upon which to base food-crops research in Africa. Owing to different cultural, environmental, political, and economic circumstances in Africa, the second-generation problems created by revolutionary changes can be expected to be quite different from those which have already arisen or have been predicted in Asia. It is manifest already that a much more intensive involvement of social scientists in the total research effort must be arranged.

This book, recording as it does the subject matter discussed in a seminar series intended to prepare for an attempt to launch a green revolution in Africa, may equally well provide a background for agricultural evolutionists. Perhaps some areas of the continent will evolve while others revolve, but areas where neither rapid evolution nor revolution occur will not stand still but regress.

2. Rice in West Africa

R. CHABROLIN

2.1. Introduction

2.1.1. *The economic importance of rice in West Africa*

Rice growing in Africa, and especially West Africa, was long considered to be of little economic importance. The crop was produced almost entirely by subsistence farming on small acreages, and exports were virtually nil. It is difficult to give exact production figures as the sources are not consistent. Table 2.1, however, shows production trends as given in the *FAO Yearbook* for the whole of Africa as well as the West African countries separately (Food and Agriculture Organization 1972). The increase in production has generally been fast and appears to be due more to an increase in the acreage cultivated than to the improvement of unit yields. Imports (according to the FAO) vary irregularly but show a tendency to increase (Table 2.2).

The rapid increase in rice consumption in West Africa may be attributed to three factors:

(1) population growth at the rate of about 2·5 per cent per annum;

(2) increasing urbanization: town dwellers tend to substitute imported foodstuffs, especially rice, for traditional foods (perhaps because of the inadequate systems for the marketing of local products or a certain snobbery value, but mainly because of the ease with which rice may be prepared);

(3) the increase in the standard of living of consumers, which has the same effect as noted in (2).

As a result, in some of the countries of West Africa rice imports are rising, which constitutes a heavy burden on the countries' foreign trade balances. According to USAID (1968) expenditure on rice imports amounted to almost US $60 million a year for the region as a whole between 1964 and 1967, while some countries such as Senegal spent up to 16 per cent of their available foreign currency on rice.

During the 1960s increased consumption of rice became, moreover, widespread in all developing countries in tropical regions. This resulted in an increase in the world price of rice, which in turn aggravated the situation, as also did the world drought during the late 1960s. In the early 1970s, however, the world price of rice began to fall again. This was certainly attributable to the effects of the Green Revolution in Asia. The general price index for rice was 88 in 1960, rose to 140 in 1967, and then dropped back to 106 in 1970. However, after the poor harvests in 1972 (Willis 1973), this trend was reversed again and prices are rising sharply, having increased in Bangkok from US $97 per t in July 1972 to US $166 per t in December for no. 1 milled rice with 10 per cent broken grain.

Estimates, such as those prepared by the World Indicative Plan, suggest that the consumption of rice within West Africa will exceed 5 million t by 1985. This represents a very fast rate of increase in the amount consumed. Production in most countries in this region is already insufficient to meet demand, and imports, about 250 000 t in 1961, rose to over 300 000 t in 1970. In future years, therefore, either imports will increase rapidly, and will give rise to serious balance of payments difficulties, or there will have to be a rapid increase in rice production. If the production of rice in West Africa continues to increase only at its present rate it will never catch up with consumption.

2.1.2. *The development of research on rice*

Agricultural research on rice has been in progress for almost 40 years, but yields in recently developed large-scale rice-growing operations demonstrate the very great possibilities of much higher productivity. However, average yields are still generally less than 2 t/ha. Great effort will have to be made in all areas (organization of production, marketing, prices, supplies to producers, dissemination of technical knowledge, etc.) to obtain results of the required magnitude.

TABLE 2.1

Production statistics for rice in West African countries 1948–71 (from FAO 1972)

	Dahomey	Gambia	Ghana	Guinea	Ivory Coast	Liberia	Mali	Niger	Nigeria	Senegal	Sierra Leone	Togo	Upper Volta	Total for African continent
1948–52														
Area†	2	11	20	340	202	261	182	4	171	57	316	11	12	2534
Production‡	—	20	23	208	104	178	148	3	250	52	274	7	11	2630
Yield§	1·9	18·8	11·5	6·1	5·1	6·8	8·2	6·6	14·6	9·0	8·7	6·8	8·7	10·4
1961–5														
Area	2	26	34	258	249	214	178	9	220	76	273	23	47	2824
Production	1	33	35	318	220	161	173	11	348	100	336	19	34	3878
Yield	4·5	12·8	10·3	12·3	8·9	7·5	9·7	11·9	15·8	13·1	12·3	8·1	7·1	13·7
1967														
Area	2	18	44	250	300	190	192	12	224	101	350	28	35	3088
Production	2	20	52	330	340	152	172	33	332	138	468	28	44	4486
Yield	12·2	11·3	12·0	13·2	11·3	8	8·9	28·3	14·8	13·6	13·4	10·0	12·3	14·5
1968														
Area	2	39	46	250	299	190	170	15	190	64	320	29	46	3045
Production	2	40	65	346	365	152	94	39	251	58	426	17	40	4408
Yield	7·7	13·7	14·3	13·8	12·2	8	5·6	25·5	13·2	9·1	13·3	6·0	8·7	14·5
1969														
Area	2	35	49	267	288	190	172	15	334	106	310	26	40	3248
Production	5	66	61	368	303	153	119	39	563	163	407	21	34	4817
Tield	23·0	18·9	12·4	13·8	10·5	8·1	6·9	25·2	16·9	15·4	13·1	8·1	8·4	14·8
1970														
Area	2	28	55	300	289	190	166	16	300	91	315	26	40	3296
Production	6	50	69	400	316	153	138	37	550	98	425	22	34	4912
Yield	26·7	17·9	12·6	13·3	10·9	8·1	8·3	22·6	18·3	10·7	13·5	8·5	8·5	14·9
1971														
Area	3	32	55	300	290	190	171	16	300	120	320	26	41	3364
Production	10	60	70	400	310	153	150	40	550	170	450	22	37	5082
Yield	33·3	18·8	12·7	13·3	10·7	8·1	8·8	25·0	18·3	14·2	14·1	8·5	9·0	15·1

† Area (1000s ha).   ‡ Production (1000s t).   § Yield (q/ha).

World, and particularly African, opinion is more and more aware of the need for increased productivity in rice growing areas, as is shown by the establishing, in September 1969, of the West African Rice Development Association (WARDA) (see p. 23), which has recently been intensively active. The role that agricultural research can and must play in making the development of higher productivity possible is obviously important. The authorities know this and, immediately after the Second World War, rice research institutes were established by the colonial powers: these were the Centre de Recherches Rizicoles du Koba in Guinea, whose activities extend to all the countries of former French West Africa; and the Rice Research Stations of Rokupr (Sierra Leone) and of Badeggi (Nigeria), for the Commonwealth countries. After the countries concerned became independent, progress in these establishments varied. In francophone West Africa, IRAT, established in 1960, devotes a large part of its activities to rice research. This institute has gradually been made responsible for the management of most

TABLE 2.2

Rice imports into West African countries (in 1000s t)
(from USAID 1968)

	Mean 1960–4	1966	1967	1968	1969	1970
Dahomey	4·6	6·0	7·0	4·0	7·6	—
Gambia	9·1	7·7	8·6	10·0	12·5	14·2
Ghana	42·7	48·3	40·1	30·6	28·1	53·1
Guinea	13·0	16·7	18·7	—	15·4	11·7
Ivory Coast	39·2	83·2	24·1	47·2	55·6	78·8
Liberia	30·0	46·3	34·4	45·5	27·8	49·0
Niger	1·4	1·0	1·4	0·4	0·1	—
Nigeria	1·5	1·3	1·5	0·3	0·7	1·8
Senegal	119·1	159·3	153·4	185·2	145·9	119·2
Sierra Leone	22·0	35·1	24·2	17·2	12·6	86·9
Togo	3·0	3·7	2·7	1·3	2·6	3·1
Upper Volta	2·0	4·1	3·8	1·3	1·5	—
Total	287·6	412·7	319·9	343·0	310·4	417·8

of the existing stations and has also set up some new ones. Thus it has a well-organized network covering all conditions likely to be encountered in tropical rice growing.

Rice research in West Africa has borne fruit, and its results are now being disseminated successfully. The rice productivity operation by SODERIZ in the Ivory Coast, by Opération Riz in Mali, and by SATEC in Senegal are bringing about significant increases in production in the areas concerned, by popularizing simple progressive ideas. Advisory services, which have been gradually developed in the region, have as their aim to pass on to producers, on a large scale, the progress made possible by the work of the agronomists. These have succeeded in finding ways of eliminating or attenuating most of the factors that limited production under the previously unsatisfactory prevailing conditions, whether production was ancient and traditional or recently introduced into an incompletely adapted environment. Adequate cultivation practices have been developed for tilling the soil, row-seeding at the right time of year and at the correct density, managing nurseries, transplanting, carrying out early and regular weeding, harvesting, and drying.

Improved cultivars have been selected or bred. Their potential yield is about 10 t/ha in paddy fields and 6 t/ha for upland rice. Such yields are very much higher than those of traditional cultivars or land races. The importance of correcting mineral deficiencies in the soil has been demonstrated. Appropriate fertilizer application often enables production per unit area to be more than doubled.

The work of the advisory services is obviously made very difficult because rice producers are scattered over a wide area and the farmers' level of education is generally very low. Nevertheless, considerable progress has been made since the Second World War, helped by numerous development operations, based on irrigation schemes, and through local training operations carried out by rice-growing teams from Taiwan (Wei Lang 1970). These training schemes have made a great contribution to the dissemination, in Africa, of the advanced cultivation techniques used in Asia.

2.1.3. *The history of rice cultivation in Africa*

Rice has been grown in Africa since ancient times. Portères (1956) traces back the cultivation of *Oryza glaberrima* in the central Nigerian delta to 1500 B.C. This species, which is associated with the megalithic cultures of Sine Saloum, was found there from the level corresponding to 800 B.C. Portères considers the introduction of *O. sativa*, of Asiatic origin, to be much more recent (thirteenth century), and the species only to have attained importance during the eighteenth century, owing mainly to the increasing level of trade using the western-coast ports.

2.1.4. *Systems of rice cultivation in Africa*

The extreme variation and adaptational flexibility of rice as a cultivated crop has enabled it to become established in a wide range of ecological and ethnic situations. The history of the development of its present-day culture from that in the 1930s is outlined later. This variation has in turn led to efforts by agronomists to develop a classification of the various types of rice and rice growing and to establish a corresponding terminology. Agreement has now been reached on the following bases of classification (Abifarin, Chabrolin, Jacquot, Marie, and Moomaw 1972; Chabrolin 1971; IRAT 1970*a*). The first criterion is the presence or absence of a free water surface over the soil of the rice field during cultivation. In the absence of free water the term used is 'upland rice growing'. More strictly, this implies that there is no soil water table high enough to provide the plant with its water requirements without rainfall. When there is a water table of this type the systems are described as 'paddy rice' (*rizculture de nappe*), 'swamp rice' (*de boue*), or 'lowland rice'. The latter term may also be used when the ground is only temporarily submerged.

Lowland rice can be subdivided into categories. When there is a complete and persistent flooding of water, i.e. a water table above soil level, the general term used is swamp rice. If the water depth is between 0·5 m and 1·0 m we speak of deep-water rice, and the term floating rice is used when the water is more than about 1 m deep. The term irrigated rice growing should only be used if the grower has control of the water, or in other words, if he can, at will, modify the water level at any time either by irrigation or drainage. This type of rice growing allows the highest yields to be attained, and this is the aim, to a more-or-less developed degree, of all existing irrigation development projects.

There are many types of swamp-rice cultivation because of the different topography of the rice fields and of water-supply regimes (major river basins, floodable basins, etc.). Mangrove rice growing is a special system related to the existence of salt in the soil. Saline soil may be cultivated only with an excess of water, which enables it to be brought temporarily to a salt content low enough to permit the rice to grow. Flood-water rice growing refers to a system in which rice is transplanted into pools left as flood water recedes or the water level drops by evaporation. Controlled submersion (see also Chapter 9, p. 171) implies a system in which bunds, equipped with valves, can delay the inflow or outflow of water from a paddy from an external source of water.

2.2. Upland rice

2.2.1. *Ecology* (Abifarin *et al.* 1972)

It is possible to practise upland (rain-fed) rice growing only in the wettest part of West Africa. Upland rice however accounts for almost 75 per cent of the West African acreage (Food and Agriculture Organization 1972) and is practised largely in areas receiving more than 1000 mm per annum rainfall.

Yields are generally low (600–800 kg/ha), but much higher yields have been obtained experimentally, e.g. 2·6 t/ha in Ghana (Aryeetey and Khan 1970), 2·5 t/ha in Sierra Leone (Will 1970), 5·4 t/ha in Nigeria (Moomaw 1970), 4·4 t/ha in Senegal (Birie-Habas, Aubin, Sene, and Poisson 1970), and recently, more than 7 t/ha in Cameroun (IRAT 1970*a*).

The rainfall distribution in West Africa follows a very marked seasonal cycle. There is only one wet season which occurs during the northern-hemisphere summer. It is very brief in the north and becomes longer and longer towards the south. Further south still the rain is interrupted at the end of summer for a short dry season and each of the two wet seasons there is of too short duration to enable most cultivars of upland rice to be grown successfully. Consequently, cultivars of shorter maturity period are required in the north of the region. As the days are longer in the summer as we move north, the flowering of photo-sensitive rice cultivars with a short-day type of response tends to be delayed. It is therefore essential for cultivars to have as short a basic vegetative period† as possible and to be as insensitive as possible to the inhibition of flowering by long hours of daylight. For adaptation to anything but a very narrow latitudinal band this is best achieved by selecting day-neutral cultivars.

The rainfall distribution in West Africa is extremely variable from year to year, especially in the timing of the start of the wet season. Selection of the sowing date by the calendar is therefore a hazardous matter and probably quite unrealistic. Generally speaking it is better to sow as soon as one can do so with the expectation of good germination.

Water requirements for rice represent only about half of the potential evapotranspiration at the beginning of growth and after grain filling (Franquin 1973) but its full amount during the remainder of the plant's life. It is therefore possible (Franquin 1973), by calculating the running hydrological balance between potential evapotranspiration and rainfall, to determine the effective length of the useful rainy season at any site. A study of this kind was made by Cochemé (UNDP 1971), who provided a map showing the annual rainfall, its variation, and the expected length of the effective wet season. As a rule of thumb, it is generally accepted that upland rice requires a rainfall of about 50–60 mm over each running ten-day period during a 3–4 months growing season.

The best soils for this crop are those with good properties of moisture retention. Fineness of texture is therefore an advantage, as is a high base-exchange capacity.

2.2.2. *Cultivation practices*

Primitive traditional techniques for growing upland rice are widespread throughout the region. The land is normally incompletely cleared, often by burning,

† (or juvenile period)—the length of time between planting and the earliest time that flowering can occur under the most favourable photoperiods. Ed.

without grubbing up the roots, the soil is lightly scraped with the hoe, and seeds are broadcast at a seedrate of about 30-40 kg/ha. Other crops are often interplanted or intersown with rice (yams, cassava, maize, etc.). Generally no manure is used, and weeding, if done at all, is carried out too late to be effective. The land is abandoned after 2-3 years of cultivation and then left fallow to revert to bush. This almost always results in uncontrolled and abundant sources of weed seeds from land near to that being cultivated.

To obtain high yields, these techniques must be changed, no matter what cultivars are used. Good ploughing is essential to allow the plant to root sufficiently quickly and deeply to survive periods of drought that may occur at the beginning of its growth cycle. Seguy, Nicou, and Haddad (1970) have suggested that ploughing should be to a depth of 15-20 cm, producing a ridged and furrowed surface so as to withstand the erosive effect of the first rains. Such ploughing can be done perfectly well with animal traction if the working width of the plough is reduced in proportion to the strength of the draught animals. It also has the effect of greatly facilitating seedbed preparation and weed control. Using this technique a plot has been cultivated in western Cameroun 5 times running without any weeding, and its yield has never fallen below 5 t/ha (Seguy *et al.* 1970). Ploughing at the end of the season, i.e. just after the rice has been harvested, has been shown to give better results than leaving this until the beginning of the next rainy season. This allows sowing to be done as soon as possible, depending on local rainfall. The optimum depth of sowing is 2 cm; the seed then has better moisture conditions than occur nearer the surface and is also protected from birds. A greater depth than this delays sprouting and therefore slows down plant growth. Row-seeding is recommended as it encourages a more orderly distribution of the plants in the soil, allowing early and efficient manual or mechanical weeding. The rows can be as much as 30-40 cm apart without apparently reducing yield in comparison with closer row spacings.

Optimum seedrate with good seed on suitably prepared soil is 50-70 kg/ha. This figure must be considerably increased wherever conditions deviate from the optimum.

Any shortage of the major nutritional elements (phosphorus and potassium) and trace elements in the soil must be detected and corrected. Nitrogen is almost always in short supply. However, it must be

added cautiously as it encourages both lodging and the growth of weeds. With regard to levels of nitrogenous fertilizers to use on upland rice, practical experience suggests that the optimum quantities and times of application can be determined best by careful observation of the growing rice (colour, height, vigour, turgidity, etc.) and that adherence to a predetermined timetable and level of application should not be encouraged if expensive and pointless waste of scarce resources is to be avoided.

In comparison with swamp-rice growing, the weed problem in the growing of upland rice is very great. Although good ploughing can inhibit weed growth, ploughing is often of inadequate quality, or available weed seeds too many, and weeding then becomes a critical field operation. Whether by mechanical or manual means, weeding must be done early, and this is impossible with broadcast sowing. It is greatly facilitated by sowing in widely spaced rows. A second cultivation is sometimes useful and may help to conserve water by reducing evaporation. Chemical weedicides are still very little used. They are very expensive and under tropical conditions their effect is usually not very long lasting. Also their behaviour is extremely erratic under very high soil moisture conditions, depending on the topography and texture of the soil, and therefore success has so far been unreliable.

Insect infestation varies, the most frequent pests being stem-boring Lepidoptera of the genera *Chilo*, *Maliarpha*, and *Sesamia*. Insecticide treatments have only rarely been found to be economically advantageous however. (Further information on rice pests may be found in Chapter 10.)

2.2.3. *Species and cultivars of upland rice* (Abifarin *et al.* 1972; Arnoux 1973; IRAT 1970*a*; USAID 1968)

The species used for traditional rice growing are *O. sativa* and *O. glaberrima*, the former having become far the more widespread. There are numerous cultivars of *O. glaberrima*, the grain of which is often red (Portères 1956).

The first phase of selection work done with this locally available germplasm resulted in the development of a number of cultivars which, although still sufficiently hardy and having adequate grain quality, had substantially improved yield potential. They included Agbede 16/56, OS6 in Nigeria, E425, R67, and 617A in Senegal (Casamance), Soavina in Gambia, Parawan in Ghana, Anethoda in Sierra

B

Leone, Morobérékan and Paté Blanc in the Ivory Coast.

Subsequently upland-rice selection programmes increasingly turned to cultivars introduced from abroad into Ghana, Senegal, Nigeria, and Sierra Leone. Quite recently, two multi-disciplinary teams have been established, at Bouaké (Ivory Coast) by IRAT and at Ibadan (Nigeria) by IITA. Both these teams are expanding the introduction of cultivars and breeding lines from abroad, and particularly from Brazil, Madagascar, IRRI (Philippines), Taiwan, and Vietnam. Collections of several hundreds of cultivars have been formed.

2.2.4. *The IRAT programme for upland-rice breeding*

The hybridization programme for upland rice is carried out at Bouaké in the Ivory Coast. Lines obtained are tested for adaptation in the Comoro Islands, Cameroun, Dahomey, Madagascar, Mali, Senegal, and Upper Volta. Pedigree selection is practised in segregating populations. Introductions having advantageous complementary characteristics are crossed and use has also been made of mutation breeding to try to eliminate certain defects, e.g. the tallness of 63-83. Pedigree selection is then made, on the basis of criteria (that will be examined later), from descendants of hybrid or irradiated plants. Among the parents used, mention may be made of OS6, IR400-5-12, IR20, IR22, IR154-61-1 (IITA), Azucena, Faya, Anethoda, Tikiri Samba (Rokupr), Miro-Miro, R67, Taichung Native no. 1, Iguape Cateto, OS6-RT-1031-69 (Bouaké), IR8, Tunsart, 1031-69, and 63-83 (Sefa).

A programme for crossing *O. sativa* and *O. glaberrima* intended to make use of the hardiness of the latter species has been started at Bouaké. The F_1s are, however, almost always totally sterile.

The first cultivars of the short ideotype† introduced from Taiwan or IRRI (type IR, TW1, I Kon Pao) can exhibit the same superiority over the long-stemmed cultivars in upland rice growing as they do in irrigated paddy cultivation. However, this superiority is expressed only under conditions of good water supply or water control; severe attacks of blast (*Pyricularia oryzae*) develop on these culti-

† The term 'ideotype' that has passed into general use among agricultural scientists was coined by C. M. Donald in Australia (Donald 1968). It is a term used to describe an idealized concept of plant form considered capable of maximizing yield under given conditions, whether any particular existing cultivar has this form or not. Ed.

vars if there is any water stress to which their shallow root system makes them particularly sensitive.

The selection objectives. The following résumé of objectives is based on information given by Abifarin *et al.* (1972), IRAT (1970a), and Will (1970).

A starting point for improvement can be defined by the characteristics of the existing cultivars. Cultivars of the traditional or only slightly improved type (OS6, Iguape Cateto, Morobérékan) are generally more than 1·3 m in height. They are relatively resistant to short periods of drought and fungus diseases, especially *Pyricularia* blast. Their grain quality is considered satisfactory by local criteria. Their potential yield is generally less than 5 t/ha except under exceptionally favourable conditions such as obtain in western Cameroun. Because of their height, their grain/straw ratio is low, they are liable to lodging, and respond poorly to nitrogen applications. A number of cultivars of *O. glaberrima* have remarkably vigorous growth and drought resistance, but these advantages are accompanied by low yield potential and a very high tendency to shattering, with consequent losses of up to 40 per cent.

The varietal improvement programmes now under way are based on pedigree selection in segregating populations. Testing is carried out under the different ecological conditions of Senegal, Upper Volta, Cameroun, etc. The selection criteria are high productivity, average tillering, plant height of less than 1·2 m, semi-erect flag leaves, high speed of growth and rooting after germination, resistance to diseases and pests, ability to withstand short periods of drought, and maximum maturity period consistent with use under the local rainfall regime. (Provided yield is not adversely affected, it is usually advantageous to have cultivars with the shortest possible maturation period.)

Since tillering is only average (i.e. there is no selection for number of panicles per unit area) yield selection must depend mainly on the weight of grain per panicle (IRAT 1970a). A high tillering rate is not desired since it is believed that this is absolutely correlated with a longer ripening period.

Reduction in the height of the plants to about 1·2 m is intended to increase their resistance to lodging. No effort is being made to further reduce plant height because very short plants are believed to be less well able to compete with weeds. It is also because of this essential ability to compete that a type with semi-erect flag leaves is sought, rather than with fully erect flag leaves typical of the latest releases from IRRI.

Speed of growth and rooting after germination is a vital factor. There is at this time a very great risk of the plant's water requirements not being satisfied because of the irregularity of the rain at the beginning of the season. The first few inches of soil dry out first and the plant will obtain a better water supply as its roots go down more deeply.

Resistance to diseases and pests is also important, since the high cost of crop-protection chemicals makes it almost impossible for the grower to use them profitably. A cultivar originating from Madagascar, known as 1345, has good resistance to stemborers because of the abundance of sclerenchyma in its vascular bundles. Several cultivars, amongst which the 63–83 is outstanding, seem to possess horizontal resistance to blast (*Pyricularia oryzae*) and are not significantly affected by races of this pathogen in any of the places where they have been cultivated in Africa.

2.2.5. *The programme for rice breeding at IITA (upland and swamp rice)*

At IITA, rice-improvement work was initiated as a sub-programme in the Cereal Improvement Programme and the Farming Systems Programme. At present about 70 per cent of its rice research is on upland and about 30 per cent on irrigated rice.

The rice research at IITA has as its main objectives: the raising of grain yield, especially of upland rice; improvement of plant type; maintenance of high physical and chemical grain quality; and resistance to the pests and diseases known to present hazards in West Africa. The programme is a unified one for both upland and swamp rice.

Grain-yield improvement. A yield-component approach is taken to improve grain yield of rice. Selection is made for medium to high number of panicles; medium to high number of grains per panicle; medium to high grain weight; and also freedom from shattering but with easy threshing of grain.

Plant-type improvement. The present unimproved plant types of rice appear to have several undesirable traits. Their height and leafiness tend to make them liable to lodging, non-responsive to nitrogen fertilizer (Tanaka and Kawano 1965, 1966), and poor yielders

on account of low grain/straw ratio. Selection criteria for plant-type improvement include: short-to-medium height, stiff-strawed plants; tough, slowly senescent, and moderate-length leaves; fairly upright leaves (for upland rice semi-droopiness is satisfactory); good tillering; superior root development; good seedling vigour; and well-exserted panicles (Tanaka, Kawano, and Yamaguchi 1966).

There are many physiological factors to be considered while attempting to achieve IITA's principal objective of yield improvement for rice. These are either directly or indirectly related to yield and include medium-to-early maturity, nitrogen responsiveness, tolerance of moisture stress, tolerance of iron and manganese toxicity under irrigation and iron deficiency under upland conditions, and satisfactory embryo dormancy in order to avoid sprouting in the ear.

Physical and chemical factors affect the appearance of the grain, the milled rice out-turn, the quantity and quality of protein, and the amylose content. Selection for quality at IITA is based on positive scoring for (1) medium over long grain; (2) clear and translucent over opaque grain appearance; (3) high percentage of milling out-turn; (4) 20–25 per cent amylose content; (5) a medium gelatinization temperature; (6) a high protein content (11–13 per cent) and a favourable amino-acid balance.

Resistance to African pests and diseases. Main emphasis is placed on resistance to the following pathogens and the associated diseases: *Pyricularia oryzae* (blast), *Rhynchosporium oryzae* (leaf scald), and *Helminthosporium oryzae*.

Rice breeding and varietal improvement procedures. To achieve the programme objectives the following steps are taken.

1. Large-scale introduction of cultivars and breeding lines, especially those showing some promise for upland conditions. IITA accessions numbered more than 4000 entries in 1974.
2. Carrying out screening and yield trials of this material under both upland and irrigation conditions.
3. Crossing between promising parents for pedigree selection and population improvement.
4. Growing multiple-entry observation nurseries and yield trials of the selected lines.

IITA already has many institute-bred cultivars performing well under upland conditions. Seven high-yielding cultivars have also been screened from introductions of rice grown under irrigated conditions.

2.2.6. *Constraints on expanding field production of upland rice*

The main factors have already been referred to above, i.e. the short duration and irregularity of the effective wet season; competition from weeds throughout the growing season; diseases and insect pests; and the low level of soil fertility. Some other, less obvious factors nevertheless play an important role and may sometimes limit production. First, there are attacks by domestic animals, rodents (various species of rats, grass-cutters, or agoutis), birds (various grain-eating passerines such as the Sudan dioch (*Quelea quelea*)), and nematode pests. Although domestic beasts can be shut out by fencing the fields, little can be done about the others except by keeping watch, a difficult and not very effective task, throughout the period when the grain is ripening. Secondly, clearing the land also raises serious problems. It should be done completely, including the removal of stumps, to permit ploughing, without which satisfactory yields cannot possibly be obtained. If clearing is done with the tools generally available to the grower (axe, cutlass, fire) it is often 10 years before opened land is properly clear of obstructions. During all this time the soil will be poorly tilled and will become infested with weeds and yields will remain low, discouraging the grower. The use of mechanical means would speed up this work, but it has several disadvantages.

The first of these is that any such procedure is very costly (50 000–100 000 CFA francs per ha) which, even if depreciated over a 10-year period, represents an annual charge equivalent to the value of 250–500 kg/ha of paddy. Secondly, it is work which, if not carried out with all due care, may well endanger the soil by accelerating erosion, with irrecoverable loss of top soil and the consequent creation of highly infertile patches. It seems unlikely that the necessary precautions will always be observed in large-scale mechanical clearing operations carried out by contractors. An advantageous solution, therefore, might be to make available to the growers themselves, by hire or sale, those tools which enable them to deal with the natural vegetation more efficiently such as chain saws, winches, and explosives. They would, of course, have to be trained to use them.

Drying of the harvest only rarely gives rise to practical problems as the weather in most of the cultivation areas is normally dry at that time. However, if threshing is done in the field with drum threshers of the Japanese type, the grain at 20–22 per cent moisture content may be too damp for immediate storage, for which a moisture content of about 14 per cent is required. Drying in the sun, even if possible, is not recommended as it causes cracking of the grain and reduces milling out-turn. According to Séguy (personal communication) it is possible to construct cribs quite cheaply which are similar to maize cribs, but in which the wire mesh normally used is replaced by mosquito netting. Several layers of paddy, each some 10 cm thick, can be dried in such cribs, even in a humid climate.

2.2.7. *Advisory organizations for upland-rice farmers*

Complex technical and human problems are raised by the transition from traditional shifting upland-rice growing, with very low yields for subsistence only, to more intensive cultivation, producing marketable surpluses and involving the use of such purchased inputs as seed, chemicals, implements, etc.

Those responsible for developing agricultural production in many countries of West Africa are becoming increasingly interested in intensifying upland-rice growing. It has the advantage over irrigated rice growing of requiring almost negligible capital investment in comparison to the very high cost of rice development based on irrigation schemes ($\frac{1}{2}$–1 million CFA francs per ha). In addition, the techniques used for intensive upland rice are simpler than those for paddy rice.

However, even the practices for intensifying upland-rice production must be taught to the growers, who must be encouraged to use them, provided with the necessary materials, and often helped to market their harvests. This is the task of the advisory organizations. Some countries prefer to have this work carried out by their own officials in the agricultural services. In such circumstances, activity tends to be dispersed because of the limited number of such officials and because such people usually have inadequate facilities. Generally such activities are not very effective and make little contribution towards increasing production. In other instances governments have set up concentrated campaigns involving intensive efforts, generally over a limited period, covering a specific area and having precise production targets. These activities may then be carried out by independent organizations, which may be state bodies (e.g. SODERIZ in the Ivory Coast) or private companies (OPR in Senegal) with either commercial financing or making use of foreign-aid funds.

In offering advice and training it is necessary first carefully to define objectives, and secondly to take note of all the relevant research information. This must be backed up, where necessary, by on-the-spot investigations. Thus a 'package' of the best available techniques and materials can be built up for any set of circumstances which may arise in rice growing. The value of a particular 'package' can then be proved by demonstrations; and finally growers can be trained to use a technique under supervision, until a new technology of rice growing has become fully established and accepted. This approach is discussed in greater detail in Chapter 14. It will be necessary to provide growers with seed, chemicals, implements, etc. at the right time; to set up and, if necessary, to supervise credit finance for this purpose; and in most situations to assist in the collecting, processing, and marketing of harvests.

2.3. Lowland-rice-growing systems

2.3.1. *Ecology of swamp-rice systems*

Because of the topography of the land and the abundant, highly seasonal rainfall, there are numerous places in West Africa that become temporarily inundated flood plains for a variable period during the wet season. Thus the flood plains of major rivers such as the Senegal, the Niger, the Mono, the Zou, and the Oueme are regularly flooded when the river is high. There are also swamps or *thalwegs* (*bolilands* in Sierra Leone) from which the flow is hindered to some extent by irregular topography, which are seasonally flooded by run-off from surrounding higher ground (IRAT 1970*a*; UNDP 1971; USAID 1968; Will 1970). It is probably in such sites that African rice growing using *O. glaberrima* first began, because this species is still found on many such sites in almost its original state, together with its close relation *O. breviligulata*. On the river flood plains the duration of the flooding obviously varies with the rainfall at or upstream of the flooded area. Likewise, at any one place the depth of flooding varies with the topography.

The water depth is generally correlated with the duration of flooding, as topographical depressions flood first and dry out last. To make complete use of a flood plain it is necessary to have a range of cultivars with different vegetative periods, those of the shortest

duration being grown at higher elevations and those with the longest vegetative periods at the lower.

It is often possible to reduce the disadvantages resulting from the uncertainty of the time of flood and ebb of the river water by constructing bunds equipped with valves to allow controlled submersion (see Chapter 9, p. 171).

Two special types of lowland rice are flood-water (Lake) and mangrove rice. Flood-water-rice growing, which is found mainly in the lake district of Mali, makes use of depressions filled by the Niger flood waters. These are filled too quickly to too great a depth to be used for controlled submersion. In these circumstances the plants are transplanted as the waters drop, starting at the periphery and working to the centre of the pools as the water recedes. This transplanting begins in October and harvesting of the first-drying areas starts at the beginning of March. The cultivars used in this system, known as Kobé rice (Portères 1956), are red-grained and races of *O. glaberrima*.

2.3.2. *Mangrove rice growing*

Mangrove rice growing is practised all along the African coast from the mouth of the Casamance to Sierra Leone, and in Nigeria. According to Will (1970), the mangrove soils, which are chemically very complex, are deep black clays and are rich in organic matter. The great majority, which formerly supported mangroves (*Rhizophora* spp.) have concentrations of sulphide built up under anaerobic conditions. When exposed to air during a prolonged period of drought, sulphide is oxidized to sulphate and the pH may fall to 2 or 3. At this pH, aluminium ions reach toxic levels in the soil water. Such problems are encountered more rarely on soils in which the *Avicenna* mangroves formed the natural climax vegetation. Mangrove soils are flooded by undiluted sea water at high tide during the dry season but during the wet season the salt concentration is sufficiently reduced to allow rice to be grown, by rain water which overflows from the rivers whenever the high tide prevents drainage. The saltiness of the soil during the dry season eliminates most of the weeds.

2.3.3. *Cultivation techniques for swamp rice*

In traditional rice growing the soil is worked as soon as it is sufficiently wetted by the first rains. Operations range from hand-hoeing to shallow ploughing with draught animals. Rice is then broadcast, at a seedrate of about 80–100 kg/ha. Early growth is rain-fed. Usually during this time the young rice is competing with many weeds. Hand-weeding is generally minimal because it is a long and tiring job and the rice fields are often far from the villages. When the fields become flooded they are left un-husbanded until harvest time, which occurs just after the waters go down. Yields under these condi-tions are 1–3 t/ha.

These traditional techniques can be improved in many ways. The construction of bunds for flood-water control, as already mentioned, enables the in-flow of the water to be delayed until the plants are of sufficient size to withstand flooding, and this partly overcomes the disadvantages of late sowing. Mineral fertilizers appreciably increase yields by making good the nutrient shortages in the soil and compensating for losses (especially phosphorus and potassium). For applications of nitrogen of the order of 40 kg/ha incremental yields of about 10–20 kg paddy per kg nitrogen are often obtained, which are extremely profitable. The practice of ridge-and-furrow ploughing can be substituted for the tradi-tional shallow tillage to enable the plants to get a better start and facilitate weed control. For the first 30–45 days from sowing, swamp rice behaves similarly to upland rice and everything already described for upland rice is appropriate also to swamp rice. Most rice-development organizations carry out ploughing using tractors on behalf of the growers (SAED in Senegal, SEMRY in Cameroun, Opération Riz in Mali, etc.).

The advantage of transplanting lowland rice is debatable. It is now rarely done except in the most long-established rice-growing areas (Guinea-Casamance, Sierra Leone). Obviously transplanting cannot be undertaken until the floods arrive, and the resulting interruption in growth could be fatal if the waters rise too rapidly. In mangrove-rice growing transplanting is normally done on large furrows as soon as the soil has lost sufficient salt to allow the rice to resume growth. The growers use their own judge-ment to determine this time, and generally leave a large safety margin so that transplanting is often done very late (September, October), a practice which adversely affects yield.

2.3.4. *Floating-rice ecology and cultivation*

Floating-rice growing is found mainly in the lower parts of the Niger river basin and also in Sierra Leone (Willis 1973). This is the traditional system of cultiva-tion in the central Niger delta. Every year in this vast region floods, caused by rainfall in the catchments of the Niger and its tributary the Benue, spread slowly over the area from the month of August to about October and then gradually diminish until the end of December. The duration and depth of flooding varies with the topography and land at different heights may be submerged for periods from 2–6 months. The indigenous *O. glaberrima* has been selected over many generations by the peasant farmers for floating rice cultivation so that numerous landraces have been evolved with precise adaptation to each area.

The rice is sown, as in upland-rice growing, at the time of the first heavy rains at the end of June or in early July. It grows under rain-fed conditions until the flood waters arrive 4–6 weeks later. To withstand the rising of the flood waters, the plant must be suffi-ciently well developed. Experiments have shown that the necessary minimum period from sowing and the arrival of the flood water is 35–40 days. This has prevented the extension of rice growing in the tradi-tional system to land on the lowest ground where the flood waters arrive too soon after the breaking of the rains.

2.3.5. *Irrigated rice by the control of flood water*

By constructing bunds around the areas to be culti-vated to form enclosures, both the maximum level of the flood and the speed at which it rises can be con-trolled to match the growth of *O. sativa*. This species is being grown increasingly because of its higher yield than *O. glaberrima* and its white grain which is preferred. The maximum water level admitted is calculated to ensure maximum safety for the harvest. This system is an irrigation system within the classi-fication outlined (pp. 9–10). The flooding of the Niger basin (see also Chapter 9, pp. 175–7), although regular in its regime over many years, varies from year to year in respect of the maximum level reached, the speed of its rise, and its duration. The further down-stream we go, the more regular becomes the flooding from one year to the next.

2.3.6. *Cultivation techniques for irrigated rice*

The method of cultivation for flooded bund-surrounded enclosures is much more like that practised under upland than under swamp condi-tions. The soil is prepared in May–June after the first rains, and the rice is then broadcast-sown. It grows under rain-fed conditions until the flood waters arrive. The speed at which flood water is admitted may be about 10 cm per day.

On behalf of Opération Riz in Mali, IRAT has studied the possibilities of improving these techniques. A change from manual power to the use of draught animals is preferable since cultivation requires considerable manpower. Tractor-powered cultivation was considered to be too expensive under conditions prevailing in Mali, which is obliged to import equipment and fuel under circumstances that are often difficult. The only motorization† adopted is for threshing, other than in pilot-scale schemes and for once-and-for-all elimination of wild rice. Cultivation by draught animals is already practised by peasant farmers in the region and the use of the plough is widespread; 95 per cent of the rice fields are ploughed by animals.

Husbandry operations for irrigated rice as recommended by IRAT in Mali. The stubble is ploughed under as soon as the harvest has been gathered and before the soil dries up to such an extent that working the land becomes difficult. A ploughing depth of about 10–15 cm is normal. The soil is left bare to permit the elimination of any *O. barthii* rhizomes which are collected and removed during the dry season. The seed bed is made by a second ploughing, at right-angles to the previous ploughing, as soon after the break of the rains as possible, to a depth of 10–15 cm. This destroys any weeds that have germinated with the first soil moisture. This is followed by harrowing, only a few days before sowing, to break up the clods and level off the rice field. The seed is broadcast at a seedrate of about 100 kg/ha or in continuous rows 30–40 cm apart, depending on the nature of the land and the availability of a drill. Row-seeding both reduces seedrate by a half and enables weeding to be done with an animal-drawn hoe, allowing easier control of weeds. Harrowing is carried out immediately after broadcast-sowing. Covering the seed with a thin layer of earth protects it from marauding birds and creates favourable conditions for germination. Weeding should be done 15 days after emergence of the rice. If necessary, a second weeding should be carried out before the plots are flooded.

Although soil analyses suggest that the application of phosphate fertilizers should be necessary on certain soils, their use cannot yet be recommended because experiments with high levels of phosphorus conducted by agronomic research institutes have not

in fact shown any response by rice. Nitrogenous fertilizer use has been studied by IRAT, which found that an application of 100 kg/ha urea, in a single application, 1 week before the flooding is usually highly advantageous. The use of fertilizer is, however, only recommended on rice fields on which other recommended practices, especially weeding, have been properly carried out. There is no risk at present of soil impoverishment, even if no fertilizer is used, as the plant nutrients are at present adequate to ensure harvests of acceptable yield levels without exhausting the soil.

Cutting must be done when the grain is ripe, which is 30 days after the emergence of the ears. Threshing can be carried out as soon as the grain has lost sufficient moisture, usually about 8–10 days later, depending on climatic conditions at the time of reaping.

2.3.7. *Limiting factors for rice cultivation with flood water control in Mali*

The most important limiting factor is the uncertainty about the extent of rainfall and the time of arrival of flood waters. Pest infestation, particularly by stem-borers, may cause 30–70 per cent damage to the crop (Bidaux 1970). The most important in near-by Northern Nigeria, according to Akinsolah (1970), are *Maliarpha separatella*, *Chilo* spp., and *Adelpherupa flavescens*, which also occur on various wild grasses including *O. barthii*. Although various insecticides used against these insects appeared to be efficient (Sevidol, Gammalin 20, Heptachlor, Azodrin), yield was not increased significantly.

Apart from blast, *Helminthosporium oryzae* also causes a damaging leafspot disease in swamp rice cultivation with incomplete water control (Aluko 1970*a*), causing often up to 15 per cent loss. High humidity appears to aggravate the disease. Few cultivars are completely resistant, and chemical treatments have been erratic in performance. SML140/10, SML18-B, and Maliong appear to be moderately resistant cultivars.

2.4. Irrigated rice growing with total water control

2.4.1. *Ecology*

Fully irrigated rice growing, in which the grower has complete control over the water and can drain or flood his plot at will at any time, is still very rare in West Africa. This is because it is a non-indigenous system and it requires, in addition to an assured

---

† For definitions of mechanization and motorization see Chapter 13.

water supply, which is found only near large water-courses, very high capital investment. The costs for irrigation schemes of this type may amount to $\frac{1}{2}$-1 million CFA francs (US $2000-4000) per ha. In theory, at least, rice can be grown under irrigation wherever fresh water is available (from rivers, groundwater, dams, etc.) regardless of the climate. Two crops a year can be obtained with this system provided that day-neutral rice cultivars are used for at least one season each year.

The agricultural missions from Taiwan have greatly helped in introducing fully irrigated rice growing to Africa (Wei Lang 1970) and the productivity under this system is such that the rice requirements of most states could be satisfied on a few thousand hectares producing more than 10 t/ha per year from two crops. In 1971 there were 57 Taiwan missions employing 487 technicians in 10 states, controlling in all 11 200 ha.

2.4.2. *Cultivation techniques*

Sites are selected according to irrigation possibilities (assuming a continuous flow of 1-2 l/ha per day). The land is formed to level, bund-enclosed fields that can be irrigated and drained at will.

In view of the high price of this development, involving very heavy depreciation costs (1 million CFA francs per ha, written off over 20 years, corresponds, including maintenance, to more than 3 t/ha per annum of paddy in most cases), plus pumping costs where appropriate, this type of rice growing is economically feasible only if yields can be maintained at a very high level. All the cultivation techniques must therefore be brought to a high degree of precision. Transplanting is the general rule, the plants being planted in rows with a spacing of 22-25 cm × 20-22 cm. Irrigation is controlled so that the depth of water is kept at 5 or 6 cm, which is adequate for maximum tillering while limiting competition from weeds. Weeding is done in keeping with a strict code of practice. Fertilizing is generous, especially with nitrogen. The missions working in the various countries have determined by experiment the formulae to be applied to each soil type. They cover a fairly wide range: nitrogen, 56-110 kg/ha; phosphorus, 20-60 kg/ha; potassium, 30-70 kg/ha.

2.4.3. *Limiting factors*

Apart from lack of available capital, which obviously restricts the extent of this type of scheme, the rigidity of the operating system adopted appears to be un-attractive to many African growers, despite the substantial profit obtainable. Wei Lang (1970) estimated that in Liberia the gross income of the rice grower and his assisting colleagues is US $4400 per ha per annum compared with US $800 for the traditional rice grower in that country.

Such results as these appear to be achievable only by the persevering work of very strict supervisory teams. As soon as they withdraw, as has happened several times following political changes in West Africa, growers almost immediately reverted to much less intensive forms of cultivation.

Irrigated rice cultivation does not necessarily require the operation of transplanting. Direct seeding can be used and can be mechanized if soil levelling has been correctly carried out. In addition to the amount of tedious and tiring work which hand labour entails, the practice of transplanting increases the risk of bilharzia from the necessary prolonged contact with infected water.

2.5. **History of development of present-day lowland-rice cultivars**

2.5.1. *Up to 1946—adapted landraces*

In the 1930s studies began on the types of rice traditionally grown in Africa. Cultivated rice belongs to two separate botanical species: *O. glaberrima*, which is found in the greatest quantity in the delta of the Niger; and *O. sativa*, introduced earlier from Asia. *O. glaberrima* is at present widely distributed in West Africa and several thousand different types can be recognized, of which some are suited to upland-, some to swamp-, and others to floating-rice growing. However, cultivation of this species is declining in favour of *O. sativa*. This is because of the low productivity of *O. glaberrima* and the high level of spontaneous seed-shattering at maturity which leads to unavoidable losses.

In countries where the production and distribution of rice seeds are centrally organized, only *O. sativa* is at present handled. Consequently the use of *O. glaberrima* may be expected to disappear within a few years. *O. glaberrima* cultivars have been collected by certain research establishments, and programmes for the hybridizing of these with *O. sativa* are in progress, e.g. at Bouaké (Ivory Coast), in order to try to obtain in a rice of *O. sativa* type the benefit of their qualities of vigorous growth and resistance to drought and certain pests.

Landraces of *O. sativa* that have long been grown in Africa almost all belong to the indica group. These genetic populations are the result of a long period of natural selection, and rudimentary selection by farmers at a low level of fertility, with no water control and in the absence of any effective crop protection practices. Genotypes that have survived to comprise these landraces possess a number of characteristics which confer great hardiness including vigorous growth, resistance to fungus diseases, and adaptation to toxic or salty soils.

Under unimproved agronomic conditions these landraces yield less than 1000 kg/ha. They respond poorly to any improvement of these conditions and lodge freely when the level of nitrogen is raised with fertilizer. Finally, their maturation period is generally long (150–180 days), short-season strains being both less productive and very prone to attack by grain-eating birds.

The first efforts made to improve rice growing mainly concerned cultivation techniques (ploughing, seeding) and schemes to improve water control. Agronomists, realizing that the low potential of the available landraces was a significant limiting factor, very quickly sought the breeding of more productive cultivars.

2.5.2. *Selection from 1946–66—selection of pure lines from landraces*

Selection work became important in the years following the Second World War and has lasted some 20 years. It consisted of selecting pure-line cultivars from the available landraces and of introducing other cultivars from tropical rice-growing countries where research had been in progress for longer (India, Sri Lanka, Indochina, Indonesia, Guyana, Madagascar, etc.). Some crosses were made and segregants selected for improved size and quality of the grain.

During the period 1946–66 little attention was paid to improving soil fertility. The prevalent idea at that time was that the unfavourable ratio between the price of fertilizer and that of paddy, the growers' poverty, and the very limited resources available to the advisory services made the economic use of fertilizers seem improbable in the medium term. Thus the selection work at the stations was mainly carried out at low or variable levels of fertility. Nevertheless, remarkable results were attained, and certain cultivars dating from that time are still successfully cultivated today. Some of the best known are the following selections from local landraces: Gambiaka, Bintoubala, Morobérékan, Ebandioul, Sintane Diofor, Malobadian, IM 16, and Oma Rosso. Intro-

ductions included: BG79 and D52–37 from Guyana; Segadis from Indonesia; Neang Veng, Phar Com En, and Kading Thang from Cambodia; Makalioka from Madagascar; Indochine Blanc and Doc Phun Lun from Vietnam; Iguape Cateto from Brazil; and SR26 from India. Cultivars bred by hybridization among local rice cultivars included: HKG98 bred at Kogoni, Mali; 63–83 bred at Séfa, Senegal; H. 821–3 bred at Richard Toll, Senegal; OS6 bred at Yangambi, Zaire.

The potential yield level has been raised appreciably, and these cultivars, still grown on a large scale under controlled water conditions, commonly give yields of about 3 t/ha. The quality of their grain is generally very good. Their sensitivity to fungus diseases, in particular blast, has not until recently caused any problems. However, because of the conditions under which they were selected, these cultivars generally have a very disappointing response to nitrogen fertilizers.

2.5.3. *Cultivar development since 1966—concentration on responsiveness to fertilizer nitrogen*

Since the mid-1960s agronomic work in West Africa, especially by IRAT, has shown how greatly the very low fertility level of most African soils can be improved by using chemical fertilizers, and their distribution and use on a large scale no longer appears to be completely Utopian. At the same time cultivars of the semi-dwarf plant type, the distribution of which initiated the Green Revolution in tropical Asia (Barker 1970a; Chandler 1970), were also introduced into West Africa. These cultivars have short, stiff straw and tiller heavily. They are efficient in the use of solar radiation and are highly responsive to nitrogenous fertilizers (up to levels of 100 kg and more nitrogen per ha), giving them a potential yield of about 10 t/ha in good conditions under irrigation. These cultivars are also virtually insensitive to day-length, which means that the date of sowing is immaterial, and have a relatively short maturation period (110–130 days), so that several crops a year can be grown (Barker 1970b).

The Taiwan technical assistance missions, established in most West African countries, have played an important part in distributing this new type of rice (ADRAO 1973a) as well as introducing the sophisticated cultivation techniques needed to exploit its potential (ADRAO 1973b). Staff of agricultural research stations are rapidly taking over from the missions and continuing this work.

The cultivars introduced from Taiwan included Taichung Native no. 1, Taichung 178, Tung Lu, I Kong Pao, Tainan 2, and TS123. IR8 was the first cultivar widely distributed by IRRI in the Philippines, then IR5, IR6, followed by IR20 and IR22, then IR24. CICA4, also of the same type came from Colombia, Jaya, and Padma from India and, more recently, SML Alupi and others from Surinam. These introductions are continuing at an ever-increasing rate from overseas breeding centres.

In addition to these introductions from abroad, there are the results of work on hybridization and mutation breeding being carried out in Africa. There has been a vast increase in the material to be evaluated, and research establishments in Africa will be hard-pressed, if not overwhelmed, to make effective use of this wealth of material with their existing facilities.

Indeed, although they represent a decisive break-through as regards potential productivity, the new cultivars already released appear to have created almost as many new problems as they have solved. There are difficulties in extending the work because of the high level of technology required to grow the cultivars. But the most worrying problems concern crop protection. Because the new cultivars have been created from parents completely foreign to the African continent, they are likely to carry little resist-ance to local strains of pathogenic fungi. In particular they show a varying, but often very high, degree of sensitivity to *Pyricularia oryzae* (causing blast disease), which makes it very risky to cultivate them without costly plant-protection measures. Some cultivars have been totally destroyed within a few years, to the great detriment of farmers who had put their faith in them (e.g. C4–63 in Ghana). Moreover, it is not impossible that the growing of sensitive cultivars has helped the spread and development of the pathogenic organisms. For example, an old cultivar that had long been acclimatized, such as D52–37, was wiped out in 2 years in Upper Volta by a new race of *Pyricularia*, to which it had never before appeared susceptible (Barat 1970). It is open to question whether there is not a vast potential danger of a similar type from the accidental introduction of viruses and bacteria able to cause diseases not yet present in Africa (Aluko 1970*a*). Large-scale cultiva-tion of cultivars becoming newly exposed to these could cause a sudden and disastrous rise in losses from disease.

Great vigilance must therefore be exercised. One way of avoiding a risk of this kind is to speed up the breeding work aimed at combining the high-yield potential of introductions with the adaptability and disease- and insect-resistant qualities of the local cultivars. Such new lines, and not the introductions themselves, should then be used for commercial release as recommended cultivars after careful field testing.

2.6. Rice research in Africa—institutions, infra-structure, and programmes

2.6.1. *Institutions and ecological coverage*

Modern studies of rice in Africa appears to date from about 1930 in Sudan and Sierra Leone (Morgan 1970), but rice research only began generally with the introduction of rice growing in areas where it was not traditional.

At the end of the Second World War the authorities began to give serious attention to food production; this included rice growing, already developed to a considerable extent without significant Government support. The new interest was reflected in the estab-lishment of the following research stations: Kankan and Koba in Guinea; Djibélor in Senegal; Ibétémi in Mali; Rokupr in Sierra Leone; Banfora in Upper Volta; Kolo in Niger; Badeggi in Nigeria; Bouaké and Ferkessédougou in the Ivory Coast; Toukou in Cameroun; and Kpong and Nyankpala in Ghana. Later, for a short period only, a Federal Rice Research Centre was set up in Guinea to bring together and co-ordinate the activities of the stations in franco-phone countries. In the 1960s IRAT was progressively entrusted, by various national agreements, with rice research and with the management of their research stations in most francophone African countries. A little later the Ford and Rockefeller Foundations founded IITA at Ibadan, which carries out rice research in addition to other research activities (Moomaw 1970). These research stations and centres now provide an excellent coverage of the various ecological conditions under which rice growing is practised in lowland tropical Africa. Very recently the West African Rice Development Association was formed. It is responsible for an extensive programme of co-operative research for member countries.

2.6.2. *IRAT's facilities and current research pro-gramme*

IRAT provides effective co-ordination between the responsible teams in individual countries and the central technical services of IRAT; co-ordination is both interdisciplinary and international. Researchers specialize in different subject areas chosen as far as possible according to the order of the limiting production factors in each area; the number of staff varies roughly according to the economic importance of rice growing in the country concerned. About 20 IRAT scientific staff are engaged on rice research and are assisted by the same number of trained technicians. A number of general agronomists and subject specialists working on a range of crops also take part in the rice programmes. The main activities of the IRAT programme are plant breeding, especially for pest and disease resistance, and agronomy and crop protection. There has emerged a common approach between all stations of IRAT to the problems to be faced. In every case the objectives and projects to be undertaken are selected by discussion between IRAT and the national government concerned.

Plant breeding. The programme is aimed primarily at producing cultivars that will make the best use of greatly improved environments as these are created. There seems little point in trying to breed swamp-rice cultivars to overcome problems that should be solved in other ways.

The basic breeding work is done at Madagascar, mainly at the Ivoloina station, and is supplemented or extended by programmes in progress in Senegal at the Richard Toll and Djibélor stations and in Mali at the Logoni and Mopti stations (floating rice). Experimental testing sites exist in most countries in which IRAT is active, i.e. in Cameroun at the Toukou station, in the Ivory Coast at the Gagnoa and Ferkessédougou stations, in Dahomey at the Ina station, in Madagascar at the Lake Alaotra and Mahitsy stations, and, at multiple-site experimental plots, in Niger at the Kolo station and in Upper Volta at the Saria, Kamboinse, and Banfora stations. The method used is pedigree selection from descendants of hybrids or artificial mutants.

Fertilization. Investigations into the following aspects of fertilization are being carried out by IRAT.

1. The determination and correction of mineral deficiencies (P, K, Ca, Si, trace elements, etc.); optimum use of maintenance fertilizers.
2. Nitrogen in plant–soil relationships; the response of rice to nitrogen (a study carried out with ^{15}N in conjunction with the French Atomic Energy Authority).

3. The role of organic matter.
4. Fixing and reversion of phosphorus in inundated soils, a study carried out with ^{32}P.

Study of rice soils. The following soil studies have been undertaken by IRAT.

1. Study and classification of rice soils; definition of optimum conditions for cultivation.
2. Investigation of the role of physical properties of the soil and tillage methods in the cultivation of upland rice, in conjunction with rooting studies. This work is being carried out in Casamance, north Cameroun (improvement of ironpan soils), and the Ivory Coast.
3. Study of the chemistry of flooded soils; toxicity phenomena. Work on this subject is being done at the Djibelor station in Senegal. The results are expected to be of wide application, and testing of the conclusions will be carried out in most of the areas where lowland rice is cultivated under inadequate drainage conditions, e.g. north Cameroun, Mali, Senegal, the west coast of Madagascar, etc.
4. Study of the desalination of soils at the Richard Toll station (Senegal).
5. A study on silica uptake in irrigated rice fields in Madagascar by multiple-site experiments.
6. Study of the water–soil–plant relationships in upland-rice growing (the use of neutron probes).
7. Studies of soil erosion in upland-rice fields in the Ivory Coast and Casamance.

Physiology. In upland-rice growing the most important limiting factor is usually the irregularity of the water supply. Improved tolerance to periods of drought can be achieved by modifying the plant through breeding and by improved cultivation techniques. Work of this nature necessitates co-operation between plant breeders and physiological agronomists. Work in this field is as yet embryonic.

Other subjects for study that make use of physiology are the connection between resistance to disease and mineral nutrition, particularly with nitrogen, and silicon of the rice plants.

Crop protection. Extensive research on crop protection has been done by IRAT. This work falls under two headings.

1. *The pathology of rice.* Five main investigations are being undertaken.

(a) Studies mainly on blast. Selection of resistant

cultivars with co-operation between plant breeders and pathologists (Bouaké).

(b) Study of *Rhynchosporium* (leaf scald), *Corticium sasakii*, and *Ophiobolus oryzae*.

(c) Studies of the use of fungicides which are carried out locally when problems arise, especially in Casamance and the Ivory Coast, in conjunction with the national plant protection services.

(d) Study of the incidence of disease in terms of plant nutrition and the state of redox potential of the soil (brown patches on rice, *Helminthosporium* disease) at Bouaké and Casamance in conjunction with soil chemists.

(e) Studies on the control of white-tip disease caused by the nematode *Aphelenchoides besseyi* are in progress in Madagascar (east coast) and are planned in the Ivory Coast (IRAT 1970*b*).

(f) Studies on methods of controlling ufra disease caused by the nematode *Ditylenchus angustus* are in progress in Madagascar. This nematode has not been found on the African mainland.

2. *Entomology*. Complementary research is in progress, mainly in Madagascar and the Ivory Coast in co-operation with the Ivory Coast's Crop Protection Department. It is concerned with the ecology of the insects, chemical and biological methods of controlling them, and the selection of resistant cultivars. The pests of particular importance are stemborers (predominant in dry areas), hispids (very widespread in Madagascar), and diopses (encountered mainly in humid areas). Other studies involve the chemical control and radioactive sterilization of the rice hispid (Madagascar), the biological and chemical control of *Maliarpha separatella* (Madagascar), and the assessment of the economic importance of lepidopterous and dipterous rice-borers (Casamance and the Ivory Coast). The main emphasis, especially in the Ivory Coast, will be on biological control research, with particular regard to the use of predatory insects.

Weed control. The main IRAT research programme is being conducted at the Bouaké station in the Ivory Coast (IRAT 1970*b*). It comprises the study of alien flora and cultivation techniques and the systematic examination of weedicides with a view to efficiency, phytotoxicity, and selectivity. The methods followed are those developed by the Commission des Essais Biologiques. Applied research is done primarily in Senegal, Mali, Cameroun, and Madagascar. The problem of eradicating wild rice is being studied in Senegal and Mali, particularly the annual species *O. breviligulata* and, mainly in Mali, the perennial rhizomatous species *O. barthii*.

Agricultural machinery. Studies of machinery are in progress at the IRAT station in Senegal, and in conjunction with the appropriate national organization in the Ivory Coast. Investigations into the use of small manual implements are to be extended in conjunction with development personnel. A study of farm mechanization is planned.

Economic studies. A study of family rice farms is included in the research programmes of IRAT agencies in Madagascar and Senegal. In the latter country the various parameters of production systems in which upland rice is included are being studied (Badinand 1969). In addition, efforts are being made to take advantage of all the experiments carried out by IRAT in order to record regularly the economic data relating to them. These records then provide reference documentation on economic aspects of rice growing (especially production costs).

2.6.3. *IITA's research programme on rice (additional to breeding)*

Rice breeding at IITA has already been described in section 2.2.5. All rice research at IITA is multi-disciplinary, involving, is addition to breeding, pathology, entomology, physiology, agronomy, weed control, soil physics, soil pedology, and rice engineering. Some of the research in these other aspects being undertaken is as follows.

Agronomy.

1. Time and rate of nitrogen application.
2. Spacing trials.
3. Trials to study the interaction between cultivars and dates of planting (planting × cultivar interaction).
4. Long-term fertility trials involving the study of N, P, K, S, and Zn under irrigation.
5. Laboratory experiments of nutritional imbalance on clayey, loamy, and sandy soils, with emphasis on the status of manganese.

Physiology.

1. Investigations on mechanism, traits, and characteristics that enable a rice plant to withstand drought.
2. Screening for drought tolerance.

3. Effect of drought stress at different stages of growth on yield and yield components.
4. The sensitivity of rice at different growth stages to different levels of water deficit.
5. The effect of mulching on growth and development of rice.

Entomology. Screening for resistance to *Chilo* spp., *Sesamia calamistis*, whorl maggot (*Hydrellia* spp.), *Nymphula depunctalis*, and *Diopsis thoracica*.

Pathology.
1. Horizontal resistance to *Pyricularia oryzae* (blast).
2. Resistance to *Rhynchosporium oryzae* (leaf scald).
3. Resistance to *Helminthosporium oryzae* (brown leafspot).

Engineering.
1. Investigation of tillage methods.
2. Investigations into the effectiveness of tillage practices in controlling bush regrowth.
3. Tillage-systems and weed-control investigations.
4. Determination of optimum time of harvesting and threshing rice.

Weed control. Investigations on weedicide combinations that give good weed control without damaging rice under upland and lowland conditions.

Soil physics. Studying rice root systems in order to develop an effective screening method for good root systems for upland rice.

Pedology.
1. Determination of various soil series suitable and unsuitable for upland rice.
2. Studying hydromorphic lands and their suitability for rice production.

2.6.4. *The West African Rice Development Association (WARDA)*

The West African Rice Development Association (P.O. Box 1019, Monrovia) was set up following a conference on the development of rice growing in West Africa held in Monrovia (Liberia) in September 1969, organized by USAID, FAO, and UNDP. The implementation document was signed in Dakar on 4 September 1970 by representatives of the Ivory Coast, Gambia, Ghana, Upper Volta, Liberia, Mali, Mauritania, Niger, Senegal, Sierra Leone, and Togo. Nigeria and Dahomey have since joined, to bring the 1974 membership to 12 (Upper Volta having left its membership open).

The Association organized a conference on research and development in rice growing in Rome in March 1972 (Arnould 1972), and seminars on improving rice cultivars (January 1973 in Monrovia) and on the fertilization of rice (January 1973 in Monrovia). Seminars on disease and pest control were also held in May 1973. Co-ordinated trials on cultivars, mineral fertilizers, weedicides, and crop production products were planned at the seminars.

2.7. Future prospects and suggestions for expanding the rice programme

Rice research in Africa carried out so far has related mainly to solving the immediate problems of agricultural development. Agricultural research facilities are scattered over a large number of countries and research is largely dependent on these countries' own funds. Research staff have only just started to undertake more fundamental studies on detailed biological phenomena determining yield and adaptation. For such investigations highly competent research teams are required. Studies of this sort (Tanaka, Kawano, and Yamaguchi 1966), which are best carried out in major global or regional research stations may lead to great further improvements in tropical rice growing.

There is sufficient evidence of the production potential of intensive rice growing from improvements achieved in South-East Asia, especially through the work done by IRRI. However, IRRI's work is directed mainly towards Asiatic rice growing, which differs in many ways from rice growing in West Africa. Asian rice cultivation was already far more intensive than that in Africa, being based on a number of working days per hectare that would rightly be considered excessive in countries with a low population density. The countries of South-East and South-West Asia could be called 'rice civilizations', and have for a very long time been familiar with the handling and control of water in agriculture—whether by gravity or by various pumping methods. There is almost no similar cultural tradition in Africa, the exceptions being a few ethnic groups on the west coast: the Bagas, Balantes, Diolas, etc. One of the main characteristics of African rice growing which must be taken into account in any development project is the variability of rainfall between seasons and reliability within seasons leading in turn to surplus or shortage of the water supply available for agriculture (Franquin 1973). It is true that there are technical solutions to this problem, but their application will always be much more costly in Africa since, in the medium term at least, little can be expected in the way of individual initiative, intuitive knowledge, or understanding from the farmers concerned, and almost everything will have to be financed and taught from outside. This view accounts for the approach adopted by most of the rice research workers in Africa; whose studies are generally carried out under very imperfect water-supply conditions. Most researchers are properly anxious that the results of their work should be applicable by the rice growers concerned, without first requiring considerable capital investment in irrigation, and thus they have chosen to conduct their research under conditions which are normal for the ordinary farmer.

However, in the long term it will be impossible to make any decisive progress in African agriculture without radically changing the attitudes of the farmers themselves towards irrigation and the use of costly inputs (fertilizers, implements, pesticides, etc.). We still all too often see practices that are quite obviously wrong, such as very low plant populations in the fields or excessively dense populations in nursery beds, weeding done poorly or not at all, and inefficient use of irrigation networks. It cannot be over-emphasized that agriculture is a whole, and that it is pointless to improve certain factors if others remain unchanged.

Consumer tastes differ rather widely from one continent to another, which makes some of the very high-yielding cultivars of an improved botanical type created in Asia unacceptable and hence unmarketable in Africa.

Consequently, although rice research done outside Africa is extremely competent and varied (Barker 1970*a*, *b*), it cannot be immediately applied within Africa. The time has come to set up rice research teams, in Africa, of a multi-disciplinary nature, provided with the necessary facilities, capable of undertaking both basic and applied research needed to improve rice growing. The staff conducting basic research should take as their starting point the results of research carried out in Asia and America (it is obviously pointless to start from scratch again) but continue it in an African context.

Several excellent research projects on rice have already been carried out in Africa, but too often the conclusions of these projects have not been applicable beyond the limits of the main research stations or of

a few restricted multiple-site trials. Trials on a broad scale are needed. An economic appraisal could be made of each trial to determine whether the proposed changes are appropriate and feasible in particular local situations.

2.7.1. Conclusions of the rice research seminar (Ibadan 1970)

During the seminar on rice research organized at Ibadan in March 1970, the participants examined in detail the need to increase rice production in West Africa. They took stock of the existing situation of rice research and pointed to problems that require additional study. The following is a summary of their conclusions.

All the participants agreed on the need to increase rice production in West Africa. In addition, they found that the research already carried out in West Africa had demonstrated the great potential for increasing production, both by improving unit yields and by expanding the acreages cultivated. The participants considered that particular importance should be given to improving irrigated rice, since it had been shown that it is possible to reach and even exceed yields of 5 t/ha. Certain research subjects were agreed as requiring priority attention:

1. To define drought-resistance criteria to be used in selection programmes.
2. To define an 'ideotype' for upland rice.
3. To organize regional variety trials for the whole of West Africa for upland rice.
4. To establish an inventory of soils suitable for upland- or lowland-rice growing.
5. To establish an inventory of water resources which would permit the extension of irrigated rice growing.
6. To assemble bioclimatological data to determine ecological areas suitable for rice growing.
7. To study the use and improvement of relatively cheap implements, combined with the use of additional sources of energy.
8. To give some attention, at the local level, to plant spacing and rates of sowing.
9. To study the use of efficient and cheap weedicides.

Acknowledgements
Grateful acknowledgement is made to A. O. Abifarin who has provided text material relating to the rice research programme of the International Institute of Tropical Agriculture, Ibadan.

References

ABIFARIN, A. O., CHABROLIN, R., JACQUOT, M., MARIE, R., and MOOMAW, J. C. (1972). *Upland rice improvement in West African rice breeding*. The International Rice Research Institute, Los Baños.

ADRAO (1973a). Papers presented at the ADRAO seminar on the improvement and breeding of rice. ADRAO SC/2/73/4.

—— (1973b). Papers presented at the ADRAO seminar on fertilizer use in rice. ADRAO SC/2/73/5.

AKINSOLAH, E. A. (1970). The incidence of stem borer damage on swamp rice in Northern Nigeria and the control of the species concerned. *Ford Foundation/IITA/IRAT seminar on rice research, Ibadan.*

ALUKO, M. O. (1970a). Plant quarantine restrictions on rice in Nigeria. *Ford Foundation/IITA/IRAT seminar on rice research, Ibadan.*

—— (1970b). Helminthosporioses of rice in Nigeria. *Ford Foundation/IITA/IRAT seminar on rice research, Ibadan.*

ARNOULD, J. P. (1972). *Proceedings of the 16th session of the inter-government group on rice of FAO*. FAO, Rome.

ARNOUX, L. (1958). Principales varietes de riz. *Riz Rizic. Cult. vivr. trop.* February Issue, 41-7.

ARYEETEY, A. N. and KHAN, E. J. A. (1970). Status of rice research in the Accra plain. *Ford Foundation/IITA/IRAT seminar on rice research, Ibadan.*

BADINAND, B. (1969). Economie du riz par les chiffres. (IRAT mimeo.)

BARAT, H. (1970). Rice protection problems in French speaking Africa. *Ford Foundation/IITA/IRAT seminar on rice research, Ibadan.*

BARKER, R. (1970a). Probable agricultural changes in rice producing and consuming nations of Asia. *Ford Foundation/IITA/IRAT seminar on rice research, Ibadan.*

—— (1970b). Environmental factors influencing the performance of new high yielding varieties of rice and wheat in Asia. *Ford Foundation/IITA/IRAT seminar on rice research, Ibadan.*

BIDAUX, J. M. (1970). Rice research in Mali. *Ford Foundation/IITA/IRAT seminar on rice research, Ibadan.*

BIRIE-HABAS, J., AUBIN, J. P., SENE, D., and POISSON, C. (1970). Rice research in Senegal: results obtained: present and future research programmes. *Ford Foundation/IITA/IRAT seminar on rice research, Ibadan.*

CHABROLIN, R. (1969). Les récherches rizicoles en Afrique tropicale francophone et à Madagascar. *Agron. trop., Nogent* 24, 1-24.

CHANDLER, R. (1970). The technology of tropical rice: present state and future development. *Ford Foundation/IITA/IRAT seminar on rice research, Ibadan.*

DONALD, C. M. (1968). The breeding of crop ideotypes. *Euphytica* 17, 385-403.

FOOD AND AGRICULTURE ORGANIZATION (1972). *Rice production statistics, 1948-1971*. FAO, Rome.

FRANQUIN, P. (1973). *La climatologie frequentielle en agriculture tropicale technique et developpement 1973*. Bulletin No. 5.

IRAT (1970a). The rice research programme of IRAT—results and outlook. *Ford Foundation/IITA/IRAT seminar on rice research, Ibadan.*

—— (1970b). Rice research in the Ivory Coast. *Ford Foundation/IITA/IRAT seminar on rice research, Ibadan.*

MOOMAW, J. (1970). The current status of rice research in anglophone West Africa. *Ford Foundation/IITA/IRAT seminar on rice research, Ibadan.*

MORGAN, H. G. (1970). Research on rice in Sierra Leone. *Ford Foundation/IITA/IRAT seminar on rice research, Ibadan.*

PORTÈRES, R. (1956). Taxonomie agrobotanique des riz cultivés. *J. Agric. trop. Bot. appl.* 3, 341-84, 541-80, 629-700, 821-56.

SEGUY, L., NICOU, R., and HADDAD, G. (1970). The root system of four varieties of upland rice in tilled and non-tilled soils. *Ford Foundation/IITA/IRAT seminar on rice research, Ibadan.*

TANAKA, A. and KAWANO, K. (1965). Leaf characters relating to nitrogen response in the rice plant. *Soil Sci. Pl. Nutr.* 11, 251.

—— —— (1966). Effect of mutual shading on dry matter production in the tropical rice plant. *Soil Sci. Pl. Nutr.* 24, 128-44.

—— —— YAMAGUCHI, J. (1966). *Photosynthesis, respiration and plant type in the tropical rice plant*. Tech. Bull. No. 7. The International Rice Research Institute, Los Baños, Philippines.

UNDP (1971). *Note on the ecology of rice and the suitability of soils for rice cultivation in West Africa*. AGS/WARDA/71/5.

USAID (1968). *Rice in West Africa*. USDA/USAID, Washington D.C.

WEI LANG, Y. (1970). The work of the Taiwanese rice missions in Africa. *Ford Foundation/IITA/IRAT seminar on rice research, Ibadan.*

WILL, H. (1970). The status of rice research in Sierra Leone. *Ford Foundation/IITA/IRAT seminar on rice research, Ibadan.*

WILLIS, J. W. (1973). Tight world rice economy may continue in 1973. *Foreign Agric.*

3. Sorghum and pearl millet[†]

C. ÉTASSE

3.1. Introduction

Cropping systems for food crops are complex (see also Chapter 14). Many farmers still commonly employ shifting cultivation, whereby the soil is exploited for a few years and then left to lie fallow for a long period, and mixed cropping is practised, providing safeguards against adverse climatic conditions and pests and, above all, offering protection for the soil against the torrential rains if ground cover can be maintained at the time these occur. The central problem for the modern agronomist, therefore, is to find a system giving greatly enhanced overall productivity that can also maintain or improve soil fertility and protect the soils equally well or better against erosion than can traditional systems.

Cultivated area and yield statistics for sorghum and millets from 1948–52 to 1966 are presented in Table 3.1, and for 1971 in Tables 3.2 and 3.3. In

TABLE 3.1

Area and yield statistics for sorghum and sorghum and millet (from FAO *Production Yearbook* 1967)

	Sorghum 1948–52	1966	Sorghum+Millet 1948–52	1966
Far East (mostly India)	16 484	18 646	51	105
Africa (largely Nigeria and Ethiopia)	7 072	8 987	9 540	11 547
North America	3 087	5 185	—	—
Latin America	312	1 951	—	—
Near East	1 103	1 643	129	68
Australia	57	183	—	—
Europe	29	96	—	—
World total	28 144	36 682	9 720	11 720

1966, countries with average yields estimated at over 2000 kg/ha were: Argentina, 2120 kg/ha; Egypt, 3920 kg/ha; France, 2840 kg/ha; Israel, 3210 kg/ha; Italy, 3550 kg/ha; Mexico, 2480 kg/ha; USA,

[†] The term 'millet' used without qualification is a general one applied to many different small grain cereal crops produced in different places.

TABLE 3.2

Production of sorghum in 1971 (from FAO *Production Yearbook* 1972)

	Area (1 000s ha)	Yield (kg/ha)	Production (1 000s t)
Developed nations			
North America, Western Europe, Oceania, and miscellaneous	6 640	3 493	23 193
Developing nations			
Africa	10 292	733	7 543
Latin America	2 854	1 735	4 953
Near East	3 453	945	3 264
Far East	16 574	460	7 628
Other developing nations	1	1 926	2
Centrally planned nations			
USSR and China	114	1 099	126
World	39 929	1 170	46 709
Average yield	Developed nations	3 493	
	Developing nations	705	
Total production	Developed nations	23 193	
	All others	23 516	

3500 kg/ha; Yugoslavia, 2330 kg/ha. Whereas in the USA, Mexico, and Argentina there had been rapid growth of both cultivated areas and yield in the period 1948–52 to 1966, in tropical Africa the area under cultivation had increased but the yields had not.

For *Pennisetum* millet alone the area in West Africa is estimated at about 10 000 000 ha, of which the distribution by countries according to FAO (1966) was: 5 000 000 ha in Nigeria; 1 700 000 ha in Niger; 600 000 ha in Chad; 500 000 ha in Mali; 400 000 ha in Upper Volta; 400 000 ha in Senegal; 100 000 ha in the Ivory Coast. The crop is also grown extensively in India and fairly widely in Togo, Dahomey, Ghana, the Republic of Sudan, and in several East African countries, including Malawi, Uganda, Zambia and Swaziland.

Whereas yields in some countries such as the USA regularly exceed 2000 kg/ha, yields of less than

TABLE 3.3

Production of millets† in 1971 (from FAO
Production Yearbook 1972)

	Area (1 000s ha)	Yield (kg/ha)	Production (1 000s t)
Developed nations			
Western Europe, Oceania, and miscellaneous	75	1 047	78
Developing nations			
Africa	13 846	612	8 467
Latin America	405	815	330
Near East	985	1 434	1 412
Far East (largely India)	17 598	464	8 164
	32 834	560	18 373
Centrally planned nations			
Asia (largely China)	29 454	759	22 355
USSR and Europe	2 726	789	2 150
	32 180	761	24 505
World	65 089	660	42 956

† Millets comprise several distinct plant species including the following: pearl millet (*Pennisetum typhoides*); finger millet (*Eleusine coracana*); Italian millet (*Setaria italica*); little millet (*Panicum miliare*); kodo millet (*Paspalum scrobiculatum*); common millet or proso (*Panicum miliaceum*); barnyard millet (*Echinochloa frumentacea*); tef (*Eragrostis tef*).

1000 kg/ha are normal in many countries in tropical Africa. Climatic factors are largely responsible for this. It is easier to farm when there is sufficient and well distributed rainfall, where the climate is essentially similar from one year to the next, or where irrigation is available, than it is where the rainfall is unreliable in both quantity and seasonal distribution. However, yields should be able to be increased to levels of 2500 kg/ha to 4500 kg/ha with only 500 mm of annual rainfall if improved husbandry and cultivars of high genetic potential are used (Webster 1970*a*).

3.2. Sorghum

In Asia and Africa sorghum grain is used as human food. The best-known preparations made from sorghum in Africa are a type of gruel, biscuits, or couscous made from whole grains or hulled grains.

For use as a food vitreous and white grains are generally preferred, but there are many different preferences in different areas. In India sorghum is used to make chapatis and the hulled grain is boiled and eaten in the same way as rice. Sorghum flour can also be used mixed with corn flour to make bread. In Africa the grain is very commonly used to make a low-alcohol beer, which is fairly nutritious. Local malting is often carried out in the preparation of beer.

For animal feed, sorghum grain is practically as good as maize. The plant can also be used for grazing and conserved as hay or silage.

3.2.1. Outline of research on sorghum in the tropics

The objective of most research has been to increase yield per unit area per unit of time. The highest yields on this basis can be obtained by using early-maturing cultivars which are day-neutral. The most striking advances in this direction have come from the release of hybrid sorghums as, for example, those developed and released in India between 1960 and 1964 (Doggett 1970*b*).

3.2.2. Agronomy

With traditional crops low plant populations are normal and the appearance of vegetation often suggests nitrogen deficiency, though this is sometimes also due to a lack of iron or zinc. In many tests in the Indian sub-continent economic levels have been worked out for nitrogen and phosphorus application, but use of potassium, iron, and zinc has rarely been economically justified. In West Africa potassium is the major limiting element in some soils (Vaille 1970).

Fertilizer use needs to be determined for each soil series and in relation to cropping history. Some cultivars of sorghum are able to respond, for the same application levels of nitrogen fertilizers, much more favourably than others (Goldsworthy 1967). For example, whereas a local cultivar may give an extra 6–10 kg of grain yield per kilogram of nitrogen applied, a more responsive one may give an extra 20–45 kg under the same conditions (unpublished data, Bambey Research Station).

Sorghum is largely grown in hot, dry areas, but it is also better adapted than maize in heavy soils liable to waterlogging (Doggett 1970*a*). Sorghum may, however, suffer serious loss of yield as a result of water deficit occurring during anthesis. The use of early-maturing cultivars in areas where the length of the rainy season is short is a means of escaping this problem. In West Africa the landraces of un-improved sorghum (Curtis 1968*a*) adapted to each ecological zone have been selected naturally in such a way as to ensure that the seed is produced and matured at the end of the rainy period (Curtis 1968*b*).

Sorghum is often grown in mixed cropping systems with other associated crops. Andrews (1972) investigated such systems in Nigeria and concluded that mixtures can provide a better overall use of resources, particularly with long growing seasons and where different components of mixtures have peak demands for light and moisture at different times. Sorghum cultivars differ in their suitability for cultivation in mixed cropping systems (Andrews 1974). In Uganda, Osiru and Willey (1972) have also investigated the comparative productivity of systematic mixtures of sorghum with beans in comparison with the same amounts of the crops grown in separate pure stands. Their work provides further evidence for the advantages of mixtures over pure stands.

3.2.3. *Weed control*

Prolonged wet periods make hand-weeding difficult, and the use of weedicides under tropical conditions has proved rather difficult since levels effective in controlling weeds are usually also toxic to the sorghum crop.

Webster (1970*a*) considered that far more research should be devoted to control of the parasitic weed *Striga* spp. in West Africa. No cultivar so far has proved to be sufficiently tolerant to *Striga* for the problem to be regarded as one to be solved by breeding alone. *Striga* can be pulled out by hand: to avoid a serious yield loss in the sorghum this must be done before the *Striga* flowers. However, visits to farmers' fields show that this is practically never done. Too-early weeding by hand is ineffective, as it merely encourages the growth of new shoots of the parasite.

Damage by *Striga* is less when the sorghum receives adequate fertilizer throughout its growth. Some other crops, such as cotton, can stimulate germination of *Striga* seeds and not themselves become parasitized. Suitable rotations therefore offer the opportunity to reduce *Striga* damage on sorghum by planting this crop after one such as cotton.

Although weedicides have reduced the incidence of *Striga* no reports are available to show that the yield of sorghum has been increased by the treatment. Foliar applications of linuron and ametrine kill *Striga* at application rates which are well tolerated by other plants, including sorghum. Granular applications of 2,4-D and MCPA, have been used at Samaru since 1967. 2,4-D does not remain effective for long enough for weed control in late-maturing cultivars (Ogborn 1970).

3.2.4. *Crop protection against diseases, pests, and predators*

Of these, insect pests are undoubtedly the most important. Diseases, while numerous (Tarr 1962), seem not to cause severe losses (Delassus 1970). Losses due to nematodes are still quite ill-defined. Birds and rodents are also formidable predators. In India, the central shoot-fly (*Atherigona soccata*) is the most severe of three important sorghum pests, the others being the stem-borer *Chilo partellus*, and grain midge *Contarinia sorghicola*. Soil application of granular 10 per cent active phorate is effective, but this chemical is too dangerous for widespread use. There is no genetic immunity to shoot-fly, but good tolerance has been found in a number of cultivars (Doggett 1970*b*) (10 per cent of the plants destroyed as against 45-55 per cent for the sensitive cultivars).

Several species of lepidopterous stem-borers damage both sorghum crops and maize (see Table 10.7). Of these, *Chilo* sp. and *Busseola fusca* are particularly important in West Africa. Most serious damage occurs when the larvae develop in the peduncle of the inflorescence causing the whole panicle to wither. Two or three sprayings with endrin or carbaryl provide an effective chemical control but may also destroy the coccinelids which would otherwise effect a degree of biological control.

No immunity has been found against borers, but there is tolerance in some breeding materials which seems able to be combined genetically with good responsiveness to fertilizers (Brenière 1970*a*). The sorghum midge, *Contarinia*, is less of a problem when large, fairly uniform areas are simultaneously sown with a single cultivar so that flowering is synchronous than it is in less well-managed cultivation (Bowden 1965). Chemical control of this pest is possible by spraying, but in breeding plots the variation in flowering time necessitates frequent applications. Doggett (1970*b*) referred to resistance in some materials being associated with cleistogamy.

Doggett (1970*a*) estimated losses of sorghum in Africa attributable to bird damage at about 600 000 t. Bitter-grained types of sorghum are less attractive to birds than sweeter types, and some cultivars can be found which lose a good deal of their bitterness on ripening and thus only suffer damage over a brief pre-harvest period. On experimental plots, if meaningful data is to be obtained, protection from birds has to be achieved by the use of mist nets and by various types of bird scarers.

3.2.5. Seed production

Plant breeding has no practical value unless the products of the breeders can become the cultivars actually grown by farmers. To make this likely to occur requires an adequate organization for seed production. The release by breeders of the first Indian hybrid sorghum made this need abundantly clear, and as a result, a seed production and control organization was set up there (Doggett 1970*a*).

3.2.6. Basic research in support of breeding

A good deal of basic research in the field of quantitative genetics has recently been carried out with sorghum (Doggett 1970*a*). Populations with a wide genetic base have been constituted by hybridizing and compositing, in order to provide the possibility for genetic recombination from diverse sources. Selection can then be carried out recurrently (Doggett and Eberhart 1968) to overcome the apparent plateaux reached in selecting from within the pre-existing narrow genetic base.

Advances are also being made in understanding the way in which seed yield is physiologically determined (Goldsworthy 1969; Willey 1970; Willey and Basime 1973). Research topics include studies of anatomy and morphology of the inflorescence and apical meristem; the migration and redistribution of sugars and starch resources resulting from photosynthesis; the relationship between reproductive and vegetative growth affecting maturity and the distribution of dry matter between seed and the remainder of the plant (crop index); responses to light and associated morphological changes; the effects of plant population and spatial arrangements on the morphology and other characteristics of individual plants; and the effect of drought and high temperatures.

Of the direct components of yield, grain number, and grain weight, it is well known that the most easily modifiable is the grain number. The time at which this is determined in ontogeny is important, because it is at this period in the growth of the plant that the yield will be most affected by different agronomic treatments.

3.2.7. Germplasm resources† and their assessment

Two projects are under way to maintain, develop, assess, and distribute world sorghum germplasm (House 1972). The first is centred on India, where more than 16 000 accessions have been assembled from around the world. This collection has been studied for resistance to many insects and diseases and has been extensively botanically characterized (Murthy, Arunachalam, and Saxena 1967). The second, carried out mainly in Texas and Puerto Rico (Miller and Rosenow 1967) aimed at converting the greatest possible range of diversity in other types into dwarf and day-neutral analogues for evaluation. This is done by back-crossing using Dwarf Martin as the dwarf day-neutral parent. In the conversion programme the letter 'C' after any sorghum accession number indicates that it has come from a 'converted' line developed in the back-cross programme.

In 1965 the All-India Sorghum Improvement Programme (AICSIP) began distributing uniform nurseries of cultivars and breeding lines to interested countries (House 1970). This function is now being resumed by the International Crops Research Institute for the Semiarid Tropics (ICRISAT) at Hyderabad. In 1970 R. C. Pickett undertook to send 300–400 lines from the programme (Pickett 1970) at Purdue, USA, to some 60 countries under a USAID contract.

3.2.8. The status of current research

OAU/Standing Technical Research Committee (STRC) joint project no. 26. Following a recommendation of the 1960 conference on African cereals, this co-operative project between African states was initiated in 1963. The three principal objectives are:

(1) to promote breeding and agronomic research on the basic cereals produced in dry-land cultivation with a view to increasing food production in Africa;

(2) to promote co-operation between research organizations;

(3) to train senior national agricultural research personnel.

† The term 'germplasm' is widely used and understood as the sum of gene assemblies carried by the widest possible range of actual genotypes. Gene assemblies of high adaptive value are likely to occur in primitive landraces of crops (according to the views of Vavilov). The conservation and study of collections of wild and primitive material of crop plants has become an important aspect of modern crop study. For a fuller account of such work see Frankel and Bennett (1970).

Until 1970 the first two subjects received most attention (Webster 1970*b*), but it is now increasingly realized that training must take priority.

Progress with new cultivars. In Nigeria selection from local populations achieved 15–20 per cent yield improvement in experimental plots over local unselected populations. But when farmers actually cultivated these selections in the traditional way these small increases were not sufficient to be noticeable, nor were there any striking differences from the local cultivars. This sort of experience has led to the view that in order to make a real impact on the farmers at least a 50 per cent increase in the yield of any new cultivar or package of practices including a new cultivar should be aimed at; and for additional effect it is helpful if the new material has quite different vegetative characteristics from the old, while at the same time preserving the grain quality and any other characteristics that determine acceptability. It was thought at first that it would be relatively easy to develop new cultivars in Africa of greatly increased potential, but new obstacles have constantly arisen.

In order to achieve populations of from 80 000 plants to 100 000 plants per ha and to use nitrogen fertilizer without risk of lodging, dwarf genes have to be present in the selections. Dwarf sorghums were bred at Samaru in Northern Nigeria using an X-ray dwarf mutant of a local 'Kaura' type (Curtis 1967) of sorghum. Breeding populations based on ms_7 genetic male sterility (Andrews and Webster 1971), also used introduced dwarfs such as 3-dwarf Combine Kaffir 60 to supply the dwarfing genes. Even by this method, however, yields have not exceeded 4500 kg/ha even in the best conditions. Only the use of hybrids seems to make it possible to go beyond this level. Dwarf cultivars appear to be more susceptible to stemborers than tall ones.

Other factors that need to be considered in developing new cultivars are as follows.

1. Traditional cultivars have acquired a good level of tolerance to *Striga* by natural selection, which is not present in many of the foreign cultivars. Care needs to be taken to preserve this tolerance in any new releases.
2. Early-maturing cultivars sown late are not suitable alternatives to longer-maturing cultivars sown earlier, unless they contain sufficient resistance to withstand the increased levels of central shoot-fly attack.

3. Day-length-sensitive cultivars have very limited latitudinal adaptation. The traditional distribution into two zones—the Guinea savanna and the Sudan savanna—is too crude to define distinctions between the adaptations of various day-length-sensitive cultivars.
4. If hybrids are to be grown, methods will have to be developed for local seed production.

3.2.9. Current sorghum research in the IRAT programme

Objectives of the IRAT breeding programme. The objectives are as follows:

(1) to reduce the height of the sorghum stalk to between 1 m and 2 m;
(2) to achieve panicles that are not too compact, so as to prevent moulds;
(3) to achieve specific adaptation to different environments, particularly in terms of the length of the vegetative period and resistance to pests;
(4) to produce a plant with large, flinty grains, giving a white flour.

At present only open-pollinated cultivars are released, but hybrids will be ready in the near future.

Methods and stages of breeding in the IRAT programme. Classical breeding methods are adopted, but are made more or less complicated according to the resources available and the length of time that the work in each country has been under way (Le Conte 1970). The methods consist of:

(1) systematic collection and testing of local germplasm;
(2) the introduction of cultivars and breeding lines and segregating populations;
(3) hybridization between local types and introductions. Some hybridizations have been facilitated by the use of introduced male sterile lines;
(4) selection within and between composites of different genetic background to improve each composite while maintaining high combining ability between them;
(5) creation of F_1 hybrids employing local selection for progressively improving the parents for maturity and grain characteristics.

Cross-pollinating populations are obtained by compositing seed from derivatives of cultivars obtained by back-crossing to a source of genetic male sterility. This population is maintained for three generations by applying a 10 per cent selection intensity for seeds from male sterile plants. In such populations repeated randomized crossing favours recombination and the breaking of linkages. By encouraging repeated genetic recombination over a number of generations, this method solves the problem of the very large population size that would otherwise be required to obtain rare recombinants in conventional breeding where crosses are made by hand. When apparently useful recombinants are observed in still largely heterozygous individuals, lines can be obtained by the pedigree method among male fertiles.

The outbreeding populations rapidly respond not only to the pressures of breeders' selection but also to natural selection. If variability is to be maintained, therefore, steps have to be taken to form each successive season's population in such a way as to remove bias in favour of particular fitness characteristics.

For the breeding of F_1 hybrid sorghums using cytoplasmic male sterility and restorer genes it is possible to use population breeding in each of two isolated populations. At Serere, Uganda (Doggett 1970*b*) eight populations were constituted and recurrently recombined for three generations of random mating. Three selection procedures were then applied to compare these for efficiency (Doggett 1970*b*).

3.2.10. *Fertilizer policy*

Research workers in all sorghum-growing countries have investigated the use of fertilizers. As with fertilizer use for all crops, the type and amount of fertilizer to use in sorghum growing cannot be generalized but needs to be considered for each individual case in relation to the soil type, previous cropping history, the rotation being practised, the particular cultivar of sorghum being grown, and the costs and returns based on current prices.

3.2.11. *Tillage of the soil*

In temperate countries agronomists generally agree on the positive value of tillage in creating and preserving a satisfactory agricultural soil. The situation needs more critical attention in a tropical zone, where erosion can be aggravated by the effect of intense rain on tilled, bare soil. By taking precautions regarding the period for this work and to obtain a rapid vegetation cover after tilling, some very favourable

results have been obtained with the sorghum crops in dry tropical zones, notably in Senegal (Nicou 1970).

Tillage at the start of the rainy season is not always possible without delaying sowing. In order partly to escape this limitation it is possible to hold the tillage over until the end of the rains provided that the rainfall period is of slightly longer duration than the vegetative cycle of the crop preceding the sorghum.

3.2.12. *Sorghum physiology*

Plant type. Tall sorghums predominate among the wild and unimproved crop populations in West Africa. In contrast, in the USA, where the crop has been most highly developed, the commercial grain sorghums have been dwarfed by the use of dwarfing genes. Increasing importance is now being attached to studies of plant morphology or ideotype (Donald 1968) as a prime determinant of potential productivity. At Samaru, Nigeria, the comparative agronomic and physiological characteristics of chosen examples of tall and dwarf sorghums (Farafara and NK300) have been studied (Goldsworthy 1970; Curtis 1970; Goldsworthy and Tayler 1970). The extent to which the contrast between these chosen genotypes can be interpreted as differences attributable to plant type needs further clarification.

Very little is known concerning the rooting behaviour of sorghum. In Bambey a careful study under two conditions of fertility and tillage of the soil has enabled us to determine the pattern of development of the root system from sowing until ripening. Rooting is not very extensive from seedling emergence until the start of tillering at about 33 days. Roots then develop very quickly up to the time of floral initiation at about 50 days. From then until the emergence of the inflorescences there is a slight increase in root depth. By ripening there are hardly any active roots left at all (Chopart 1970).

Tillage favours deep rooting but fertilizer application tends to make it more superficial. Maximum root depth for different conditions ranges from about 110 cm to 140 cm; 75 per cent of all roots occur in the upper 20 cm of horizon.

Relationship between root and aerial growth and reproduction. The ratio of above ground to below ground growth varies a good deal during the course of vegetation as the timing of each is different. There is a close correlation between weight of root and weight of grains.

Physiological characteristics influencing grain yield.
In Nigeria (Goldsworthy 1969) and at Makerere
University in Uganda, Basime and Willey (Willey
1970; Willey and Basime 1973) have studied the
applicability to sorghum of the growth and development models developed for temperate cereals. Three
principal phases can be defined:

(1) the purely vegetative phase (until the initiation
of the inflorescences);
(2) the phase of development of the panicle;
(3) the phase of grain filling.

The prime object of the work was to examine the
relative importance of the panicle capacity (assimilate sink) and the supply of carbohydrates (assimilate
source) as determining factors of yield.

In the Uganda work, a shading test on cultivar
Dobbs of intermediate height and on a dwarf breeding line from the Texas conversion programme (see
p. 30) showed that a 30 per cent reduction of
incident light reduced the yield of Dobbs by 22·6
per cent but only by 5·2 per cent for the dwarf.
A 50 per cent defoliation treatment also showed that
the grain size and yield were far more affected with
Dobbs than occurred with the dwarf line.

These findings were interpreted as implying that
whereas the panicle capacity is the limiting factor in
the dwarf, it is the availability of assimilates and not
ear size that is limiting in Dobbs. Thus it is apparent
that, in sorghum, physiological determination of
yield may differ between cultivars and that there is
scope for renewed physiological study for each new
selection. In later studies (Willey and Basime 1973)
five cultivars including Dobbs were compared.

3.2.13. Improving the quality of sorghum

Delicate pieces of apparatus, such as automatic
amino-acid analysers, do not lend themselves to
extensive use in countries very short of technical
skill and financial resources. However, a number
of partial studies on nutritional quality have been
carried out in Africa.

In Senegal research has shown that a 20 per cent
increase in the protein content of the grains can be
obtained with nitrogen applied at seeding, and
an 88 per cent increase with additional nitrogen
fertilizer dressing applied at emergence of the panicle
(Abifarin 1970). In the most extreme set of data (but
bearing in mind the increased grain yield from the
treatments), the protein production per hectare was
529 per cent of the control. Nitrogen effects both net

photosynthesis as assessed by the production of dry
matter and metabolism of the nitrogen in the plant
as assessed by the protein content of the grain.

Most of the work on improving the nutritional
quality of sorghum has been carried out at Purdue,
USA by Pickett and his associates (Pickett 1970).
They found large variations in the grains studied,
both in total proteins (from 8 per cent to 20 per cent)
and in the proportions of amino acids in the proteins
(0·85–3·36 per cent lysine; 2·84–5·53 per cent
threonine; 0·34–4·51 per cent tryptophan).

The protein level and lysine content of the embryos
were found to be higher than that of the endosperm.
Selection for a relatively large embryo in relation to
total seed weight can therefore be used as an indirect
approach to selecting for nutritional quality. Studies
have shown that the world collection contains genotypes that have both an acceptable yield level and
good protein and lysine content.

3.2.14. Documentation

A sorghum bibliography of 2000 references was prepared for the period 1930–63 by a group from the
George Washington University. Doggett's book
(Doggett 1970a) on sorghum and Tarr's book (Tarr
1962) on sorghum diseases are important major
works. An annual Sorghum Newsletter is produced;
the sixteenth was published in 1973.

3.3. Pearl millet

All cereal species of pearl millet are classified in the
section *Penicillaria* of the genus *Pennisetum*. All
species of this section have a basic chromosome
number of $x = 7$. Hutchinson and Dalziel's classification (1958) is unsatisfactory for agricultural botanical
use since there is apparently no limit to the recombination of characteristics involved in taxonomic
differentiation when crosses are made. *Pennisetum
typhoides* Stapf and Hubbard comprises the broadest
concept of the species (Stapf and Hubbard 1934) that
includes all the cultivated pearl or bulrush millets.
Subspecific separation within this species is possible
by placing all cultivated forms into the subspecies
Pennisetum typhoides spp. *cereale* and the wild forms
of the 'violaceum' type into *Pennisetum typhoides* spp.
violaceum.

Genetic analysis of crosses between the spp.
violaceum and *cereale* have suggested that the transformation of the wild into the cultivated forms has
occurred by a number of mutations involving the

following changes: persistence of spikelets until ripening (elimination of abscission); and far larger grains with non-adherence of the glumellae. The occurrence in crops of individual plants that look more or less like a wild variety of *Pennisetum*, having deciduous spikelets and medium-sized grains, suggests either that introgression from wild relatives or back mutation are occurring.

The study of stabilized local landraces enables us to get a good idea of the species' natural variability, but once these are brought together in collections at a single site, the cross-fertilization due to protogyny, makes it difficult to maintain these as distinct entities.

3.3.1. The general objectives of pearl-millet research

Improvement of yield. In traditional conditions and with unimproved cultivars yields rarely exceed 500 kg/ha. A prime objective therefore is yield increase. This may come about by either agronomic or genetic improvement or both.

Production agronomy. Pearl millet is a plant which responds well to tillage, especially so in a very sandy soil. The use of nitrogenous fertilizer often gives spectacular results, but too much fertilizer will promote rank vegetative growth at the expense of grain yield. For example, in some instances up to 20 t/ha dry straw have been obtained for only 2–3 t grain (CNRA 1964, 1965).

Genetic improvement. Genetic improvement has so far given discouraging results in Africa. However, in India F_1 hybrids, which were first released in 1966, have attained an established position in commercial cultivation (Andrews, D. J., personal communication). In Africa the lack of success so far may be either because the selections have been based only on the genetic base of local populations without bringing in alien germplasm by hybridization, or because mass-selection methods have been used on characteristics with low heritability instead of using components of yield that are more highly heritable, such as the size, length, and compactness of the panicles.

Attempts to use selections for general combining ability have not been made under conditions of sufficient experimental precision and the very strong genotype × environment interaction has not been sufficiently taken into account.

Physiological approach to grain yield improvement. To obtain a substantial gain in grain production,

straw production may have to be reduced, i.e. achieving an improvement of the grain/straw ratio and hence more favourable partitioning of assimilates. One of the most important factors may be the reduction in the weight of the straw through the use of dwarf cultivars, coupled with the use of high-density plant populations.

Recent physiological work at Bambey (Jacquinot 1970*a*) has shown that the dry matter that is used for grain filling is synthesized after flowering; only the products of two or three upper leaves are concerned with supporting the early stages of floral development, and these do not subsequently respire more than they assimilate.

Improvement in the nutritional quality of the grain. Recent work at Bambey (Jacquinot 1970*b*) has shown that variability in nutritional value of the grain is extensive. If the grain protein is unusually rich in tryptophane, it is often correspondingly deficient in lysine.

Principles and methods of grain improvement. It is possible to achieve self-fertilization by bagging the ears of pearl millet. After a few generations there are effects on vigour and fertility, but this inbreeding depression is not so marked as in some other heterozygous plants such as maize.

Lines can be obtained through selection in the partially inbred segregants, whose grain production is equal to, and in some instances even higher than that of local open-pollinated populations. It is possible therefore to envisage creating stable synthetics by crossing between selected partially inbred lines of high combining ability. D.174 was obtained in this way in Coimbatore from the progeny of the $D_2 \times IP.81$ cross.

According to Bilquez (1970) studies of the genetic mechanisms of heterosis for yield in the Indian sub-continent (Burton 1968) have shown that heterosis is closely associated with the degree of heterozygosis and that additive effects only contribute 40 per cent to the total variance. This indicates clearly the value of using F_1 hybrids, where the non-additive gene interactions can be exploited (Burton and Powell 1968). If, instead of considering the total grain yield, selections are made for yield components, then general combining ability (Murthy, Tiwari and Harinarayana 1967) (depending upon additive genetic effects) becomes a far more useful selection criterion than specific combining ability. High values of different components may be selected in separate groups of lines between which recombination will subsequently be allowed to occur so as to combine two or more favourable yield components. Fortunately, three different forms of cytoplasmic sterility are known in pearl millet (Burton and Athwal 1967), which makes it possible to obtain F_I hybrids quite easily.

To obtain constantly improved F_I hybrid combinations, reciprocal recurrent selection can be used in paired populations, each providing a restorer for the cytosterility of the other. Nevertheless, until now in Africa it has been considered wiser to provide for the creation of synthetic populations for which new seed would only have to be provided to cultivators every three or four years instead of each year, as would be desirable for F_I hybrids.

3.3.2. *Improvement objectives for pearl millet for adaptation to intensive agriculture*

Agronomical imperatives and recent physiological studies both indicate the same steps that must be taken to correct the principal defects in the local cultivars of pearl millet. Of these, the need to reduce straw production is definitely the most important requirement. Straw length affects the plants' requirements in mineral elements, especially for nitrogen. Reduced length enables the stand to be increased without risk of lodging and makes mechanical farming easier, especially combine harvesting, interim cultivation, and the burying of residues after harvest.

The following type of pearl millet is suggested as an ideotype (Étasse 1970): dwarf stature; medium tillering capacity, to allow phenotype adaptation to more or less favourable environmental conditions; a favourable leaf structure so as to avoid self-shading; a short maturity period achieved by reducing the number of nodes; disease resistant; having long, compact heads; and drought and heat tolerant.

However, to improve our knowledge of selection criteria further, a number of studies must be carried out to give this general model more precision. We need to know which morphological characteristics of the plant have the greatest influence on the grain/total dry matter ratio, what should be the maximum head density in relation to the structure of the plant, and just how much tillering capacity is desirable.

3.3.3. *Diseases and pests of economic importance*

Principal diseases of pearl millet. Three diseases can be considered as being of economic importance, and

these, moreover, vary according to the annual climatic conditions and the country (King 1970).

1. *Downy mildew caused by Sclerospora graminicola.* It is not uncommon to find symptoms on 50 per cent of the plants in a field and yield losses have been estimated at as much as 10 per cent some years in Nigeria (King 1970). The symptoms vary enormously, the loss of yield depending on the stage in the plant's life at which they first appear. Oospores in the soil infect the plant from the seedling stage. Secondary infections occur through conidial dispersal. The degree of infection depends very closely upon soil and atmospheric moisture. The only effective method of controlling the disease is by the use of resistant cultivars; good sources of resistance can be found in photoperiod-sensitive West African cultivars (Andrews, D. J., personal communication). Breeding for resistance can be facilitated by increasing the probability of infection by mixing infected dead plant materials into the soil where segregating material is to be grown.

2. *Smut caused by Tolyposporium penicillariae.* In this disease the flowers become infected before the stigmas appear. Generally speaking the damage is not particularly serious, but in breeding plots the use of bags to ensure self-fertilization creates conditions very favourable to the disease. Genetic resistance is the most appropriate means of combating the disease, although fungicides have also been effective (King 1970).

3. *Honeydew and ergot caused by Claviceps microcephala (conidial Sphacelia sp.).* Most of the local cultivars of pearl millet in West Africa are very susceptible. No adequate source of resistance is yet known, and other methods of attempted control have not been effective.

Principal insect pests of pearl millet. There are two principal pests which attack pearl millet.

1. *The millet midge, Geromyia penniseti.* Damage by this pest was first recognized only in 1966 in Senegal (Brenière 1970b) where the millet midge completely destroyed the production of all millet plants flowering between 10 September and 15 October. This parasite in fact multiplies very rapidly at that time of year, then decreases under the joint effects of parasitism and larval diapause.

2. *Borers.* The most common stem-borer in pearl millet in West Africa is *Acigona ignefusalis*. Traditional millet cultivars are tolerant in that the stalks can contain several larvae without any visible

C

damage. With short and slender-stalked cultivars that might be developed in future, it is possible that damage for comparable levels of infestation might be much greater. Late-maturing millet is more heavily infested than early maturing millet, as it may be attacked by three successive generations of larvae. Damage, however, in these vegetatively vigorous forms is comparatively less.

3.3.4. *Agronomy and physiology of millet*

Research has been carried out in all countries of West Africa on techniques of cultivation and on fertilizer response in pearl millet following which some generalizations can be made.

1. Tilling is worthwhile, even in very sandy soils where it might seem unnecessary.
2. Fertilizer applications to rectify recognized deficiencies often produced spectacular improvements, especially in long-term trials. In single-season trials extraneous factors often mask the effect of fertilizer treatments.
3. Nitrogenous fertilizer is very important but largely increases vegetative growth, often bringing about imbalance if not accompanied by changes in other factors, such as increasing plant population.

In experiments with the use of nitrogenous fertilizers, grain yields exceeding 2000 kg/ha have not been obtained. At nitrogen application levels beyond 60 kg/ha the excessive growth appears to depress grain production. However, a late dose of nitrogen at the end of the season increases both the grain production and its protein content (Blondel 1970).

It is suggested that the presently used ammonia-based fertilizers (ammonium sulphate or urea), although satisfactory for slow-growing plants, are not at all suitable for rapid-growing dry-land plants such as pearl millet. The nitrogen effectiveness is almost doubled if we change from a fertilizer with a predominance of ammonia to one that has nitrogen at higher oxidation levels (Jacquinot 1970*a*).

Water requirements of early-maturing pearl millet crops. Experiments were carried out during two seasons of very different rainfall pattern, the first was characterized by a very high PET (potential evapotranspiration) and only just sufficient precipitation to support short-period crops, and the second by a low PET and sufficient rainfall. The Souna cultivar was grown normally and provided

grain yields of 1700 kg/ha and 1500 kg/ha for 300 mm and 320 mm of water consumption respectively. It thus appears that the crop requires 350 mm of soil moisture, and if rainfall is well distributed—taking into account losses due to run-off, percolation, etc. which can be estimated at 20 per cent—suitable rainfall for a rain-fed crop would be about 440 mm for a cultivar with a maturation period of 90–100 days (Dansette 1970).

The multiplication of similar and more detailed studies would enable us to learn in greater detail the crop's water requirement at different periods and to explain and possibly to forecast the yield, but above all, to determine from agroclimatological information the zones most favourable to each cultivar.

3.3.5. *Documentation*

A review of pearl millet has recently been published (Ferraris 1973).

3.4. Future research and the application of existing information

3.4.1. *Sorghum*

Genetic improvement. Much work remains to be done in determining the value of physiological parameters to breeding and selection. In particular whether the number of leaves and their orientation is as important as is suggested by work on other cereal crops. The role of erect leaves in maize is still the subject of great controversy, and similarly in sorghum, net photosynthesis and the partitioning of dry matter need further critical study before ideotypes based on research in other crop species are uncritically adopted. It is generally acknowledged that dwarfism is desirable, but exactly how far this characteristic should be taken is not known. Dwarfism reduces straw production, but when it is exaggerated there tends to be a density of leaves which is unfavourable to light penetration and likely to impede photosynthesis.

The maturity period of cultivars needs to be related to climate and photoperiod. To produce good-quality grain, breeders need to select cultivars whose flowering approximately coincides with the end of the rains. Cultivars adapted to zones with a relatively long rainy season tend to be photosensitive. The nature of this characteristic is still too little studied, and observations must therefore be made of the reaction of all germplasm under well-controlled conditions.

The expression of resistance to drought during early stages of vegetation probably depends upon soil physical properties and the tillage methods used. There is little evidence of genetic variation between cultivars in early drought resistance; resistance to drought at the end of the cycle however is dependent on genotype. This feature is very complex because of the interactions between the soil, root system, plant population, and cultivars. It is considered that it is safer for the breeder to attempt to develop new cultivars adapted to cultivation at low plant population than to breed for high yield per unit area, but having this depending on establishing and maintaining high plant populations. In the former approach we may produce a very satisfactory yield but the total leaf area and, therefore, evaporation may be reduced.

Damage caused by leaf diseases is rarely considered by physiological agronomists, but, in especially humid conditions this damage may be very considerable. Differences in the susceptibility to such damage of the various cultivars should be catalogued. Moulds that destroy grain in the ear are a great problem which must be solved. If cultivars could be found which were more resistant to such damage, then earlier maturity and more efficient types could be more widely recommended.

Selection for the total protein content and especially for a good amino-acid balance is an important objective. Analyses are difficult and expensive to carry out, and no country in West Africa is at present in a position to consider such a programme on its own. The most appropriate way to work in the direction of quality improvement will be through international collaboration, and the workers at Purdue in the USA are recognized as being in the forefront of this movement (see p. 30).

Further emphasis should be given to the preservation of germplasm in a 'gene bank' or 'world collection' and a tropical country where sorghum has a long growing period should be used to house the collection.

Selection techniques. The theory of a population approach to genetic improvement has been very well defined by Doggett and Eberhart (1968), but the practical implementation of their method raises a certain number of problems which need to be overcome. The first results of tests of F_1 hybrid breeding in Africa are very promising; this breeding method should be intensively followed up. Although there are plenty of restorer (R) lines available, very few lines adapted to tropical Africa have the cytoplasmic male sterility which is required for the female parents of hybrids.

Plant protection. The use of genetic resistance is the cheapest way of limiting damage by pests and diseases, but it is not likely to be possible in all cases. Some other approaches are also very easy; one example of these is the treating of seeds against smuts.

In order to control the main insect pests research must be continued on this front, especially on the use of systemic insecticides against borers and shoot-fly.

The *Striga* problem must be studied in depth from the chemical-treatment point of view and also from the agronomical (trap plants, rotations) or biological points of view (feeding substances capable of promoting the germination of *Striga* seeds). This is being undertaken at the University of Sussex in England.

Agronomy. Studies must be made on some neglected problems. Topics include the variations in rooting from the agronomical and genetic points of view, and the reasons for the reduction in yield of other crops following one of sorghum on the same land.

Technology. As sorghum in Africa is still essentially a product for family consumption, it is marketed only on a very small scale, and research into its use is still very limited. To promote increased production, a marketing organization and regular outlets, whether for milling, brewing, or cattle feed, would be essential. Although work is in progress at the Institute of Food Technology in Dakar on the use of both sorghum and millet in baking, there has so far been no large-scale achievement.

3.4.2. *Pearl millet*

Most of the problems of sorghum growing are also problems of pearl-millet growing. The most important advance to be expected with pearl millet is the transfer of true dwarfism to early cultivars, to obtain a plant type resembling a temperate cereal. Conversion of late cultivars to dwarfs is also envisaged.

Further research is urgently needed to obtain genetic resistance to downy mildew (*Sclerospora graminicola*) and ergot (*Claviceps purpurea*) in acceptable cultivars. Emphasis should be placed on obtaining vigorous and homogeneous cultivars. We may look forward to the eventual production of F_1 hybrids of this crop in West Africa.

References

ABIFARIN, A. O. (1970). Quality improvement in sorghum. *Ford Foundation/IITA/IRAT seminar on sorghum and millet, Bambey*.

ANDREWS, D. J. (1970). Breeding and testing dwarf sorghums in Nigeria. *Exp. Agric.* **6**, 41–50.

—— (1972). Intercropping with sorghum in Nigeria. *Exp. Agric.* **8**, 139–50.

—— (1974). Responses of sorghum varieties to intercropping. *Exp. Agric.* **10**, 57–63.

—— WEBSTER, O. J. (1971). A new factor for genetic male sterility in *Sorghum bicolor* (L.) Moench. *Crop Sci.* **11**, 308–9.

BILQUEZ, A. F. (1970). Aspect général des récherches sur les mils en Afrique. *Ford Foundation/IITA/IRAT seminar on sorghum and millet, Bambey*.

BLONDEL, D. (1970). Recent results in altering the protein content of pearl millet grain by the use of nitrogen fertilizers in Senegal. *Ford Foundation/IITA/IRAT seminar on sorghum and millet, Bambey*.

BOWDEN, J. (1965). Sorghum midge (*Contarinia sorghicola* Coq.) and other causes of grain sorghum loss in Ghana. *Bull. ent. Res.* **56**, 169–87.

BRENIÈRE, J. (1970a). Les problèmes d'importance économique aux insectes du sorgho en Afrique occidentale. *Ford Foundation/IITA/IRAT seminar on sorghum and millet, Bambey*.

—— (1970b). Millet pests of economic importance. *Ford Foundation/IITA/IRAT seminar on sorghum and millet, Bambey*.

BURTON, G. W. (1968). Heterosis and heterozygosis in pearl millet forage production. *Crop. Sci.* **8**, 229–30.

—— ATHWAL, D. S. (1967). Two additional sources of cytoplasmic male sterility in pearl millet and their relationship to Tift 23A. *Crop Sci.* **7**, 209–11.

—— POWELL, J. B. (1968). Pearl millet breeding and cytogenetics. *Adv. Agron.* **20**, 48–89.

CHOPART, J. L. (1970). Morphology and root development of sorghum in contrasting fertility conditions. *Ford Foundation/IITA/IRAT seminar on sorghum and millet, Bambey*.

CNRA (1964, 1965). Sections on plant breeding in *Annual Reports for CNRA, Bambay, Senegal*.

CURTIS, D. L. (1967). The races of sorghum in Nigeria. *Exp. Agric.* **3**, 275–86.

—— (1968a). *The races of sorghum in Nigeria: their distribution and relative importance*. Institute for Agricultural Research, Bull. No. 86. Samaru, Nigeria.

—— (1968b). The relation between the date of heading of Nigerian sorghums and the duration of the growing period. *J. appl. Ecol.* **5**, 2–5.

—— (1970). The canopy structure of tall and short sorghums. *J. agric. Sci. Camb.* **75**, 123–31.

DANCETTE, C. (1970). Use of the neutron probe to study the water relations of souna millet under contrasting moisture regimes. *Ford Foundation/IITA/IRAT seminar on sorghum and millet, Bambey*.

DELASSUS, M. (1970). The main diseases of sorghum in West Africa. *Ford Foundation/IITA/IRAT seminar on sorghum and millet, Bambey*.

DOGGETT, H. (1970a). *Sorghum*. Longmans, London.

—— (1970b). Sorghum breeding. *Ford Foundation/IITA/IRAT seminar on sorghum and millet, Bambey*.

—— EBERHART, S. A. (1968). Recurrent selection in sorghum. *Crop Sci.* **8**, 119–21.

DONALD, C. M. (1968). The breeding of crop ideotypes. *Euphytica* **17**, 385–403.

ÉTASSE, C. (1970). Breeding aims for pearl millet for intensive cultivation. *Ford Foundation/IITA/IRAT seminar on sorghum and millet, Bambey*.

FERRARIS, R. (1973). *Pearl millet*. Commonwealth Agricultural Bureaux Review Series No. 1. Commonwealth Bureau of Pastures and Field Crops, Hurley, U.K.

FRANKEL, O. H. and BENNETT, E. (1970). *Genetic resources in plants—their exploration and conservation*. I.B.P. Handbook No. 11, Blackwells, Oxford and Edinburgh.

GOLDSWORTHY, P. R. (1967). Responses of crops to fertilizers in northern Nigeria—I—sorghum. *Exp. agric.* **3**, 29–40.

—— (1969). The source of assimilates for grain development in tall and short sorghum. *J. agric. Sci. Camb.* **74**, 523–31.

—— (1970). The growth and yield of tall and short sorghum in Nigeria. *J. agric. Sci. Camb.* **75**, 109–22.

—— TAYLER, R. S. (1970). The effect of plant spacing on growth and yield of tall and dwarf sorghum in Nigeria. *J. agric. Sci. Camb.* **74**, 1–10.

HOUSE, L. R. (1970). A world view of sorghum research including the economics of production. *Ford Foundation/IITA/IRAT seminar on sorghum and millet, Bambey*.

HUTCHINSON, J. and DALZIEL, J. M. (1958). *Flora of West tropical Africa*. The Crown Agents, London.

JACQUINOT, L. (1970a). La nutrition carbonée du mil. I—Migrations des assimilats carbonés durant la formation des grains. *Agron. trop. Nogent* **12**, 1088–95.

—— (1970b). The potential of pearl millet. *Ford Foundation/IITA/IRAT seminar on sorghum and millet, Bambey*.

KING, S. B. (1970). Millet diseases. *Ford Foundation/IITA/IRAT seminar on sorghum and millet, Bambey*.

LE CONTE, J. (1970). Current state of research on sorghum in the IRAT programme. *Ford Foundation/IITA/IRAT seminar on sorghum and millet, Bambey*.

MILLER, F. R. and ROSENOW, D. T. (1967). Sorghum conversion project. *Sorghum Newsl.* **10**, 100.

MURTHY, B. R., ARUNACHALAM, V., and SAXENA, M. B. L. (1967). Classification of a world collection of sorghum. *Ind. J. genet. Pl. Breed.* **27**(1), Special number.

—— TIWARI, J. L., and HARINARAYANA, G. (1967). Line × Tester analysis of combining ability and heterosis for yield factors in *Pennisetum typhoides*. *Ind. J. genet. Pl. Breed.* **27**, 238–49.

NICOU, R. (1970). The effects of tillage practices on sorghum productivity in the dry tropical zone. *Ford Foundation/ IITA/IRAT seminar on sorghum and millet, Bambey.*

OGBORN, J. (1970). Methods of controlling *Striga hermonthica* for West African farmers. *Ford Foundation/IITA/ IRAT seminar on sorghum and millet, Bambey.*

OSIRU, D. S. O. and WILLEY, R. W. (1972). Studies of mixtures of dwarf sorghum and beans (*Phaseolus vulgaris*) with particular reference to plant population. *J. agric. Sci. Camb.* **79**, 531–40.

PICKETT, R. C. (1970). *Inheritance and improvement of protein quality and content in Sorghum vulgare* Pers. Progress Report to January 1970 USAID/Purdue University contract AID/csd-1175.

STAPF, O. and HUBBARD, C. E. (1934). *Pennisetum.* In *Flora of tropical Africa* (ed. D. Prain), Vol. 9. L. Reeve, Ashford, England.

TARR, S. A. J. (1962). *Diseases of sorghum, sudangrass, and broomcorn.* Commonwealth Mycological Institute, Kew.

VAILLE, J. (1970). Fertilizer practices for sorghum in Northern Cameroun. *Ford Foundation/IITA/IRAT seminar on sorghum and millet, Bambey.*

WEBSTER, O. J. (1970*a*). Current status of sorghum research in the OAU/STRC joint project. *Ford Foundation/ IITA/IRAT seminar on sorghum and millet, Bambey.*

—— (1970*b*). Report on 3rd conference of the OAU standing technical research committee at Zaria, October 13th–16th, 1969. *Sorghum Newsl.* **13**, 61–3.

WILLEY, R. W. (1970). Agronomy of sorghum. *Ford Foundation/IITA/IRAT seminar on sorghum and millet, Bambey.*

—— BASIME, D. R. (1973). Studies on the physiological determinants of grain yield in five varieties of sorghum. *J. agric. Sci. Camb.* **81**, 537–48.

4. Grain legumes

K. O. RACHIE AND P. SILVESTRE

4.1. Importance and production of grain legumes (pulse crops and oilseeds)

4.1.1. Present status

Grain legumes are a major component of cropping systems in the lowland tropics of Africa and ecologically similar regions around the world. Several species have great potential for contributing to both protein and calorie intake of humans and domestic animals. Dry legume seeds are frequently the most practical source of storable and transportable protein in regions lacking refrigeration facilities. Their ability to grow vigorously in diverse environments and especially on poor, nitrogen-deficient soils is particularly advantageous in subsistence agriculture in remote areas. The consistently high yields of groundnuts, the rapid vegetative growth of cowpeas and dry beans, the extended reproductive period of viny species (yam, lima, and velvet beans), and the long survival of woody perennials such as pigeon peas and locust beans offer complementary advantages in complex farming systems.

4.1.2. Food uses and value of legumes

Food legumes often have a distinct advantage over many other crops in the simplicity of their preparation and in the multiplicity of edible forms. Tender green shoots and leaves, unripe whole pods, green seeds, and dry seeds are all commonly eaten, while some species, notably the yam bean (*Sphenostylis stenocarpa*) and winged bean (*Psophocarpus tetragonolobus*), produce edible tubers as well. The nutritional value of most food legumes is particularly important in complementing the non-cereal, low protein, starchy staple foods of many tropical diets, such as the roots and tubers and edible bananas. However with cereal staples also they provide a well-balanced diet. Grain-legume proteins are usually the least-expensive protein source for both rural and urban populations in Africa.

According to Tahir (1970) approximately 70 per cent of the world's human protein needs in 1968 were met directly from plant sources and only 30 per cent from animal sources, including fish. Of the human intake of vegetable protein, cereals contribute about 66 per cent, grain legumes about 20 per cent, and other sources, such as roots, tubers, nuts, fruits, and vegetables, make up the remainder. But while the overall average consumption of plant protein was 43 kg *per caput* per annum, that in Africa was only 26 kg *per caput* per annum.

In a survey carried out in 1963–4 by FAO (1966) in Nigeria it was found that available supplies of calories at 2719 cal *per caput* per day and proteins at 80 g *per caput* per day provided adequate nutrition in the semi-arid northern region, but intake was only 1909 calories and 40 g of protein in the subhumid western region and 1774 cal and 33 g of protein in the humid eastern region. These were considered inadequate.

Overt nutritional imbalance with relative protein deficiency in relation to energy availability may be acute in some humid and subhumid areas at intermediate elevations, as in Uganda (1000–1200 m) where although carbohydrate sources are more than adequate at 3000–4000 cal *per caput* per day protein supplies are inadequate. Death from malnutrition in Uganda reaches 5 per 1000 births and occurs primarily in post-weaning children.

4.1.3. Constraints to increased production

Tropical food legumes are subject to many natural hazards, especially pests and diseases, and are often not genetically capable of responding sufficiently to improved growing conditions. Some legumes have consumer-restricted use because of undesirable flavours, flatus factors, metabolic inhibitors, and other toxic substances. Other species that are well adapted, high yielding, and nutritious are not yet widely eaten because of ignorance or unfamiliarity with their culture and methods of preparation.

4.1.4. *Production of grain legumes in the lowland tropics*

The tropics (defined as countries with the majority of their territories lying between the Tropics of Cancer and Capricorn) of southern Asia produced in 1970 20·8 million t of grain legumes on 33·1 million ha—or 51·8 per cent of total tropical grain legumes production; tropical Africa produced 8·0 million t on 12·6 million ha (or 20·1 per cent); and Latin America 7·9 million t on 9·8 million ha (or 28·1 per cent).

Production statistics for 1970 for leguminous oilseeds and pulses have been published by FAO (1971, 1972). Four categories of grain legumes, including unspecified Asian grams, which FAO include as a single category, constitute the majority of lowland tropical pulse and leguminous oilseeds production. These, in descending order of importance are indicated as being (1) groundnuts, (2) pigeon peas, (3) cowpeas, and (4) Asian grams. Trends apparent over the period 1961–70 show that soya-bean production has increased the most rapidly at intermediate to high elevations, with a fivefold increase in production being recorded. Whilst at low elevations cowpea growing increased about 2·6 times during the same period, at a rate substantially faster than that of green and black grams.

The most important grain-legume crops of the African lowland tropics are groundnuts (4·4 million t in shell from 5·4 million ha or 61·4 per cent of the total lowland tropics legume crop), cowpeas (1·1 million t; 22·8 per cent), bambara groundnuts (roughly estimated at 150–250 thousand t; 3–5 per cent), and pigeon peas (50–100 thousand t or about 1–2 per cent of the total production). Lima beans, African yam beans, haricot beans, velvet beans, hyacinth beans, African locust beans, and soya beans are of undetermined secondary importance in the African lowland tropics.

4.1.5. *World production of pulse crops and oilseeds by regional areas*

Groundnuts. India produced 25·3 per cent of the 1970 world crop. In Africa the three biggest producers were Nigeria with 6·5 per cent, Senegal with 3·6 per cent, and the Sudan with 2·1 per cent of total world production. However, production declined by 17·1 per cent in Nigeria and 33·8 per cent in Senegal between 1961 and 1970, owing possibly to quality restrictions in Europe.

Cowpeas. Africa produced 93 per cent of the world crop of 1·18 million t in 1970. Major growing countries were Nigeria (60·2 per cent), Niger (12·7 per cent), Upper Volta (8·1 per cent), and Uganda (5·9 per cent). However, it is quite possible that this production was underestimated by 10–15 per cent, because unreported plantings may result from the largely subsistence mode of cultivation.

Bambara groundnuts. The role of bambara groundnuts in the nutritional economy of tropical Africa is difficult to determine, because of the lack of production estimates. It is possible that this species together with the very similar Kersting's groundnut are third in importance after groundnuts and cowpeas in the lowland tropics. They are mainly confined to the hot, drier regions between the desert and sub-humid savannah-forest lands along the southern fringe of the Sahara, and in scattered areas in eastern Africa. They contribute as much as 200 000 t from 300 000 ha. Niger alone produces 30 000 t on 40 750 ha (Nabos 1970).

Pigeon peas. India produced 1·84 million t of dry grain on 2·67 million ha in 1970, comprising 94 per cent of the world crop. Other producers were Uganda (1·7 per cent) and Malawi (1·0 per cent) in Africa; and Burma (1·3 per cent) and the Dominican Republic (1·1 per cent). Allowing for unreported and 'kitchen garden' plantings for home use, these statistics are likely to be underestimated by as much as 15–20 per cent.

4.2. Grain-legume species in relation to the ecological regions of Africa

In the African lowland tropics, the four most important species of grain legume are groundnuts, cowpeas, bambara groundnuts, and pigeon peas. Lima beans, yam beans, and velvet beans are also locally important. Some of these 'secondary' species could become major contributors to the protein food supply with a minimal effort towards improving their production. As shown by the list below, several grain legumes can be cultivated in the semi-arid to subhumid African tropical lowland, but fewer are well adapted to the more extreme arid and humid regions.

1. Semi-arid regions:

    (a) short-duration cowpeas (*Vigna unguiculata*);†
    (b) short-duration groundnuts (*Arachis hypogaea*);†

† Major species.

(c) bambara groundnuts (*Voandzeia sub-terranea*);†

(d) Kersting's groundnut (*Kerstingiella geocarpa*);

(e) locust beans (*Parkia* spp.).

2. Semi-arid to subhumid regions:
   (a) medium- and long-duration groundnuts;†
   (b) medium- and long-duration cowpeas;†
   (c) locust beans (*Parkia* spp.);
   (d) pigeon peas (*Cajanus cajan*).†

3. Subhumid to humid regions:
   (a) pigeon peas (*Cajanus cajan*);†
   (b) lima beans (*Phaseolus lunatus*);
   (c) cowpeas, medium- and long-duration;†
   (d) haricot beans (*Phaseolus vulgaris*).

4. Humid to very humid regions:
   (a) lima beans;
   (b) yam beans (*Sphenostylis stenocarpa*);
   (c) mucuna beans (*M. pruriens* var. *utilis* and *M. sloanei*);
   (d) pigeon peas.†

4.3. Leguminous oilseeds

4.3.1. Anticipated change in the relative importance of groundnuts and soya beans

The two important leguminous oilseed crops are groundnuts and soya beans. Whereas groundnuts are already of major importance in the lowland tropics, soya beans play at present only a very minor role and are found mostly in south-eastern Asia. In Africa, not more than 50 000 t of soya beans are produced annually, as a consequence of lack of an established demand and problems of management. Nevertheless, the soya bean may have considerable potential in the lowland tropics. There is increasing demand for industrial protein sources for both food and feed in developing tropical countries, and it is anticipated that, in the near future, increasing emphasis will be placed on adapting and improving soya beans under lowland tropical conditions.

4.3.2. *Groundnuts* (Arachis hypogaea L.)

Groundnuts are a major crop of long-standing cultivation in Africa. They are considered to have been introduced by the Portuguese to their port settlements in Africa in the sixteenth century from Brazil, and germplasm from Western Mexico may have reached Madagascar and Eastern Africa via the

† Major species.

Spanish route through the Philippines and then by trade and traffic through Asia (Purseglove 1968). Thus continental Africa became the meeting place for the genetic descendants through easterly and westerly routes around the world from their undoubted centre of origin in the New World. Africa may now be regarded as an important secondary centre of genetic diversity for the species.

Ecological adaptation and constraints. Groundnuts have become an important African food crop as well as an industrial crop grown for export.

Groundnuts are somewhat drought tolerant and are adapted to light and well-drained soils but also thrive on black or red clay soils provided that the rainfall is not so high as to cause any waterlogging. The most favourable climatic conditions are high solar radiation with moderate rainfall during growth but with a reliable period of soil drying as the pods mature so as to harvest with rapid drying of the lifted pods. Although the crop is grown in all zones from sub-arid to fully equatorial it is best adapted and also most important in the Sudan–Sahel region. In more humid regions its use is mainly as a subsistence crop.

The problem of ensuring freedom from aflatoxin contamination (see p. 48) in groundnuts destined for export has recently become an important consideration in selecting areas for continuing expansion of commercial production of the crop.

Production areas. According to Purseglove (1968), although India, China, and the USA produce more groundnuts, Nigeria is the world's largest exporter. In West Africa Senegal, Niger, and Gambia also have well-established exports. France buys about half of the trade in groundnuts which amounted altogether in 1968 to about 1·5 million t.

Botany. The morphology of the species has been studied by many workers including Gregory, Smith, and Yarborough (1951), Bunting (1955, 1958), Krapovickas and Rigoni (1960), Krapovickas (1968), and Gibbons, Bunting, and Smartt (1972). All authors recognize the variation in the mode of distribution of vegetative and reproductive branches and the extent of branching as the basis of sub-specific classification. The most modern treatment (Gibbons *et al.* 1972) classifies *Arachis hypogaea* L. as follows:

Subspecies *hypogaea* Waldron. A group with alternate branching or Gregory's Virginia type: the inflorescences are simple and never borne by the main stem;

the first bud of the cotyledon axes is always vegetative; the branches have in turn two vegetative buds and two reproductive buds; the plant may be spreading or erect; in the latter case, because of the abundant branching, it has a bushy appearance.

> Var. *hypogaea*: runner or erect in habit; the main stem is less than 50 cm in the runner forms; the branches are rarely very hairy; the growing season is fairly long.
> Var. *hirsuta* Kohler; runner form in habit; the main stem is often more than 1 m long; the branches are very hairy; the season is very long and the plants are very sensitive to attack by *Cercospora* spp.

Subspecies *fastigiata* Waldron. A group of cultivars with sequential branching (Gregory's Valencia and Spanish types). Inflorescences always occur on the main stem; the first buds on the cotyledon axes are reproductive; reproductive and vegetative buds then follow each other irregularly; the plant structure is little branched and is always erect.

> Var. *fastigiata* Valencia type. Inflorescences are simple; branches arising from the main stem do not rebranch, or if so only at the tip; pods contain 2, 3, or 4 seeds.
> Var. *vulgaris* Harz. Spanish type. Inflorescences are compound; the branches arising from the main stem are irregularly secondarily branched; the pods generally contain 2 seeds only.

The variety *hirsuta* is not widespread and appears to be only of botanical interest. The distinctions between all groups are important agriculturally because of the genetic or agronomic characteristics generally closely associated with the classification features given. To simplify matters however, and adhere to the most widely known terminology, the designations, following Gregory *et al.* (1951), of Virginia, Valencia, and Spanish will be used hereafter.

Comparisons and contrasts between Virginia and Valencia and Spanish types of cultivar. The Virginia group generally have smaller leaflets, are darker green in colour, and of longer growth season than those of the Valencia and Spanish groups. Seeds of the Virginia cultivars generally have a period of dormancy while those of Valencia and Spanish cultivars can germinate immediately and indeed often germinate in the earth before they are harvested if the soil has remained moist. On ripening, maximum

oil content is attained before maximum dry weight in Virginia cultivars, whereas oil content and dry weight reach maximum levels simultaneously in the Spanish and Valencia cultivars (Schenk 1961).

The non-saturated fatty acid component is higher in the oil of Virginia cultivars than Spanish and Valencia cultivars (Worthington and Hammons 1972). The crude protein content appears generally to be lower in Virginia cultivars and higher, particularly in Valencia cultivars (Mauboussin 1970). Virginia cultivars are more tolerant to *Cercospora* leaf-spot diseases and also it is only in this group that cultivars resistant to rosette disease have been found. Virginia cultivars appear to have a higher production potential than those of the sequential groups, but are more ecologically demanding, especially with regard to solar radiation and water supply.

The requirements become most stringent for good-quality production for types with large nuts. The large-fruited Virginia bunch types, which comprise the main dessert groundnut, can only be well grown in the most favourable rain-fed regions such as Georgia (USA), parts of China, Malawi, and the south of Senegal, or in areas where groundnuts are irrigated such as Egypt, Sudan, and Israel.

Although a good knowledge of the distinctive features of the two groups appears useful for an understanding of the behaviour of traditional cultivars of groundnut, it must be pointed out that recent breeding has created new cultivars from inter-group hybridization and hence some of the characteristics of these do not conform to one or other of the groups.

Groundnut breeding. The groundnut is a largely self-pollinating plant. However, outbreeding has been assessed at 0·20 per cent in Senegal (Mauboussin 1968), 0·01–0·55 per cent in the USA (Culp, Bailey, and Hammons 1968) for Virginia cultivars, and Bolhuis (1951) indicated 6·6 per cent out-crossing for Spanish cultivars in Java.

As a result, natural populations may be somewhat heterogeneous and heterozygous despite an apparent uniformity and a majority of stable types. Such populations comprised landraces adapted to and cultivated in different areas before the start of modern selection and hybridization.

Breeders first produced improved populations by mass selection from these landrace populations, e.g. Mwitunde from Tanzania was used for selection work in South Africa, Madagascar, and India. The selection of pure line cultivars (based on single plant selection of supposedly homozygous plants) was first made from natural or improved populations; currently it is practised in populations in which greatly increased genetic variability has generally been obtained by hybridization. Often single plant selections are made in 'early' generations before homozygosity has been attained (F_4-F_5) to become the parent selections of breeding lines for selection as new cultivars after trials. At Samaru a series of successive crosses is being made between F_1 plants in order to increase the opportunity for recombination between the parental genomes (IAR 1969).

The use of radiation-induced mutations has given some practical results in groundnut breeding (Gregory and Cooper, 1959; Martin 1968; Mukewar 1966; Patel 1968; Madhava Menon, Raman, and Krishnaswami 1970) and has also been used in various genetic and physiological studies (Gregory 1956).

Shchori and Ashri (1970) and Ashri (1972) have studied chemically induced mutants.

Great genetic variability occurs in the descendants of hybrids of parents of divergent genetic origin, such as between Virginia and Valencia or Spanish cultivars. The possibility of using interspecific hybridization for improving disease resistance is also being studied (Smartt and Gregory 1967; Raman, Sree, and Rangaswamy 1972).

Hybridizations between genetically remote parents tend to give rise to some quite new recombinations of characters, other desired recombinations of characteristics belonging to different groups are difficult to obtain. For example, breeders would like to recombine the early maturity of the Valencia and Spanish with the resistance to rosette disease of certain Virginia cultivars, but this seems difficult to achieve by conventional backcross methods. The same appears to apply to dormancy, the control of which is not well understood because of practical difficulties in studying it (Mauboussin 1968).

Considerable advances have been made in groundnut breeding in recent years. An attempt will be made to take stock of them briefly for the main selection aims.

The commercial Virginia cultivars at present available have oil contents of about 50 per cent. Various research projects show that this characteristic is highly heritable (Martin 1969). Oil contents of 57–8 per cent have been obtained in Upper Volta in segregants from Virginia × Spanish crosses. In India, according to Gopalswamy and Veerannah

(1968), TMV1, a Virginia cultivar has a higher oil content than TMV2 (Spanish) when grown on dry land, but vice versa when grown on irrigated land.

Protein content has been little studied. According to Holley and Hammons (1968) a negative correlation is observed between oil content and protein content. The gross protein contents, as a percentage of dry weight, of groundnuts of the Bambey collection vary from 24 per cent to 35 per cent (Mauboussin 1970).

Improvement of groundnuts intended for the confectionary trade has been carried out mainly in the USA and Israel. Generally large-fruited groundnuts of the Virginia type with constricted pods giving the nuts a regularly rounded shape have been selected. These include NC2, NC4, GA 119-20, Florigiant, and NC17 from the USA, Valencia 247 from Madagascar, A124/B from Central Africa, and Acholi White from India.† To shorten the season of Virginia types is a frequent aim of breeders. In Senegal (IRAT 1969) line 57-422, of Virginia character, obtained in Senegal from a Virginia × Valencia cross has a season of 105 days. However, it appears difficult to obtain as short a maturity time in alternate branching types, as commonly exists in sequential branching types. A number of lines combining dormancy with short season have been obtained in Senegal and Sudan (Gillier and Silvestre 1969). Correlations between the volume of soil exploited by the root system and drought resistance and between the possession of short internodes and drought resistance have been noted in India (IARI 1970).

In Senegal, Gautreau (1970) has been unable to show any relationship between drought resistance and the number of stomata but rather it appears to depend upon the efficiency and speed of stomatal regulation following changes in the water potential in the tissues. Germination tests of a collection of cultivars under high osmotic pressure has allowed systematic screening of numerous lines and selection of new resistant lines. A suction pressure test measurement allows the drought resistance of individual plants to be assessed and hence a study to be made of the heritability of this characteristic.

The naturally occurring rosette virus resistant selections found so far in Africa (IITA 1970) were all the Virginia type and of long season. New cultivars

† Acholi White is in fact an old landrace from the Acholi district of Uganda from which a selection may have been made in India. Ed.

of good agronomic quality such as 48–37 and 1040 have been obtained from them by selection in Bambey and Niangoloko stations. These are well adapted in Upper Volta, the north of Dahomey, Togo, Nigeria, and in southern Niger.

The inheritance of resistance to rosette disease is based on two pairs of major recessive genes modified in expression by the presence of other genes and perhaps also some cytoplasmic action. Resistance has been introduced by back crosses in Senegal in the population 28-206 which is a Virginia type of wide ecological adaptation and medium maturity of about 120 days. From this stock new lines such as 28.206 RR have been obtained (Mauboussin, Laurent, and Delafond 1970).

Several different crosses are being evaluated at Niangoloko in Upper Volta in an attempt to obtain lines of different maturity length of high agronomic quality and resistant to rosette disease. Virginia selections 55–460, 48–37, and 52–14 have shown reasonable resistance to *Cercospora* spp. in Nigeria (Fowler 1970a). Resistance to *Cercospora* spp. has been reported in the wild species *A. repens*, *A. glabrata*, *A. hagenbeckii*, and *A. villosa* (Abdou 1966; Gibbons and Bailey 1967; Anon. 1964). Some resistance appears to be due to factors inhibiting sporulation by the fungi (Abdou 1966) and other resistance is due to the small dimension of the stomata (Gibbons and Bailey 1964).

Resistance to *Aspergillus flavus* is a most important problem. Suryanarayana Rao and Tulpule (1967) and Kulkarni (1967) reported some resistance in cultivars Kaboka and Mwitunde, but Mixon and Rogers (1973) could not confirm this with the strains of *Aspergillus flavus* that they used to test the resistance of 2000 lines of several species of *Arachis*. Two lines of the Valencia type originating from Argentina PI 337394 and PI 337409 proved fairly resistant. The resistance of these is thought to be due to the resistance of the testa to invasion by the pathogen. These authors are now studying the inheritance of this resistance.

A selection from Mwitunde and another selection from Schwartz 21 are claimed to be tolerant to *Verticillium dahliae* (Franck and Krikun 1969). Punjab 1 and TMV3 are said to be tolerant to *Sclerotium bataticola* (Mathur, Singh, and Joshi 1967). Garren (1964) has shown differences between cultivars in sensitivity to *Sclerotium rolfsii*. Schwartz 21 (Bolhuis 1955), CES 101 (Anon. 1968), PI 341884, PI 341885, and PI 341886 (Simbwa-

Bunnya 1972) have been reported resistant to *Pseudomonas solanacearum*.

Physiological research on the groundnut. Temperature of 32–34 °C is optimum for germination, and 15–45 °C are the extreme limits between which germination can occur. During growth the optimum temperatures have been reported for different cultivars between 24 °C and 33 °C (Catherinet 1956; Montenez 1957; Bolhuis and de Groot 1959). Large diurnal temperature differences seem to be unfavourable; differences of 20 °C inhibit flowering (Fortanier 1957) and night temperatures below 10 °C prevent ripening of the fruit (Shear and Miller 1955).

Although groundnuts do not appear to show photoperiodic responses in flower initiation (Prévot and Ollagnier 1956) both photoperiod and total radiation affect the development of the flowers (Prévot 1949).

The groundnut resists drought well. In the sandy soils of Senegal, Dancette (1970) has determined that the soil moisture below which the groundnut suffers is 70 per cent of field capacity and the permanent wilting point is at about 40 per cent. Water requirements have been studied by various authors (Billaz 1959; Devez 1959; Ilyana 1958, 1959; Mantez and Goldin 1964; Ochs and Wormer 1959) and were reviewed by Salter and Goode (1967). The flowering period is the time when the plant is most sensitive. Total requirements are of the order of 500–600 mm; they rise in the first month of growth from 1·5 mm to 3–4 mm per day and then reach a peak of 5–7 mm, dropping in the last month of the season to 4·3 mm and finally to about 2 mm per day.

Mineral nutrition of groundnuts. Foliar diagnosis of mineral-nutrient deficiency has been developed mainly in West Africa (Prévot and Ollagnier 1961; Martin 1965; Anon. 1969). This technique makes it possible to evaluate deficiencies of the main elements or unbalanced mineral nutrition of the plant. The plant's uptake of the major elements for each 1000 kg of pods produced can be taken approximately as:

	Foliage and stems kg	*Fruit (pods and nuts)* kg
N	10–12	30–35
P_2O_5	1·5–2	6
K_2O	10–12	6–10
CaO	8–12	1–2
MgO	8–10	2

Nitrogen requirements are high, but they are normally met by bacterial symbiosis. In a wide-ranging series of studies in francophone countries applications of nitrogenous fertilizers have generally only proved to be economic in the relatively low level of 10 kg/ha applied during the first 3 weeks of growth (starter nitrogen) when symbiosis has not yet become functional.

Phosphate fertilizers are often effective because the soils of Africa are so short of available phosphate.

Potassium deficiencies occur on leached ferrallitic and overcultivated soils and are most easily recognized by an abundance of single-seeded fruits. Foliar diagnosis can be used to avoid the wasteful application of more potash than is required to avoid this.

Calcium is one of the most important elements for fruit formation and development, especially in large-fruited cultivars. This element is absorbed both by the roots and by the gynophores (Brady 1947). Calcium deficiency can be overcome by top-dressing if necessary, preferably in the form of calcium sulphate ($CaSO_4$) at the time of flowering. However, larger applications of lime throughout the profile during land preparation can be useful in correcting the acidity of soils if the pH is below 4·5 or 5·0 depending on the type of soil.

Sulphur is a most important nutrient element in groundnut production. It can be absorbed both by the aerial parts and by the roots (Brzozowska and Hanower 1964; Hanower, Brzozowska, and Prévot 1963; Cas 1972; Bromfield 1973). It both promotes and prolongs flowering. Sulphur requirements of about 12–15 kg/ha have been demonstrated in many parts of Africa. If nitrogen and phosphate fertilizers containing sulphur are used they generally satisfy the sulphur requirement.

Deficiencies of boron and molybdenum and manganese toxicities have been observed in Africa (Vaille and Gona 1972; Anon. 1964). Boron deficiencies, like calcium deficiency, produces 'hollow heart' (Harris and Brolmann 1966). Blackening of the embryo and sometimes the splitting of the stems associated with shortage of boron may appear as a secondary effect of drought (Gillier 1969). Boron deficiency is easily corrected by application of only 5–10 kg/ha of borax.

Sufficient molybdenum is required for effective bacterial symbiosis, increasing both the weight and number of nodules and consequently is important for nitrogen nutrition. Deficiencies can be corrected by small applications of 28 g/ha (Gillier 1966; Martin and Fourrier 1965) of ammonium molybdate.

Manganese toxicity in leached acid soils in which Mn becomes exchangeable has been observed in the

Congo (Prévot, Ollagnier, Aubert, and Brugière 1955; Anon. 1964). It aggravates the effects of calcium deficiency.

Because the groundnut has a great capacity for growing and producing some crop in poor soils the practice of fertilizing has not developed in most of the countries where it is cultivated. In Africa fertilizer applications recommended are often low and limited to dressings of phosphates and sulphur by the use of single superphosphate, e.g. in northern Nigeria (Goldsworthy and Heathcote 1964).

In Senegal, however, Carrière de Belgarric and Bour (1963) recommended ternary formulae at the level of 150 kg/ha of a compound fertilizer containing small quantities of nitrogen and sulphur and quantities of phosphate and potassium, the actual composition varying according to the type of soil and cultivation system, since potassium requirements become high in overcultivated areas and in the permanent and intensive agricultural systems currently being taught.

Crop protection. 1. *Pests.* The groundnut is susceptible to several pathogens. Storage pests include *Caryedon fuscus*, *Tribolium* spp., and above all *Trogoderma granarium*. Field pests are generally only of economic importance in so far as they encourage the development of certain pathogens (see Chapter 11) and in particular, this applies to the aphid vectors of the rosette virus complex and the julid (*Peridontopyge spinosissima*), which provides infection sites allowing fungi to grow on the seeds during germination and damage seedlings. This same insect and termites also encourage development of mould on the fruit, and in particular *Aspergillus flavus* the cause of aflatoxicosis.

2. *Diseases.* The most important disease problems in Africa are the leaf spots caused by *Cercospora* spp., damping-off diseases of seedlings, and the virus diseases, especially the rosette virus complex.

Leaf spots are caused by *Cercospora arachidicola* and *C. personata* and, to a lesser extent, by *C. canescens* (Fowler 1970). Leaf spots cause losses of 10–60 per cent, depending on conditions. Sulphur deficiency greatly increases disease severity and the use of tolerant cultivars is generally to be recommended. The disease can, however, be controlled by fungicidal treatments. McDonald (1970) has advised that, for seed multiplication schemes, weekly spraying with Mancozeb 45% W.P. at 2·2 kg/ha in 400 l of water from the third week after sowing is economic in Nigeria. In Cameroun, with very sensitive Valencia types having a high yield potential, fungicide treatment is also economic (IRAT undated). In Malawi, Corbett and Brown (1966) recommended fungicide treatment after every period in which 8–10 days of rain had accumulated.

The most frequently encountered soil organisms associated with seedling blights are species of *Aspergillus, Rhizopus, Penicillium, Fusarium, Pythium, Sclerotium*, and *Macrophomina*. However, crown rot caused by *Aspergillus niger* is often mainly responsible for these losses (Gibson 1953). Seed and soil-borne fungi can be easily and generally very cheaply controlled by treating the seed with an insecticide-fungicide mixture (e.g. thiram/BHC). Numerous experiments in Senegal show that improvements in yield from seed disinfection may be 40–50 per cent in conditions unfavourable for rapid germination, due not only to reduced plant mortality but also to stronger surviving plants. *Macrophomina phaseoli*, which is common in West Africa in light soils, causes wilting of the adult plant, breakage of gynophores, and forms a mycelial felting in the intercotyledon space of the seeds (Bouhot 1967).

Aspergillus flavus growing on the seeds forms aflatoxins. The importance of aflatoxins became apparent in 1960 in Britain in intensive turkey and duck farms (Bampton 1963). *A. flavus* may develop on numerous food grains but it finds a particularly favourable environment in the groundnut. On the nuts, the fungus develops at moisture contents between 9 per cent and 35 per cent, particularly between 15 per cent and 25 per cent (Moreau 1968). Before harvesting, the pods in the ground are at a humidity favourable to the pathogen, but they have natural protection against infection when the plant is alive and healthy. If pods are damaged by insects and the plants wither in the field during the growth period *Aspergillus* gains entry more easily. If harvesting is delayed too long after the nuts are ripe or harvesting ripe pods is carried out when the atmospheric humidity is high without this being followed by rapid drying then this will also encourage the development of *A. flavus*. (McDonald and Harkness 1963, 1964, 1965, 1967; McDonald and A'Brook 1963; McDonald, Harkness, and Stonebridge, 1964.)

4.3.3. *Soya beans* (Glycine max (*L.*) Merr.)

The rather limited investigations on soya beans in the lowland tropics have demonstrated the extraordinarily high yielding potential of this crop

under favourable conditions of moisture, soil fertility, and plant protection. But there are some major constraints to be overcome, including rapid loss of seed viability, poor establishment in tropical soils, occurrence of diseases, and attack by insect pests. Soya beans are frequently more responsive to high levels of soil fertility than are some other tropical legumes, but may also be more sensitive to soil acidity. Moreover, most tropical soils—particularly in Africa—are devoid of effective strains of the specific organism (*Rhizobium japonicum*) required for symbiotic nitrogen fixation.

Producing areas for soya beans. Soya beans were domesticated in China and Manchuria, where the use of the crop is believed by some historians to date back to the Chinese Emperor Shen Nung (2500 B.C.), though good evidence of the existence of cultivated soya beans is lacking beyond the eleventh century B.C. (Hymowitz 1970). Soya beans were introduced and first cultivated in the Americas by the turn of the twentieth century. Production in the USA rose from practically nothing to 31·8 million t on 17·2 million ha between 1907 and 1971. Introductions were made into Tanzania as early as 1907 and into Uganda in 1913, but the crop did not really become established until the early 1940s (Auckland 1970). Through the work of Auckland (1966, 1967) and others production had reached 2400 ha in the vicinity of Nachingwea, Tanzania (elevation 425 m, latitude 10° 25′ S; 800 mm annual rainfall) by 1963. Production in Uganda increased from about 1000 t to more than 8000 t by 1968 (Leakey 1970a), and in Nigeria it has increased from 4000 t in 1948–52 to 33 000 t on 38 000 ha in 1971. Scattered production has also occurred in Ethiopia, Rwanda, Zaire, and some francophone countries as well.

Botanical aspects. There are two principal species belonging to the subgenus *Soja* (Moench) F. J. Herm.: the cultivated *Glycine max* (L.) Merr. and wild soya bean *G. ussuriensis* Regel and Maack. Both species have $2n = 40$ chromosomes compared with 40 and 80 chromosomes for the six species belonging to the subgenus *Glycine* Willd.; and 22 and 44 chromosomes for the five subspecies of *G. wightii* in the subgenus *Bracteata* Verdc. (Hymowitz 1970). Most highly developed modern cultivars are predominantly monopodal, they range in height from 30 cm to 120 cm, they produce pods profusely at the nodes, range from 85 to 140 days in maturity, are more or less physiologically determinate,† and produce dry-seed yields in the range of 1000 kg/ha to more than 3000 kg/ha under favourable circumstances in the tropical lowlands. However, many South Asian cultivars grow much more vigorously, have very large leaves and stems, branch profusely, and are often partially decumbent. Most cultivars of soya beans are sensitive to photoperiod, but the known exceptions provide an opportunity to achieve extended adaptation.

Genetic improvement of soya beans in Africa. The most important activities in Africa have been the introduction and testing of cultivars of soya bean from other areas. In francophone countries these activities have been centred in Madagascar, with further testing in Cameroun and Centralafrique (Silvestre 1970*b*; Marquette 1970*a*). However, genetic recombination has been employed in breeding programmes at Nachingwea, Tanzania (Auckland 1966, 1967), Makerere University in Uganda (Hawtin 1973), and in Nigeria (Van Rheenen 1972; Ebong 1970*a*; IITA 1973). The best low-elevation lines, developed in Tanzania near sea level and considered well adapted from latitude 5° S to 10° S, were derived from a cross between Hernon 237 and Light Speckled. These matured in 110–20 days and had small yellow seeds of comparatively high oil and low protein content. This type of bean was considered particularly well suited to the Japanese market. However, at higher elevations, with lower night temperatures, these cultivars became too late in maturity, and could only be used as cover and green manure crops (Auckland 1970). In Uganda, selections were made in local landraces (surviving from early introductions) near Kampala by Leakey (Rubaihayo and Leakey 1970) towards the development of suitable commercial cultivars. Bukalasa 4 and 6, which were based on selected single plants, matured in about 125 days at that elevation (1200 m).

Breeding objectives. Some of the disincentives to increasing the cultivation of soya beans in the West African lowland tropics might be overcome through breeding with the following aims:

(1) Wide adaptation, through hybridization and selection for insensitivity to photoperiod and night temperature differences;

(2) elimination or reduction of metabolic inhibitors, flatus factors, and lipoxidase in the seed testa;

† Physiological determinism in crops means that vegetative growth ceases when reproductive growth begins. Ed.

(3) resistance to insect pests including foliage feeders like thrips (*Sericothrips occipetalis* and *Taeniothrips sjostedti*) and *Spodoptera* spp., and pod-boring pests (*Maruca testulalis*, *Laspyresia* sp., and various Hemiptera);

(4) resistance to diseases, primarily bacterial pustule (*Xanthomonas phaseoli* var. *sojense*), soil-borne pathogens (*Pythium* and *Rhizoctonia* spp.), and *Cercospora* leafspots;

(5) resistance to nematodes—primarily root-knot types (*Meloidogyne incognita*);

(6) resistance to pod-shattering and stem-lodging.

Soya-bean cultivar testing. The importance of predictable and stabilized growth in soya beans is widely recognized, and considerable screening for these characteristics has been carried out to attain this. Among cultivars of soya beans tested that show photoperiod and temperature insensitivity are the narrow-leaved cultivars Tokachinagaha from Japan, Fiskeby V from Sweden, Hshi Hshi from Taiwan, and Grant, SRF 300, and Clark 63 from the USA (Leakey and Rubaihayo 1970; Radley 1971; Summerfield and Huxley 1973).

In Madagascar yield tests have been conducted for many introduced cultivars including Davis, Dare, Clark, Kent, Pelican, and Acadian from the USA and Chung Hsing and Wakashima from Taiwan, two selections from South Africa and the local cultivars Hatho and Morissonneau (Silvestre 1970*b*; Marquette 1970*a*; IRAT 1972). Table 4.1 records

TABLE 4.1

Yield data from soya bean trials in Madagascar at Lac Alaotra (750 m altitude) (source IRAT)

Cultivar	Yield (kg/ha)		
	1970	1971	1972
Davis	4377	1759	1755
Pelican		3014	3375
Acadian	2267	3089	3724

the yields from the three most consistently successful cultivars from the USA at 750 m elevation. The results of uniform and advanced trials carried out in West Africa at Ibadan, Western State, and Bende, Eastern Central State in Nigeria, and at Kpong in south-eastern Ghana in 1972 were reported by IITA (1973), and are shown in Table 4.2.

Growth of soya beans in relation to temperature regime and day-length. The effects of day-length, altitude, and temperature have been investigated over a range of tropical conditions. Studies at IRAT showed an increase of about 3 days in growth duration for each 100 m higher elevation in plantings made in Madagascar, Cameroun, Centralafrique, and Senegal (Silvestre 1970*b*). Summerfield and Huxley (1973) demonstrated the existence of genetic differences between cultivars in sensitivity to day-length and night temperature changes, using growth cabinets at Reading University, England. Increasing day-lengths from 11 h 40 min to 13 h 20 min, while

TABLE 4.2

Data from soya bean yield trials in West Africa (source: IITA *Annual Report* for 1973)

Variety	Mean yield (kg/ha) ††	Time taken to flower (days)	Time taken to maturity (days)	Plant height† (cm)	Efficiency (kg/day)
Imp. Pelican	2997	37	103	90	29·1
Bossier	2684	34	105	52	25·6
CES 486	2563	37	113	87	22·7
Kent	2413	27	90	50	26·8
Chung Hsing	2327	37	91	68	25·6

† From IITA planting made on 6 April 1972.
†† Maximum yields in 1972 trials were 3009 kg/ha for CES 486 at IITA, 3653 kg/ha for Bossier at Bende, and 3137 kg/ha for Kent at Kpong, Ghana.

maintaining the same temperature regime, delayed the maturity of Improved Pelican by 22 days and up to 34 days for CES 407. Reducing night temperatures from 24 °C to 19 °C, while maintaining day-length conditions, increased maturities by 20 days and 18 days respectively for the same two cultivars. Other cultivars that are highly sensitive to both day-length and night temperatures include Bukalasa 4 (Rubaihayo and Leakey 1970), Hernon 237, HLS 223, Yellow Kedele, and 3H/55F (Auckland 1970). Strains insensitive to these treatments included Grant, Clark 63, Tokachinagaha, and SRF 300.

Plant nutrition. The use of nitrogenous fertilizers for soya beans has seldom been proved economic in tropical soils if seeds have been inoculated with an effective strain of *Rhizobium*, although early growth may benefit. However, lime, phosphate, and potash may be beneficial if not essential under many conditions. Radley (1971) obtained increases up to 395 kg/ha of grain (18·4 per cent) from 112 kg/ha applications

of P_2O_5† in Trinidad. In Madagascar, 40–60 kg/ha of P_2O_5 have generally given a good economic response and the application of 60 kg of P_2O_5, 90 kg/ha of K_2O together with inoculation of seeds is recommended for newly cleared lands (Silvestre 1970*b*). At Ibadan fertilizer experiments carried out on the Egbeda soil series (pH 6·3; organic carbon 0·92 per cent; clay 13·4 per cent; sand 84·6 per cent) indicate the optimum levels of nitrogen and phosphorus to be 45 kg nitrogen per ha and 45 kg phosphorus per ha.

Symbiotic nitrogen-fixing process. Rhizobial inoculation is almost always essential in tropical soils where soya beans have not previously or recently been grown. Experiments conducted in Madagascar on degraded ferralitic soils of hilly lands resulted in yields 4·35 times those of the uninoculated control on granitic-gneiss soils of Ampangabe and 6·02 times the uninoculated plots on magmatite at Sakay and Kianjosoa (Marquette 1970*b*). Special sources of inoculum (*Rhizobium japonicum* in a pulverized peat carrier) are available, primarily from American and Australian sources. One of the better strains for tropical conditions, according to Dart (1973), appears to be CB 1809.

Populations and spatial arrangements of plants. Commercial seedrates of 50–65 kg/ha in rows 92–105 cm apart used to be recommended for mechanized cultivation. However, in the tropics, optimal plant populations and spatial arrangements for the tall and narrow type of soya beans are similar to those now advocated in temperate regions. Radley (1971) concluded that 200 000–275 000 plants per ha in 61 cm rows maximized economic returns under Trinidadian conditions. Yields in southern Nigeria (IITA 1973) were highest with the closest spacings tried of up to 5 cm ×40 cm (5 000 000 plants per ha). Plant populations advocated in Madagascar, however, are only 250 000 plants per ha (Marquette 1970*a*). In the USA recent studies on the effects of population and spacing have been made by Lehman and Lambert (1960) and Shibles, Weber, and Byth (1966) and in East Africa by Gray (1967), Auckland (1970), and Leakey and Rubaihayo (1972). Most of these authors point out that factors other than the

† P_2O_5 and K_2O equivalents: this expression of levels of application of phosphorus and potassium respectively is conventional in agricultural literature. Applications are usually made in the form of single or double superphospate and as muriate or sulphate of potash.

theoretical maximization of yield may in fact have to determine the field population and arrangement chosen.

Weeds. Even in the absence of any other constraints weed growth can seriously limit yield. Recent and continuing developments in weedicide technology offer practical solutions to this problem if inter-row cultivation is impractical. Among the most practical pre-emergence herbicides are amiben (up to 3 kg/ha), Trifluralin (1·5 kg/ha), linuron (1·5 kg/ha), diphenamid (4·4 kg/ha), and chorobromuron (up to 2 kg/ha) or various combinations of these chemicals (Radley 1971).

More recently, alachlor (2 kg/ha) together with metribuzin (0·6 kg/ha) has been recognized as a generally effective pre-emergence treatment and bentazon (2 kg/ha) as a useful early post-emergence treatment against broad-leaved weeds. However, soya-bean cultivars vary in susceptibility to weedicides and the weather may also affect their activity.

Insect pests. Foliage-feeding pests like thrips and *Spodoptera* spp. may occur throughout the growing period and are effectively controlled by applications of lindane, endosulfan, or dimethoate, singly or in combination. Flower- and pod-borers like *Maruca* and *Laspyresia* spp. are well controlled by Gardona in experiments carried out in southern Nigeria (IITA 1973). Experiments on control of the pest complex carried out in eastern Tanzania by Robertson (1969) during 1965-7 showed that two applications of endosulfan 35 per cent E.C. applied at 1·41 l/ha gave economic control of the soya-bean pest complex in that area. An increase of 100 kg/ha which was easily exceeded would cover the full costs of protection.

Soya-bean diseases. The most serious diseases of soya beans in the lowland tropics of Africa are probably bacterial pustule, caused by *Xanthomonas phaseoli* var. *sojense*, and damage by root-knot nematodes (*Meloidogyne incognita*). Other diseases observed in East Africa included leaf spots caused by *Cercospora* sp. and *Ascochyta sojaicola* (Weiss 1967). In southern Nigeria pod and stem blights occur that are caused by *Diaporthe phaseolorum* var. *batatis*, and virus-like symptoms similar to those on several other legumes occur on soya beans in the Ibadan area, but do not spread aggressively. Resistance to bacterial pustule has been shown in 52 entries of 351 soya-bean lines screened at Ibadan in 1972 (IITA 1973).

Further tests will be required to determine whether this resistance is 'stable' and is readily transferred by breeding.

Soya-bean rust caused by *Phakopsora pachyrhyzae* which is a disease limiting productivity of soya beans in Asia does not yet occur in Africa or in the New World. Vigilence to exclude this pathogen is therefore important.

Uses of soya beans. There is widespread interest in soya beans in tropical Africa as a new source of plant proteins and oil for manufacturing human food, animal feed, or for other industrial purposes. Some institutions are even planning to utilize soya beans more directly. Two such schemes are the use of soya beans in preparing bread flour for a school bun programme in Uganda (Leakey 1970a) and also at the Agricultural College (ETSA) at Tshibashi, Kasai Province, Zaire (Vanneste and D'Heer 1972).

Conclusions. In the lowland tropics soya beans have an enormous, but largely unrealized, potential for production of food, animal feed, and industrial sources of protein and oil. However, constraints to rapidly increasing the use of this species include:

(1) unfamiliarity with the crop and its uses;
(2) lack of primary markets at a reasonable price;
(3) difficulty in obtaining satisfactory field populations under high soil and air temperatures and because of rapid loss of seed viability under ambient storage;
(4) poor soil conditions in most tropical soils—low fertility, high or low pH, inadequate or excessive moisture, and poor physical structure.
(5) lack of control of pests and diseases;
(6) need for special rhizobial inoculation;
(7) competitive effects of weeds;
(8) agronomic deficiencies—primarily shattering and lodging.

Nevertheless, most of these problems have straightforward solutions through better management and genetic improvement.

4.4. Major pulses

The three major and two secondary species of low-oil grain legumes grown in the African lowland tropics are compiled in Table 4.3. There are no accurate or complete production statistics on some of these species, but rough approximations can be made based on partial or fragmentary information

TABLE 4.3
The production of five principal pulse crops in West Africa

Crop	Area (1 000s ha)	Production (1 000s t)	Remarks
Major pulses			
1. Cowpeas	3 000	1 100	Nigeria: 700 000 t on 1 400 000 ha (1971) Niger: 150 000 t on 920 000 ha (1970)
2. Bambara groundnuts	300–450	300–350	Nigeria: 100 000 t in 1966–7 Niger: 29 970 t in 1964–8 Ghana: 20 000 t.
3. Pigeon peas	200–400	90–200	Present official estimates (50 000 t) may be low by a factor of 2–3 times
Minor pulses			
4. Lima beans	120–200	50–100	Grown predominantly in kitchen gardens
5. African yam beans	150–250	60–150	Nigeria alone may produce 50 000–100 000 t annually

and general experience. Bambara groundnuts are mainly confined to semi-arid regions; cowpeas are grown in the semi-arid to subhumid areas; yam and lima beans occur in more humid areas; and pigeon peas are ubiquitous from humid to semi-arid regions. All species produce better-quality seeds when ripening occurs after the heavy rains are over. Local management practices and the adaptive features of most cultivars normally combine to achieve synchronization with the major period of fruiting and ripening as the rains diminish or even after the onset of the dry season. Generally, as would be expected, landraces at higher latitudes have more pronounced photoperiod and/or temperature response than those adapted nearer the equator.

4.4.1. *Cowpeas* (Vigna unguiculata (*L.*) *Walp.*)

The cowpea—southern pea, black-eyed bean, lubia, or niebe—appears to have originated in West Africa, where a profusion of weedy and wild species of *Vigna* abound in both savannah and forested zones. The species *V. unguiculata* exists as herbaceous, erect, semi-upright, prostrate-spreading, and twining-climbing forms, some of which were given specific status in the past. Cowpeas were gathered or cultivated in prehistoric times in Africa and must have reached Egypt, Arabia, and India at a very early date, since they are known from sanskritic times. The early Greeks and Romans knew about cowpeas, and they were introduced by the Spaniards into the West Indies in the sixteenth century, reaching the USA about 1700.

Botanical aspects. The cowpea and its closely related weedy and wild relatives have $2n = 22$ and 24 chromosomes, but 22 is the more common condition. Outcrossing is infrequent, depending on season and activities of pollen vectors. In subhumid parts of West Africa, the first rains are often characterized by a high level of bee activity, resulting in out-crossing to the extent of 10 per cent or more; whereas in the second-season rains, insect activity is reduced and out-crossing may be less than 1 per cent. Seed germination is epigeal, rapid (48–72 hours), and usually a very high proportion of seeds germinate.

Adaptation and production. The cowpea is predominantly a hot-weather crop, well-adapted to the semi-arid and forest-margin tropics. It is frequently grown in mixtures with other crops like maize, sorghum, pearl millet, and cassava, but sometimes as a pure crop. Cowpeas are grown on a wide range of soil types from sands to heavy, expandable clays. Most cultivars of cowpea do not tolerate waterlogging as well as soya beans. However, some forms, like the yard-long bean, can tolerate, and may even require, higher rainfall than other forms. Some cultivars are insensitive to day-length, while others require short days to mature within a reasonable time. Maturities range from less than 60 days up to 7 or 8 months, depending on genotype and the environment. Cowpeas, except the most erect forms, grow and spread very quickly, thereby quickly forming a cover which prevents soil erosion.

The producing areas. Cowpeas are grown extensively throughout the lowland tropics of Africa, in a broad belt along the southern fringe of the Sahara, and also in eastern Africa, from Ethiopia to the Republic of South Africa. They are mainly produced in the hot semi-arid to subhumid areas, with significant production in Nigeria, Niger, Upper Volta, Uganda, and Senegal. Nigeria alone produces about 61 per cent of the recorded world crop—about 760 000 t annually.

The need for improvement of the cowpea cultivation in Africa. Yields of cowpeas in West Africa are very low as a consequence of several constraints.

1. *Climate* is often characterized by insufficient,

poorly distributed, or excessive moisture; lack of sufficient insolation; and extremes of temperature.

2. *Soils* are usually of poor physical structure; low water-holding capacity; have a deficiency of organic matter and extremes of low or unbalanced fertility; and often unfavourable microbiological conditions.

3. *Plant protection* measures to control large numbers of insect pests at all stages of growth are usually lacking and a complex of diseases occurs including those of fungal, bacterial, viral, and nematode causation.

4. *Weeds* are largely uncontrolled and compete for moisture, nutrients, and light.

5. *Cultural practices* are primitive for land pre-paration, planting method, planting dates, populations, spatial arrangements and fertilizer applications (mainly phosphorus and potassium) are inadequate.

6. *Genetic factors* include low productivity efficiency; limited range of adaptation; a tendency to lodging; susceptibility to pests and diseases; and poor palatability and nutrient values.

Breeding strategy. A strategy for breeding cowpeas has been described by Ebong (1970a, 1970b). He stresses the importance of assembling and maintaining collections of genetically diverse materials, and breeding for (1) high yield, (2) acceptable quality, (3) day-neutrality, (4) erect growth habit, (5) long peduncles (above foliage), and (6) resistance to diseases (anthracnose, seedling blights, stem rot, viruses, and leafspot). In francophone Africa, research on cowpeas was started in 1953, when variety and fertilizer trials and population experiments were begun and germplasm collected (734 accessions— working collection of 222 entries). Most intensive efforts on niebe were made during 1962-6 (Sene and N'Diaye 1970). Bambey, Senegal has been the centre for hybridization and breeding, with emphasis on erect, determinate plant types, while observations and selection are also being carried out in Niger, Upper Volta, North Cameroun, and Dahomey (Silvestre 1970c).

Élite strains. Among the improved cultivars developed in Senegal are N. 59-25, an erect, early cultivar with grey spotted seeds, and which is best adapted at Bambey and central Senegal, and N. 58-111, a late, procumbent strain with black and white pied seeds, and adapted for eastern and southern Senegal. N. 58-57 and N. 58-75 grow well in northern Senegal. Nabos (1970) described two improved cultivars developed in Niger having satisfactory yields, uniform flowering, and good seed quality (cream coloured; 15 g per 100 seeds) (see Table 4.4). Other erect or semi-erect short-duration (76-90 days), medium-large (10-12 g per 100 seeds) cultivars with grey-brown to cream-coloured seeds,

TABLE 4.4

Strain	Plant type	Days of maturity	Grain yield 1964-9 (kg/ha)
TN 98-63	Procumbent	130	1300
TN 54-63	Semi-erect	110	1300

yielding as high as 2150-460 kg/ha over a three-year period included TN 62-64G (Bambey N. 58-134), TN 65-64C (Bambey N. 59-7), TN 3-65G (Bambey N. 65-3), TN 58-161G, and TN 36-34C. Crosses were also made in Niger using Bambey N. 58-75, which has a short growth duration and high-quality, large, creamy seed. Yield-testing trials run by IITA at three different locations and seasons in Nigeria (Ibadan, Umudike, and Samaru) and Kpong, Ghana during 1971 and 1972 have led to the identification of five high-yielding and broadly adapted cultivars (Table 4.5). These are primarily semi-erect or semi-upright in growth habit, and medium-early maturing (mean time to 50 per cent first flowers = 45·6 days;

TABLE 4.5

Pedigree	Mean rank	Mean yield (kg/ha)
IVu 37 Pale Green	2·7	1993
IVu 57 New Era	3·2	1884
IVu 72 Bambey N. 59-25	3·5	1883
IVu 354 Hanbru 58-123	4·5	1850
IVu 335 Jebba Pea A	5·5	1717

and mean time to 50 per cent first ripe pods = 69·2 days), but do not have the grain qualities generally preferred in the region. Maximum yields of the best entries were obtained at Kpong (2493 kg/ha) and Samaru (3008 kg/ha) (IITA 1973).

These results illustrate what can be achieved by effective pest control. In much of tropical Africa, cowpeas in normal cultivation produce highly erratic yields which can be less than 100 rising to, at most, 400 kg dry seeds per ha (Rounce 1949; Ebong 1965;

Ojehomon 1970*a*). However, effective pest control combined with favourable growing conditions can increase seed yields by 10–30 times (IITA 1973; Ebong 1965; Sellschop 1962). Comparatively smaller increases in grain yields had been obtained through the use of fertilizers or by cultivar selection by the early 1970s (Ojehomon 1970*a*). In fact, fertile soils, or too liberal use of fertilizers, may tend to increase vegetative growth and reduce seed yields (Johnson 1970; Sellschop 1962).

Genetics and inheritance. Among the self-pollinated legumes hybridization is possibly easiest in cowpeas. The flowers are large and very easily emasculated and seed setting can be high when conditions are favourable (Ojomo 1970). Reasonably high humidity and moderate temperature appear to favour setting in hand-manipulated flowers, but there is also a strong genetic component, since some parents are much easier to cross than others (IITA 1973). Naphthalene acetic acid in talc dusted into emasculated flowers reduced blossom drop and resulted in 30 per cent setting in hand-crosses (Barker 1970). Nevertheless, wide crossing with wild or other cultivated species of *Vigna* has been largely unsuccessful as either pollen germination fails or union of gametes does not occur. Sometimes the embryos tend to collapse very soon after apparent fertilization has taken place.

Components of yield. Recent investigations have tended to emphasize the number of pods per plant as the major component of yield; the number of seeds per pod tend to be negatively correlated with pods per plant (Steele 1972; IITA 1973). Therefore, seeds per pod is not an effective selection index; nor is flower buds per plant, since the rate of abortion is very high, depending on the interaction of genotype with environment (Ojehomon 1970*a*, 1972).

Cross-pollination in cowpeas. The amount of cross-pollination is of direct importance in choice of breeding methods. Cowpeas can be classified as predominantly self-pollinated, but in the more humid regions of south-eastern USA up to 10 or 12 per cent out-crossing may occur (Purseglove 1968). Generally, out-crossing is less than 5 per cent. The principal vectors for transfer of pollen are bees—both honey, bumble, and other wild types—gathering pollen or nectar. Other insects (thrips) feeding on flowers could also contribute to out-crossing.

Insect pests of cowpeas. Insects attack cowpeas in all stages of growth and in storage and are considered the major limiting factor in cowpea production in the low humid tropics (Silvestre 1970*c*). Effective control of insect pests in these circumstances often results in 10–30 times the productivity of unprotected crops (IITA 1973). In southern Nigeria about 15 major and more than 50 minor pests attack the cowpea crop. The most serious control problems involve the pests listed in Table 4.6. Other species, in addition

TABLE 4.6

Principal pests of the cowpea in humid areas of West Africa

Stage of growth	Insect species
1. Early seedling growth, foliage, flowers, and pods	*Taeniothrips sjostedti* (Trybom) *Sericothrips occipitalis* Hood
2. Green foliage (chewing, rasping)	*Ootheca mutabilis* (Sahlb.) *Zonocerus* spp. *Spodoptera* spp.
3. Flowers and floral buds	Thrips (same as above)
4. Floral buds and pods	Various Hemiptera *Coreid* spp. *Maruca testulalis* Geyer. *Laspeyresia ptychora* Meyr. *Melanagromyza vignalis* Spenser *Heliothis* spp.
5. Stored seeds	Bruchideae *Maruca* and *Laspeyresia* Others

Source: information supplied by IRAT and IITA entomologists.

to those in Table 4.6, which are important in the drier regions of West Africa, have been noted by Delassus (1970):

(1) *Melangromyza phaseoli*—the bean-fly attacks young developing shoots;

(2) *Sphenoptera* sp. (Buprestidae)—also attacks the young stems;

(3) species of coreids which puncture the floral buds and developing pods include species of *Anoplocnemis*, *Acanthomia*, and *Tassidedes*;

(4) Pod- and seed-borers—including *Piezotrachelus varium*, *Dendorix* sp., and *Lampides* sp.

Direct loss in grain yields resulting from uncontrolled insect attack have been estimated in carefully controlled experiments carried out in Ibadan by

Whitney (IITA 1973). Individually, these losses may be as high as the following:

(1) thrips: 50 per cent damage;
(2) *Maruca* (flower damage): 20 per cent damage;
(3) *Maruca* (pod damage): 20 per cent damage;
(4) Hemiptera (seed damage): 35 per cent damage;
(5) *Laspeyresia* (seed damage): 50 per cent damage.

Collectively these estimates exceed 100 per cent, and productivity is virtually nil without some control in certain seasons in the humid tropics.

Chemical insecticides. The use of insecticides is still the most practical solution to the insect problems of cowpea production. Whitney, in Chapter 10 of this book, reviews much of the more important literature.

For practical use the best insecticides or combinations of insecticides are, from several points of view, endosulfan (Thiodan 50% WP) at 0·15% a.i. (active ingredient) in water or 0·9 kg a.i. per ha; lindane 50% WP (Gammalin) at 0·14% a.i. concentration or 0·6 kg a.i. per ha (if emulsifiable compound is used, concentration should not exceed 0·05% concentration); Azinphosmethyl 25% WP (Gusathion M) at 0·14% a.i. concentration or 0·6 kg a.i. per ha; dimethoate 30% EC (Rogor 40) at 0·03% a.i. concentration or 0·3–0·6 kg a.i. per ha. Combinations of Thiodan and Rogor 40 applied 6–8 times during the growth cycle provides a high level of control; but often as few as two or three sprays are highly profitable under commercial practice. Gardona, alone or mixed with Thiodan, may be more effective against podboring species during the post-flowering period (IITA 1973).

Soil- or seed-treatments with formulations of several insecticides provide insect protection for 6–7 weeks. Some applications are comparatively inexpensive and, when combined with two or three post-flowering foliar applications (Gardona or Gardona+Thiodan), should provide a high level of economical plant protection throughout the growing period of the crop. Finally, the harvested, threshed, and properly dried seeds can be safely stored in airtight plastic bags (0·3 mm thickness) holding 40–50 kg of grain and to which 18 g of carbon tetrachloride have been added (Caswell 1968).

Cultural and biological controls. The possibilities for cultural and biological controls and host-plant resistance must not be overlooked and should be investigated intensively in the future. There already appears to be a genetically controlled mechanism for low-level tolerance or resistance to thrips; but the possibility of resistance to podborers has not yet been established. The control of pests of cowpeas in West Africa is vitally important to the protein resources of the region.

Diseases and nematode pests of cowpeas. There are several major diseases of cowpeas grown in the lowland tropics, although the overall reduction in yield which they cause may be less than that arising from insect predations—at least in West Africa (IITA 1973). The major problems from disease and nematode pests include the following:

1. Fungal and bacterial diseases
    (a) Seedling blights and wilts caused by:
        (i) *Rhizoctonia solani*;
        (ii) *Pythium aphanidermatum*;
        (iii) or, of probably secondary etiology associated with *Colletotrichum* sp., *Fusarium* sp., and *Botryodiplodia theobromae*.
    (b) Stem blight caused by *Colletotrichum lindemuthianum*;
    (c) Necrotic leaf spots caused by:
        (i) *Cercospora cruenta*;
        (ii) *Cercospora canescens*;
        (iii) *Xanthomonas vignicola* (bacterial pustule).
    (d) Rust caused by *Uromyces appendiculatus*.

2. Virus diseases
    (a) Cowpea green mottle virus (green blister).
    (b) Cowpea yellow mosaic virus (or yellow flecks).

3. Nematodes
    (a) Root-knot nematode (*Meloidogyne incognita*).
    (b) Root-lesion nematode (*Pratylenchus* sp.).
    (c) Spiral nematode (*Helicotylenchus pseudorobructus*).
    (d) Sting nematode (*Belonolaimus gracilis*).

4. Phanerogam parasites
    Striga gesnerioides is a parasitic weed growing from the roots of cowpeas in tropical Africa.

Delassus (1970) also mentions several other pathogens causing significant disease in the drier parts of West Africa. *Neoscosmopora vasinfecta* is the cause of a vascular wilt and unidentified species of *Cercospora*, *Helminthosporium*, *Leptosphearula*, and *Choanephora*, together with *Rhizoctonia bataticola* cause a miscellany of symptoms on stems, leaves, and pods.

Loss in production of cowpeas due to disease. In southern Nigeria losses in stand and grain productivity have been estimated (IITA 1973) to be as high as the following:

(1) seedling blights (fungus): 75 per cent of stand;
(2) anthracnose: 50 per cent of yield;
(3) *Cercospora* leaf spot: 30 per cent of yield;
(4) bacterial pustule: 12 per cent of yield;
(5) yellow mosaic virus: 50 per cent of yield;
(6) root-knot nematode: 25 per cent of yield.

Seedling blights do not necessarily reduce yields, for if the loss in stand occurs early, adjacent plants will tend to compensate for the dying plant by increased and extended branching and fruiting (IITA 1973). In studies on *Cercospora* leaf spots at Ibadan, Schneider (1973) attributed direct loss in yield from *C. canescens* and *C. cruenta* as 18 per cent and 42 per cent respectively.

Chemical seed-treatments and host-plant resistance. The most practical and promising approaches to the control of diseases of cowpeas generally are by the use of chemical seed-treatments and the use of resistant cultivars. The systemic fungicide, chloroneb (Demosan 65 W) at the rate of 2 g per kg seeds, either alone or in combination with a standard thiram dressing, has given excellent results in experiments carried out at Ibadan (IITA 1973). Other fungicides, such as mancozeb (Dithane M-45), benomyl (Benlate), carboxin (Vitavax), or copper oxide or oxychloride formulations can be used as foliage sprays to provide some control of several diseases, but have seldom been shown to be economic on a commercial scale. Therefore major emphasis should be given to the search for and incorporation of host-plant resistance. Preliminary evaluation of a comprehensive collection of cowpea germplasm indicates that several sources of tolerance or resistance to all major pathogens are available.

Reproductive ontogeny of the cowpea. Flowers are borne in racemes at the distal ends of peduncles carried in leaf axils. However, 70–88 per cent of the flowers are shed before anthesis. Of the remaining 12–30 per cent on which fruits are set, up to half abort prematurely, so that only 6–16 per cent of the total flower buds develop to mature fruits (Ojehomon 1968a, 1968b). Additional buds may occur on secondary and tertiary branches. Ojehomon (1970b) demonstrated a 43 per cent reduction in seed yield per plant by removing all flowers for 12 days after anthesis. However individual flowers represent very little loss in dry weight (0·01 g) so that their loss in

terms of the over-all carbohydrate economy of the plant is unimportant (Summerfield and Huxley 1973).

The effect of light: photoperiod. Cultivars of the cowpea adapted in the higher tropical latitudes have a short-day type of photoperiodic response. This serves to regulate flowering in such a way that pods are formed and mature towards the end of the rainy season. Short-day types often produce excessive vegetation in relation to grain yield when planted earlier than is optimal for that particular cultivar (Summerfield, Huxley, Dart, and Hughes 1975). Photo-insensitive (day-neutral) cultivars can be grown in both low tropical latitudes and in temperate regions as summer crops.

The range of optimal photoperiod for induction of flowering in 14 cultivars studied by Wienk (1963) was from 8 hours to 14 hours. Many cultivars show habit changes in response to light quantity (irrespective of photoperiod), becoming etiolated and bearing twining terminal and lateral shoots under reduced light. This promotes the climbing, as contrasted with a sprawling, habit when grown in association with other crops and weeds.

The effect of temperature on the growth of the cowpea. Recent experiments at Reading University (Summerfield and Huxley 1973) have demonstrated the profound effects of night temperatures on both vegetative and reproductive development in terms of growth, days to first flower, and seed yields in 30 cultivars which were studied. Diurnal temperature change influences *Rhizobium* activity and nodulation as shown in experiments conducted by Dart and Mercer (1965). Maximum dry-matter production occurred with the combination of 27 °C day and 22 °C night temperatures when combinations of 21–36 °C day temperatures with 16–31 °C night temperatures were imposed. Dart and Mercer concluded that air temperature is of considerably greater importance than either light intensity or nitrogenous fertilizers in determining the efficient functioning of the symbiotic system.

Water requirements of the cowpea. Cowpeas are highly drought-resistant, but may also be reasonably tolerant of high soil moisture (IITA 1973). Most cowpeas are grown under rain-fed conditions, but they may also be grown with surface or sprinkler irrigation. Cowpeas may be cultivated without rainfall by growing them after swamp rice on the residual moisture of soils of high water-holding capacity.

According to Huxley and Summerfield (1973), moisture stress can reduce productivity considerably during the period from emergence to first flower, but with determinate cultivars, may not significantly affect yields when water stress occurs thereafter. Doku (1970) found nodulation to be reduced by water stress, particularly when combined with experimentally lengthened days (to 16 hours).

Mineral nutrition. The requirements of cowpeas for nitrogen, phosphorus, potassium, calcium, magnesium, and sulphur have been partially established under certain conditions and for some genotypes. However, there is very little information for this species on minor nutrient requirements. By analogy with what is known for other legumes it may be assumed that molybdenum, manganese, copper, zinc, and boron are required for effective nodulation and correspondingly increased productivity. It has been estimated at IITA, that each tonne of cowpeas removed as crop results in the loss from the soil of 40-50 kg nitrogen, 17 kg (P_2O_5), 48 kg (K_2O), 16 kg (CaO), 15 kg (MgO), and 4 kg S.

Information available on the physiological effects and requirements of major nutrients other than nitrogen is summarized as follows:

1. Phosphorus uptake occurs mainly towards the end of the vegetative growth period and is largely transported to the seed (Jacquinot 1967).
2. Potassium taken up is accumulated mainly in the stem during early growth and later in the seeds (Jacquinot 1967). Field response to potassium by cowpeas in Africa has generally been low, which is a reflection of the generally high content of potash in the soil, but potassium applications equivalent to 40 kg/ha K_2O increased nodulation in eastern Nigeria (Tewari 1965*b*).
3. Calcium stimulates nodulation, perhaps because the calcium status in acid soils effects the availability of molybdenum. Most calcium is taken up during the first 40 days of growth, but it may accumulate in the leaves in replacement of potash during later growth (Jacquinot 1967).
4. Magnesium uptake is maximal during the last third of the period of growth and foliar concentrations are slightly higher than in other plant organs (Jacquinot 1967).

The symbiotic nitrogen-fixing process. Very little response to nitrogenous fertilizers is observed when seeds are properly inoculated or the appropriate rhizobial cultures occur in the soil. Therefore, it is usually more efficient to improve conditions tending to maximize the rhizobial process than to use nitrogenous fertilizers. Ways of doing this include the following:

1. Inoculate with efficient strains of *Rhizobium* if nodulated cowpeas or related species have not been recently grown on the land. Efficient strains may double yields in comparison with some indigenous *Rhizobium* strains (IARI 1971).
2. Improve soil moisture and mulching and avoid excessive cultivation (Masefield 1957).
3. Use temperatures of 24 °C, which are optimum for primary root nodulation according to Dart and Mercer (1965). The number of plants nodulating as well as numbers of nodules produced decrease linearly as temperatures increase from 31 °C to 42 °C (Philpotts 1967).
4. For optimum nodulation ensure photoperiods are not longer than 16 hours (Doku 1970).
5. Apply phosphate to the seedbed to increase nodulation (Tewari 1965*b*); avoid high soil nitrogen levels, which would inhibit nodulation during early growth (Ezedinma 1964).

Fixation of nitrogen by a well nodulated cowpea crop was estimated by Nutman (1971) at 73-240 kg/ha. Nitrogenous substances accumulate in leaves during vegetative growth and migrate to the seeds during grain filling. Each tonne of cowpeas harvested is estimated to remove about 40 kg of nitrogen (Jacquinot 1967). If Nutman's and Jacquinot's estimates are correct it follows that, providing conditions for nodulation are favourable, nitrogen fixation provides adequate nitrogen to sustain cowpea production at at least current productivity levels.

Management practices. Productivity in cowpeas grown in Africa is only 100-300 kg/ha of dry seeds. This is because it is cultivated in subsistence agriculture as a secondary crop in association with cereals like sorghum, pearl millet, or maize. The cowpeas are sometimes (not in Nigeria) planted broadcast at 22-33 kg/ha when the cereal crop is about 50 cm tall. After germination the seedlings are often thinned out (and used as a pot herb), depending on the cultivator's judgement of the stand he has obtained in relation to the availability of moisture. In this system, the cultivator and his dependants regard the cereal as the more important food crop and give it first priority in the season's activities. If additional land beyond that needed for the farmer's subsistence is cultivated, it is usually planted to a cash crop such as cotton or groundnuts.

Associated cropping. In Niger, mixed cropping has been demonstrated to be more highly productive than sole cropping of pearl millet and cowpeas. The mixture grown on one-hectare plots produced yields of 682 kg and 1525 kg of cowpeas and pearl millet, respectively, compared with 1072 kg of cowpeas and 905 kg of pearl millet from pure stands of ½ ha each. However, the profits from sole cropping were greater as a consequence of the higher market value of cowpeas (Nabos 1970).

Date of planting. Day-neutral cowpeas can be planted at any time of the year in low tropical latitudes, when moisture and fertility are adequate and if satisfactory pest control can be practised (IITA 1973). However, it is highly desirable for planting time to be restricted so that maturation occurs during bright, sunny weather. This helps to reduce pod and seed damage from both insects and diseases (McDonald 1970). Most cultivars begin to flower 35–70 days from germination. The date of planting should be so timed that protracted rainy periods are over by the time the crop begins flowering. Thus, at Ibadan, late May and late August are the best periods for planting of day-neutral types, whereas at higher latitudes, in monomodal or monsoon-type climates, a late June or early July planting may be preferable. Day-length sensitive cultivars are not well adapted to growing in the first season in bimodal rainfall regions.

Populations and spatial arrangements. In mechanized agriculture, cowpeas are usually planted in rows 75–100 cm apart, 7–10 cm apart within the row, and at a seedrate of 17–28 kg/ha. In African mixed cropping systems cowpea seeds are frequently planted at a rate of 22–33 kg/ha. In francophone Africa, hill plantings (2–3 seeds per drop) are recommended at spacings of 50 × 50 cm or 50 × 60 cm for early cultivars, and wider for late or spreading cultivars (Silvestre 1970c).

Fertilization. Low but significant responses have been obtained in fertilizer experiments in francophone West Africa to the three major nutrients, nitrogen, phosphorus, and potassium (Nabos 1970). A common recommendation for phosphorus is 20–60 kg of P_2O_5 per ha. Potash, at the rate of 30–60 kg

of K_2O per ha if the soil is known to be deficient, and possibly a light application of nitrogen, 15–30 kg/ha, may be included. However, where possible, all fertilizer use should take account of the soil series, its known characteristics, and the previous history of any particular piece of land.

Weed control. Although the cowpea is a good competitor when well established, weed competition can nevertheless be a constraint on yield when other factors are not limiting. Clean weeding for only the first month after planting for short-duration, determinate cultivars has given as good yields as plots clean-weeded throughout growth in plots in southern Nigeria (IITA 1973). Mechanical cultivation or hoeing is probably the most practical means of control under most conditions, but several weedicides have been tried with varying degrees of success in different places. Trifluralin at 0.56–1.12 kg/ha applied presowing and immediately harrowed or rotovated in has given good control in USA (Ogle 1967). Chloramben (Amiben) has also generally given good results when used at the manufacturers' recommended rate.

Cover cropping. The effectiveness of cowpeas in rapidly covering the soil surface and preventing loss of topsoil has been amply demonstrated in a run-off experiment in southern Nigeria. Cowpeas proved superior to maize and other cereals for this purpose (IITA 1973).

Utilization. The primary use of cowpeas is as a dry pulse, but the green pods, green seeds, seedlings, and tender young leaves are also often used as pot herbs. If the tender green leaves are plucked before the reproduction phase begins, the plant continues to produce new leaves. Mehta (1971) demonstrated that it was possible to remove all tender leaves up to a maximum of 3 times at weekly intervals during the vegetative stage of growth without reducing the final seed yield. The vegetation also makes excellent hay, and the surplus culled and broken seeds can be used as a protein concentrate for domestic animals. Cowpeas cook more easily and quickly than *Phaseolus* beans, and are therefore favoured in areas where both beans and cowpeas are grown when fuel is scarce. Cowpea hay is high in nutrients and its fibre is more easily digested than lucerne fibre. It is also excellent for grazing by milk-producing animals.

Cowpeas are preferred to other pulse crops in many regions, particularly in Africa. This is fortunate as they also provide an important source of proteins, caloric energy, and other nutrients and require a minimum amount of cooking or preparation.

Cooking. In Africa cowpeas are consumed in three basic forms, among which there are many minor variations. Most frequently they are cooked together with vegetables, spices, and other ingredients to make a thick soup or gruel, which is eaten in association with the basic staple such as preparations of cassava, yams, plantain, or cereals. The second method of preparation is as deep-fried cake (akara balls), prepared from a dough of flour made from ground-up shelled cowpeas to which onions and seasonings are added. Cowpeas are also eaten as steamed bean cakes (moin-moin in Nigeria), prepared from cowpea flour to which chopped onions and seasonings have been added. In preparing the flour the testas are removed by soaking the dry seeds in water for a short period and rubbing. Rough or wrinkled testas are preferred as they soak quickly and are more easily removed.

Nutritive qualities.

Constituent	Percentage	Constituent	Percentage
Water	11·0	Fat	1·3
Protein	23·4	Fibre	3·9
Carbohydrate	56·8	Ash	3·6

Contents of calcium (90 mg per 100 g), iron (6–7 mg per 100 g), nicotinic acid (2·0 mg per 100 g), and thiamin (0·9 mg per 100 g) are high and contribute substantially to these requirements in the African diet (Platt 1962).

Crude protein levels are highly variable, ranging from 19 per cent to 35 per cent on a dry-weight basis, depending on genotype, seed yield, management, and environment (Boulter 1972). The amino-acid spectrum is excellent except that methionine and cystine tend to be sub-optimum for monogastric nutrition, as in most other grain-legume species. A range of essential amino-acid content is reported by several investigators as follows:

Amino acid	Percentage total protein	Average percentage total protein
Lysine	5·7–9·6	6·6
Cystine	0·7–1·7	0·9
Methionine	0·7–1·6	0·9
Histidine	2·7–4·0	3·3
Threonine	3·4–5·3	4·1
Tryptophan	0·6–1·6	0·9

According to Liener (1969) levels of toxic substances and antimetabolites like the trypsin inhibitor,

haemaglutinnins, and flatus factors are 'minimal' in the cowpea. Nevertheless, cowpeas have been shown to contain trypsin and chymotrypsin inhibitors (Ventura and Filho 1967), and may have a cyanogen, with a titre as high as 2 mg per 100 ml extract (Montgomery 1964). Therefore, cooking is needed to inactivate these undesirable principles.

Potential. The cowpea probably has the greatest potential of all food legumes in the semi-arid to sub-humid lowland tropics. It has many important advantages in terms of quick, early growth, short duration, wide adaptation, drought resistance, grain quality, acceptability, a broad range of genetic diversity, and ease of hybridization. By mid-1973 a world germplasm collection of 4200 accessions had already been assembled (including recent systematic collecting within Nigeria) and is being studied at IITA in Nigeria. However, it would be highly desirable to explore systematically and collect more intensively, particularly in other parts of West Africa.

Dry-seed yields of 1500 kg/ha within 70 days from planting and 2000–3000 kg/ha within 95 days have been recorded in experimental plots (IITA 1973). These have been obtained under generally favourable conditions with adequate fertility levels, where there is reasonable control of the pest complex and weeds.

4.4.2. *Bambara groundnuts* (Voandzeia subterranea (L.) Thou.)

The bambara groundnut is also called voandzu, congo goober, groundbean, earth pea, kaffir pea, jugo bean, Madagascar or stone groundnut, haricot pistache, and many other names in local dialects. Bambara is a district on the Upper Niger near Timbuktu, and the crop is probably of West African origin (Johnson 1968). In recent times it has been largely replaced by groundnuts (*Arachis hypogaea*) except on the poorest and sandiest of soils or in areas of uncertain rainfall (Stanton 1966). It is cultivated as a pulse; freshly harvested seeds are boiled directly but very hard matured and dried seeds have to be pre-soaked. Sometimes the seeds are roasted and ground into flour (Purseglove 1968).

Botanical. The bambara groundnut is an annual herb with short, prostrate, profusely branched stems rooting at the nodes, usually with very short internodes. There are both compact, bunch and open, or spreading types, which can be categorized according to the ratio of petiole length to the internode length. This

D

ratio, according to Doku's (1969) classification, varies for bunch varieties from 8·2 to 11·0, for semi-bunch varieties from 7·4 to 7·7, and for open varieties from 4·4 to 6·5.

The pinnately trifoliate leaves are borne on long, erect, grooved petioles up to 22 cm long, thickened at the base, and growing at right-angles to the main stem. The leaves are stipulate and leaflets are lanceolate to narrowly elliptic and glabrous. The larger terminal leaflet is 5–10 cm × 1·5–3·0 cm with a rounded, emarginate apex. One, two, or three flowers develop in each axil; the flowers are smaller than those of cowpeas. The tip of the peduncle is swollen, with a brush of hairs behind. After pollination this bends downward, excavates a tunnel in the soil, and draws in the developing pods (Purseglove 1968). Doku and Karikari (1970a) found that flowering commenced about 28 days after germination, and the peduncle tip or aerial disc had touched the soil surface within 1–2 days after syngamy; soil penetration required 5–7 days; further penetration (to about 5 cm) required 10–12 days; and full maturation took up to 40 days. Some cultivars continue producing flowers indefinitely, whereas others cease flowering about 2 weeks before the growth period is completed.

The fruits are more-or-less spherical or lenticular in shape and wrinkled when mature, about 2 cm in diameter, and usually contain one, but sometimes two, seeds. Seeds are round, smooth, very hard, and vary in size from quite small up to 1·5 cm in breadth and from about 14–98 g per 100 seeds. Seeds take up to 15 days to germinate in the cultivated forms but up to 31 days in wild relatives (var. *spontanea*) (Doku and Karikari 1970b). Seeds may be black, red, creamy white, or variously blotched, and have a white eye, sometimes surrounded by a black eye. They usually contain 16–21 per cent protein, 4·5–6·5 per cent fat, and 50–60 per cent carbohydrates (Purseglove 1968).

Evolution of cultivated strains. Doku and Karikari (1970b) compared wild strains of bambara groundnut (*V. subterranea* var. *spontanea*) with the open, semi-bunched and compact or bunch types. They postulated that the cultivated forms were derived from the wild types through a series of gradual changes including:

(1) more compact or bunchy growth habit;
(2) more highly self-pollinated with less dependence on pollinating ants;
(3) increased stem number;

(4) reduced leaf area;
(5) reduced shell thickness.

Earlier Doku (1968, 1969) had observed that wild and more open or semi-bunched types are largely out-crossed and ant-pollinated, whereas the compact or bunch types are predominantly self-pollinated. Ant activity in the soil at the base of the plants also helped loosen the earth aiding penetration of the pods.

Producing areas and adaptation. Bambara groundnuts are widely distributed in tropical Africa and probably reached Madagascar with the earliest peoples from the mainland. The groundnuts were transported to Brazil and Surinam early in the seventeenth century and later taken to the Philippines and Indonesia. Harlan (1972) found this crop widely grown in the savanna fringe south of the Sahara from Senegal and Guinea to Sudan and southwards to the Republic of South Africa. The crop is usually grown in small patches as a 'garden crop', being especially important in Mali, Upper Volta, northern Ghana, northern Togo, northern Dahomey, Nigeria, Chad, northern Cameroun, western Tanzania, Malawi, eastern Zambia, eastern Rhodesia, the Republic of South Africa, and southern Zaire. It is also grown to a lesser extent in Senegal, the Ivory Coast, Niger, Sudan, Kenya, Uganda, Mozambique, and Angola.

The bambara groundnut is widely adapted from arid to subhumid tropical regions. Although it grows very well on poor soils in hot, dry climates, it withstands high rainfall, except at harvest. It also tolerates drier conditions than most other pulses and is believed to require high temperatures and adequate sunshine for optimum growth.

Plant improvement. Very little breeding or other investigations have been carried out on the bambara groundnut, with the exception of some selection and varietal testing. There is a wide diversity of plant and seed characteristics and in growth habit. Rassel (1960) stated that accessions from Zaire matured in 5 months, whereas introductions from Senegal and Madagascar matured in 3–4 months. Several countries have recommended cultivars, some of which are listed by Stanton (1966):

Country	Cultivars
Malawi	Zambian and Barotseland cultivars
Congo (D.R.)	Kahemba brune
Upper Volta	Souma nianga
Basaga	Nantasemberwa, Kibuzu, and Kikol (large seeds)
Niger	TV21, TV12, TV37, and TV83 (TV12 best)

Yields are frequently quite high in comparison with reputed cowpea yields in subsistence agriculture. In Niger during 1964-8 it was estimated at 742 kg/ha; average yields frequently range between 600 kg/ha and 800 kg/ha. Nabos (1970) described collections maintained at Tarna (Niger) including 62 local types, 20 from Bambey, and 2 from Dahomey, and reported that 4 cultivars mentioned above (TV numbers 12, 21, 37, and 83) out-yielded local types by 20-50 per cent, averaging 1018 kg of dry seeds per ha. Exceptional yields of shelled nuts of 2600 kg/ha have been reported from Malawi and 3000 kg/ha from Rhodesia (Stanton 1966).

Cultural practices. Bambara groundnuts are frequently planted as a mixed crop with cereals (usually pearl millet), at wide spacings of 100-200 cm, or as a sole crop at much higher populations and closer spacings of 15×30 cm, 25×30 cm, 20×50 cm or 40×40 cm. Nabos (1970) states that planting is usually done by women and at a plant population of 222 000 plants per ha. The seeds are usually planted at a depth of 3-6 cm. Light earthing-up may be done to promote fruiting, and the crop is harvested by digging and pulling the whole plant to which the pods adhere by tough, wiry stalks (Purseglove 1968). Inoculating with soil from an old bambara groundnut field is frequently desirable when *Voandzeia* is grown for the first time or after several years. Moreover, fertilizing with super-phosphate at 60 kg/ha at planting time and with ammonium sulphate at 40 kg/ha has been demonstrated to be economic (Stanton 1966).

Plant protection. The bambara groundnut is reputed to be extraordinarily free of diseases and pests and is much less susceptible to pod damage both in the field and in storage than are groundnuts. However, under certain conditions several pathogens do attack the crop (Stanton 1966):

(1) stem and foliage rots caused by *Sclerotium rolfsii*, *Pythium* spp., and *Rhizoctonia solani*;
(2) root rot caused by *Phytophthora manihotis*;
(3) foliage diseases—powdery mildew, *Sphaerotheca voandzeias*, and leafspots associated with *Cercospora canescens* and *Ascochyta phaseolorum*.

Potential. The bambara groundnut should be rated much more highly than at present as a pulse crop adapted to the hot, semi-arid to arid lowland tropics. The underground fruiting habit and extremely hard seeds are favourable characteristics for adverse conditions, reducing liability to attack by flower and pod insects and by storage pests. Moreover, the nutritive value of the green and ripe seeds appears to be excellent, with virtual freedom from metabolic inhibitors and toxins.

The high yields that are already sometimes attained occur without any serious scientific efforts on genetic improvement having been made. Increasing pressures on cultivating marginal lands should now focus attention on the importance of collecting and improving this already popular, productive, and nutritious species.

4.4.3. Pigeon peas (Cajanus cajan *Millsp.*)

The pigeon pea (red gram, arhar, congo bean, or gandul) is probably a native of Africa as an apparently truly wild form occurs in the sub-Saharian region. Seeds of the pigeon pea have been found in Egyptian tombs of the twelfth dynasty, and it was possibly cultivated in the Nile valley before 2000 B.C. Pigeon peas are thought to have been cultivated in Madagascar from very early times; they were possibly brought there—and to India—by people from the African mainland. Previously, India was thought to be the country of origin of the pigeon pea: *C. indicus* is a synonym of *C. cajan*. Pigeon peas reached the New World only after 1492, and there is no reason to suppose that they spread into the Pacific regions until much later. They are recorded as having been introduced in Guam, which was a standard port of call between Mexico and the Philippines, in 1772. The crop is now very widely spread throughout the tropics.

Production. On a world-wide scale, pigeon pea is by a considerable margin the most important pulse (non-oilseed grain legume) crop of the lowland tropics, with an annual production of about 2 million t produced on 2·9 million ha (Food and Agriculture Organization 1972). About 95 per cent of the crop is produced in southern Asia, and there mainly in India. Statistics for the African tropics may be erroneously low by a factor up to 4 times, since it is largely grown in small amounts at any one location, in compounds or kitchen gardens, but occurs in this way throughout the humid to semi-arid low- and intermediate-elevation tropical areas of Africa. Field plantings, however, are recorded mainly in Malawi and Uganda, but also occur in most other countries in East and West Africa (Rachie 1970).

Morphology. The pigeon-pea plant is a woody, short-lived perennial shrub growing to more than 4 m tall, which can be grown either as a seasonal or a perennial crop (some bushes survive more than 10-12 years). It has a deep taproot with longer laterals in the spreading than in erect types. The narrowly lanceolate and finely pubescent trifoliate leaves are spirally arranged on the main axis and on branches. Some cultivars tend to produce long primary branches which overtop the main axis. Such primary branches are leafy over their entire length but flower and fruit along the terminal one-third to one-half of their length. In other cultivars there is profuse secondary and tertiary branching. Yet others branch very little, producing leaves and fruits directly on the main axis, and are similar in structure to the modern soya bean. There are both erect and spreading types with branch angles ranging from about 30° to 60° from the main stem.

Inflorescences are either terminal or axillary, and the racemes vary from about 4 cm to 12 cm long in different cultivars. Flowers are about 2·5 cm in length and are either yellow in colour or with the dorsal side of the standard red, purple, deep orange, or may be veined with red or purple. The pods are somewhat flattened and 2–8 seeded (commonly 4) and measure 4–10 cm × 0·6–1·5 cm. Unripe pods may be plain green (recessive), purple, or maroon, or green blotched with purple or maroon. Seeds vary in size and are usually globular in shape. They may be white, greyish, red, brown, purplish, or speckled in colour, have a small white hilum, and when mature commonly weigh 10–15 g per 100 seeds.

Adaptation. Pigeon peas are widely adaptable to climatic and soil conditions, but thrive best if the soils are not markedly deficient in lime and in which water-logging does not occur. They are highly drought- and heat-resistant. As already stated, it is more widespread throughout the range of lowland tropical conditions than most other leguminous crops.

Most pigeon peas are highly sensitive to daylength, though some lines are insensitive. Maturities for different cultivars range from 90 days to more than 240 days, depending on the time of planting and climatic factors. Early growth is very slow, and pigeon peas are frequently interplanted with other, shorter-term crops, including both cereals and other grain legumes. They continue to grow and fruit after the shorter-duration companion crop has ripened and been harvested. Competition with weeds and with the companion crops is rather poor during the first 4–6 weeks, but the plants are very hardy once a canopy has been established which overtops the companion crop. A leaf litter builds up beneath them which not only suppresses weeds but also reduces erosion. High yields can be obtained with seasonal cultivars when there are good rains during the first two months, but no further precipitation occurs during the remaining 2–4 months before harvest.

Composition. The green pigeon pea is a popular vegetable in several tropical countries as well as being canned. The fresh seeds comprise 45 per cent of the weight of the whole pod. In this form they contain about two-thirds water (they are normally harvested for processing when alcohol soluble solids reach 25 per cent), 20 per cent carbohydrates, 7·0 per cent protein, 3·5 per cent fibre, 1·5 per cent fat, and 1·3 per cent mineral material. Dry, ripe seeds contain about 10 per cent water, 23 per cent protein, 56 per cent carbohydrate, 8·1 per cent fibre, and 3·8 per cent mineral matter. The protein is of reasonably good quality but, like most grain legumes, is deficient in sulphur amino acids and tryptophan in comparison with animal protein. However, the seeds are comparatively low in metabolic inhibitors, the flatus-inducing sugars, and the testa is free of lipoxidase which can cause off-flavours in soya beans and other legumes. Normally the seeds are split and testas removed in the preparation of Indian dhal.

Plant improvement. Pigeon peas are highly cross-pollinated (5–40 per cent, averaging 20 per cent) in comparison with most other grain legumes. A high proportion of flowers shed before setting fruit, particularly in hand cross-pollinated flowers. Blooming occurs over several weeks and flowers normally open between 11 a.m. and 3 p.m., and remain open for about 6 hours. It is important to emasculate before 9 a.m. on the day before the flowers open as pollen is shed later that morning (the day before opening). Rain reduces the success achieved in hand-pollination. Germination is hypogeal. Seed dormancy occurs in at least some cultivars (Purseglove 1968).

Plant types. Early botanists recognized two botanical varieties: var. *flavus* DC comprises the *tur* cultivars of peninsular India, which are short in stature, early in maturity, and have predominantly yellow flowers, green pods, light-coloured seeds, and usually 3 seeds per pod. The second variety, var. *bicolor* DC, comprises the *arhar* varieties of northern India and similar types of pan-tropical distribution. These are large, bushy, late-maturing perennials with red or purple or darkly veined flowers and hairy, maroon, or purple unripe pods with 4–5 dark coloured or speckled seeds when ripe. Most permutations of these characters have been observed in the world collection currently being studied at IITA and in breeding lines, as there appear to be few, if any, barriers to recombination save the physical ones of time and space. The continuation of the distinction into botanical varieties on the basis of a small number of characters has little to recommend it.

A more useful classification of the cultivated species was developed by Akinola and Whiteman (1972), who studied 95 accessions. They used 31 attributes, including plant and leaf morphology, growth, flowering patterns, disease tolerance, and components of seed yields to form 15 classes which fall into three major groups according to a hierarchical computer program (MULTCLAS). The group categorization included:

Group A. inflorescences terminal on shoots (basipetalous), comparatively early maturity;

Group B. inflorescences axillary (acropetalous), early maturity, and pod ripening over a protracted period;

Group C. inflorescences axillary (acropetalous), very late maturity, and pod ripening concentrated into a short period.

Management.

1. *Cropping systems and spatial arrangements.* As already pointed out, pigeon peas are often intersown with other crops like maize, sorghum, or pearl millet and left to mature on residual moisture after the cereal has been harvested. Row spacings vary widely depending on the companion crop and the plant type of the cultivar used. In mixed cropping, pigeon peas may be planted after every 2–4 rows of the main crop. In pure stands spacings normally vary from 30 cm to 90 cm in the row and from 90 cm to 300 cm between rows. Pigeon peas can be used as a forage crop; defoliating and slashing back to 90 cm stubble height at 3–5 month intervals produces the highest vegetative dry-matter yields.

2. *Fertilizer response.* It is often difficult to demonstrate response to fertilizers or rhizobial inoculation in tropical soils having a near-neutral pH and good drainage. However, moderate applications of phosphate and potash could be expected to produce economic returns on soils deficient in those elements. As always, responses cannot be generalized but depend upon the soil and its previous cropping history.

3. *Plant protection.*

(*a*) *Diseases*: The pigeon pea is generally less liable to damage by diseases and pests than other grain legumes in the lowland tropics. However, root and stem rots caused by *Macrophomina phaseoli* and *Phytophthora manihotis* have been reported in various parts of Africa; and *Fusarium udum* Butl. causing a wilt has been considered the most serious pathogen in India (Stanton 1966). Collar and stem canker caused by *Physalospora cajanae* is a serious disease in the Caribbean, as is rust caused by *Uromyces* sp. Leafspots caused by *Cercospora* sp. and *Colletotrichum cajani* may also occur in some areas. The most practical approach to these problems is to achieve host-plant resistance and to choose the optimum planting dates so as to escape disease as far as possible.

(*b*) *Insects*: Caterpillars of *Heliothis armigera* Hubn. and *Exelastis atomosa* and the fly *Agromyza obtusa* M. are serious pests of pigeon peas in India (Purseglove 1968), whereas *Heliothis* sp., *Maruca testulalis*, and *Laspeyresia* sp. occur in Africa. The pod-borers *Elasmopalpus rubedinellus* (Zell.), *Ancylostomia stercorea* (Zell.), and *Heliothis virescens* (F.) are serious pests in the West Indies (Purseglove 1968). Insecticide sprays are the only control measure for these pests at present, but these problems have not been very thoroughly investigated with respect to host-plant resistance or cultural and biological controls.

4. *Potential.* The pigeon pea has exceptional potential for use over a wide range of tropical conditions from subhumid to semi-arid regions. It is particularly valuable in mixed cropping and in bush-fallow systems of agriculture, where a perennial crop of 3–4 years is desirable. It can be used either as a field or a garden crop for producing dry seeds, green seeds as a vegetable, green leaves for cooking and for forage, or it can be used as a cover crop. Green-pod yields of 1000–8000 kg/ha (conversion ratio of green pods to dry peas is 3·3:1) have been recorded, and dry-seed yields of 500–1000 kg/ha are realizable. However, favourable growing conditions can result in high yields of 1600–2500 kg/ha; the exceptional yield of 5000 kg/ha of dry seeds has been reported from India (RPIP 1967). Akinola and Whiteman

(1972) obtained highest dry-seed yields per year (7600 kg/ha) based on two major harvests from a single planting of the cultivar UQ 50.

4.5. Secondary pulses

There are two secondary and at least nine minor pulses grown in the African lowland tropics on which information is largely unavailable or fragmentary. However, some of these may have exceptional productivity potentials, unique characteristics of adaptation, resistance to pests and diseases, and good nutritive qualities. Production statistics, if available are usually crude extrapolations from rough estimates obtained from limited areas.

4.5.1. *Lima beans* (Phaseolus lunatus *L.*)

Lima beans are also known as butter, Madagascar, Burma, or sieva bean in English, and *pois du cap, haricot du kissi, pois savor*, or *feve creole* in French. They are grown both for their green and dried shelled beans. It is essential that the mature bean should be cooked, as some cultivars—particularly those with coloured testas—may contain dangerous levels of cyanogenic glucocides, and hence produce hydrogen cyanide on hydrolysis. This poison is readily dissipated in the cooking process.

Origin and spread of lima beans. The lima bean is considered by Mackie (1943) to be indigenous to Central America, where wild forms occur. From there large white-seeded types were spread southward to Peru by the Incas; small-seeded types spread northward through Mexico to the southern USA; and tropical, perennial types spread eastward to the West Indies and from there to Brazil. The latter were short-day and highly cyanogenic plants. The large-seed types (*Ph. lunatus* var. *macrocarpus*) have been found in Peruvian excavations probably dating from 6000 B.C. to 5000 B.C. (Purseglove 1968). Early Spanish explorers carried the Caribbean-type limas with them across the Pacific to the Philippines and southern Asia; but African limas trace their origin back to Brazil, although the large-seeded types of Madagascar came originally from Peru.

Production. The lima bean is a major pulse in a few countries of the humid tropics in Africa. Up to 30 000 ha are planted annually with lima beans in Madagascar. One estimate from Nigeria, where it is grown mainly in the eastern and western Regions, places the crop second in importance to cowpeas

(FAO/CCTA 1958). Burma is the major producer of lima beans in Asia. The beans have also become an important vegetable crop for canning and freezing being grown extensively in the USA for both commercial and home use. A considerable number of improved cultivars, both pole and bush types, are available.

Identification. Lima beans can be readily distinguished from *Ph. vulgaris*. The bracteoles beneath the flower are shorter than the calyx, whereas they are longer than the calyx in *Ph. vulgaris*. *Ph. lunatus* has generally smaller white or cream flowers 1·2 cm or less in length, distinctive oblong, recurved, 2–4 seeded pods, and, to some extent the bean can also be distinguished by the flattened shape of the seeds, which vary in size from 45 g per 100 seeds to 200 g per 100 seeds. The seeds are rounded in 'potato' types, and are white, ivory, red, purple, brown, or black in solid colours or variously mottled and speckled. Germination is epigeal.

Adaptation. There are two major plant types—bush types 30–80 cm tall and twining and climbing herbs 2–4 m tall; both types are more or less perennial and indeterminate in nature. They tolerate wet weather and humid conditions better than most cultivars of *Ph. vulgaris*, but require dry weather to produce good-quality dry seeds, although the pods tend to shatter on maturation under dry conditions. The small-seeded sieva types are more resistant to hot, arid conditions than the large-seeded types. There appears to be a positive correlation between seed size and maturity.

Diseases and pests. Under humid tropical conditions in West Africa, lima beans are less affected by diseases and insect pests than most other grain legumes (IITA 1973), though podborers and leafhoppers attack the crop. The most seriously limiting diseases are those caused by viruses, downy mildew (*Phytophthora phaseoli*), pod blight (*Diaporthe phaseolorum*), anthracnose (*Colletotrichum lindemuthianum*), root rots (*Rhizoctonia solani* and *Fusarium oxysporum*), and rust (*Uromyces appendiculatus*).

Flowering and crossing.

1. *Out-crossing.* Allard (1954c) observed 1–15 per cent out-crossing in California, depending on cultivar, wind direction, and spacing. Out-crossing declined rapidly beyond 75 cm of isolation and was almost nil

at 10 m. Thrips of the genus *Frankliniella* appeared to be the principal pollinator.

2. *Controlled crossing*. Genetic male sterility is available and can be used to facilitate genetic recombination. Wester and Jorgensen (1950) showed that a very high rate of successful cross-pollination can be achieved without emasculation by forcing the stigma of the female parent out of the keel by means of pressure with forceps on the wings. Since a high rate of out-crossing occurs naturally (Magruder and Wester 1943), controlled selfing may be needed, and this can be achieved by covering the row or plants in the row just prior to flowering with a thin 28/20-mesh cotton plantbed cover material of suitable width (Magruder and Wester 1942). Wester and Marth (1949) were able to reduce excessive drop of flowers and pods in breeding materials by the use of growth substances; light scratches were made at the base of the flowers with a needle to facilitate absorption.

3. *Seed set*. A major cause of inadequate fertilization of ovules in the lima bean is the sensitivity of pollen to moisture stress and high temperatures. Some cultivars, such as Fordhook 242, have greater tolerance to excessive soil moisture and high temperatures than others (Lambeth 1950). Mackie (an upright bush type of 110-120 days duration) was released in California as a selection for tolerance to high temperatures (Allard 1954*a*).

Improved cultivars. Cultivars adapted to the tropics include both bush and pole types. Among bush types are Fordhook 242 and Burpee Bush; high-performing pole types include King of the Garden and Florida Speckled Butter. In southern Nigeria several small-to-large, white-to-variously coloured pole types have been collected or introduced, generally with good results, except that some showed susceptibility to nematodes. The bush cultivars produced upwards of 800 kg dry seeds per ha, whereas the pole lines have produced dry seed yields in excess of 3000 kg/ha from multiple pickings over a 5-month span (IITA 1973).

Management. The great range of plant types, growth habits, duration, and genotype–environment interactions occurring in lima beans makes it difficult to generalize about cultural practices. Perhaps the major limitations on productivity are moisture availability and diseases and pests. Bush beans are usually planted in rows 70-92 cm apart with 5-20 cm between plants in the row, whereas pole cultivars are planted in hills 92-125 cm apart with 3-4 seeds per hill for training on poles. Seeds of bush cultivars are usually planted 1·5-5·0 cm deep and at the rate of 132-165 kg/ha for large-seeded cultivars, and 55-80 kg/ha for small-seeded cultivars. Dry beans are usually harvested when pods have turned light yellow or brownish, after about 3 months for early cultivars or 7-9 months for long-duration, photosensitive types (e.g. for large-seeded Peruvian strains). If maturation occurs during humid weather the beans can be harvested green for immediate use of the ripe seeds, and will remain wholesome for up to 11 days at about −1 °C (Purseglove 1968).

Potential. Lima beans may well be one of the most promising pulses for the wetter parts of the lowland humid tropics. Under West African conditions the crop is extraordinarily free from serious disease and pest problems, with the possible exception of nematodes. However, good resistance to root-knot nematodes has been reported by Allard (1954*b*) and has also been observed in breeding lines in Nigeria (IITA 1973). Recent reports on chick-feeding experiments in the USA (McGinnis 1973) suggest that some cultivars of lima beans if autoclaved had feeding values better than some cultivars of *Ph. vulgaris* and as good as cowpeas. Supplementation with synthetic methionine (and penicillin) improved the nutritive values and made them equivalent to soya beans supplemented with methionine when fed at equivalent protein levels.

Perhaps the major deterrent to greater production of lima beans in West Africa is the taste preference for cowpeas and the fact that lima beans take longer to cook than cowpeas. Nevertheless, lima beans are already quite commonly used, and their use should increase with the development of cultivars with better seed qualities and the dissemination of more information on methods of preparation.

4.5.2. *African yam beans* (Sphenostylis stenocarpa Harms.)

This pulse is a slow-growing, herbaceous climber, frequently grown in association with the common yam (*Dioscorea* spp.) and other beans in humid and forested regions of the African lowland tropics. It is grown both for its seeds and edible tubers, which somewhat resemble sweet potatoes in size and shape and Irish potatoes in taste.

Botanical. The plant may be procumbent or grow upright by twining on supports. The leaflets are lanceolate or ovate. The axillary inflorescences are racemose. The calyx is broadly cupular and shortly, undulately lobed. The stigma is spatulate. Fruits are linear, up to 30 cm long, and glabrous, becoming tough and hard when ripe. The seeds are ellipsoidal, smooth, shiny, and hard when ripe (Hutchinson and Dalziel 1958).

Utilization. The tough pods and hard seeds appear less susceptible to pest attacks either in the field or storage than either beans or cowpeas. The seeds, which require several hours of soaking before cooking, contain exceptionally high levels of methionine (up to 1·92 g per 16 g nitrogen) and cystine (1·44 g per 16 g nitrogen) in comparison with other grain legumes.

Nicol (1959*a*, 1959*b*) studied the utilization of the yam bean in Nigeria. In some places it was cultivated mainly for its edible tubers and in others for its seeds. It is also known to be cultivated in the Central African Republic, Cameroun, Zaire, Ethiopia, and various other parts of East and Central Africa.

4.6. Minor pulses

There are several minor pulses grown to a limited extent in different parts of the African lowland tropics. Some of these may have high potential, but little attention has been paid to their improvement. Others have hardly been studied at all, except for basic taxonomic descriptions. The more widely grown of these less important species are now very briefly described. The information is from several sources: the description of the species and the recorded distribution of their cultivation have been obtained mainly from Hutchinson and Dalziel (1958), FAO/CCTA (1958), Stanton (1966), and Purseglove (1968).

4.6.1. Common beans (Phaseolus vulgaris L.)

The common (dry, dwarf, kidney, French, navy, snap, runner, salad, or string) bean is of world-wide distribution and the most widely grown and best known of all *Phaseolus* species. Although extensively grown at intermediate and higher elevations, it is grown only to a minor extent in the lowland tropics, usually as a fresh market vegetable. Its use in this capacity is reviewed by Tindall (see Chapter 7). However, the use of beans as a grain legume has also been studied, especially in the francophone countries of West Africa (d'Arondel de Hayes 1971). Different cultivars are adapted to growth in the wet and dry

seasons; in particular dwarf or bush beans are considered as suitable as a pulse crop for cultivation only in the dry season.

Dry-season cultivation. During the dry season of 1967–8 trials were conducted with six commercial dwarf cultivars. Sowing was done in mid-November; the plants flowered in late December and were harvested in mid-February, a growth cycle of 90 days. The three highest-yielding cultivars were Coco Rose Iran at 2843 kg/ha, Lingot Bean at 2333 kg/ha, and Michigan Pea Bean at 1704 kg/ha.

For climbing cultivars an average yield of 3414 kg/ha was obtained from all plots for the cultivar Sossogbe in a trial of 90 days' duration which was carried out in the dry season of 1967–8. The expense and labour involved in erecting supports made the growing of climbing cultivars uneconomic in the dry season, despite the appreciable increase in yield over the dwarf cultivars. It was calculated that supporting sticks, each about 2 m high and set 40 cm apart in the row with 0·6 m between rows (40 000 sticks were required per ha), cost approximately 10 000 CFA francs per ha. The sticks could, however, be used for two crops over a period of 2 years. The actual cost was 13 325 CFA francs per ha. At a price of 0·25 CFA francs per kg of beans to the producer, 535 kg of the harvest are required to cover the cost of the labour involved just to erect the supports.

Wet-season cultivation. Dry beans can be cultivated in the wet season but only climbing cultivars, such as Sossogbe, are suitable. The dwarf cultivars are flattened by heavy rain and are sensitive to diseases.

Trials with Sossogbe in the wet season of 1968 produced yields of 2170 kg/ha; four dwarf cultivars gave no harvest since they failed to survive the excessive moisture during the months of July and August.

4.6.2. Mucuna spp.

There are two species of *Mucuna* grown as pulses in various parts of tropical Africa. Both are vigorous, herbaceous climbers with long duration, often requiring 8–12 months to fruit.

The fruits of velvet beans (*M. pruriens* var. *utilis*) are smaller than those of the horseye bean (*M. sloanei* Adars = *M. urens*). The pods of both species are distinctive in appearance and frequently covered with profuse, stiff, golden, urticating hairs, giving them a velvety appearance. This feature, together with extreme hardness of testa in horseye bean, render them unpleasant to harvest and thresh. The seeds are cooked by boiling or roasting and, after the testas have been removed, are eaten in soups and stews. Apparently the flour has good thickening properties, as less than 10-12 large seeds are reported to make a gallon of soup. The beans are cultivated in such diverse regions as southern Nigeria, Dahomey, Senegal, Upper Volta, Sudan, and Mozambique. The seeds of *M. pruriens* contain the unusual amino acid l-dihydroxyphenylalanine, which is of pharmaceutical interest (Leakey, personal communication).

Dioclea reflexa Hook. is a woody climber whose seeds have extremely hard testas and resemble those of *M. sloanei*. It has been reported to occur from Sierra Leone and Guinea eastward to Cameroun. There is comparatively little evidence of its use for food, and it may be only an occasional 'gathered' crop used sporadically or in times of scarcity.

4.6.3. Hyacinth beans or lubia (Lablab niger *Medik*)

This species is also known as the bonavist, dolichos, lablab, Indian butter, and Egyptian kidney bean (Syn. *Dolichos lablab* L.: *Lablab vulgaris* Savi). The young pods and fresh ripe green beans are used as a vegetable and the dry seeds as a pulse. It is also grown for forage and is a major crop in rotation with sorghum and cotton in the Gezira irrigation area of Sudan. It is a hardy, drought-resistant dry-land crop, well adapted to low rainfall areas in the tropics. Although it is a herbaceous perennial herb it is frequently grown as an annual. It is often twining, but bush forms also occur. The pod resembles that of a lima bean and the seeds are also similar in size to a medium lima bean, but plumper, and have a characteristic aril or projection from the hilum, extending one-third of the circumference of the seed, making it easy to identify. Two botanical varieties are recognized:

(1) var. *lablab* (var. *typicus* Prain.): short-lived, twining perennial herb; pods are longer, more tapering with long axes of the seeds parallel to the suture; grown mainly for green pods;

(2) var. *lignosus* (L.) Prain.: longer-lived, semi-erect, bushy perennial also called Australian pea. Pods are shorter, more truncated, and the long axes of the seeds at right-angles to the suture. The plants have a strong unpleasant smell and are used mainly as a dry pulse and fodder.

The hyacinth bean is rather widely (but not extensively) grown throughout the Sudano-Sahelian zone, and also in Kenya (the Kikuyu bean), Malawi, Angola, and Mozambique.

4.6.4. Locust beans (Parkia *spp.*)

Fernleaf, nitta trea, nere, or nele are some of the common names for *Parkia* spp. particularly *P. filicoides* Welw. (syn. *P. clappertonia*) and *P. biglobosa* Benth. (also *P. oliveri*), which become large trees. Parkias are seldom, if ever, planted and occur throughout the African savannah zones. The seeds are not used as a dietary staple, but are cooked and fermented to make a condiment widely used as a food flavouring. The fruit pulp from immature pods is sweet, a popular flavouring in desserts and drinks, and is particularly nourishing by virtue of its high content of sulphur amino acids—up to 2·9 g methionine and 3·8 g cystine per 16 g nitrogen on a dry-weight basis (Busson, Toury, and Bergeret 1958).

4.6.5. Jack and sword beans (Canavalia *spp.*)

Jack beans (*C. ensiformis* L. and *C. plagiosperma*) and sword beans (*C. gladiata* Jacq.) are used as a dry pulse or a green vegetable (immature pods), and the vegetative portion may be grown for forage, green manure, or cover crop. However, dry *C. gladiata* seeds may contain toxic substances. All three species are very hardy, deep-rooted, and exceptionally drought-resistant, but also tolerate water-logging, shade, and saline soils better than many other grain legumes. The extremely large, tough pods and hard seeds (2-3 cm long) germinate epigeally and very quickly (48-72 hours), and the developing plant is exceptionally free from insects, pests, and diseases under African conditions.

The large, dry seeds are hard to cook and can be somewhat toxic, requiring boiling in salt water for several hours, with a change of water. The jack bean is an important source of the enzyme urease. The lectin concanavalin-A occurs at levels of 2·5-3·0 per cent by weight in dry seeds and is being used in medical research (Sharon and Lis 1972). This particular lectin has a high specific agglutinating activity in cells transformed by DNA tumor viruses or carcinogens.

Dry-seed yields of jack beans can be considerable even on poor soils—where 2000-2500 kg/ha are reported. Production is scattered generally throughout West Africa, and also in Zaire and Angola.

4.6.6. *Mung beans and black grams* (Vigna radiata (L.) *Wilczek*)

Mung beans (green gram) and black gram *Vigna radiata* (L.) Wilczek vars. *aureus* and *mungo* respectively, are highly relished by Asians and have been grown to a limited extent in eastern Africa as far south as Natal, to satisfy a portion of the local market. They are much used for preparing 'bean sprouts' as a vegetable. The seed and its nutritive qualities are excellent, but cultivation and utilization of these crops are largely unfamiliar to Africans.

At Ibadan dry-seed yields in experimental plots approached 2000 kg/ha.

4.6.7. *Rice beans* (Vigna umbellata† (Thunb.) *Ohwi & Ohashi*

This species, formerly named *Phaseolus calcaratus* Roxb., is a vigorously climbing or sub-erect annual producing long, slender, glabrous, shattering pods, with variously coloured, medium–small (8 mm long) seeds, and is used mainly as a dried pulse. It is more important in Asia, where it is frequently grown in rotation with rice, and can make a crop in as little as 60 days. It performs well under humid conditions and has been reported to be cultivated together with *Phaseolus angularis* in the Zairean southern savannah and river basin. The rice bean may have great potential for the humid tropics as it is less susceptible than many other pulses to diseases and pests in West Africa.

4.6.8. *Tepary beans* (Phaseolus acutifolius *var.* latifolius)

This species, originating in north-western Mexico was introduced into Africa in fairly recent times and is grown to a limited extent in the drier cultivated sub-Sahara regions. It has been recorded from both West and East Africa as far south as Lesotho, Botswana, and the Republic of South Africa.

4.6.9. *Kersting's groundnuts* (Kerstingiella geocarpa *Harms*)

Kersting's groundnut is grown in West Africa in similar ecological conditions to the bambara groundnut, but on a much smaller scale. Kersting's groundnut can be distinguished from *Voandzeia* by its deeply divided calyx with narrow lobes and its glabrous style. The axillary inflorescences are sub-

† A substantial reallocation of species previously included in *Phaseolus* to the genus *Vigna* is made and justified in Verdcourt (1969).

sessile and the pods are about 2 cm in length and usually contain two seeds. The fruits penetrate the soil by a carpophore and elongation of intercalary meristem tissue at the base of the ovary. This is similar to the mechanism of geocarpy in *Arachis hypogaea* and different from that in *Voandzeia*.

4.7. Conclusions

4.7.1. *The need for improvement*

Five species, groundnuts, cowpeas, bambara groundnuts, pigeon peas, and lima beans, are considered of major importance in West Africa and occupy between them most of the cultivated land devoted to grain legumes in the lowland tropics. In addition to these, soya beans must be considered to have exceptional potential for contributing to both vegetable oil and plant protein needs and for processing in the developing agricultural industries of the region. Groundnuts in the tropics have received the greatest improvement emphasis in the past owing to their industrial and export potential.

4.7.2. *Future trends in research and development*

Various difficulties oppose increasing groundnut production. According to FAO (Tahir 1970) the expansion on the world market for oil and oilseeds is unlikely to exceed 3·2 per cent per annum in the future, and the aflatoxin hazard limits the desire to use groundnut proteins, even though world protein requirements are growing very fast. The necessary research for the maintenance of at least existing levels of groundnut production should be directed at two main objectives; increasing the productivity of farmers and developing outlets, in particular by obtaining the best return for nuts and oilcake by improved quality control. On both these two points there is a considerable gap between the knowledge already acquired through research and its application (see also Chapter 14).

It now appears feasible to draw on both extensive knowledge and improved genetic materials from soya bean improvement programmes in temperate regions to bring this crop to a position of comparable importance with groundnuts. Soya beans and groundnuts can be considered complementary rather than competitive, as they have rather different soil and moisture requirements. Comparatively little in-depth study on plant improvement has so far been done for cowpeas, pigeon peas, or lima beans, and virtually nothing has been done on bambara groundnuts.

It is both surprising and encouraging that other 'secondary' pulses, like pigeon peas, mung beans, and jackbeans, have occasionally been shown to have high productivity potential, often without serious problems or need for substantial management inputs. This suggests that comparatively modest investments on improvement might pay off handsomely and quickly. Nevertheless, it would probably be unrealistic to activate major improvement efforts on more than three or four of these crops at the present time, at least at the international level. However, it may become expedient over the longer term, as demand increases, that more marginal lands be brought under cultivation and resources be made available to mount programmes on some of the more promising of these 'secondary' species. Furthermore, localized interests may well decide that emphasis on some of the minor species is relevant in national programmes.

Since minor species tend to be quickly lost with the expansion of more sophisticated farming systems, there is an urgent, immediate need thoroughly and systematically to collect and maintain indigenous germplasm.

Acknowledgements

The editors would like to thank H. D. Tindall for the section on *Phaseolus vulgaris*.

References

ABDOU, Y. A. M. (1966). The source and nature of resistance in *Arachis* sp. to *Mycosphaerella arachidicola* and *M. berkeleyii* and factors influencing sporulation of these fungi. Thesis, University of North Carolina, Raleigh, USA.

AKINOLA, J. O. and WHITEMAN, P. C. (1972). A numerical classification of *Cajanus cajan* (L.) Millsp. accessions based on morphological and agronomic attributes. *Aust. J. Agric. Res.* 23, 995-1005.

ALLARD, R. W. (1954a). New heat tolerant lima bean. *Calif. Agric.* 8, 5 et seq.

—— (1954b). Source of root knot nemetode resistance. *Phytopathology* 44, 1-4.

—— (1954c). Natural hybridization in lima beans in California. *Proc. Am. Soc. hort. Sci.* 64, 410-16.

ANON. (1964). Annual Report of the Indian control oilseeds Committee. No. 18.

ANON. (1968). UCPA recommends new improved peanut variety. *Mon. Bull. UP Coll. Agric.* 34, 4.

ANON. (1969). (Laboratoires membres du Comité Inter-Instituts d'Études des Techniques Analytiques du Diagnostic Foliaire.) Méthodes de références pour la détermination des éléments minéraux dans les végétaux: Azote, Phosphore, Potassium, Sodium, Calcium, Magnésium. *Oléagineux* 24, 497-504.

ANON. (1970). *A new technology for dry land farming.* Indian Agricultural Research Institute, New Delhi.

ASHRI, A. (1972). *Mutations and physiological reaction to several chemical mutagens in peanuts,* Arachis hypogaea L. Proc. Latin Amer. study group on induced mutations and plant improvement, Buenos Aires, 1970. International Atomic Energy Authority, Vienna, Austria.

AUCKLAND, A. K. (1966). Soybean in Tanzania. I—The exploitation of hybridization for the improvement of soyabeans. *J. agric. Sci. Camb.* 67, 109-19.

—— (1967). Soybean in Tanzania. II—Seasonal variation and homeostasis in soyabeans. *J. agric. Sci. Camb.* 69, 455-64.

—— (1970). Soybean improvement in East Africa. In *Crop improvement in East Africa* (ed. C. L. A. Leakey). Commonwealth Agricultural Bureaux, Farnham Royal.

BAMPTON, S. S. (1963). Growth of *Aspergillus flavus* and production of aflatoxin in groundnuts.—I. *Trop. Sci.* 5, 74-81.

BARKER, L. N. (1970). A review of cowpea breeding at the University of Ife (1966-1970). *Ford Foundation/IITA/IRAT seminar on grain legume research in West Africa, Ibadan.*

BILLAZ, R. (1959). L'alimentation en eau de l'arachide dans les sols diors du Sénégal. *IRHO, Rap. mult.*

BOLHUIS, G. G. (1951). Natuurlijke bastaardering bij de Cardnoot (*Arachis hypogaea*). *Landbouwk. Tijdschr.* 63, 447-55.

—— (1955). La culture de l'arachide en Indonésie. *Oléagineux,* 10, 157-60.

—— DE GROOT, W. (1959). Observations on the effect of varying temperature on the flowering and fruit set temperature in three varieties of groundnut. *Neth. J. agric. Sci.* 7, 317-26.

BOUHOT, D. (1967). Contribution a l'étude des maladies des gousses et des graines d'arachides dues au *Macrophomina phaseoli. Agron. trop., Nogent* 22, 864-87.

BRADY, N. C. (1947). The effect of period of calcium supply and mobility of calcium in the plant on peanut fruit filling. *Proc. Soil Sci. Soc. Am.* 12, 336-41.

BROMFIELD, A. R. (1973). Uptake of sulphur and other nutrients by groundnut (*Arachis hypogaea*) in northern Nigeria. *Exp. Agric.* 1, 55-8.

BRZOZOWSKA, J. and HANOWER, P. (1964). Absorption et distribution du soufre 35 chez quelques arachides tropicales. *Oléagineux* 19, 663-72.

BUNTING, A. H. (1955). A classification of cultivated groundnuts. *Emp. J. exp. Agric.* 23, 158-70.

—— (1958). A further note on the classification of cultivated groundnut. *Emp. J. exp. Agric.* 26, 254-8.

Busson, F., Toury, J., and Bergeret, B. (1958). *Composition in acides amines des graines des principales légumineuses tropicales utilisées dans l'alimentation.* Document No. 5 Laboratoire de Récherches du Service de Santé de la France d'outre Mer, Marseilles; et Organisme de Récherche sur l'Alimentation et la Nutrition Africaines, Dakar.

Carrière de Belgarric, R. and Bour, F. (1963). Le développement de la productivité de l'arachide au Sénégal. *Agron. trop., Nogent* **18**, 863-75.

Cas, S. (1972). Absorption radiculaire et distribution du 35 S dans les organes de l'arachide. Influence de la déficience en soufre. *Oléagineux* **27**, 545-51.

Caswell, G. H. (1968). The storage of cowpea in the northern states of Nigeria. *Proc. agric. Soc., Nigeria* **5**, 4-6.

Catherinet, M. (1956). Quelques données sur la germination de l'arachide. Étude de la température optimum. *Ann. C.R.A. Bambey, Bull. Agron.* **16**, 93-8.

Corbett, D. C. M. and Brown, P. (1966). Fungicidal control of *Cercospora* leaf spots of groundnuts in Malawi. *Rhod. Zam. Mal. J. agric. Res.* **4**, 13-21.

Culp, T. W., Bailey, W. K., and Hammons, R. O. (1968). Natural hybridization of peanuts, *Arachis hypogaea* L. in Virginia. *Crop. Sci.* **8**, 109-11.

Dancette, C. (1970). Détermination au champ de la capacité de rétention après l'irrigation dans un sol sableux du Sénégal. Intérêt agronomique de cette mesure et application à la culture d'arachide. *Agron. Trop., Nogent* **25**, 225-40.

D'Arondel de Hayes, J. (1971). Complementary crops which could be of interest in the Sudanian zone. *Ford Foundation/IITA/IRAT seminar on vegetable crops, Ibadan.*

Dart, P. J. (1973). Effects of temperature on activities of soybean rhizobial strains. Unpublished results of experiments carried out at Rothamstead Experiment Station, Hertfordshire, U.K.

—— Mercer, F. V. (1965). The effect of growth temperature level of ammonium nitrate and light intensity on the growth and nodulation of cowpeas. *Aust. J. agric. Res.* **16**, 321-45.

Delassus, M. (1970). Crop protection: major problems observed and studied on edible legumes in various French speaking countries. *Ford Foundation/IITA/IRAT seminar on grain legume research in West Africa, Ibadan.*

Devez, J. (1959). La culture de l'arachide dans la plaine de Ruzizi. *Bull. Inf. INEAC* **8**, 219-30.

Doku, E. V. (1968). Flowering, pollination and pod formation in bambara groundnut (*Voandzeia subterranea*) in Ghana. *Exp. Agric.* **4**, 41-8.

—— (1969). Growth habit and pod production in bambara groundnut (*Voandzeia subterranea*) in Ghana. *Ghana J. agric. Sci.* **2**, 91-5.

—— (1970). Effect of daylength and water on nodulation of cowpea (*Vigna unguiculata* (L.) Walp.) in Ghana. *Exp. Agric.* **6**, 13-18.

—— Karikari, S. K. (1970*a*). Flowering and pod production of bambara groundnut (*Voandzeia subterranea* Thouars). *Ghana J. agric. Sci.* **3**, 17-26.

—— —— (1970*b*). Operational selections in wild bambara groundnuts. *Ghana J. Sci.* **11**, 47-56.

Ebong, U. U. (1965). *Cowpea production in Nigeria.* Nigerian Federal Department of Agricultural Research, Memorandum 80. Ibadan, Nigeria.

—— (1970*a*). On the Nigerian grain legume gene bank. *Ford Foundation/IITA/IRAT seminar on grain legume improvement, Ibadan.*

—— (1970*b*). Strategies for breeding cowpea crop for improvement and high productivity. *Ford Foundation/IITA/IRAT seminar on grain legume improvement, Ibadan.*

Ezedinma, F. O. C. (1964). Effects of inoculation with local isolates of cowpea *Rhizobium* and application of nitrate nitrogen on the development of cowpeas. *Trop. Agric., Trinidad,* **41**, 243-9.

Food and Agriculture Organization (1966). *Agricultural development in Nigeria 1965-1980,* pp. 392-400. FAO, Rome.

—— (1971). *Agricultural production yearbook.* No. 24, pp. 153-74, 228-35. FAO, Rome.

—— (1972). *Agricultural production yearbook.* No. 25, pp. 229-35. FAO, Rome.

FAO/CCTA (1958). *Report of the FAO/CCTA technical meeting on legumes in agriculture and human nutrition in Africa.* FAO, Rome.

Fortanier, E. J. (1957). De beinvloeding van de Bloei big *Arachis hypogaea* L. *Med. Landb. Hoogesch. Wageningen* **57**, 1-116.

Fowler, A. M. (1970). The epidemiology of *Cercospora* leaf spot diseases of groundnuts. *Samaru Agric. Newsl.* **12**, 66-9.

Franck, Z. R. and Krikun, J. (1969). Evaluation of peanut (*Arachis hypogaea*) varieties for *Verticillium* wilt resistance. *Plant Dis. Rep.* **53**, 744-6.

Garren, K. H. (1964). Inoculation potential and differences among peanuts in susceptibility to *Sclerotium rolfsii.* *Phytopathology* **54**, 277-81.

Gautreau, J. (1970). Étude comparative de la transpiration relative chez deux variétés d'arachide. *Oléagineux* **25**, 23-8.

Gibbons, R. W. and Bailey, B. E. (1967). Resistance to *Cercospora arachidicola* in some species of *Arachis.* *Rhod. Zamb. Mal. J. agric. Res.* **5**, 57-9.

—— Bunting, A. H., and Smartt, J. (1972). The classification of varieties of groundnut (*Arachis hypogaea* L.). *Euphytica* **21**, 78-85.

Gibson, I. A. S. (1953). Crown rot, a seedling disease of groundnut caused by *Aspergillus niger.* *Trans. Brit. Mycol. Soc.* **36**, 198-209, 324-34.

Goldsworthy, P. R. and Heathcote, R. G. (1964).

Fertilizer trials with groundnuts in Northern Nigeria. *Emp. J. exp. Agric.* **31**, 351–66.

GOPALSWAMY, A. and VEERANNAH, L. (1968). Studies on oil, protein and free fatty acid content (titrable acidity) of bunch and spreading varieties of groundnut. *Madras agric. J.* **55**, 209–11.

GREGORY, W. C. (1956). Induction of useful mutation in the peanut. In *Genetics in plant breeding. Brookhaven Symposia in Biology.* No. 9, pp. 177–90.

—— COOPER, W. E. (1959). Atomic peanut. *Res. farming* **17**, 3.

—— SMITH, B. W. and YARBROUGH, J. A. (1951). Morphology, genetics and breeding. Chapter III in *The peanut, the unpredictable legume.* A Symposium of the National Fertilizer Association, Washington D.C., pp. 28–88.

HANOWER, P., BRZOZOWSKA, J., and PREVOT, P. (1963). Vitesse d'absortion et de translocation du 35 S chez l'arachide. *C.r. hebd Séanc. Acad. Sci. Paris* **257**, 496–8.

HARLAN, J. V. (1972). Report of genetic resources in Africa. Unpublished report of exploration and survey made by the author during 1967–72. Department of Agronomy, University of Illinois.

HARRIS, H. C. and BROLMANN, J. B. (1966). Comparison of calcium and boron deficiency of peanut. *Agron. J.* **58**, 575–82.

HAWTIN, G. C. (1973). The exploitation of genetic variation in *Glycine max* (L. Merrill.). Ph.D. Thesis, Cambridge University, England.

HOLLEY, K. T. and HAMMONS, R. O. (1968). Strain and seasonal effects on peanut characteristic. *Univ. Georgia Res. Bull.* **32.**

HUXLEY, P. A., SUMMERFIELD, R. J., DART, P., and HUGHES, A. P. (the late) (1975). The effect of nitrogen, water and shade stress on the seed yield of cowpea cv. Prima. *Pl. Soil.* (In press.)

HYMOWITZ, T. (1970). On the domestication of the soybean. *Econ. Bot.* **24**, 408–21.

IAR (1969). Institute for Agricultural Research, Samaru (Nigeria). Annual reports.

IARI (1970). *New vistas in pulse production.* The Indian Agricultural Research Institute. New Delhi.

ICOC (1964). Indian Control Oilseeds Committee Annual Report, 1964–5.

IITA (1973). *Grain legume improvement program of the International Institute of Tropical Agriculture.* Published reports of the IITA for 1971 and 1972. Ibadan, Nigeria.

ILYANA, A. I. (1958). Definition of the periods of high sensitivity of peanut plants to soil moisture (Russian). *Fiziol. Rast.* **5**, 253–8.

—— (1959). Étude des periodes de haute sensibilite de l'arachide aux differences d'humidite du sol. *Oléagineux,* **14**, 89–92.

IRAT (1972). Annual reports of IRAT (Madagascar) for the years 1970–2.

JACQUINOT, L. (1967). Comparison of growth and mineral nutrition in four cowpea varieties (French). *Agron. trop.*, Nogent **22**, 575–640.

JOHNSON, D. T. (1968). The bambara groundnut—a review. *Rhodesian agric. J.* **65**, 1–4.

—— (1970). The cowpea in African areas of Rhodesia. *Rhodesian agric. J.* **67**, 61–4.

KRAPOVICKAS, A. (1968). Origen, variabilidad y diffusion del mani (*Arachis hypogaea* L.). *Actas Mems XXXVII. Congr. internac. Am.* **2**, 517–34.

—— RIGONI, V. A. (1960). La nomenclatura de las subspecies y variedades de *Arachis hypogaea* L. *Rev. Invest. agric.* **14**, 198–228.

KULKARNI, L. G. (1967). Asiriya Mwintunde: groundnut variety that will surpass many others. *Ind. Farming* **17**, 4–7.

LAMBETH, U. N. (1950). Some factors influencing pod set and yield of lima bean. *Res. Bull. agric. Exp. Stn. Univ. Wis.* **466**, 60 *et seq.*

LEAKEY, C. L. A. (1970). Brief summary of work on soyabeans in East Africa. *Ford Foundation/IITA/IRAT seminar on grain legume research, Ibadan.*

—— RUBAIHAYO, P. R. (1972). Soyabean agronomy in Uganda. I—Plant population and spacing. *E. Afr. agric. for. J.* **27**, 201–5.

LEHMAN, W. F. and LAMBERT, J. W. (1960). Effects of spacing of soyabeans between and within rows on yield and its components. *Agron. J.* **52**, 84–6.

LIENER, I. E. (1969). *Toxic constituents of plant foodstuffs.* Chapters 1–3. Academic Press, New York.

MACKIE, W. W. (1943). Origin, dispersal and variability of the lima bean (*Phaseolus lunatus*). *Hilgardia* **15**, 1–29.

MADHAVA MENON, P., RAMAN, V. S., and KRISHNASWAMI, S. (1970). An X-ray induced monosomic in groundnut. *Madras agric. J.* **57**, 80–2.

MAGRUDER, R. and WESTER, R. E. (1942). Prevention of field hybridization in the lima bean. *Proc. Am. Soc. hort. Sci.* **40**, 413–14.

—— (1943). Natural crossing of lima beans in Maryland during 1941. *Proc. Am. Soc. hort. Sci.* **42**, 557–61.

MANTEZ, A., and GOLDIN, E. (1964). The influence of irrigation frequency and intensity on the yield and quality of peanut (*Arachis hypogaea*). *Israel J. agric. Res.* **14**, 103–210.

MARQUETTE, J. (1970*a*). Report on grain legumes in Madagascar. *Ford Foundation/IITA/IRAT seminar on grain legumes research, Ibadan.*

—— (1970*b*). Results of bacterial inoculation of soyabean (*Glycine max*) and of haricot beans (*Phaseolus vulgaris*) in Madagascar. *Ford Foundation/IITA/IRAT seminar on grain legume research, Ibadan.*

MARTIN, G. (1965). Diagnostic foliaire de l'arachide. *Oléagineux* **20**, 287–91.

—— (1968). Evolution de la richesse en huile dans la descendance d'arachides irradiées. *Oléagineux,* **23**, 105–7.

—— (1969). Contribution à l'étude de certains caractères

d'importance agronomique chez l'arachide. *Cah. Orstom. Sèr. Biol.* **7**, 3–53.

MARTIN, G. and FOURRIER, P. (1965). Les oligo-elements dans la culture de l'arachide du Nord Senegal. *Oléagineux* **20**, 287–91.

MASEFIELD, G. B. (1957). The nodulation of annual leguminous crops in Malaya. *Emp. J. exp. Agric.* **25**, 139–50.

MATHUR, S. B., SINGH, A., and JOSHI, L. M. (1967). Varietal response in groundnut (*Arachis hypogaea*) to *Sclerotium bataticola. Plant Dis. Rep.* **51**, 649–51.

MAUBOUSSIN, J. C. (1968). Études sur le taux d'allogamie chez l'arachide. IRAT (mimeo).

—— (1970). La sélection de l'arachide à l'IRAT/Senegal. *Ford Foundation/IRAT/IITA seminar on grain legume research, Ibadan.*

—— LAURENT, P., and DELAFOND, G. (1970). Les variétés d'arachides récommandées au Sénégal et leur emploi. *Cah. Agr. Pr. Pays chauds.* No. 2, 63–89.

McDONALD, D. (1970). Survey of cowpea market samples for seed borne fungi. *Ford Foundation/IITA/IRAT seminar on grain legume research, Ibadan.*

—— A'BROOK, J. (1963). Growth of *Aspergillus flavus* and production of aflatoxin in groundnuts. Part III. *Trop. Sci.* **5**, 208–14.

—— HARKNESS, C. (1963). Growth of *Aspergillus flavus* and production of aflatoxin in groundnuts. Part II. *Trop. Sci.* **5**, 143–54.

—— —— (1964). Growth of *Aspergillus flavus* and production of aflatoxin in groundnuts. Part IV. *Trop. Sci.* **6**, 12–27.

—— —— (1965). Growth of and production of aflatoxin in groundnuts. Part VIII. *Trop. Sci.* **7**, 122–37.

—— —— (1967). Aflatoxin in the groundnut crop at harvest in Northern Nigeria. *Trop. Sci.* **9**, 148–61.

—— —— STONEBRIDGE, W. C. (1964). Growth of *Aspergillus flavus* and production of aflatoxin in groundnuts. Part VI. *Trop. Sci.* **6**, 131–54.

McGINNIS, J. (1973). Feeding values of dwarf and lima beans and cowpeas in chick feeding experiments. Personal communication, Washington State University.

MEHTA, P. N. (1971). The effect of defoliation on seed yield of cowpeas (*Vigna unguiculata*). An analysis of leaf harvest for dry matter and nitrogen content. *Tech. Common. Int. Soc. Hort. Sci. N.Y.* **21**, 167–71.

MIXON, A. C. and ROGERS, K. M. (1973). Peanuts resistant to seed invasion by *Aspergillus flavus. Oléagineux* **28**, 85–6.

MONTENEZ, J. (1957). *Récherches experimentales sur l'écologie de la germination chez l'arachide.* Dir. Agric. Forêts, Elevage, Brussels.

MONTGOMERY, R. D. (1964). Observations on the cyanide content and toxicity of tropical pulses. *W. Ind. med. J.* **13**, 1–11.

MOREAU, C. L. (1968). *Moisissures toxiques dans l'alimentation.* P. Lechevalier, Paris.

MUKEWAR, A. M. (1966). Cytogenetics of X-ray irradiated groundnut (*Arachis hypogaea* L.). *Ann. agric. Res. Abstr. port-grad. Res. WK 1960–65.* Nagpur agric. Coll. Mag. 1966: Spec. Res. No. 124-5.

NABOS, J. (1970). Grain legumes in Niger—present state of research. *Ford Foundation/IITA/IRAT seminar on grain legume research, Ibadan.*

NICOL, B. M. (1959a). The calorie requirements of Nigerian peasant farmers. *Br. J. Nutr.* **13**, 293–306.

—— (1959b). The protein requirements of Nigerian peasant farmers. *Br. J. Nutr.* **13**, 307–20.

NUTMAN, P. S. (1971). Perspectives in biological nitrogen fixation. *Sci. Prog. Lond.* **59**, 55–74.

OCHS, R. and WORMER, T. H. M. (1959). Influence de l'alimentation en eau sur la croissance de l'arachide. *Oléagineux* **14**, 281–91.

OGLE, W. L. (1967). An evaluation of herbicides for Southern Peas (cowpeas) and snap beans. *Proc. Am. Soc. hort. Sci.* **90**, 290–5.

OJEHOMON, O. O. (1968a). The development of the inflorescence and extra floral nectaries of *Vigna unguiculata* (L.) Walp. *Jl. W. Afr. Sci. Ass.* **13**, 92–110.

—— (1968b). Flowering, fruit production and abscission in cowpea, *Vigna unguiculata* (L.) Walp. *Jl. W. Afr. Sci. Ass.* **13**, 227–34.

—— (1970a). A comparison of the vegetative growth, development and seed yield of three varieties of cowpea. *J. agric. Sci.* **74**, 363–74.

—— (1970b). Effects of continuous removal of open flowers on the seed yield of two varieties of cowpea. *J. agric. Sci.* **74**, 375–81.

—— (1972). Fruit abscission in cowpeas. *Vigna unguiculata* (L.) Walp. *J. exp. Bot.* **23**, 751–61.

OJOMO, O. A. (1970). Pollination, fertilization and fruiting characteristics of cowpeas (*Vigna unguiculata*). *Ghana J. Sci.* **10**, 33-7.

PATIL, S. H. (1966). Cytogenetics of X-ray induced aneuploids in *Arachis hypogaea* L. *Can. J. Genet. Cyt.* **10**, 545–60.

PHILPOTTS, H. (1967). The effects of soil temperature on nodulation of cowpeas. *Aust. J. exp. Agric. Anim. Husb.* **7**, 371–6.

PLATT, B. S. (1962). *Tables of representative values of food commonly used in tropical countries.* Medical Research Council special report, Series 20. MRC, London.

PRÉVOT, P. (1949). Croissance, dévéloppement et nutrition minérale de l'arachide. *Oléagineux coloniaux, serie scientifique* No. 4. SETCO, Paris.

—— OLLAGNIER, M. (1956). Methode d'utilisation du diagnostic foliaire. In *Analyse des plantes et problème des fumures minérales*, pp. 175–92. IRHO, Paris.

—— —— (1961). Diagnostic foliaire. Relations réciproques de certains elements minéraux. *Advances in Horticultural Sci.* (Proceed. of the XVth Int. Hort. Cong., Nice 1958.) Pp. 217–28.

—— —— Aubert, G. and Brugière, J. M. (1955). Degradation du sol et toxicité manganique. *Oléagineux* **10**, 239-43.

Purseglove, J. W. (1968). *Tropical crops: dicotyledons 1*, pp. 199-232. Longmans, Green, London and Harlow.

Rachie, K. O. (1970). Progress in *Cajanus* investigations at Makerere University (Kampala, Uganda). *Ford Foundation/IITA/IRAT seminar on grain legume research, Ibadan.*

Radley, R. W. (1971). Soybean production paper presented to the FAO/UNICEF sponsored seminar on the production, utilization and marketing of soyabeans. Kampala, Uganda.

Raman, V. S. and Rangasamy, S. R. (1972). Consideration on the phylogenetic relationships of species of *Arachis*. *Madras Agric. J.* **59**, 262-71.

Rassel, A. (1960). Le voandzu *Voandzeia subterranea* Thouars et sa culture au Kwango. *Bull. agric. Congo Belg.* **51**, 1-26.

Robertson, I. A. D. (1969). Insecticide control of insect pests of soyabean (*Glycine max* L.), in Eastern Tanzania. *E. Afr. agric. for. J.* **35**, 181-4.

Rounce, N. V. (1949). *The agriculture of the cultivation steppe of the Lake, Western and Central Province (Tanganyika)*. Longmans, Green. Capetown, South Africa.

RPIP (1967). Regional pulse improvement project Progress Report. No. 5, pp. 169-70. Indian Council of Agricultural Research, New Delhi.

Rubaihayo, P. R. and Leakey, C. L. A. (1970). Introduction, testing and selection of soyabeans in Uganda. *E. Afr. agric. for. J.* **36**, 77-82.

—— —— (1972). Soyabean agronomy in Uganda. I. Plant populations and spacing. *E. Afr. agric. For. J.* **37**, 201-5.

Schenk, R. V. (1961). *Development of the peanut fruit.* Georgia African Experimental Station Technical Bulletin No. 22.

Schneider, R. W. (1973). Epidemiology, yield-loss prediction and control of *Cercospora* leaf spot of cowpea (*Vigna unguiculata*). Ph.D. thesis, Department of Plant Pathology, University of Illinois.

Sellschop, J. P. R. (1962). Cowpeas, *Vigna unguiculata* (L.) Walp. *Fld Crop Abstr.* **15**, 259-66.

Sene, D. and N'Daye, S. M. (1970). Improvement of the cowpea (*Vigna unguiculata*) at the CNRA, Bambey. *Ford Foundation/IITA/IRAT seminar on grain legume research, Ibadan.*

Sharon, N. and Lis, H. (1972). Lectins: cell-agglutinating and sugar specific protein. *Science, N.Y.* **177**, 949-59.

Shchori, Y. and Ashri, A. (1970). Inheritance of several macromutations induced by diethyl sulfate in peanuts, *Arachis hypogaea*. *Radiat. Bot.* **10**, 551-5.

Shear, G. M. and Miller, L. I. (1955). Factors affecting fruit development of the Jumbo Runner peanut. *Agron. J.* **47**, 354-7.

Shibles, R. M., Weber, C. R., and Byth, B. E. (1966). Effect of plant population and row spacing on soybean development and production. *Agron. J.* **58**, 99-102.

Silvestre, P. (1970a). IRAT's work on groundnuts. *Ford Foundation/IITA/IRAT seminar on grain legume research, Ibadan.*

—— (1970b). IRAT's work on soybean. *Ford Foundation/IITA/IRAT seminar on grain legume research, Ibadan.*

—— (1970c). IRAT's work on various food legumes. *Ford Foundation/IITA/IRAT seminar on grain legume research, Ibadan.*

Simbwa-Bunnya, M. (1972). Resistance of groundnut varieties to bacterial wilt (*P. solanacearum*) in Uganda. *E. Afr. agric. for. J.* **37**, 341-3.

Smartt, J. and Gregory, W. C. (1967). Interspecific cross compatibility between cultivated peanut *A. hypogea* L. and other members of genus *Arachis*. *Oléagineux* **22**, 455-9.

Stanton, W. R. (1966). *Grain legumes in Africa*, p. 180. FAO, Rome.

Steele, W. (1972). Cowpeas in Africa. Ph.D. thesis, University of Reading, England.

Summerfield, R. J. and Huxley, P. A. (1973). Daylength and night temperature sensitivity screening of selected cowpea and soyabean cultivars. Reading University—IITA internal communication No. 5. Reading, England.

Suryanarayana Rao, K. and Tulpule, P. G. (1967). Varietal differences of groundnut in the production of aflatoxin. *Nature, Lond.* **214**, 738-9.

Tahir, W. M. (1970). Review of the world's plant protein resources. *Ford Foundation/IITA/IRAT seminar on grain legume research, Ibadan.*

Tewari, G. P. (1965). Effect of nitrogen, phosphorus and potassium on nodulation in cowpea. *Exp. Agric.* **1**, 257-9.

Vaille, J. and Gona, H. (1972). Mise en évidence d'une déficience en bore sur arachide dans le Nord Cameroun. IRAT, Paris.

Vanneste, P. and D'Heer, A. (1972). Report of the soybean project in comprehensive nutritional education in the Luluabourg region (unpublished). ETSA and WHO, Luluabourg, Kasai Province, Zaire.

van Rheenen, H. A. (1972). Improvement of soybeans at Mokwa, Nigeria. Personal communication. IITA, Ibadan.

Ventura, M. M. and Filho, J. K. (1967). A trypsin and chymotrypsin inhibitor from Black-eyed Pea (*Vigna sinensis*). I. Purification and partial characterization. *Anals Acad. Bras. Cienc.* **38**, 553-66.

Verdcourt, B. (1969). New combinations in *Vigna*. *Kew Bull.* **23**, 464 *et seq.*

Weiss, E. A. (1967). Soyabean trials on the Uasin Gishu (Western Kenya). *E. Afr. agric. for. J.* **32**, 223-8.

WESTER, R. E. and JORGENSEN, H. (1950). Emasculation unnecessary in hybridizing lima beans. *Proc. Am. Soc. hort. Sci.* **55**, 384-90.
—— MARTH, P. C. (1949). Some effects of a growth regulator mixture in controlled pollination of lima bean. *Proc. Am. Soc. hort. Sci.* **53**, 315-18.
WIENK, J. F. (1963). Photoperiodic effects in *Vigna* *unguiculata* (L.) Walp. *Meded. Landb. Hoogesch. Wageningen*, **63**, 1-83.
WORTHINGTON, R. E. and HAMMONS, R. O. (1972). Varietal differences and seasonal effects on fatty acid composition and stability of oil from 82 peanut genotypes. *J. agric. Food Chem.* **20**, 727-30.

5. Root and tuber crops

D. G. COURSEY AND R. H. BOOTH

5.1. Introduction

Tropical root and tuber crops are in many ways the food plants *par excellence* of the lowland tropics and in general have a high degree of ecological adaptation to humid tropical conditions. Some of the major species of edible yam were first cultivated in West Africa. Today they are still food crops of major importance. Cassava, aroids, and sweet potatoes are also extensively grown.

Food crops have until recently received much less attention from research workers than tropical cash crops. And among food crops, the tropical root crops—even in comparison with the tropical cereals—have been particularly neglected.

Agricultural science has largely developed within the cultural framework of Western Europe and has thus been deeply imbued with the cultural norms of that area. Vegetatively propagated root and tuber crops were virtually unknown in the traditional food patterns of Western Europe until the rise of the potato, introduced during the sixteenth century. European agriculture developed using grain crops as the staple carbohydrate foods. Even in remote antiquity, in the Mediterranean world and the earlier civilizations of the Middle East, vegetative root crops were virtually unknown. It is significant that FAO took for its motto *Fiat panis* (let there be bread), indicating clearly the equation of bread with food, which exists at least at an unconscious level amongst most of those educated in the traditions of Western Europe.

In many parts of the forest zone of the humid lowland tropics highly organized societies have evolved which have never used grain crops to a significant degree, and whose culture, in both social and economic senses, relates to food-production systems based on vegetatively propagated root crops. The area of West Africa described by Miège (1954) as *la civilization de l'igname*, where both culture and nutrition are based upon the yam, and the yam-, taro- or sweet-potato-based cultures of Melanesia and Polynesia are examples (Coursey 1972).

The influence of Western ethnocentric ideas upon the development of scientific thought has probably limited the progress of study of these vegetatively propagated crops, and their role in tropical ecosystems and food-production techniques. Frequently, under these ecological conditions, root crops have a far higher potential for food production than have grain crops. Rice and maize have been genetically improved under the conscious direction of the plant breeder to a state where they are considered to approach their maximum attainable yield potential (de Vries, Ferweda, and Flach 1967). Root crops, however, are still, to a large degree, unimproved by selection and breeding and should therefore be expected to have a greater potential for future improvement. In terms of calorific production per hectare per year, root crops such as cassava and sweet potato already compare very favourably with most tropical grain crops, even when these grain crops are grown under most favourable conditions. In the forest zone, grain crops, with the exception of swamp rice, are not ecologically favoured, and root crops are at a substantial advantage (de Vries, Ferweda, and Flach 1967; Coursey and Haynes 1970; Idusogie 1971; Miche 1971), and taro is able to compete with swamp rice even under the conditions in which this latter crop is best adapted.

The high biological efficiency of root crops as food producers arises partly from the 'architecture' of the crops. Strength in other parts of the plant is not needed to support bulky roots or tubers. Increased size of the edible part need not therefore be associated with increased production of non-edible tissue. In grain crops, in contrast, a substantial fraction of the total biomass is required to support the edible grains above ground. The ratio of the amount of edible grain crop to the total biomass lies between $0.2:1$ and $0.6:1$, while for root or tuber crops it may be over $0.7:1$.

Not only are root crops capable of relatively high efficiencies in converting solar energy to stored energy in the form of edible carbohydrate, but their capacity for protein production is also higher than is commonly realized (Table 5.1). Cassava, however, the most important of the tropical root crops, is normally extremely low in protein, and its use as a dominant staple carbohydrate food is commonly associated with protein deficiency. It is quite incorrect to generalize, however; this deficiency does not necessarily apply to all other root crops, some of which have protein contents comparable with those of the cereal grains when calculated on a dry-weight basis.

It has been pointed out by Idusogie (1971) that crops such as yam or Irish potato can produce substantially more protein per unit area of land under West African conditions than crops such as rice or sorghum, and more utilizable protein than some of the legumes or oilseeds such as soya bean or benniseed. However, little information exists on the biological value of the protein of most of the tropical roots, and this needs further study.

Tropical root crops provide the basic food of about 400 million people within the tropics, this estimate being based on a production (as of 1967) of some 140 million t of these crops within the tropical world (excluding mainland China), which represents a total energy value of 200×10^{12} nutritional calories (Coursey and Haynes 1970). Lucas (1968) has pointed out that it is in Africa that this group of foodstuffs is most important and, within this continent, he estimates that they provide nearly a third (29·2 per cent) of the total nourishment.

TABLE 5.1

Calorie and protein productivity of various food crops in West Africa (modified from Idusogie 1971)

	Calorific production (million cal/ha)	Protein production (kg/ha)
Cassava	8·2	37
Yam	5·7	107
Sweet potato	7·4	96
Irish potato	4·7	128
Cocoyam (taro, tannia)	4·5	80
Maize	3·2	82
Rice	3·2	72
Sorghum	2·4	70
Soya beans	0·8	78

In considering what sources of information to review in this chapter, a conscious selection has been made in favour of publications which are themselves well referenced. Thus, it is hoped that any reader approaching the subject for the first time will be able, through the references quoted, to obtain access to a very large proportion of the total literature. Because this book is orientated towards West African agriculture, there has been a deliberate concentration on writings of anglophone and francophone authors concerned with that geographical area. However, there is also a substantial volume of Spanish and Portuguese literature from Latin America, for which further reference should be made to Montaldo (1967, 1972). Concise descriptions of the botany of the various crops may be found in Purseglove (1968, 1972).

5.2. Cassava (*Manihot esculenta* Crantz)

5.2.1. *Botany and classification*

Cassava, a perennial shrubby plant of the family Euphorbiaceae, is known by many different names. In Brazil it is known as 'mandioca' or 'aipim', in Spanish-speaking America as 'yuca', in the francophone world as 'manioc', in anglophone Asia as 'tapioca', and in the rest of the English-speaking world usually as cassava, or sometimes also as manioc.

Although many different clonal cultivars are now being grown, little is known about their wild ancestors, but two supposedly wild forms, *Manihot saxicola* Lanj. and *M. melanobasis* Muell.-Arg., are so similar to cassava that it is difficult to separate them. However, the roots of these species usually show only rudimentary thickening, and neither of them is as easily propagated by cuttings as is cassava. These species and others may represent the ancestral pool from which our present-day cassavas have been selected. Selection for high yield of root and for the ability to establish quickly from stem-cuttings occurred in the very early domestication of the crop long before Columbus's contact with the New World (Doku 1969).

Various classifications of cassava have been made by botanists. The two species formerly widely recognized were a poisonous, bitter species, *Manihot esculenta* Crantz (*M. utilissima* Pohl), and a sweet, non-poisonous species variously described as *M. dulcis* Baill., *Jatropha dulcis* Rottb., *M. palmata* Muell., or *M. aipi* Pohl. This classification was based on the propensity of the roots to release hydrogen cyanide. Sweet or non-poisonous forms usually contain smaller quantities of cyanogenic glucosides or of the enzyme linamarase than the bitter or poisonous ones. Taste alone is too subjective a concept to use for a taxonomic separation, and in any case no clear line can be drawn between bitter and sweet cultivars. Because of the variation in root sugar levels, some sweet types may in fact have a higher glucoside content than some bitter types. Environmental factors also have an effect on root toxicity: drought, low soil fertility, and especially, potassium deficiency have been said to increase the root cyanogenesis. Rules of taxonomic nomenclature dictate that the taxon *M. esculenta* Crantz has precedence over *M. utilissima* Pohl (Rogers and Fleming 1973), and this is now taken to embrace both sweet and bitter cassavas.

Difficulties still arise over the classification and distinction between cultivars (Rogers 1967; Hendershott, Ayres, Brannen, Dempsey, Lehman, Obioha, Rogers, Seerley, and Tan 1972). The general morphology and development of the cassava plant from both seeds and stem cuttings are reasonably well described, but little is known of the detailed anatomy of cassava.

5.2.2. *Origin and distribution*
Cassava is native to South America and perhaps also Central America. It has long been an important staple of Amerindians, and is now widely grown from Argentina to Florida. Cassava was introduced to the African continent by Portuguese traders; first into West Africa via the Gulf of Benin and the Congo river during the second half of the sixteenth century, and secondly into East Africa via the islands of Reunion, Madagascar, and Zanzibar towards the end of the eighteenth century. Following these two coastal introductions the establishment and spread inland of the two types of cassava, to meet around Lake Tanganyika, was probably slow, and cassava did not become a widely established staple in Africa until the nineteenth and twentieth centuries (Jennings 1970; Jones 1959).

The Portuguese also introduced the South American technology for de-toxifying the roots and producing a food known as 'farina'. In this technique, cassava roots are peeled and grated and the product placed in a wicker sleeve on the end of which is hung a heavy weight. The internal compression of the flexible sleeve under vertical tension brings about rapid expression of the juice containing a large proportion of the cyanogenic glucosides. However, this original method of preparing cassava was apparently not widely adopted in Africa, possibly because the local populations had no tradition of wicker weaving and could not produce the sleeves for the extraction of the flour, or because cassava was an unfamiliar food and people were not sufficiently interested to adopt it with its associated technology, which is more complicated than that needed for other crops. By comparison, the spread in Africa of sweet potato and Asiatic yam species, thought to have been introduced at about the same time, was initially much more rapid. This reflects their immediate palatability, greater ease of storage after harvest, and ease of adaptation to the traditional African farming systems (Doku 1969).

Despite a slow acceptance, the cultivation of cassava continued to expand, largely because of its ability to withstand locust attack, drought, low soil fertility, and poor husbandry, and it slowly became established as the most important famine reserve crop in many areas. The subsequent development of cassava to the status of a staple food crop in West Africa in the nineteenth and twentieth centuries was largely stimulated by the development of a local technique for processing the roots into non-poisonous food before they deteriorated. The innovation was probably made by repatriated Brazilian slaves, who modified the traditional Amerindian process by putting the grated mass of peeled roots into jute bags and squeezing out the liquor with heavy stones or logs placed on the sacks. During the process fermentation takes place, and a product of a characteristic flavour called 'gari' is produced (see p. 81). It is possible to correlate the expansion of cassava production with diffusion of the knowledge of gari-making (Beck 1971).

By the mid-twentieth century cassava had become fully integrated into tropical African agriculture. For example, Jones (1959) estimated that for a third of the population of tropical Africa cassava was a dominant staple and for 65 per cent it was an important secondary crop. Within Africa the boundaries of its cultivation correspond roughly with the 760 mm per annum isohyet† and a belt between 15° N and 15° S (Jennings 1970). Within these limits of rainfall and latitude the dominance of any particular root or tuberous food crop is more dependent upon sociological and historical than edaphic or climatic

† *Isohyet*: Line drawn on a map through areas having the same rainfall.

conditions (Agboola 1968). Although in recent years the largest recorded rate of increase in area (480 000 ha per annum) for cassava cultivation has occurred in Africa, this has been accompanied by little or no increase in the average yield per hectare and the estimated average yields in Africa (6·7 t/ha) are less than half those of South America (14 t/ha). Thus, although the area under cassava production in Africa in 1970 was estimated as 4·6 million ha, comprising about 49 per cent of the world total, it produced only about 36 per cent of the total world production, whereas about 40 per cent of the total world production of cassava comes from South America, which contains only 25 per cent of the world cassava area (Diaz 1972).

5.2.3. *Efficiency of food production from solar energy*

In terms of calories (fixed in edible carbohydrate) per hectare per unit time, cassava can out-yield all other food crops except sugar cane. Although yields of 20–40 t/ha are usually reported under field experimental conditions, de Vries *et al.* (1967) quote figures for maximum yields as 77·0 t/ha per harvest, equivalent to 71·1 t/ha per annum or 250 kcal/ha per day. This compares with 16·4 t/ha per harvest, 26·0 t/ha per annum, and 176 kcal/ha per day for rice also grown under conditions for maximum production. Several interspecific hybrids of cassava have been quoted as giving yields of between 50 t/ha and 70 t/ha per crop under field experimental conditions (Jennings 1970).

5.2.4. *Crop physiology*

Yields of cassava are very variable and are influenced by many factors which have rarely been studied under carefully controlled conditions. Final root yield of course depends on both the number and size of the swollen roots. Although Beck (1960) claimed that the swelling is initiated only during the first 6 months of growth and that subsequent yield is due only to incremental growth of the early formed storage roots, recent studies (CIAT 1972) have shown that cultivars differ markedly in characteristics determining bulking and that the timing and relative magnitude of storage root growth components (number and size of roots) are markedly influenced by planting densities. The optimum planting density varies between cultivars, although a useful median figure for maximizing yield is about 1 × 1 m. The plant density and arrangement traditionally used in any given area has usually been arrived at empirically as a result of local experience of a number of inter-

acting factors such as cultivars, environmental conditions, and suitability for the methods of husbandry historically adopted (including intercropping systems) in that region or by a particular ethnic group.

There is a close relationship between leaf area and yield. Williams and Ghazali (1969) postulate from analogy with cereals that plants having leaves with attenuated lobes and with more vertical mid-day orientation may be associated with higher yields by allowing more light to penetrate to lower leaves. These factors, particularly the relation of leaf retention and leaf angle and movement, and their effect upon yield, are the subject of research programmes at the Centro Internacional de Agricultura Tropical at Cali, Colombia (CIAT 1972).

The ratio of the weight of the swollen roots to the total weight of roots and tops was called the coefficient of utilization by Cours (1951).

5.2.5. *Propagation of cassava*

The use of 30–45 cm cuttings as planting material has given higher yields than those obtained when shorter ones are used. Moderately thick cuttings taken from the basal parts of stems are more satisfactory than from upper stems (Jones 1959). Cuttings may be planted vertically or at an angle, buried completely or to only half of their length, or they can be buried horizontally about 10 cm deep. They can be planted on the flat or in mounds or ridges, singly or in pairs. Different methods are adopted either as a local fashion or as found best suited to local conditions, such as soil depth and drainage. Generally, planting in a mound or on a ridge is thought to be advantageous. The position of the cutting in the soil is considered to influence root distribution more than final yield; the roots tend to penetrate deeper when cuttings are planted vertically and to be distributed more widely when they are planted horizontally.

Planting is normally carried out at the start of the rains and is usually done by hand, but may also be accomplished mechanically. In some areas short-duration cultivars are planted at the beginning of the rains and long-duration ones later, but cuttings are often planted at any time when the parent plants are harvested. Cuttings are sometimes stored for a short period and storage for 10–20 days is claimed to be advantageous to strike and vigour. Longer periods of storage usually results in severe loss of viability (Wholey, personal communication).

The feeder (non-storage) roots of cassava plants

penetrate the soil to depths of 40–80 cm (Cours 1951), and deep cultivation to facilitate this may be advantageous. This deep root penetration is probably a partial explanation of cassava's tolerance of drought and low soil fertility.

As a subsistence crop cassava is frequently intercropped with such crops as maize, legumes, bananas, and vegetables. Where it is grown as a commerical crop it is more usually grown in monoculture, either grown continuously or rotated with other crops. In traditional shifting systems cassava is usually the last crop taken as fertility declines, before land is left to bush-regeneration.

5.2.6. *Responses to fertilizers*
Reported responses of cassava to fertilizer applications are very variable as would be expected when soil nutrient status and climate differs between trials. These are frequently inadequately reported. There may be also variation in the inherent nutrient responsiveness of different cultivars. Studies need to be undertaken using specified cultivars and on a range of soil types of known nutritional status under a range of recorded environmental conditions so that a consistent body of useful information, as opposed to a collection of unrelated data, can be built up.

In nutrient culture experiments, phosphorus levels over the normal range found in soils greatly influence yield, much more so than potassium. In one investigation employing both nutrient culture and field conditions nitrogen tended to increase stem growth at the expense of root yield, whereas increasing potassium tended to restore a more favourable crop index. In another experiment nitrogen, while increasing root yield, did not correspondingly increase starch yield but resulted in a higher root nitrogen content, interpreted as an increase in protein. The whole subject obviously needs further attention (Doku 1969; Jennings 1970; Silvestre and Delcasso 1971).

5.2.7. *Weed control*
It is often considered that weed control in the cassava crop is necessary only in the early stages of crop growth, until about 3–4 months after planting, when full ground cover is usually obtained. It has recently been shown in Colombia that adequate weed control is the most expensive operation in cassava production aimed at maximizing yield. Weeding is usually done manually but sometimes by ox- or tractor-drawn hoes. Recently several pre- and post-emergence chemical weedicides have been investigated (CIAT 1972). Usually at least two hand-weedings are needed to obtain sufficient freedom from competition for maximum growth of the crop.

5.2.8. *Harvesting*
Another expensive operation is harvesting. This is usually carried out by hand, often with a digging stick or other implement. Root size and distribution, as already explained, are influenced by cultivar and by cultural and environmental factors. The disposition of the roots, as well as soil compaction and moisture, affect the ease of harvesting and amount of damage caused to the roots. The use of various kinds of mechanical lifters has so far had very limited success. New designs seem to be needed if mechanical harvesting is to become more common.

5.2.9. *Pests and diseases*
Although freedom from pests and diseases and particularly its ability to withstand locust attack, was historically a factor leading to the spread of cassava in Africa, it is now considered that these negative factors, particularly diseases, are amongst the major factors limiting yields and further development of this crop.

African mosaic disease of cassava. The most important disease limiting both yield and cultivation in Africa is cassava or African mosaic. This disease, which became important in West Africa in the early 1920s, is not certainly known to exist outside Africa. Since cassava is not indigenous to Africa it must be assumed that this is a disease new to cassava, introduced to it from some other crop or wild host. The disease is still of uncertain etiology, though previously it was presumed to be of viral cause. The pathogen is transmitted by white flies (*Bemisia* spp.) and through the use of infected cuttings. Mechanical transmission has not been achieved. Acquisition feeding time for the white flies is 4–6 hours on diseased leaves, and this is followed by a latent period of 4 hours before transmission to healthy plants is possible. Effective inoculation requires at least a 15-minute feed and the white flies remain infective for more than 48 hours after becoming vectors. Susceptible cultivars develop symptoms 12–20 days after inoculation (Chant 1958).

Variations in symptom severity are reported to be caused by variation in the strain of the pathogen, growing conditions, and growth of the host. Yield loss due to African mosaic varies greatly, depending on the degree of infection and the resistance of the cultivar. Estimates of loss range from 5 per cent to 95 per cent. Losses of about 80 per cent are not uncommon, and losses of about 20 per cent attributable to the disease may be regarded as quite normal in many parts of West Africa. There have from time to time been reports that obviously infected plants may yield better than symptom-free ones, but no such reports based on reliable data are known. Symptom-free (and supposedly pathogen-free) cuttings of susceptible cultivars of *M. esculenta* can be obtained from distal parts of diseased plants if they are grown at high temperatures (35–37 °C) for 28–42 days. Although some knowledge thus exists on a possible means of control, research needs to be intensified as control of this disease is of paramount importance to the future of this crop in Africa.

Cassava bacterial blight. Also of increasing importance is cassava bacterial blight (caused by *Xanthomonas manihotis*). This disease (Lozano and Sequeira 1974), first reported and studied in Central and South America, is known to have become established in many parts of Africa, though when this occurred is most uncertain and somewhat contentious. It is one of the most devastating plant bacterial diseases known, causing both leaf-spotting and complete wilt and die-back of infected plants. Severe infection with bacterial blight results in complete loss of yield. The disease is easily carried accidentally from one area to another through lack of, or in spite of, phytosanitary control, and also from one season to another by the use of infected planting material. Locally the disease is disseminated by rain splashing, insects, and on infected clothing and implements, and can thus spread very rapidly during wet weather. Once the disease has been observed in a crop little can be done to control it while saving the crop, although the rate of spread may be reduced by severely pruning infected plants. Differences in resistance between cultivars are known. It is possible that the disease could be controlled by eradication of cassava from a given area and, after a sufficient time, using bacteria-free planting material for replanting (CIAT 1972). Crop rotation must be practised to eliminate bacteria from the soil before planting clean material. The use of tip cuttings from infected stocks, and propagation of resistant cultivars, appear to offer possible means of controlling this disease, but execution on an agricultural scale of such a programme would be a major undertaking.

Other diseases. Other cassava diseases include several *Cercospora* leaf spots, all of which, under particular environmental conditions, cause some defoliation. *Phytophthora* root rot can cause over 80 per cent loss of yield in wet, poorly drained soils. There are numerous other pathogens, which, although generally of minor importance, may, under certain circumstances, cause considerable loss. Much of the available information concerning these has recently been reviewed (Lozano and Booth 1974).

Cassava, as stated earlier, suffers less than any other crop from the ravaging invasions of the migratory African locust (*Locusta migratoria migratorioides*). However, the crop may suffer considerably from attack by other insects, in particular horn worms, though these do not occur on cassava in Africa, thrips, shoot-flies, and scale insects and also from attack by mites. Insect pests have been little studied, and it is generally considered that, as the roots remain undamaged and new shoots are produced after an attack, they are of minor importance, although in certain cases severe loss of yield occurs. The effect of such insect attacks on crop yield needs to be further investigated and suitable control measures sought.

In recent years mites appear to have become increasingly important pests on cassava in South America and East Africa. *Mononychellus tanajoa* Bondar is the pest concerned in both Brazil and in Uganda (Fennah, Director, Commonwealth Institute of Entomology, personal communication, 1974). This mite causes extensive yellowing and shrivelling of the leaves and sometimes complete defoliation.

In addition to diseases, insects, and mites, mention should be made of other predators such as rodents, pigs, monkeys, baboons, and other animals. The control of such pests is not always easy, but they are frequently combated and discouraged by surrounding plots with a perimeter row of the most bitter cultivars. It is partly because wild pigs leave bitter cassavas alone that such types continue to be preferred to 'sweet' cassavas in many areas of Africa.

5.2.10. Storage

Cassava roots cannot be kept in the fresh state for more than a few days after harvesting without serious deterioration in quality. This has imposed severe limitations on the home use, marketing, and industrialization of the crop. It is not completely understood why cassava roots decay so rapidly, but much of the relevant literature has recently been reviewed (Ingram and Humphries 1972). The normal practice for 'storing' fresh cassava is not to harvest the plants until required, but this results in the occupation at any time of $\frac{3}{4}$ million ha of land planted with already mature cassava. Post-harvest deterioration appears to be a complex process involving both pathological and endogenous enzymatic processes. The onset of deterioration, which is shown by internal streaking and discoloration, usually commences at the sites of mechanical damage. Deterioration can be delayed or even prevented if, after harvest, the roots are kept for 10 days at relatively high temperature and humidity. During this period the 'skin' of the roots becomes thickened and hardened and wounded tissue becomes suberized and then sealed off by a wound periderm. These changes retard moisture loss and provide a physical barrier against invasion by wound pathogens (Booth 1973). This curing process and the subsequent storage of fresh roots for up to 2 months can be undertaken in simple field structures similar to the clamps used for storing Irish potatoes in temperate climates.

5.2.11. *Food use of cassava*

Cassava, when used as a human staple food or as animal feed, needs to be supplemented with protein-rich foods or feed. The chemical composition of cassava roots varies with maturity, cultivar, and growing conditions. Typical values suggested for the edible portions (about 80 per cent of the total) are 62 per cent water, 35 per cent carbohydrates, 1 per cent protein, 0·3 per cent fat, and 1 per cent mineral matter, which is rich in calcium. The roots are also rich in ascorbic acid (30–43 mg per 100 g) and contain nutritionally significant amounts of thiamine, riboflavin, and niacin. However, the content of protein and of vitamins tends to be reduced by the processing required to remove cyanide. The root proteins are relatively deficient in the sulphur-containing amino acids (Oke 1968). Thus cassava must be regarded essentially as an energy source and, whenever possible, be supplemented with protein-rich foods. However, as de Vries and his co-workers pointed out, the protein content of cassava roots, when compared with other food crops such as rice on the basis of weight of protein per 100 calories is not so unfavourable as is generally believed (de Vries *et al.* 1967).

'Gari' is a popular cassava food in West Africa. It is produced by fermenting peeled and grated roots

in jute bags on top of which heavy weights are placed to squeeze out the liquor. Gari meal is then sieved, spread out to dry, and roasted on flat earthenware or iron pans with continual stirring until the meal becomes light and crisp. Gari may be stored for several weeks or even months and prepared and eaten in various ways. By the addition of small quantities of cold water it can be softened and eaten with soups or stews or, by soaking in boiling water, it can be made into a dough which can be rolled into balls and eaten alone or with various additives. Attempts have recently been made to mechanize the production of gari (Akinrele, Cook, and Holgate 1962).

The term 'fufu' is often used loosely to describe any starchy food pounded into a sticky dough. In making fufu from cassava fresh roots are 'fermented' by soaking them in running water or in water-filled vessels for 3–4 days, after which they are peeled, squeezed to express the juice, and pounded.

'Kokonte' flour is prepared by sun-drying pieces of roots which are then milled into flour. Fufu can also be prepared from kokonte instead of from fresh cassava. Several types of cake can be prepared from cassava. This usually involves peeling, washing, and grating the roots into a meal which is then dried. This meal may then either be steam-baked or fried in oil.

In some parts of Africa, cassava leaves are consumed in considerable quantities; these are richer than the roots in both quantity and quality of protein. Cassava leaves can contain between 17·8 per cent and 34·8 per cent protein on a dry-weight basis, depending on the cultivar and the age of the leaf; values of 5–7 per cent on a fresh-weight basis are usual. The amino-acid spectrum in leaf protein is good, only methionine being low, whilst lysine is particularly high. Young cassava leaves may therefore, after boiling for about 5 minutes to remove hydrogen cyanide, form a useful supplement to a diet of cassava roots (Terra 1964).

The water which is pressed out of the grated meal in the preparation of gari and similar products contains a high proportion of starch which settles out on standing for some time. This can be utilized by simply decanting the supernatant liquid, washing the starch, and drying it in the sun or in an oven. Alternatively it can be made into tapioca by roasting the washed starch, which causes the individual starch grains to burst open, and adhere together to form large granules (Grace 1971; Oke 1968).

5.2.12. Feed use of cassava

In addition to its major use as human food, cassava is used as feed for fattening pigs and feeding cattle and poultry. Cassava can be fed to animals either raw, boiled, or after drying. On account of its low protein content, it needs to be blended with maize or mixed with protein concentrates such as oilseed-cake and mineral salts. The high nitrogen content of cassava leaves† has attracted attention for animal feeding and, in some countries, the making of a cassava silage from whole plants is practised. In the last decade a growing international trade has developed in dried cassava chips or pellets for use in animal feeds. This trade has been principally between Indonesia and Thailand on the one hand and Belgium, Holland, and West Germany on the other, but a considerable expansion of this type of trade is envisaged; it has been projected that the expanded EEC alone will import 3 times its current level of 1·5 million t per annum of dried cassava by 1980, providing tariff charges and production costs do not change significantly (Nestel 1973; Phillips 1974).

Chronic cyanide toxicity in animals fed high cassava diets, as well as in humans eating a cassava staple food, is a well-recognized problem. It appears that unless effective steps can be taken to reduce toxicity, it could become an increasing problem both from the nutritional standpoint and from the retardation of domestic and possible export earnings of producing countries (Nestel 1973). The toxicity of cassava is caused by the presence of the cyanogenic glucoside linamarin, together with much smaller amounts of the closely related lotaustralin (Bolhuis 1954; Coursey 1973a). These substances are hydrolysed under the influence of the endogenous enzyme linamarase to liberate hydrogen cyanide (HCN). Little reliable information is available as to whether the glucosides themselves are toxic or whether toxicity arises only from hydrolysis of the glucosides. Similarly, little reliable information is available on the efficiency of traditional detoxication methods or on alternative improved methods. These and many other aspects of cassava toxicity were recently discussed, particularly in relationship to the development of neuropathic conditions (Nestel and Mac-Intyre 1973).

In addition to its use as human or animal food,

† Sometimes, though without adequate confirmation, assumed to be due to high protein content and expressed as 'crude' protein. Ed.

cassava finds considerable use as a source of industrial starch, which is used mainly in the textile and adhesive industries.

5.2.13. Genetic selection during the history of cultivation of cassava

Throughout the history of cassava cultivation the selection of 'suitable' cultivars has been taking place. As suggested earlier the initial selection was probably for high-yielding and easily established types. Since that time selection aims have become more specialized. In West Africa priority has been given to mosaic resistance; in Madagascar, Indonesia, and Thailand to the selection of forms with the high starch content and large starch grains particularly suitable for industrial starch production; and in other regions to cultivars adapted to such factors as particular soil types, low rainfall, and good nutritive value and freedom from toxic glucosides.

In general little is known of cassava genetics. Many of the characters which have been studied for breeding purposes, such as mosaic resistance, have shown continuous variation. Controlled breeding and screening of cultivar collections has been attempted at several centres (Magoon 1970) and, although largely cross-pollinated, cassava, like several other flowering crop plants which are vegetatively propagated, offers excellent scope to plant breeders to develop improved clones. Cassava has a somatic chromosome number $(2n = 36)$ and is probably allopolyploid, possibly a segmental allopolyploid with six duplicated chromosomes.

The aim of the plant breeder must be to attempt to incorporate into one or more clonal cultivars most, if not all, of the desirable characters required by the particular growers and users of the crop. The first necessity is to define (in collaboration with agronomists, crop protection specialists, and food and industrial technologists) more closely the desirable characteristics, and then see to what extent these can be combined. For all food crops, high yielding ability and good food qualities rank foremost, and for cassava another important character is high starch yield. Any other factors directly or indirectly affecting these characters, such as resistance to pests and diseases, ease of harvesting, storage ability, low cyanogenic glucoside, and high protein content, should also engage the attention of the plant breeder.

A programme of breeding cassava for resistance to mosaic disease was carried out under EAAFRO (The East African Agriculture and Forestry Research

Organization) in collaboration with national programmes in the East African region (Nichols 1947;
Jennings 1957; Doughty 1958; Jennings 1970).†
Although after screening a large collection of clones
(including Kru from Ghana, Binti Misi from Uganda,
and Aipin Valença from Brazil) a few were found
which possessed adequate resistance, it was possible
to obtain improved resistance from intraspecific
hybrids between clones. These intraspecific hybrids
were so successful that some of them (notably 3724-4,
4023-4 (Kru×Binti Misi) and 4440-6) became
standard clones in Uganda (Jameson 1964) and were
never succeeded by the interspecific hybrids, which
are generally better known. The interspecific hybrids
were bred between cassava and two other species,
Manihot glaziovii Muell-Arg. (ceara rubber) and
Manihot dichotoma Ule (jaquie manicoba rubber).

A similar back-cross programme was conducted
with *Manihot dichotoma* but no clones as good as
those from the cassava × ceara cross were obtained.

5.3. Yams (*Dioscorea* spp.)

5.3.1. Agro-ethnology

Of all the tropical food crops, few are as closely
associated with a particular cultural area as are
certain species of yam with West Africa. Yams are
extensively grown and used in other parts of the
world, but nowhere else, except perhaps in Melanesia, are they of such central importance. In much
of West Africa yams are still today the preferred
staple food among many of those inhabiting the
forest and the wetter parts of the Guinea savanna
zones, although they were relatively more important
before the introduction of American food crops such
as maize, cassava, and tannia (*Xanthosoma*). It is
perhaps no coincidence that the great states of the
second millennium A.D. in the forest zone of West
Africa–Ashanti, Ife, and Benin arose in areas where
yam was the staple food (Coursey 1967; Ayensu and
Coursey 1972). Yams are still a crop of major
importance, although the paramount position of
the yams has been affected to some extent by the
introduction and spread first of the Asiatic taro
(*Colocasia*) probably about 1000 years ago, then of
cassava and, most recently, within living memory,
of Asiatic rice. African cultural patterns tend to be
strongly oriented towards the staple carbohydrate
food items in the food–agricultural complex (Uchendu

1970) and, in much of West Africa, yams are still
prominent in cultural life as well as in the material
subsistence of the population. Comparable situations
exist also in Melanesia (Barrau 1956, 1970; Coursey
1972).

The economically important yams belong to the
genus *Dioscorea*, which is the largest genus of the
family Dioscoreaceae. It is of pantropical distribution
and contains some 600 species. The Dioscoreaceae
are monocotyledons, formerly classed with the
Liliales (Burkill 1960; Coursey 1967), although
studies on certain African species have demonstrated
the existence of a non-emergent second cotyledon.
The most recent taxonomic revision (Ayensu 1972)
establishes the order Dioscoreales and slightly revises
the lower orders of classification.

It is not necessary here to discuss in great detail
all the past work on the edible yams, as adequate
recent monographic treatments covering most aspects
exist (Waitt 1963; Coursey 1967; Coursey and
Martin 1970). However, in summary, recent research
suggests that the yams have been selected as crop
plants in several tropical areas where there were
sharply demarcated wet and dry seasons and that
it is mainly in such regions that they are still plants
of economic importance today.

5.3.2. Botany of yams

During the wet season the plant in the wild or primitive agriculture grows as a 'vine', which twines
through trees or undergrowth, but, in cultivation,
it is usually trained on stakes, strings, or wires. The
vines bear racemes of inconspicuous unisexual, male,
or female flowers which usually occur on separate
plants (i.e. most species are dioecious); the female
flowers are succeeded by dehiscent trilocular capsular
fruits which are winged and wind-dispersed. At the
end of the wet season, the vine dies completely and
the plant survives the dry season as dormant tubers.
It is these tubers which are economically useful as
human food.

The tuber may be an annually renewed organ,
which is the case in the edible species, or may be
perennial, becoming larger and progressively more
lignified from year to year, as occurs in many other
species. A few of the more primitive members of the
genus are rhizomatous rather than tuberous. The
tuber has generally been regarded as being morpho-
logically a stem tuber. Recent studies (Lawton and
Lawton 1969) show that in certain West African
species the tuber arises from the hypocotyl, while in

---

† The paragraphs on breeding cassava for mosaic resistance in
East Africa were added by the editors.

E

other species (Archibald 1967) this is not so: tuber genesis in yams awaits further research.

Yams are propagated vegetatively by small tubers or pieces of tuber known as 'setts'. Vegetative propagation for a prolonged period has reduced the necessity of sexual fertility, and many cultivars evolved and selected under this system flower comparatively seldom and even less frequently set fertile seed (Doku 1966). Most wild species are fertile.

5.3.3. *Evolution and domestication*

Assuming the evolution of man in tropical Africa, the yams are probably amongst the oldest of man's foods. Wild yams are extensively used at the present time by 'pre-agricultural' peoples in various parts of the tropical world, but such primitive people do have some concept of protecting the wild plants, which represents a first step towards crop agriculture. We may suppose that a similar use of yams as food could go back to the earliest days of man's pre-history (Barrau 1956; Coursey 1972).

True yam-based agriculture may have developed independently, though almost simultaneously, in West Africa and in South-East Asia around 5000 years ago. Different, although similar, species of yam were brought into cultivation in these two areas, while yet other different species were brought into cultivation in tropical America.

The Asiatic yams, *D. alata* and *D. esculenta*, became disseminated throughout South-East Asia, the Indian sub-continent, the Indonesian archipelago, and the Pacific Ocean in early times, probably more than 2000 years ago—long before European contact with the areas. At least the former reached Madagascar, and probably the East African littoral, about 1500 years ago or earlier, as a result of Malaysian contacts. Nevertheless, it is unreasonable to suppose that this had any influence on the development of West African cultivation. It appears that *D. alata* was probably introduced to West Africa only in the sixteenth century by Portuguese traders returning from the East. *D. esculenta* has been grown in West Africa only very recently, as a deliberate introduction in the Colonial era.

In West Africa, yam cultivation was an indigenous development, and the main species grown, the typical West African white yam, *D. rotundata*, is unknown elsewhere, except in the Caribbean, where it was doubtless taken on the slave ships as provisions. The cultivation of yams not only antedates European contact, but also the beginning of the use of iron tools in Africa, around 2000 years ago. The amount of ritual, ceremony, and superstition which surrounds the cultivation and utilization of yams also indicates the great antiquity of their use in West Africa (Coursey 1967; Ayensu and Coursey 1972). The yellow yam, *C. cayenensis*, is also of some economic importance as a crop plant throughout the West African forest zone.

5.3.4. *Taxonomy of West African yams*

The taxonomy of these West African yams is somewhat complex. It is now considered that *D. cayenensis* is a wild species, which has been substantially selected in cultivation, whilst *D. rotundata* is of hybrid origin, derived from it and a savanna species. The putative relationships of the various forms are discussed in detail elsewhere (Ayensu 1970; Coursey 1971; Ayensu and Coursey 1972).

In the New World, indigenous yams, notably *D. trifida*, were domesticated by Amerindians before Spanish contact, but never became major crop plants, perhaps because of the availability of the more easily cultivated and less seasonal cassava. The more important Asian and African yams were introduced by Europeans and today are far more extensively grown than are any of the indigenous species. Forms of *D. alata* are the most popular in the eastern Caribbean, but in Jamaica and Puerto Rico, and much of the mainland, the African species are preferred, especially *D. cayenensis*, which has undergone a substantial degree of improvement through selection (see Table 5.2).

5.3.5. *Production areas*

The main areas of yam production in the world are West Africa, the Caribbean, parts of South-East Asia, New Guinea, and the islands of the Pacific. Production of yams in West Africa is largely confined to the West African 'yam zone' (Coursey 1967; Miège 1954), which extends from the central Ivory Coast to Cameroun, spanning both the forest and the more humid parts of the Guinea savanna. Within this zone about two-thirds of the yams in the world are grown: indeed, Nigeria alone produces almost half the world's recorded yam crop. Within this area, yams were formerly the main staple and, although during recent decades they have been displaced to a substantial degree by other crops, even today the combination of production and demographic statistics indicates that the average inhabitant of the region consumes between $\frac{1}{2}$ kg and 1 kg of yams daily.

TABLE 5.2

Cultivated yam species (after Coursey 1971)

	Africa	Asia	America
Major spp.	*D. rotundata* Poir.	*D. alata* L.	*D. trifida* L.
	D. cayenensis Lamk.	*D. esculenta* (Lour.) Burk.	
Minor spp.	*D. bulbifera* L.	*D. bulbifera* L.	*D. convolvulacea* Cham. & Schlecht.
	D. preussii Pax.	*D. hispida* Dennst.	
	D. praehensilis Benth.	*D. pentaphylla* L.	*Rajania cordata* L.
	D. sansibarensis Pax.	*D. nummularia* Lamk.	
	D. dumetorum (Knuth) Pax.	*D. opposita* Thunb.	
		D. japonica Thunb.	

5.3.6. *Yields of yams*

Both in West Africa and in other yam-growing parts of the world, most yams are produced by subsistence farmers. The particular cultivars grown are planted for traditional reasons; cultural practices are simple and chemical fertilizers are rarely used. Under these conditions, yields, when recorded, are of the order of 5–10 t/ha per crop. Under better conditions, such as prevail at experimental stations, far higher yields have been obtained with similar planting material (see Table 5.3). If the approach of de Vries *et al.* (1967) is applied to the yield figure of 70 t/ha for West Irian, and a growing period of 9 months is assumed, this is equivalent to calorific production of 266 kcal/ha per day, compared to a maximum food calorie production of grains of the order of 100–200 kcal/ha per day.

This high potential capacity for food calorie production is present in species which have been subjected to little or no selection on a scientific basis. With even a limited amount of effort on genetic improvement, substantial advances might be expected.

5.3.7. *Prospects for the genetic improvement of yams*

Among the first problems to be overcome in attempts to improve yams is the lack of satisfactory records and of large collections of species and cultivars. Most clones are of localized occurrence and are known only by vernacular names. Clones vary greatly in nutri-

tional quality, yield potential, adaptation to different climatic regions, and many other factors. Although several attempts have been made in various countries to assemble and classify living collections of yams, only limited success has been achieved, and considerable erosion of the genetic resources may be occurring as yams are displaced by alien crops (Coursey 1971, 1973*b*). The collection and evaluation of existing cultivars is an essential preliminary to effective work on the improvement of yams as crop plants. Such a collection should be made on a worldwide basis. The studies on *D. alata* recently made in Puerto Rico (Martin and Rhodes 1973) are a valuable beginning but need to be extended to other species and locations.

TABLE 5.3

Selected high yield reports in yams (after Coursey and Martin 1970)

Species	Country	Yield (t/ha)†
D. alata	Malaysia	42·5
D. alata	Trinidad	46·8
D. alata	St. Vincent	58·2
D. alata	Fiji	25·4
D. esculenta	Malaysia	24·6
D. esculenta	Trinidad	32·2
D. esculenta	West Irian	70·0
D. rotundata	Ghana	20·8
D. rotundata	Nigeria	16·2
D. cayenensis	Trinidad	31·5

† The authors provide in the text information that the average growth period is 9 months.

Most cultivated forms are high polyploids and attempts to improve yams by breeding have been hampered by the infrequent flowering and poor seed-setting associated with this, and by dioecy. Many of the better cultivars are male. Male and female plants tend to flower at different times and pollen production in male flowers is often poor. However, seeds from good cultivars of *D. rotundata* have been successfully germinated (Doku 1966), and the seedlings have, themselves, flowered. Wild yams have retained their sexual fertility and are possible seed parents that might be combined with cultivated forms as pollen parents. Although the greatest natural variation of the genus *Dioscorea* occurs in the New World, only a single American species (*D. trifida*) has been brought into cultivation to any significant extent. This species sets seed freely and could be a promising starting material, as could wild American species.

The aims of any programme of selection and breeding must be clearly defined. Improved yield constitutes an obvious aim, but some known yam cultivars are already so high-yielding that secondary aims might be of more immediate practical significance. Particular attention needs to be paid to the following aspects:

(1) tuber size and form, to allow the application of mechanical harvesting techniques;

(2) industrial processing characteristics;

(3) response to fertilizer treatment—many existing forms show little positive response;

(4) enhanced nutritional value—this should concern not only protein content, but also the amino-acid spectrum of the protein;

(5) good storage life in the fresh state, which is associated with tuber dormancy;

(6) adaptation to various ecological conditions.

Conventional propagation by tuber cuttings limits the multiplication rate to only two-, three-, or at best five- or six-fold per year. Vine cuttings provide a means whereby stock can be multiplied 20- or 30-fold, or even more per year (Ferguson 1972). Meristem culture techniques may prove much more efficient and encouraging preliminary experiments have already been made with *D. bulbifera* and *D. sansibarensis*. Further development work is needed.

The yams which have been most highly selected, such as *D. alata* and *D. rotundata*, produce single or a few large tubers. Such types have clearly been favoured by cultivators using manual harvesting techniques. When grown under favourable conditions, these species respond by producing even larger tubers, which become progressively more difficult to harvest. Plants producing larger numbers of individually smaller tubers would be preferable if mechanization is envisaged (Haynes 1967). *D. esculenta* and *D. trifida* should be promising species to develop for this purpose. Since it is probable that particular cultivated species in use today are simply those that chanced to be selected by primitive farmers, more of the 600 species known could well be regarded as having food-crop potential.

5.3.8. *Applied crop physiology*

Only recently has information become available on the growth physiology of the yam plant (Gooding and Hoad 1967; Sobulo 1972; Enyi 1972). There are still broad areas of ignorance. In particular, little information is yet available on responses to fertilizer,

though such information as there is suggests little positive response in tuber yield. Most of the available data have been reviewed by Ferguson and Haynes (1970).

5.3.9. *Practical aspects of yam cultivation*

The high labour requirement for yam cultivation compared with other manually cultivated root crops is well known and accounts for the interest in mechanized harvesting of the crop. Yams will probably long continue to be grown mainly as a small-scale horticultural crop and improvements directed towards the needs of horticulturists, as well as potential large-scale producers, are needed. Staking has been identified (Haynes 1967; Wholey and Haynes 1971) as requiring a major labour input. Systems of cultivation that reduced or obviated the need for staking could effect considerable economies. Yams generally have better storage performance than most other root crops because of natural tuber dormancy. However, both physiological and pathological factors can contribute to storage deterioration. This has been reviewed by Coursey (1967) and by Thompson, Been, and Perkins (1975).

It has been suggested that yam production will decline in the next few decades, mainly owing to the high labour input required in comparison to other root crops (Anon. 1971). They are a strongly preferred food among many peoples, however, and the authors of this chapter consider that any such relative decline will be very slow, even in the face of substantial increases in price.

5.4. Cocoyams: Taro (*Colocasia* spp.), Tannia (*Xanthosoma* spp.), etc.

5.4.1. *Introduction*

These aroid root crops include representatives of two very closely similar genera of the Araceae, *Colocasia* and *Xanthosoma*, which are widespread in cultivation in the tropical world and of considerable importance as food plants, together with the localized and much less important *Alocasia*, *Cyrtosperma*, *Amorphophallus*, and *Anchomanes*. They have received even less attention from research workers, particularly in Africa, than other tropical root crops, in spite of their considerable importance as food plants: it is perhaps symptomatic that, at the 1971 Ford Foundation/IITA/IRAT seminar on root and tuber crops, no mention was made of them (Anon. 1971), although they were discussed briefly in the seminar on vegetable crops. Even the basic taxonomy of the group has been the cause of much confusion, at least part of which arises from the fact that the major species exist in a profusion of clonal cultivars, and that poplyploidy, including aneuploidy, is present. Many clones have been vegetatively propagated for many thousands of years—a situation similar to that which exists with the edible *Dioscorea*. Variation in domesticated clones will have arisen by the selection of mutants.

5.4.2. Colocasia *spp*.

The principal edible cultivars are now considered all to belong to a single species *C. esculenta* (L.) Schott. This includes the cultivated forms commonly known as taro, curcas, or dasheen, as well as the eddoe type, formerly often regarded as a separate species (*C. antiquorum* Schott) but now relegated to subspecific status as *C. esculenta* subsp. *antiquorum*.

5.4.3. Xanthosoma *spp*.

Similarly, most of the economically important edible *Xanthosoma*, commonly referred to as tannia, yautia, ocumo, or malanga, are now considered to be members of the species *X. sagittifolium* (L.) Schott, although the rarely grown *X. violaceum* Schott and the salad vegetable *X. brasiliense* Engl. can perhaps be regarded as separate species (Barrau 1957; Coursey 1968). All the aroid root crops are known in most of anglophone Africa as cocoyams, but in East Africa—confusingly—also as arrowroot: *Colocasia* is sometimes qualified as 'old' cocoyam, and *Xanthosoma* as 'new' cocoyam, referring to the relative dates of their introduction to the African continent.

Neither taro nor tannia is indigenous to Africa, taro having been introduced with the banana, probably 1000 or more years ago, through the well-known southern Arabian route. It was originally domesticated in South-East Asia, where it may be one of man's oldest cultivated plants (Barrau 1970). Tannia, on the other hand, is an introduction of the mid-nineteenth century from tropical America, where it was cultivated and originally domesticated by the Amerindians (Doku 1966; Coursey 1968). Although a more recent introduction, tannia is now almost as widespread in West Africa as taro and in many areas is preferred, particularly as it can be prepared for food in ways similar to those used for the *Dioscorea* yams. Taros and tannias now play a major part in the nutrition of the forest zone of West Africa; in Ghana, in particular (Doku 1966), production was second only to that of plantain (*Musa* sp.) amongst the staple crops of the country. Taros and tannias are also important in parts of the Ivory Coast and the eastern parts of Nigeria (Annegers 1973).

There is much variation in preference amongst different ethnic groups for one genus or the other, or even for particular cultivars (Catharinet 1965). The spread of tannia in Africa has also been assisted by its widespread recommendation for, and use as, a shade plant for young cocoa trees.

5.4.4. Taro husbandry

Taro, in common with other aroid root crops, is of much greater and more widespread importance in the Pacific Ocean area than it is in Africa. Research on its cultivation and utilization is more advanced there. Much of this work, which has recently been reviewed (Plucknett 1970; Plucknett, de la Peña, and Obrero 1970; de la Peña 1970; Sivan 1970; Plucknett and de la Peña 1971), could well be of useful application in West Africa. In particular, whereas yields reported in Africa and indeed in many other parts of the tropics are usually only of the order of 4–10 t/ha, in Hawaii they are up to 25 t/ha under dry-land conditions and over 70 t/ha under lowland irrigated and intensively farmed conditions. Hawaii is probably the one country where taro cultivation has been placed on a fully organized modern footing (Plucknett and de la Peña 1971), with a substantial degree of mechanization. The system is that selected cultivars of taro are grown on flooded paddy fields in a manner closely analogous to the techniques of swamp-rice cultivation. The fields are prepared before flooding by ploughing and discing behind suitable tractors, followed by puddling of the soil using a disc or spiked-toothed harrow. Rotary tillers are occasionally used, with two-wheeled tractors. The planting setts, which consist of cuttings from the upper tip of the corm or cormel together with the bottom 15–25 cm of the petiole, are then inserted by hand into the soft, puddled soil, the setts being spaced regularly with the aid of strings or spacers. This phase of the operation could probably be further mechanized by the use of modified cane-planting or tuber-planting machinery. A substantial degree of weed control is effected automatically through the flooding of the fields, which is done as soon as the taro is well established, but experiments in Hawaii have shown that, where necessary, weedicides may be added to the irrigation water with a high degree of success and little damage to the crop. Although, even in Hawaii, harvesting is still often done by hand, using a crowbar as a digging stick, experiments there have shown that this operation can very simply be mechanized by the use of a light two-wheeled tractor with a special horizontal auger which 'winds' the tubers out of the soil and leaves them ready to be collected in trucks. It is under cultivation conditions such as these that the highest yields have been recorded. Under dry-land conditions, but using overhead sprinkler irrigation, 50 t/ha have been achieved with other taro cultivars in Cyprus, indicating that the plant has yield potential far beyond normal yield achieved under West African conditions.

Despite little breeding work (Plucknett *et al.* 1970), it seems that there is already such a wide variation within the cultivated material available in the world, and that a rationally oriented programme of introductions and selections might provide greater returns, at least in the near future, than any large-scale breeding programme. Proper attention should also be paid to agronomic improvements and the appropriate use of fertilizers. The present established cultivars of taro have been selected by cultivators from among the spectrum of natural variation derived from mutation. Mutant forms also arise spontaneously in collections and provide new variation.

Under conditions of subsistence agriculture, taro is not commonly manured, except with household refuse, etc. It has, however, been established that the plant can respond well to nitrogen, phosphorus, and potassium under both lowland (i.e. irrigated) and upland conditions in the Pacific area (de la Peña and Plucknett 1967). de la Peña and Plucknett successfully used foliar analytical techniques for the determination of the nutrient status of this crop. It seems that investigation of fertilizer requirements should be a major item in any attempt to expand the production of taro under African conditions, especially where more intensive and possibly partially mechanized farming is contemplated.

5.4.5. *Plant diseases*

Diseases are rarely serious under conditions of subsistence agriculture in small plant populations and at low densities, although in the Solomon Islands and adjacent islands in the Pacific there have been reports of a devastating virus disease. Where attempts are made to cultivate the plant on a large scale, disease problems are likely to become more serious. In the Pacific, a leaf blight is caused by *Phytophthora colocasiae*, and this pathogen also causes losses in both taro and tannia under West African conditions (Doku 1966), although it here appeared to be associated with other organisms as well, and the symptoms of the disease were somewhat different from those described in the Pacific area. Rotting of the corms and roots during growth and consequent stunting or even collapse of the plant has been attributed to various species of *Pythium*. The incidence of root and corm rots and other diseases in the Pacific area has been reviewed by Trujillo (1967), who also indicated possible control measures, and by Buddenhagen, Milbroth, and Hsieh (1970). Work in Ghana on diseases of taro and tannia was reviewed by Doku (1966). In these studies, all introduced cultivars proved susceptible to most diseases with the exception of some inedible tannias.

5.4.6. *Tannia husbandry*

Tannia is almost as widely grown in West Africa—and indeed in many other parts of the tropics—as is taro. It is a larger and more vigorous plant than taro and perhaps better adapted to poor agronomic conditions. Tannia, being less important in the Pacific than taro, has received less attention from research workers and is less well documented. Such data as are available, however, indicate that its yield potential is at least as great, if not greater, than that of taro. Yields approaching 70 t/ha have been attained under experimental conditions (Warid 1970). Some studies of the physiological parameters of crop growth for tannia have been reported by Karikari (1971) from West Africa and Spence (1970) from Trinidad. Tannias are adapted to drier conditions than taro, and cultivation under paddy conditions is unlikely to be appropriate. Tannia has the same potential pathogens as taro and is also subject in some parts of tropical America and the Caribbean to attack by a dynastid beetle, *Ligyrus ebenus* (de G.), which has not been reported to be a significant pest in other parts of the world.

The leaves of both taros and tannias are edible and are quite extensively used as food, especially in Ghana and some other parts of West Africa.

5.4.7. *Storage*

When produced under conditions of subsistence agriculture, at least in Africa, cocoyams are not normally stored for any substantial period but are harvested as and when required. Under the more highly sophisticated conditions in Hawaii they are normally processed into 'poi' (a fermented food) shortly after harvest. There seems to be a considerable degree of confusion as to the suitability of cocoyams for storage; in certain parts of Melanesia and Polynesia they are stored successfully for considerable periods, and some of the earlier literature (reviewed by Coursey (1968)) indicated that storage for some months is possible. Recent work conducted in Melanesia (Gollifer and Booth 1973) suggests, however, that storage life is limited to a few weeks by the development of a complex of post-harvest rots. The development of larger-scale commercial production of cocoyams therefore needs to be associated with investigations directed towards the establishment of satisfactory storage conditions for the fresh material, and/or the development of industrial processes for the manufacture of prepared food, based on these crops, which would have an inherently longer storage life than the fresh material.

5.4.8. *Use of cocoyam for stock feed*

Utilization of the entire plant directly from harvesting for stock-feeding purposes is also a subject that warrants considerable attention, in view of the potentially exceedingly high calorific production that can be achieved. A limitation of the nutritional value of these crops that applies particularly to some cultivars of taro is the high content of calcium oxalate in the form of raphides, which have an irritant effect on the mucous membranes when they are consumed. In normal culinary practice this is at least partially overcome by prolonged boiling, followed by pounding or maceration. It is possible that selection or breeding for low oxalate strains would be a feasible proposition, and this would be of particular interest to enhancing the utilization of the crops for animal feed.

5.4.9. *Other aroid genera having edible tubers*

In the Indo-Pacific region a number of other genera of aroids such as *Cyrtosperma*, *Amorphophallus*, and *Alocasia* are cultivated and used for food to a greater or lesser degree, though none has achieved the importance of *Colocasia* and *Xanthosoma* (Barrau 1957; de la Peña 1970). These crops have not, however, been introduced as food plants to other parts of the tropics, though *Alocasia* is frequently met with in Africa and elsewhere as an ornamental plant. Even in the Indo-Pacific region their importance is declining, and it seems difficult to justify any major research

attention to them in other parts of the world, although it has been noted that very substantial yields can be achieved with *Amorphophallus*. There are several species of this genus, and also of the closely related *Anchomanes*, which are indigenous to Africa. Some of these species were formerly used—and may still occasionally be used—as famine or reserve foods in the remoter parts of the continent, although others are intensely poisonous. It appears unlikely that any of these have much potential for development as crop plants (Coursey 1968).

5.5. Sweet potato (*Ipomoea batatas*) (L.) Lam.

5.5.1. *Origin and distribution*

Like cassava, the sweet potato was originally domesticated in tropical America, although it is now grown almost throughout the tropical world and is an important food crop in many countries. Its cultivation is generally considered to have spread in pre-Columbian times through the Pacific Ocean area as far as New Zealand, and possibly, through the Indian ocean trade, even as far as Africa. A Central American centre of domestication is now generally accepted (Martin and Jones 1972). The genus *Ipomoea* is of pantropical distribution (Yen 1970), and a number of indigenous African species are collected from the wild as emergency foods (Irvine 1952). *I. aquatica* Forsk. is cultivated as a leafy vegetable in South-East Asia (Yen 1970).

5.5.2. *Adaptation*

The sweet potato is more tolerant than most other tropical root crops to a wide range of edaphic and climatic conditions. It is mainly a dry-field crop, intermediate in its optimal habitat between cassava at one extreme and taro at the other. It is more tolerant of cold than many other root and tuber crops within the tropics and can be grown at altitudes as high as 3000 m. Thus it has become the staple food crop of many of the highland peoples of New Guinea (Oomen, Spoon, Heesterman, Ruinard, Luyken, and Slump 1961) and is of great importance in comparatively high-altitude areas of eastern Africa, such as parts of Uganda, Rwanda, and Burundi. In these areas it has enabled people traditionally accustomed to growing other vegetatively propagated root crops to settle at high altitudes. The crop can be harvested after a growing period which is quite short compared with most other tropical root crops. The sweet potato, because of this, has been developed as a summer crop in the warmer temperate latitudes. It is a crop of great importance in the southern parts of the USA, Japan and neighbouring islands, and New Zealand, and is also sometimes grown in Mediterranean countries. Because of its importance in the USA and Japan more research has been undertaken on many aspects of this crop than on more exclusively tropical crops. It is a major industrial crop in the southern USA, and indeed the only recent text (Edmonds and Ammerman 1971) on the crop is written almost exclusively from the North American standpoint.

5.5.3. *Transfer of technology—a caveat*

In view of the comparatively high state of development of the sweet potato as a crop, and of industries based on it, special possibilities exist for the transfer of the comparatively advanced technologies from the USA and Japan to the developing world. Nevertheless, any attempts at such transfer must be handled with considerable caution because of the great ecological differences. In the tropical world the sweet potato is mainly a crop of subsistence farmers who do not have access to high-level technology (MacDonald 1963, 1967). Some of the major limiting factors on sweet-potato production under these conditions were considered by Haynes (1971) to be:

(1) low and inconsistent yields;
(2) uncertain market demand;
(3) high labour inputs for production; and
(4) absence of established production systems.

Although substantial amounts of research on the crop under tropical conditions have been undertaken in Trinidad (Campbell and Gooding 1962; Haynes 1971), in Uganda (MacDonald 1963, 1967), and in various Pacific territories (Poole 1955), attempts to develop sweet-potato production further must take full account of social and economic, as well as technical, conditions in the country concerned. The influence of traditional food habits, for example, needs major consideration; this probably accounts for the far greater importance of the sweet potato in Ghana than in Nigeria, though even in Ghana the sweet potato contributes less than 0·1 per cent of total food expenditure, compared with over 12 per cent for cassava and nearly 7 per cent for yam (Doku 1966).

5.5.4. *Breeding*

Most breeding work on sweet potatoes so far has been conducted in temperate climates. Although many newly bred cultivars have become commercially available, it is very doubtful that many of these will be both adapted to and acceptable in developing countries in the tropics. Considerable emphasis has been placed in the USA on orange-fleshed cultivars, which are considered attractive and which have a high carotene (vitamin A) content. In most parts of the forest zone vitamin-A deficiency is not a problem, and orange-fleshed cultivars, which are unfamiliar, have no advantage *per se* over white-fleshed types unless they can produce substantially more carbohydrate.

5.5.5. *Yields*

Yields reported under tropical conditions are exceedingly variable, from 5 t/ha to as much as 40 t/ha. In Trinidad it was found (Haynes 1971) that cultivars could be classified into three groups according to their yield response to nitrogen treatment of the soil, some giving strong positive responses, some giving little or no response, and some actually giving a negative response under the same conditions of other factors. The negative response was associated with production of excessive leaf and stem growth and hence apparently to a diversion of resources away from storage. Differences were associated with the period from planting to the initiation of storage roots by different clones. If systematic industrial agriculture based on the sweet potato is to be developed anywhere in the tropics, the importation of established industrial cultivars having a high response in yield to fertilizer may be desirable, as most 'local' cultivars are not responsive.

5.5.6. *Diseases*

Under conditions of subsistence agriculture, diseases of the sweet potato in Africa have received only a little attention (Sheffield 1957, 1958) but can result in substantial reductions of yield (MacDonald 1963). It may be anticipated that, if monoculture of the crop is attempted under tropical conditions, a number of diseases currently of only minor importance could become much more serious.

After initial investigation in a joint programme between workers at Yangambi (Zaire) and in Uganda in the early 1950s, Hansford (unpublished) and Sheffield (1957) described two different diseases of viral (or mycoplasmal) origin. Sweet-potato virus A is transmitted by aphids and is of widespread occurrence and serious economic importance, especially in the Zaire and Uganda. East African sweet-potato virus B is white-fly transmitted. Martin (1957) reviewed sweet-potato virus diseases known in the

USA, to none of which the diseases in East Africa seem identical. MacDonald (1963) reported that losses of 66 per cent in yield were attributed to virus (presumably virus A) in trials at Makerere University in Uganda.

More generally, the diseases of the crop which are of importance under warm temperate conditions in southern USA have been reviewed by Hildebrand and Cook (1959) and this provides information that could provide a useful background for further investigations in the tropics.

5.5.7. *Pests*

Many species of insect pests have been recorded as attacking the sweet potato. The most important, which are of pantropic distribution, appear to be *Cylas* spp., *Megastes grandalis*, and *Euscepes batatae*. Control has been achieved by the use of dieldrin, DDT, and chlordane applied as dusts on the planting material, in the soil during the tuber-forming period, or as sprays applied to the plants. However, the extensive use of such chemicals should hardly be encouraged. Under the conditions of subsistence agriculture an adequate amount of control is generally achieved through rotations and shifting cultivation, which eliminate the possibility of a serious build-up of pests to a destructive level. At the IITA genetic resistance to sweet-potato weevil has been found (Singh, unpublished), and this and other aspects of genetic pest resistance need further explanation.

Under African conditions the absence of reports to the contrary suggest that sweet potatoes suffer little from nematode attack. Indeed, Whitehead in an EAAFRO report on a nematode survey conducted in East Africa claimed that 'sweet potato roots appear to be practically immune to most known nematode attacks' (MacDonald 1963).

5.5.8. *Weeds and weed control*

Under monoculture weeds and their control are important in sweet-potato production. Investigations in the West Indies indicated that control is most important during the first few weeks after planting. Weeds of sweet potato have been satisfactorily controlled by weedicides, including amiben and diphenamid (Haynes 1971).

5.5.9. *Storage*

A limitation in the development of the sweet potato as a crop in the tropical world is its poor storage life. Much work has been undertaken, particularly in the

USA, on sweet-potato storage, but this has been in relation to storage during cold winters.

In the tropics the sweet potato is frequently harvested as and when it is required, and once harvested is generally considered to be a highly perishable crop. Under tropical conditions sweet potatoes are extremely prone to waste (Kushman and Wright 1969). However, various simple storage techniques have been evolved in different parts of the world and some of these may very well be of much wider applicability and warrant further investigation.

5.5.10. *Leaf use*

Reference has already been made to the use of the leafy tops of the sweet-potato plant as animal feed. In some areas of Africa the tops are also used as a green vegetable, though on a minor scale. The leaves have a peculiar flavour and are best dried before using for food, to reduce this.

5.6. Irish or white potato (*Solanum tuberosum* L.)

5.6.1. *Status and adaptation*

Of highland tropical origin the Irish or white potato has achieved such great prominence in the temperate world that it has become a world crop rather than merely a tropical crop. It was first domesticated at high altitude in the Andes. Most research on potatoes has been devoted to their production and use in temperate climates, but in recent decades increasing attention has been paid to the crop in tropical regions (Montaldo 1970). The potato in the temperate world has been monographed by Burton (1966). Recent ideas on its use in the tropics have been reviewed by Simmonds (1971) and French (1972).

Substantial work has been undertaken in India (Upadhya, Purohit, and Sharda 1972) on the use of the potato in tropical ecosystems. Most cultivars in the temperate world produce tubers only during periods of lengthening days. To be successful under tropical conditions, potatoes must produce tubers in short days. However, even when the day-length requirement is satisfied, the potato is usually adapted in the tropics only to areas of land at altitudes of 1000 m or higher. Potato crops have been grown successfully under experimental conditions at low altitudes in the tropics (Chapman 1967), but this would probably be an economic proposition only in circumstances where they rank as luxury vegetables for upper-income-group markets, rather than as a staple food crop. Disease problems are serious but can be partially overcome by the continual re-importation of clean stocks of seed-tubers. In many tropical countries it should be feasible to establish disease-free areas for seed-potato production and to apply strict plant quarantine restrictions to maintain disease freedom in such areas.

5.6.2. *Status in West Africa*

In West Africa, potatoes have never been a crop of great importance, as the amount of medium-altitude land suitable for their growth is extremely limited. They have been grown in Ghana (Doku 1966) in the Mampong and Amedzofe areas at altitudes between 400 m and 250 m and in Nigeria (Suchomel 1971) on the Jos and Biu plateaux at altitudes between 600 m and 1300 m, and in the Central African Republic, Upper Volta, and Cameroun.

The Jos plateau industry in particular is of considerable interest as it initially developed quite spontaneously from the private gardening activities of expatriate miners. It has supplied substantial proportions of Nigeria's potato requirements for 30 years or more. Recently, however, the industry has declined seriously owing to the increasing incidence of a number of pests and diseases, of which the following are particularly important:

bacterial wilt	*Pseudomonas solanacearum* (E. F. Smith) Smith
late blight	*Phytophthora infestans* (Mont.) de Bary
alternaria blight	*Alternaria solani* (Ell. and G. Martin) Sor.
common scab	*Streptomyces scabies* (Thaxter) Waksman & Henrici
nematode	*Meloidogyne javanica* (Treub.) Chitwood

Since the decline a substantial amount of research effort has been devoted to rehabilitation of the industry. This has involved screening a large number of cultivars (including Mexican strains of *S. phureja*, which is better adapted to the tropical environment than is *S. tuberosum*, with the aim of establishing satisfactory high-yielding, disease-resistant lines which are adapted to local conditions. Although prominence has been given in the past to the problem of late blight, another major limiting factor is probably bacterial wilt. Unless and until satisfactory commercial cultivars can be developed that are adequately resistant to this disease, there appears

little hope of maintaining the industry. A research programme being conducted on the potato in Kenya has the development of resistance to blight and wilt as major priorities. In that country potatoes are grown at altitudes mainly between 1000 m and 2500 m.

In view of the difficulties of potato cultivation in the tropics, Doku (1966) has suggested that it would be more appropriate to concentrate research work in West Africa on indigenous, or at least well-established, lowland tropical root crops rather than on potato. He emphasizes that most of these have advantages of yield, adaptation, and ease of cultivation and are already staples familiar to most of the people. The potato is not substantially superior, from a nutritional point of view, to most indigenous root crops (except perhaps in the amino-acid spectrum of its protein fraction), although it is widely regarded as having some prestige value, and for this reason tends to be popular, especially amongst higher-income groups.

5.6.3. *Storage under tropical conditions*

Storage of potatoes under tropical conditions presents special problems. As with most other technical matters concerning the crop, storage systems have been developed mainly under temperate conditions, where protection from cold is the major consideration, whereas in the tropics the avoidance of excessively high temperatures in storage is more important. Where potatoes are grown in the tropics as a luxury crop the produce is often stored in controlled-temperature buildings, but this could not be justified economically were the crop to attain wider utilization as a staple food within the tropics. Even under refrigerated conditions, careful attention needs to be paid to produce-quality at the time of harvest and to correct handling. Storage of potatoes in the tropics is discussed by Booth and Proctor (1972).

5.6.4. *The cultivation of potatoes in francophone Africa*

Information on potato growing in francophone Africa was presented at the seminar on vegetables at IITA (see also Chapter 7). Of 24 cultivars tested in the Central African Republic, the following four were selected as being adapted: Bintje, Claudia, Arran Banner, and Regale. Mean yields, over 3 years of cropping and including both wet- and dry-season growing periods, varied from 6·3 t/ha to 7·2 t/ha on ferralitic soils, without fertilizer. Phosphate was the first limiting nutrient, and there was some evidence for interaction between phosphate and potash responses. Yields in 1968 were (t/ha): Bintje, 15·7; Claudia, 24·2; Arran Banner, 17·5; Regale, 19·5.

In the Upper Volta, Claudia and Bintje were found to give the best results in the dry season, with mean yields of about 18 t/ha. Irrigation and fertilizer studies were also reported. Bacterial wilt (*Pseudomonas solanacearum*) and blight (*Phytophthora infestans*) were serious in the wet season. Some resistance to both diseases was shown by Arran Banner; Valdor, Regale, and Ackersegen were partially resistant to blight. Étoile du Leon appeared to possess some resistance to wilt.

5.7. Minor root crops

A number of root or tuber crops of minor importance are known in Africa, although it seems unlikely that any of these have any very great potential for development they nevertheless require mention in this chapter. The indigenous aroids of the genera *Amorphophallus* and *Anchomanes* have already been referred to (see p. 89).

The hausa potato (*Solenostemon rotundifolius* (Poir) J. K. Morton: Labiateae, better but incorrectly known as *Coleus rotundifolius* Chev. & Perrot, is an ancient African food plant, still cultivated to a small degree in many remote districts and known in cultivation in some parts of Asia as well (Chevalier 1946). A number of cultivars are known and the tubers have an excellent flavour, resembling that of the Irish potato. Yields are thought to be low, but little reliable information is available. In most of the more developed parts of Africa, the crop has almost disappeared from cultivation.

The African yam bean (*Sphenostylis stenocarpa* Harms), widely known in West Africa, is a leguminous plant closely resembling the better-known Asian and American yam beans, *Pachyrhizus* spp. The plant is a trailing vine which is commonly grown on mounds, in somewhat the same manner as yams. In addition to the turnip-like fleshy root, the pods can be eaten as a vegetable.

The winged bean (*Psophocarpus tetragonolobus* Neck.) is another legume which has recently been suggested (Pospisil, Karikari, and Boamah-Mensah 1971) as having great potential as a combined root and vegetable crop for the lowland tropics. Though not indigenous to Africa but to South-East Asia, it has been cultivated successfully in Ghana. The root is believed to have a higher protein content than is usual for root crops but, although edible, is not par-

ticularly palatable. The immature fruit can be eaten as a green vegetable or the mature seeds as dry beans, while the whole plant may be used as forage for animals.

The guinea potato (*Dioscoreophyllum cumminsii* Diels: Menispermaceae) is a common wild plant in the forest zone from Sierra Leone to Zaire. It is a vine-like plant, with white fleshy rhizomes 1 cm or so in diameter and up to 15 cm long, and rich in starch. These are occasionally eaten boiled, like potatoes, especially in Gabon (Chevalier 1937). The guinea potato is not cultivated to any great extent, but the rhizomes were formerly collected from the wild. There has recently been a sudden upsurge of interest in this plant because of an intensely sweet substance contained in the fruit.

Icacinas (*Icacina senegalensis* A. Juss and *I. trichantha* Oliv.: Icacinaceae) are widely distributed shrubby wild plants of the savanna and the forest savanna mosaic zones of West Africa. The plants produce large round underground organs, which can weigh up to several kilograms. They have long been used as emergency foods (Dalziel 1937). Although they have never been systematically cultivated, they are said to give surprisingly high yields from wild stands, and appear to have a potential for development as crop plants.

A number of other root and tuber plants are used occasionally in Africa as reserve or famine foods (Dalziel 1937; Irvine 1952). Some of these plants might well have a potential for development into useful crops, given the necessary research interest. There are numerous minor root crops used to a limited degree in South America (Leon 1967; Montaldo 1972), and some of these, notably *Arracacia xanthorrhiza* Bancroft, might also be of interest in other parts of the tropical world. Problems of acceptability, however, might have to be overcome.

Acknowledgements

Section 5.6.4 was prepared by H. D. Tindall, and placed in this chapter by the editors.

References

AGBOOLA, S. A. (1968). The introduction and spread of cassava in Western Nigeria. *Nigerian J. econ. soc. Studies* **10**, 369–85.

AKINRELE, I. A., COOK, A. S., and HOLGATE, R. A. (1962). The manufacture of gari from cassava in Nigeria. *Proceedings of the 1st International Congress in Food Science and Technology* Vol. 4, pp. 633–44.

ANNEGERS, J. E. (1973). Ecology of dietary patterns and nutritional status in West Africa. 1. Distribution of starchy staples. *Ecol. Fd Nutr.* **2**, 107-19.

ANON. (1971). A summary of the proceedings of the seminar. *Ford Foundation/IITA/IRAT seminar on research on root and tuber crops in West Africa, Ibadan.*

ARCHIBALD, E. E. A. (1967). The genus *Dioscorea* in Cape Province west of East London: *Jl S. Afr. Bot.* **33**, 1-46.

AYENSU, E. S. (1970). Comparative anatomy of *Dioscorea rotundata* and *Dioscorea cayenensis. J. Linn. Soc. (Bot.)* **63**, 127-36.

—— (1972). In *Anatomy of the monocotyledons* (ed. MET-CALFE, C. R.), Vol. 6: *Dioscoreales.* Clarendon Press, Oxford.

—— COURSEY, D. G. (1972). Guinea yams. *Econ. Bot.* **26**, 301-18.

BARRAU, J. (1956). Les ignames alimentaires des îles du Pacifique sud. *J. Agric. trop. Bot. appl.* **3**, 385-401.

—— (1957). Les aracées à tubercules alimentaires des îles du Pacifique sud. *J. Agric. trop. Bot. appl.* **4**, 34-52.

—— (1970). La région indo-pacifique comme centre de mise en culture et de domestication des végétaux. *J. Agric. trop. Bot. appl.* **17**, 487-503.

BECK, B. D. A. (1960). Cassava trials on Moor Plantation. *Rep. Dep. agric. Res. Nigeria. 1958-1959.*

—— (1971). Cassava production in West Africa. *Ford Foundation/IITA/IRAT seminar on research on root and tuber crops in West Africa, Ibadan.*

BOLHUIS, G. G. (1954). The toxicity of cassava roots. *Neth. J. agric. Sci.* **2**, 176-85.

BOOTH, R. H. (1976). The storage of fresh cassava roots. *Proc. 3rd int. Symp. trop. root Crops, Ibadan* **3**. (In press.)

—— PROCTOR, F. J. (1972). Considerations relevant to the storage of ware potatoes in the tropics. *Pest Artic. News Summ. (PANS)* **18**, 409-32.

BUDDENHAGEN, I. W., MILBROTH, G. M., and HSIEH, S. P. (1970). Virus diseases of taro and other aroids. *Proc. 2nd Int. Symp. trop. root Crops, Hawaii* **2**, 53-5.

BURKILL, I. H. (1960). The organography and the evolution of the Dioscoreaceae, the family of the yams. *J. Linn. Soc. (Bot.)* **56**, 319-412.

BURTON, W. G. (1966). *The potato.* Veenman and Zonen, Wageningen.

BUSSON, F. (1965). *Plantes alimentaires de l'Ouest africain.* Leconte, Marseilles.

CAMPBELL, J. S. and GOODING, H. J. (1962). Recent development in the production of food crops in Trinidad. *Trop. Agric., Trin.* **39**, 261-70.

CATHARINET, M. (1965). Note sur la culture du macabo et du taro au Cameroun. *Agron. trop., Nogent* **20**, 717-24.

CHANT, S. R. (1958). Studies in the transmission of cassava mosaic virus by *Bemisia* spp. (*Aleyrodidae*). *Ann. appl. Biol.* **46**, 210-15.

CHAPMAN, T. (1967). Some experiments with potatoes (*Solanum tuberosum*) in Trinidad, 1963-4. *Proc. 1st Int. Symp. trop. root Crops, Trinidad* **1**, 171-9.

CHEVALIER, A. (1937). Legumes curieux de l'Afrique tropicale. *Revue int. Bot. appl. Agric. trop.* **17**, 444-6.

—— (1946). Un legume tropical à repandre: la petit pomme de terre d'Afrique (*Coleus rotundifolius*). *Revue int. Bot. appl. Agric. trop.* **26**, 296-330.

CIAT (1972). Cassava production systems. *Rep. Centre int. Agric. trop. Annual report for 1972*, 43-82.

COURS, G. (1951). Le manioc à Madagascar. *Mém. Inst. scient. Madagascar, Serie B* **3**, 203-400.

COURSEY, D. G. (1968). *Yams.* Longmans, London.

—— (1968). The edible aroids. *Wld Crops* **20**, 25-30.

—— (1971). The history and possible future of yam cultivation in West Africa. *Ford Foundation/IITA/IRAT seminar on research on root and tuber crops in West Africa, Ibadan.*

—— (1972). The civilizations of the yam—interrelationships of man and yam in Africa and the Indo-Pacific region. *Archeol. phys. Anthropol. Oceania* **7**, 215-33.

—— (1973a). Cassava as food: toxicity and technology. In *Chronic cassava toxicity* (eds. B. L. Nestel and R. MacIntyre). International Development Research Centre, Ottawa.

—— (1973b). Genetic erosion in West African yams. *Crop genetic resources in their areas of diversity.*

—— HAYNES, P. H. (1970) Root crops and their potential as food in the tropics. *Wld Crops* **22**, 260-5.

—— MARTIN, F. W. (1970). The past and future of yams as crop plants. *Proc. 2nd Int. Symp. trop. root Crops, Hawaii* **1**, 87-90; 99-101. (Reprinted (1972). *Pl. Fds hum. Nutr.* **2**, 133-8.)

DALZIEL, J. M. (1937). *The useful plants of west tropical Africa.* Crown Agents, London.

DE LA PEÑA, R. S. (1970). The edible aroids in the Asian-Pacific area. *Proc. 2nd Int. Symp. trop. root Crops, Hawaii* **1**, 136-40.

—— PLUCKNETT, D. L. (1967). The response of taro (*Colocasia esculenta* (L.) Schott) to N, P and K fertilization under upland and lowland conditions in Hawaii. *Proc. 1st Int. Symp. trop. root Crops, Trinidad* **1**, 70-85.

DE VRIES, C. A., FERWEDA, J. D., and FLACH, M. (1967). Choice of food crops in relation to actual and potential production in the tropics. *Neth. J. agric. Sci.* **15**, 241-8.

DIAZ, R. O. (1972). *World cassava production and yield trends, 1960-1968*, Centro International de Agricultura Tropical, Colombia.

DOKU, E. V. (1966). Root crops in Ghana. *Ghana J. Sci.* **6**, 5-36. (Reprinted (1967), *Proc. 1st Int. Symp. trop. root Crops, Trinidad* **1**, 39-65.)

—— (1969). *Cassava in Ghana.* Ghana Universities Press, Accra.

DOUGHTY, L. R. (1958). *Cassava breeding for resistance to mosaic and brown streak viruses—a review.* Contribution to 7th Meeting of the East African Specialist Committee for Agricultural Botany. EAAFRO, Nairobi. (Mimeo)

DUMONT, C., RENAUT, G., and VANDEVENNE, R. (1971). Integration de la culture de l'igname dans un système de culture modernisé. *Ford Foundation/IITA/IRAT seminar on research on root and tuber crops in West Africa, Ibadan.*

EDMONDS, J. B. and AMMERMAN, G. R. (1971). *Sweet potatoes: production, processing, marketing*. Avi Publishing Co., Westport, Connecticut.

ENYI, B. A. C. (1972). Effects of staking, nitrogen, and potassium on growth and development in lesser yam, *Dioscorea esculenta. Ann. appl. Biol.* **72**, 211-19.

FERGUSON, T. U. (1972). The propagation of *Dioscorea* spp. by vine cuttings: a critical review. *Trop. Root Tuber Crops Newsl.* **5**, 4-7.

—— HAYNES, P. H. (1970). The response of yams (*Dioscorea* spp.) to nitrogen, phosphorus, potassium and organic fertilizers. *Proc. 2nd Int. Symp. trop. root Crops, Hawaii* 1, 93-6.

FRENCH, E. R. (1972). *Prospects for the potato in the developing world*. Centro International de la Papa, Lima.

GOLLIFER, D. E. and BOOTH, R. H. (1973). Storage losses of taro corms in the British Solomon Islands Protectorate. *Ann. appl. Biol.* **73**, 349-56.

GOODING, E. G. B. and HOAD, R. M. (1967). Problems of yam cultivation in Barbados. *Proc. 1st Int. Symp. trop. root Crops, Trinidad* 1, 137-51.

GRACE, M. (1971). *Processing of cassava*. FAO, Rome.

HAYNES, P. H. (1967). The development of a commercial system of yam production in Trinidad. *Trop. agric., Trin.* **44**, 215-21.

—— (1971). Recent research on sweet potato (*Ipomoea batatas* (L.) Lam.) at the University of the West Indies. *Ford Foundation/IITA/IRAT seminar on research on root and tuber crops in West Africa, Ibadan.*

HENDERSHOTT, C. H., AYRES, J. C., BRANNEN, S. J., DEMPSEY, A. H., LEHMAN, P. S., OBIOHA, F. C., ROGERS, D. J., SEERLEY, R. W., and TAN, K. H. (1972). *A literature review and research recommendations on cassava* (Manihot esculenta *Crantz*). University of Georgia, USA.

HILDEBRAND, E. J. and COOK, H. T. (1959). *Sweet potato diseases*. Farmers Bulletin, U.S. Department of Agriculture 1059. Government Printing Office, Washington, D.C.

IDUSOGIE, E. O. (1971). The nutritive value per acre of selected food crops in Nigeria. *Jl W. Afr. Sci. Ass.* **16**, 17-24.

INGRAM, J. S. and HUMPHRIES, J. R. O. (1972). Cassava storage—a review. *Trop. Sci.* **14**, 131-48.

IRVINE, F. R. (1952). Supplementary and emergency food plants of West Africa. *Econ. Bot.* **6**, 23-40.

JAMESON, J. D. (1964). Cassava mosaic disease in Uganda. *E. Afr. agric. for. J.* **29**, 208-13.

JENNINGS, D. L. (1957). Further studies in breeding cassava for virus resistance. *E. Afr. agric. J.* **22**, 213-19.

—— (1970). Cassava in Africa. *Fld Crop Abstr.* **23**, 271-8.

JONES, W. O. (1959). *Manioc in Africa*. Stanford University Press.

KARIKARI, S. K. (1971). Cocoyam cultivation in Ghana. *Wld Crops* **23**, 118-22.

KUSHMAN, L. J. and WRIGHT, F. S. (1969). *Sweet potato storage*. Agricultural Handbook 358. Government Printing Office, Washington, D.C.

LAWTON, JUNE R. and LAWTON, J. R. S. (1969). The development of the tuber in seedlings of five species of *Dioscorea* from Nigeria. *J. Linn. Soc. (Bot.)* **62**, 223-32.

LEON, J. (1967). Andean root crops: origin and variability. *Proc. 1st Int. Symp. trop. root Crops, Trinidad* 1, 118-23.

LOZANO, J. C. and BOOTH, R. H. (1974). Diseases of cassava (*Manihot esculenta* Cranz). *Pest Artic. News Summ. (PANS)* **20**, 30-54.

—— SEQUEIRA, L. (1974). Bacterial blight of cassava in Columbia: epidemiology and control. *Phytopathology* **64**, 83-8.

LUCAS, J. W. (1968). The role of plant foods in solving the world food problem. 1. Energy requirements. *Plant Foods hum. Nutr.* **1**, 13-21.

MACDONALD, A. S. (1963). Sweet potato with particular reference to the tropics. *Fld Crop Abstr.* **16**, 219-25.

—— (1967). Some aspects of the sweet potato and its agronomy in Uganda. *Proc. 1st Int. Symp. trop. root Crops, Trinidad* 1, 112-23.

MAGOON, M. L. (1970). Problems and prospects in the genetic improvement of cassava in India. *Proc. 2nd Int. Symp. trop. root Crops, Hawaii* 1, 58-61.

MARTIN, F. W. (1967). The sterility-incompatibility complex of the sweet potato. *Proc. 1st Int. Symp. trop. root Crops, Trinidad* 1, 3-15.

—— JONES, A. (1972). The species of *Ipomoea* closely related to the sweet potato. *Econ. Bot.* **26**, 201-15.

—— RHODES, A. M. (1973). Correlations among greater yam (*Dioscorea alata* L.) cultivars. *Trop. Agric., Trin.* **50**, 183-92.

MARTIN, W. J. (1957). The mosaic and similar diseases of the sweet potato: review. *Pl. Dis. Reptr* **41**, 930-45.

MICHE, J. C. (1971). Les usages possibles des plantes à racines ou à tubercules. *Ford Foundation/IITA/IRAT seminar on research on root and tuber crops in West Africa, Ibadan.*

MIÈGE, J. (1954). Les cultures vivrières en Afrique occidentale. *Cah. d'outre-mer* **7**, 25-50.

MONTALDO, A. (1967). *Bibliografia de raices y tuberculos tropicales*. Universidad Central de Venezuela, Maracay.

—— (1970). The potato in Latin America. *Proc. 2nd Int. Symp. trop. root Crops, Hawaii* 1, 107-19.

—— (1972). *Cultivo de raices y tuberculos tropicales*. Instituto Interamericano de Ciencias Agricolas, Lima.

NESTEL, B. L. (1973). Current utilization and future potential for cassava. In *Chronic cassava toxicity* (ed. B. L. Nestel and R. MacIntyre). International Development Research Centre, Ottawa.

NESTEL, B. L. and MACINTYRE, R. (1973). *Chronic cassava toxicity*. International Development Research Centre, Ottawa.

NICHOLS, R. F. W. (1947). Breeding cassava for virus resistance. *E. Afr. agric. for. J.* **12**, 184-94.

OKE, O. L. (1968). Cassava as food in Nigeria. *Wld Rev. Nutr. Diet.* **9**, 227-50.

OOMEN, H. A. P. C., SPOON, W., HEESTERMAN, J. E., RUINARD, J., LUYKEN, R., and SLUMP, P. (1961). The sweet potato as the staff of life of the highland Papuan. *Trop. geogr. Med.* **13**, 55-6.

PHILLIPS, T. P. (1974). *Cassava utilization and potential markets*. International Development Research Centre, Ottawa.

PLUCKNETT, D. L. (1970). Colocasia, Xanthosoma, Alocasia, Cyrtosperma and Amorphophallus. *Proc. 2nd Int. Symp. trop. root Crops, Hawaii* **1**, 127-35.

—— DE LA PEÑA, R. S. (1971). Taro production in Hawaii. *Wld Crops* **23**, 244-9.

—— DE LA PEÑA, R. S., and OBRERO, F. (1970). Taro (*Colocasia esculenta*). *Fld Crop Abstr.* **23**, 413-26.

POOLE, C. F. (1955). The sweet potato in Hawaii. *Circ.* **45**, Hawaii agric. Exp. Stn.

POSPISIL, F., KARIKARI, S. K., and BOAMAH-MENSAH, E. (1971). Investigations of winged bean in Ghana. *Wld Crops* **23**, 260-4.

PURSEGLOVE, J. W. (1968). *Tropical crops: dicotyledons*. Longmans, London.

—— (1972). *Tropical crops: monocotyledons*. Longmans, London.

ROGERS, D. J. (1967). A computer-aided morphological classification of *Manihot esculenta* Crantz. *Proc. 1st Int. Symp. trop. root Crops, Trinidad* **1**, 57-78.

—— FLEMING, H. S. (1973). A monograph of *Manihot esculenta*. *Econ. Bot.* **27**, 1-113.

SHEFFIELD, F. L. M. (1957). Viruses in sweet potatoes. *Phytopathology* **47**, 582-90.

—— (1958). Transmission of sweet potato virus to alternative hosts. *Phytopathology* **48**, 1-6.

SILVESTRE, P. and DELCASSO, G. (1971). Le manioc dans la region maritime du Togo. *Ford Foundation/IITA/IRAT seminar on research on root and tuber crops in West Africa, Ibadan*.

SIMMONDS, N. W. (1971). The potential of potatoes in the tropics. *Trop. Agric., Trin.* **48**, 291-9.

SIVAN, P. (1970). Dalo growing research in the Fiji Islands. *Proc. 2nd Int. Symp. trop. root Crops, Hawaii* **1**, 151-4.

SOBULO, R. A. (1972). Studies on white yam (*Dioscorea rotundata*). 1. Growth analysis. *Exp. Agric.* **8**, 99-106.

SPENCE, J. A. (1970). Growth and development of tannia (*Xanthosoma* sp.). *Proc. 2nd Int. Symp. trop. root Crops, Hawaii* **2**, 47-51.

SUCHOMEL, D. R. (1971). The joint Institute for Agricultural Research USAID Nigerian Irish potato project. *Ford Foundation/IITA/IRAT seminar on research on root and tuber crops in West Africa, Ibadan*.

TERRA, G. J. A. (1964). The significance of leaf vegetables, especially of cassava, in tropical nutrition. *Trop. geogrl Med.* **2**, 97-108.

THOMPSON, A. K., BEEN, B. O., and PERKINS, C. (1975). Reduction of wastage in stored yams. *Proc. 3rd Int. Symp. trop. root Crops, Ibadan*. (In press.)

TRUJILLO, E. E. (1967). Diseases of the genus *Colocasia* in the Pacific area and their control. *Proc. 1st Int. Symp. trop. root Crops, Trinidad* **2**, 13-19.

UCHENDU, V. C. (1970). Cultural and economic factors influencing food habit patterns in sub-Saharan Africa. *Proc. 3rd Int. Congr. Fd Sci. Technol., Washington*.

UPADHYA, M. D., PUROHIT, A. N., and SHARDA, R. T. (1972). Breeding potatoes for sub-tropical and tropical areas. *Wld Crops* **24**, 314-16.

WAITT, A. W. (1963). Yams, *Dioscorea* species. *Fld Crop Abstr.* **16**, 145-57.

WARID, W. A. (1970). Production and improvement of edible aroids in Africa. *Proc. 2nd Int. Symp. trop. root Crops, Hawaii* **2**, 39-44.

WHOLEY, D. W. and HAYNES, P. H. (1971). A yam staking system for Trinidad. *Wld Crops* **23**, 123-6.

WILLIAMS, C. N. and GHAZALI, S. M. (1969). Growth and productivity of tapioca (*Manihot utilissima*). 1. Leaf characteristics and yield. *Exp. Agric.* **5**, 183-94.

YEN, D. E. (1970). Sweet potato. In FRANKEL, O. H. and BENNETT, E. (Eds.) *Genetic resources in plants—their exploration and conservation*. IPB Handbook no. 11. Blackwells, Oxford and Edinburgh.

6. Bananas as a food crop

O. J. BURDEN AND D. G. COURSEY

6.1. Introduction

The banana is an important food crop of the humid lowland tropical areas of the world, and has received, in relation to its use as a staple food, far too little attention by agricultural scientists. Bananas are also grown extensively as plantation crops for sale as dessert fruits, and the cultivars suited to this use are well known and researched. Of the global production of bananas, however, the export to the temperate world of dessert bananas probably accounts for little more than a quarter of the total production of bananas, and the greater part of the small-farm production is utilized within the producing countries. Much of this is of cultivars eaten cooked or made into beer. This chapter is concerned only with bananas in the context of their production as a food crop. Bananas as an export crop will not be considered.

6.2. Taxonomy

All the edible bananas, whether cooking or dessert, are derived from the two wild species *Musa acuminata* and *M. balbisiana* of the section Eumusa of the genus *Musa* (Musaceae). These both originated in the Indo-Malaysia region, and were taken by man to the rest of the wet tropics (Kurz 1865; Cheeseman 1948; Simmonds and Shepherd 1955; Simmonds 1966). The most widely accepted classification of the edible banana cultivars (Simmonds 1966) is based on a scoring method which estimates, from an index based on multiple characters, the relative contribution of the two parent species, *M. acuminata* (A) and *M. balbisiana* (B), to the genetic make-up of the cultivar under consideration (Simmonds and Shepherd 1955), where A and B represent the haploid genomes of each species. This system has been followed by De Langhe (1964) and Karikari (1973) in considering the origins of bananas and plantains in the Congo and Ghana respectively. Problems of classification for *M. acuminata* cultivars have been discussed by Hari (1968). Most edible bananas are triploids with the combinations AAA, AAB, or ABB.

No edible forms of pure diploid *M. balbisiana* (BB) or of *balbisiana* polyploids are known to occur naturally, though the studies of induced polyploid forms of *M. balbisiana* by Vakili (1967) suggests that these could occur and be misinterpreted as ABBs. Some diploid forms are edible and widely distributed, and several edible tetraploids have been recorded as occurring naturally in South-East Asia (Vakili 1967).

There is no clear botanical separation between dessert cultivars and starchy or cooking cultivars. Sometimes dessert types are called bananas and cooking types are called plantains, but usage is not regular. Indeed, all edible types are termed plantains in India and in some Latin American countries. The word 'banana' is possibly of West African origin. In the remainder of the chapter the word plantain will be used to designate bananas that are produced to be eaten cooked.

Bananas used as staple foods are derived from three triploid groups:

AAA Mainly comprise the Cavendish group of cultivars which have a low starch and high sugar content when ripe and are used for cooking only in the green state. AAA also includes a well-known red banana which is used for cooking both in East Africa and South America.

AAB These have been considered by some to represent the true plantains. Bananas of this genome complex are generally starchy even when ripe and edible only after cooking. De Langhe (1964), discussing the origin of 56 'plantain' cultivars grown in Zaire concluded that at least five different syntheses of the AAB genome, through hybridization, may have occurred in the development of the whole plantain sub-group. Plantains are considered by others as falling into two natural groups (Simmonds 1966).

  1. French plantain: with 7–10 or more hands and numerous fingers to each hand, and having a persistent male axis.

2. Horn plantain: usually with only 3–5 hands and few fingers per hand, and having a deciduous male axis.

There are many intermediate forms, however, and it is possible that the two forms are of common origin.

ABB The group known in the Caribbean as 'bluggoes'; these are starchy cooking bananas which are found in all banana-growing areas under a wide variety of local names. The centre of origin of the ABB triploidy is thought to be in southern India. This group tends to be adapted better than AAA or AAB bananas in drier climates and more exposed situations (Simmonds 1966).

6.3. Utilization

The importance of these *Eumusa* cultivars as food crops varies with different areas of the tropics. Simmonds (1966) considered that about half the world production of bananas is eaten uncooked in the ripe state whilst the remainder are cooked by boiling, steaming, baking, roasting, or frying. A small proportion is used in the preparation of other products such as sweetmeats or alcoholic drinks. Bananas (plantains) are a major component of the local diet in lowland tropical Central and West Africa (Zaire, Ghana, and southern Nigeria): they are eaten boiled, boiled and pounded, baked, fried, or occasionally sun-dried or smoked for preservation of the dried product (Dalziel 1937).

Both bananas (AAA), including the red bananas, and bluggoes (ABB) (generally eaten steamed) form a major part of the diet in East Africa, especially in Uganda and adjacent parts of Tanzania on the western shore of Lake Victoria.

In India there is a large consumption of bananas (AAA) in the uncooked ripe condition, but the use of cooked plantains does not appear to be widespread or of major importance. Moreover, in South-East Asia the starchy cultivars of *Eumusa* do not apparently form a significant part of the staple diet. In Hawaii and other South Pacific Islands they are more widely used.

Plantains are a major item in diets in Colombia and Venezuela, being consumed either fried or baked, or as preparations made from plantain flour. In Venezuela, the intake of plantains and of bananas together is said to account for about 7 per cent of the daily calorie intake, 20 per cent of vitamin C, and 48 per cent of vitamin A (Jaffé, Chavez, and De Koifman 1962).

In Colombia the annual intake per person is put at 41 kg, but this figure is said to vary considerably between rural and urban populations (Cardenosa-Barriga 1961). Plantains and other cooking bananas are similarly important in other humid tropical areas of Latin America.

In the West Indies all three main types of cooking bananas (green bananas (AAA), plantain (AAB), and bluggoe (ABB)) contribute to the staple diet. The bluggoe (ABB) is extensively used, particularly in Grenada in the Windward Islands. Plantains, either fried or boiled when ripe, are eaten on all islands of the West Indies.

Unripe dessert bananas (AAA) are important in the diet of villagers in the banana-exporting islands of the Caribbean (Windwards and Jamaica), where they may be grown as a backyard crop or be bunches rejected from the packing of the export crop. They are commonly eaten boiled with salt fish.

6.4. Agronomy

Haarer (1964) presented a broad view of the agronomy of commercial banana plantations. The major proportion of the crop is grown in mixed cultivation, often on very small holdings. Thus yields are difficult to estimate and not very meaningful in comparison with those of commercial pure stands. In several countries they are grown as shade trees for coffee or cocoa (Lassoudière 1973; Cardenosa-Barriga 1961). A comprehensive account of banana and plantain cultivation in Colombia has been provided by Garcia Reyes (1968).

The conditions of cultivation to maximize production per unit area of plantains can be expected to be essentially as for dessert bananas, but when crops are grown mainly for local consumption maximizing the yield to land for one crop may not be the sensible objective of the farmer. In any case, bananas do not normally receive the same level of cultural care as crops grown for export. Often only minimal work is put into the production of the crop, and the return for labour may be high and make for easy living. Bananas tolerate a wide variety of soil types, provided that rainfall and temperature are adequate. Temperatures with no greater fluctuation than between 20 °C and 35 °C are desirable and rainfall should be distributed so that soils are never dry in the root zone, or if so only for short periods. Under low-altitude tropical conditions 200 cm of well-distributed rainfall is adequate. Salter and Goode (1967) include a useful review of water requirements for bananas in their monograph. Where the precipitation is markedly seasonal, the time of planting in relation to rainfall is critical. Well-drained soil is necessary as the crop will not tolerate waterlogging.

The yield potential in most plantain-growing areas is rarely achieved since fertilizer is rarely applied and cultural care is minimal. In some areas animal manure is used. Some research into the use of commercial fertilizers has been carried out in West Africa as well as elsewhere. The use of foliar diagnosis is well developed in commercial production, and symptoms of deficiency have been usefully illustrated by Charpentier and Martin-Preval (1965).

Garcia Reyes (1968) recommends that three equal applications of 150 g each of complete fertilizer (12:6:22:2) (N:P:K:Mg) should be made to each plant every year. Other recommendations are made for Ghana (Karikari 1972), the Ivory Coast (Lassoudière 1973), and Puerto Rico (Vicente-Chandler 1973). Melin (1972), in a preliminary report on trials with a cultivar of French plantain in Cameroun, found that maximum yield was obtained in phosphate-rich volcanic soils by the use of heavy dressings of nitrogen, sulphur, and potash.

A considerable amount of research remains to be done in the various production areas to determine suitable fertilizer practices for different cultivars under varying soil and climatic and economic and social conditions.

6.5. Diseases and pests

The diseases of AAA bananas important in world trade are well known and have been extensively discussed (Stover 1972; Wardlaw 1972). Many problems are common to these and to plantains (AAB) and bluggoes (ABB), but the specific problems of the latter crops have not, however, received a similar level of attention. The most serious field disease of bananas, Sigatoka disease or leaf spot (*Mycosphaerella musicola*), also attacks AAB and ABB cultivars, although there is a degree of tolerance in these latter types which appears to be related to the relative contribution made by the *Musa balbisiana* component (Stover 1972). Plantains (AAB) are susceptible to damage from oil sprays at concentrations used to control leaf spot in bananas (Stover 1972).

Bluggoe bananas are particularly susceptible to bacterial wilt caused by *Pseudomonas solanacearum*. Fortunately the virulent Race 2 of this widespread pathogenic species, which causes Moko disease, is limited in distribution (Buddenhagen 1961). The tomato Race 1 of this pathogen, which is also pathogenic on some *Musa* species, is virulent only on a few edible cultivars (Vakili 1965). All three cultivar types are susceptible to severe attack of black leaf streak (*Mycosphaerella fijiensis*).

Cordana leaf spot, which is rarely severe in AAA dessert bananas, can cause severe defoliation of plantains. On the other hand, pitting disease, caused by *Pyricularia grisea* (which causes severe damage in bananas of Cavendish varieties, requiring costly control measures), is not known to cause severe damage to plantains.

Panama disease caused by *Fusarium oxysporum* f. sp. *cubense*, which has devastated commercial plantings of Gros Michel bananas and brought about a general conversion to Cavendish cultivars, does not cause extensive damage on plantain cultivars. Bluggoe (ABB) bananas are said to have a degree of tolerance, but they may suffer severely under some adverse environments and conditions.

Plantains are also subject to severe damage from nematodes of those species which attack commercial bananas (Stover 1972). According to Gorenz (1963) in the western region of Ghana nematode infestation is so severe as to be the first limiting factor in plantain production. Losses of production attributable to nematodes may be as high as 25 per cent in the first year of planting, rising to 50 per cent or even 90 per cent in ratoon crops.

The diseases, pests, and weeds of bananas and their control have been the subject of a recent and most useful manual (Feakin 1971).

6.6. Post-harvest factors

Interest in cultivation of cooking cultivars of banana for commercial marketing has been stimulated in recent years by the demand for these among immigrant populations in North America and Britain, and also by rapid urbanization which is occurring in countries where plantains are a staple food. This has brought about problems in handling, preservation, and ripening.

Plantains in general remain in a satisfactory condition when given shipping and ripening environments used for dessert bananas in world trade. The very wide range of cultivars and the varying conditions of production, however, indicate that much detailed research work is still needed to determine optimum handling conditions in particular situations.

The minimum temperature at which the plantains can be held after harvest without suffering cold damage is 11 °C (IIT 1967), but Hernandez (1973) has indicated that even at a storage temperature of 13 °C, the normal shipping temperature for green bananas, plantains can be impaired in quality. Ripening studies in Puerto Rico, using the local plantain cultivars Guayamero and Maricongo, showed that even ripening could be achieved in 4-5 days at a temperature of 22 °C, and a relative humidity of 95-100 per cent, using a single application of ethylene 1/1000 v/v to initiate ripening (Sanchez Nieva, Hernandez, and Bueso de Vinas 1970). The effect of humidity on plantains ripened under ambient tropical conditions, without the application of ethylene or temperature control, is discussed by Thompson, Been, and Perkins (1974).

There are several published references to the composition of plantains, some compared with bananas at varying stages of ripeness, but the figures given are very variable and reflect the wide range of cultivars studied (von Loesecke 1949; van den Berghe 1958; Ketiku 1973; Anon. 1972). Although the texture of ripe AAB or ABB fruit is firmer than that of AAA fruit, it is not yet clear how far this is due to incomplete starch–sugar conversion.

References

ANON. (1972). *Plantains* 32 Frutales. Folleto No. 7 Mexico. Commission Nacional de Fruticultura.

BERGHE, L. VAN DEN (1958). La valeur nutritive de la banane plantain. *Folia scient. Afr. cent.* 4, 85.

BUDDENHAGEN, I. W. (1961). Bacterial wilt of bananas: history and known distribution. *Trop. Agric., Trin.* 38, 107-21.

CARDENOSA-BARRIGA, R. (1961). Platanos y bananos en Colombia. *Agric. Trop.* 17, 665-77.

CHARPENTIER, J. M. and MARTIN-PREVAL, P. F. (1965). Carences et oligo-elements chez la bananier. *Fruits* 20, 521-57.

CHEESEMAN, E. E. (1948). Classification of the bananas, III. Critical notes on species. *Kew Bull.* (2), 145-53.

DALZIEL, J. M. (1937). *The useful plants of West Tropical Africa*. The Crown Agents, London.

DE LANGHE, E. (1964). Origin of variation in the plantain banana. *Meded. Landb Hoogesch. Opzoek Stns Gent.*

FEAKIN, S. (ed.) (1971). *Pest control in bananas*. PANS Manual No. 1 (2nd edn). PANS, London.

GARCIA REYES, F. (1968). El cultivo del platano y el banano. *Revta Cafet. Colomb.* 17, 83-98.

GORENZ, A. M. (1963). Preparation of disease-free planting material of banana and plantain. *Ghana Fmr* 7, 15-18.

HAARER, A. E. (1964). *Modern banana production*. Leonard Hill, London.

HARI, P. C. (1968). Bract imbrication as a taxonomic character in *Musa acuminata. Trop. Agric., Trin.* 45, 99-108.

HERNANDEZ, I. (1973). Storage of green plantains. *J. Agric. Univ. P. Rico* 57, 100-6.

INSTITUTO DE INVESTIGACIONES TECNOLOGICAS, BOGOTA (1967). Algunas anotaciones sobre el almaceniamento del platano 'Dominico Marton'. *Technologia* 9, 41-3.

JAFFÉ, W. G., CHAVEZ, J. F., and DE KOIFMAN, B. (1962). Sobre el valor nutritivo de platanos y cambures. *Archos venez. Nutr.* 13, 9-23.

KARIKIRI, S. K. (1972). Plantain growing in Ghana. *Wld Crops* 24, 22-4.

—— (1973). Some taxonomic assessment of the contribution of *Musa acuminata* and *Musa balbisiana* to the origin of plantains and bananas in Ghana. *Ghana J. agr. Sci.* 6, 9-19.

KETIKU, A. O. (1973). Chemical composition of unripe (green) and ripe plantain (*Musa paradisaica*). *J. Sci. Fd Agric.* 24, 703-7.

KURZ, S. (1865). Note on the plantains of the Indian archipelago. *J. agri-hort. Soc. Madras* 14, 295-301.

LASSOUDIÈRE, A. (1973). Le bananier plantain en Côte d'Ivoire. *Fruits* 28, 453-62.

LOESECKE, H. W. VON (1949). *Bananas*. Interscience, New York.

MELIN, P. H. (1972). Potential de productivité d'un cultivar de 'French plantain'. Resultats preliminaries. *Fruits* 27, 591-3.

SALTER, P. J. and GOODE, J. E. (1967). *Crop responses to water at different stages of growth*. Research Review No. 2. Commonwealth Bureau of Horticulture and Plantation Crops, London.

SANCHEZ-NIEVA, F., HERNANDEZ, I., and BUESO DE VINAS, C. (1970). Studies on the ripening of plantains under controlled conditions. *J. Agric. Univ. P. Rico* 54, 517-29.

SIMMONDS, N. W. (1966). *Bananas* (2nd edn). Longmans, London.

—— and SHEPHARD, K. (1955). The taxonomy and origins of the cultivated bananas. *J. Linn. Soc. (Bot.)* 55, 302-12.

STOVER, W. (1972). *Diseases of the banana*. Commonwealth Mycological Institute, Kew.

THOMPSON, A. K., BEEN, B. O., and PERKINS, C. (1974). Effects of humidity on ripening of plantain bananas. *Experientia* 30, 35-6.

VAKILI, N. (1965). Inheritance of resistance in *Musa acuminata* to bacterial wilt caused by the tomato race of *Pseudomonas solanacearum. Phytopathology* 55, 1206-9.

—— (1967). The experimental formation of polyploidy and its effect in the genus *Musa. Am. J. Bot.* 54, 24-36.

VICENTE-CHANDLER, J. (1973). Plantains—a versatile crop with commercial potential. *Wld Fmg* 15, 18-19.

WARDLAW, C. W. (1972). *Banana diseases* (2nd edn). Longmans, London.

7. Vegetable crops

H. D. TINDALL

7.1. Introduction

7.1.1. Historical background

Vegetable crops from Europe were probably introduced into West Africa by Portuguese traders on early visits to the several natural harbours of the west coast. Other introductions may have come from the north and east, via overland trading routes which existed from very early times. Many fruits and vegetables were also introduced in the late eighteenth century, notably by Afzelius, who established a vegetable garden in Freetown, Sierra Leone. Many early nineteenth-century introductions came from the West Indies, as a result of the establishment of the settlements of freed slaves. Several of these introductions were New World food crops such as maize, tomatoes, peppers, and sweet potatoes.

Several flourishing market gardens were established in West Africa in the nineteenth century, growing mainly vegetables of temperate origin to supply ships calling at west coast ports.

The cultivation methods employed were probably based on those currently used in Europe, modified to meet the demands of the local environmental conditions. Cultivation on raised beds became an established practice early in the development of vegetable cultivation in the more humid areas of West Africa.

7.1.2. Indigenous vegetables

The only vegetables which can be considered to be truly indigenous to West Africa are water melon, bambara groundnut, various kinds of spinach, the fluted pumpkin or oyster nut, water leaf, and some yams. Over a long period of cultivation and adaptation, naturally occurring botanical varieties and cultivars of these vegetables have become widely established in West Africa, although their range of genetic variability is still largely unexplored.

Local vegetables provide an essential component of the daily food intake in most rural and urban areas in West Africa and contribute to the nutritional balance of the diet. Their production is limited to small-scale operations, either on a domestic level in kitchen or home gardens or as minor crops in indigenous mixed-cropping farming systems. Research on the production aspects of local vegetable crops in West Africa has been on a limited scale, although some species have been used in plant physiological research and other specialized studies.

7.1.3. Diversification

In recent years the need to diversify crop production has encouraged research into the problems related to new crops and cropping patterns in areas where economic crop production was previously limited or non-existent. However, there has been relatively little investigation into the possible inclusion of vegetable crops, whether exotic or of indigenous origin, in these programmes.

There is, however, considerable evidence that this situation may now be changing in some parts of West Africa. The main factors which are likely to promote an increase in vegetable production to supply urban areas, for example, are the expansion in the consumer population and, with rising incomes, a simultaneous demand for improved quality and a greater choice in vegetable produce. The development of more efficient means of communication and a higher educational standard also contribute to a changing approach to food consumption. The growth of local tourist industries has added, to a limited extent, to this demand for a higher standard of commodity and a more varied supply of produce.

Finally, the development of a fresh-vegetable export industry, notably in Senegal, has demonstrated what it is technically possible to achieve. Rural areas are relatively unaffected by the social and economic changes and pressures to which urban communities are subjected, and it appears that the attitude to vegetable consumption of rural communities is unlikely to change significantly unless radical alteration in the rural style of living results from the introduction of some form of industrial development or commercial agricultural activity.

There is insufficient data for any realistic quantitative assessment of vegetable production in West Africa to be made, although some countries can provide omnibus statistics for total exports of vegetables and fruits where an export industry has become established. In addition, the data concerning yields, the amount and types of fertilizer in use, and costs of production of specific crops are generally inadequate.

7.1.4. *Categories of production*

There are four main categories of vegetable producer in West Africa. The first is the domestic producer, living in a predominantly rural area, who grows only enough for his family needs. Most of this production is by women and the crops grown are generally limited to those which are well adapted to local climatic and soil conditions and are easily propagated. For these producers, the types of vegetable grown are often influenced by tribal or religious customs. Surplus produce may sometimes be exchanged or bartered with neighbours but is rarely sold for cash.

The second main category comprises the part-time grower who grows more than is required for the family needs and who expects to be able to sell most of his produce either to a middleman or through a regular market trader. Such growers usually live on the fringes of the urban areas, where transport is not a major constraint. In this category there is practically no attention to grading, packing, or presentation of produce.

The third type of producer is the small-scale grower for whom commercial farming, including vegetable production, is the only productive occupation. He frequently keeps pigs or poultry and may also have fruit trees on his holding. He grows only crops which are assured of a ready sale in the local market and usually sells his produce to a middleman or wholesaler with whom he has a contract. Contracts sometimes involve the collection by the entrepreneur of the produce from the holding. Some attempt is usually made by this class of grower to grade and pack his produce into containers which will preserve it during the journey to market but his storage and packing facilities are usually extremely rudimentary.

The fourth class of producer is becoming increasingly important, particularly in the francophone areas

of West Africa. This is the agri-business type of enterprise which is often linked with some form of horticultural co-operative. The cost of the production of vegetables in Europe during the winter period has escalated in recent years. This is mainly due to rises in production costs in the glasshouse industry, particularly those associated with heating fuels. There is, therefore, an increasing interest in many parts of Africa in the possibility of producing the temperate and subtropical types of vegetable which are in demand in Europe. This type of operation is now well established in some areas of both East and West Africa.

7.2. The nutritional value of vegetables

7.2.1. *Agriculture and nutrition*

A great deal of emphasis has been given by nutritionists, but less by agricultural development planners, to the value of vegetables in maintaining adequate health standards. In West Africa there is a continuing need for research and investigation into the interaction between agriculture and nutrition, because alterations to traditional ways of living have introduced new problems. The development of urban areas has, for example, introduced an increasing dependence on food from rural areas. Haswell (1971) referred to the fact that, in many tropical countries, changes in income level can be correlated with changes in patterns of consumption from a diet composed almost exclusively of cereals to a more varied diet, with significant increases in the consumption of vegetables.

7.2.2. *Basis of assessment of nutritional value*

Munger (1971) referred to the fact that the percentage of protein in vegetables, which is usually expressed on a wet-weight basis, is not very impressive when compared with that of dry cereals, which is normally expressed on a dry-weight basis. The fact that cereals are seldom eaten dry is often overlooked. The protein content of many fresh vegetables, on a comparable basis, is favourable with that of cooked cereals, which may lose up to 75 per cent of their protein content during cooking (see Table 7.1). Munger also suggested that the choice of crops for protein production should not be based solely upon the percentage of protein they contain and that it is also important to consider the yield of protein per unit time. This is not always correlated with the percentage of protein produced in any one part of the crop.

TABLE 7.1

Some vegetables with higher protein percentages than cooked rice (2·2 per cent)

Food	Percentage crude protein (wet-weight basis)
Soya bean, boiled (*Glycine max*)	12·8
Mung bean, boiled (*Vigna aureus*)	11·0
Pigeon pea, boiled (*Cajanus cajan*)	9·4
Chickpea, boiled (*Cicer arietinum*)	9·3
Papaya leaves (*Carica papaya*)	8·0
Bago leaves (*Gnetum gnemon*)	7·4
Lima beans, boiled (*Phaseolus lunatus*)	7·4
Cassava leaves, boiled (*Manihot esculenta*)	7·2
Beans, boiled (*Phaseolus vulgaris*)	7·2
Garlic (*Allium sativum*)	7·0
Malunggay leaves, boiled (horseradish tree) (*Moringa oleifera*)	6·5
Cashew nut, young leaves (*Anacardium occidentale*)	5·2
Ampalaya leaves (*Momordica charantia*)	5·0
Patola (sponge gourd) tops (*Luffa cylindrica*)	4·6
Kulitis, boiled (spineless amaranth) (*Amaranthus gracilis*)	4·4
Broccoli (*Brassica oleracea* var. *italica*)	3·5
Kang kong, boiled (water convolvulus) (*Ipomoea aquatica*)	3·4
Gabi leaves, boiled (taro) (*Colocasia esculenta*)	3·3
New Zealand spinach, boiled (*Tetragonia expansa*)	3·3
Cowpeas, edible pods, boiled (*Vigna unguiculata*)	3·0
Sitao, edible pods, boiled (*Vigna unguiculata* var. *sesquipedalis*)	2·9
Winged beans, edible pods, boiled (*Psophocarpus tetragonolobus*)	2·9
Sugar pea, edible pods, boiled (*Pisum sativum*)	2·7
Sweet potato, boiled (*Ipomoea batatas*)	2·6
Gabi (taro) (*Colocasia esculenta*)	2·4

Except for broccoli, values are extracted from *Food composition Tables*, Handbook I (4th revision), Food and Nutrition Research Center, Manila (1968). A number of the values for boiled vegetables are for a single analysis but are not greatly different from values for the fresh vegetables, which are based on more analyses.

An additional point made by Munger was that more information is needed on the protein quality of many vegetables, since crude protein determinations, based on the total nitrogen obtained from Kjeldahl estimations multiplied by 6·25, leave much to be desired. Amino-acid analyses or biological assay methods would give more relevant information. For example, Phansalker (1960) showed that the addition of red amaranth to cereal or pulse protein produced a mixture which was almost comparable with skimmed milk in protein efficiency ratio. The most important yield data for tropical vegetables could, therefore, be the amounts of sulphur-containing amino acids produced per hectare per day. Investigations based on this factor could well be rewarding and should include both exotic and indigenous kinds of vegetables. Munger also suggested that more data were required on the protein yield of leaves of cocoyam and cassava and the effect that harvesting the leaves had on the yield of the roots. Terra (1967) stated that the protein content of the old leaves of cassava is about 4 per cent while that of the young leaves is 6–10 per cent. He also emphasized the fact that young leaves of cassava were rich in methionine and that the tubers contained only about 0·8–1·0 per cent of protein.

Platt (1965) reported that dark green leaves of some types of vegetable were high in carotene and contained about 5·0 per cent protein (wet-weight basis); examples cited were *Corchorus*, *Basella*, *Ipomoea*, and *Amaranthus* species. Light-green leaves contain approximately 2 per cent of protein. The protein content of vegetables was also referred to by Samson (1971), who mentioned that lysine and methionine were two of the essential amino acids which were likely to be present in limited quantities in vegetables. He quoted some examples of *per capita* daily consumption levels of vegetables, such as 400–500 g per day in southern Europe, 200 g per day in most developed countries, but less than 50 g per day in most tropical countries. Samson also referred to available statistics for the number of grams of protein which could be derived from vegetables, linked with the consumption of vegetable protein per day for various countries (Food and Agriculture Organization 1969*b*).

Discussing the nutritional values of vegetables, Oomen (1971) made a strong plea for the use of leafy vegetables, particularly in the diets of young children. Leaves contained the most actively growing tissues, and those which were dark green, with a high content of chlorophyll, were probably the most nutritious. He cited, as examples, sweet potato, pumpkin, and water leaf. These vegetables contain many valuable supplements to the diet such as protein, calcium, iron, carotene, and ascorbic acid. They also contain more thiamin, riboflavin, and niacin than many cereal food crops. Oomen also drew attention to the low calorific value and high water content of most leaf crops and the relatively large volumes which would need to be consumed if they were the main source of nutrition. An adult would require some 4–5 kg per day, which would be excessively high since it is seldom that the daily rate of consumption exceeds 500 g per day. The consumption of leaf vegetables in powder form, or as protein extract, could possibly prove acceptable in the future.

Xanthosoma brasiliense Engl., is cultivated for its leaves specifically as a salad crop in some parts of South America (Coursey, D. G., personal communication). It is so far unknown in West Africa and its introduction could well be considered. According to Busson (1965), the protein in the leaves of edible aroids is less deficient in sulphur-containing amino acids than the protein of the storage-organs.

7.2.3. *Consumption of leafy vegetables*
In European countries, the consumption of vegetables may be as high as 60–70 kg per year per person. Consumption of vegetables in tropical areas is often much lower and intermittent, owing to seasonal variations, but vegetables are seldom proscribed by traditional taboos. In Dahomey, the consumption of leaf vegetables was estimated to be as low as 6 kg per year per person while that of tomatoes, okra, and chillies was 11 kg per annum.

Terra (1967) reported that some 25 investigations from Africa showed that the average *per capita* consumption of leafy vegetables varied from 0 g per day to 47 g per day whereas the *per capita* consumption of other vegetables varied from 6 g per day to 83 g per day. Many tropical vegetables are much richer in proteins and vitamins than the temperate ones. An extensive list giving the composition of tropical vegetables has been compiled by Wu Leung, Busson and Jardin (1969).

7.2.4. *Poisonous substances in leafy vegetables*
Reference was made by Samson (1971) to the need for more research on the presence of substances in leaf vegetables which could be harmful to the consumer. He cited the well-known example of the occurrence of cyanogenic glucosides in cassava and also the presence of enzymes in all parts of the plant which liberate hydrocyanic acid. After crushing, the enzymes present in the leaves separate gaseous hydrogen cyanide from the sugar moiety of the glucoside, and the gas is easily removed by a few minutes of boiling (de Bruijn 1971). The presence of similar cyanogenic glucosides in other leafy vegetables has been reported by Oke (1966).

The occurrence of poisonous substances in the family Solanaceae was also referred to by Samson, who cited examples of botanical varieties within single species which were reported to have either poisonous leaves and edible fruits or edible leaves and poisonous fruits, the occurrence of either possibly being dependent on environmental conditions. Further investigation of these reports appears to be necessary; it may be possible for plant breeders to produce new cultivars which have both edible leaves and fruits.

Oke (1966) has also reported that high levels of oxalates have been found in the leaves of *Talinum*, *Celosia*, *Corchorus*, and *Amaranthus* species which are eaten. The amounts present appeared to vary with the cultural methods employed. Selection and breeding techniques were suggested as possible means of reducing the levels of oxalates in these leaf vegetables. Sadik (1971) has given an outline of recent work concerned with the content of oxalate found in some leafy vegetables. Andrews and Viser (1951) found the oxalate content of 13 out of 45 fruits and vegetables to be more than 0·1 per cent on a fresh-weight basis. Munro and Bassir (1969), working in Nigeria, analysed 30 fruits and vegetables for their calcium and oxalate content. The latter, on a dry-weight basis, ranged from traces to 19·7 per cent. The authors considered 3 g of oxalic acid to be a toxic dose and concluded that, of all the vegetables they analysed, only spinach could become toxic. An intake of 100 g fresh weight of spinach could give a toxic dose of oxalate, assuming that there was no interaction between calcium and oxalic acid. The presence of soluble and insoluble oxalate in some Nigerian vegetables is presented in Table 7.2.

From Table 7.2 it is noted that a large percentage of the total oxalate in some vegetables is in a soluble form which could induce oxaluria if taken in excess. From recent work in Nigeria, it has been found that only negligible amounts of soluble oxalates are removed by washing and rinsing. It appears necessary, therefore, to pursue further investigations into the residues of oxalates left in leafy vegetables following upon various methods of preparation and on the nutritional consequences of a high daily but non-toxic level of oxalate intake from leafy vegetables.

7.3. Home gardens
Detailed accounts of the principles and practical measures applied in a scheme involving the establishment of home gardens have been given by van

TABLE 7.2

The oxalate content (dry weight) of some green vegetables from the Ibadan-Ilesha area which have been recently analysed at IITA (from Sadik 1971)

	Percentage oxalate		
	Total	Soluble	Insoluble
Amaranthus hybridus	14·33	5·60	8·73
Talinum triangulare	14·21	13·37	0·84
Basella alba	12·94	4·52	8·42
Celosia argentea	11·92	7·55	4·37
Basella rubra	11·14	6·59	4·55
Telfairia pedata	7·99	5·08	2·91
Solanum sp.	7·73	3·74	3·99
Colocasia antiquorum	7·37	5·32	2·05
Manihot esculenta	5·29	2·66	2·63
Crassocephallum crepidioides	4·89	2·69	2·20
Solanum nigrum	3·78	2·41	1·37
Struchium sparganophora	3·68	1·21	2·47
Vernonia amygdalina	2·82	2·26	0·56

Eijnatten (1969). In a general paper on some aspects of this work van Eijnatten (1971) discussed the indiscriminate planting of trees, shrubs, and herbaceous plants in a typical West African home garden and drew a comparison with the vegetation existing in a well-established rain forest; both afford a desirable protection to the soil from the excessive radiation and from heavy rainfall. The various stages in development of home-garden projects in western Nigeria (Ilesha) and South-Eastern Dahomey (Porto Novo) were also outlined, and active collaboration between horticultural, nutritional, and home economics workers was stressed as essential for the success of any scheme of this nature.

Munger (1971) also emphasized the need for research into the specific problems of the home producer who is responsible for most vegetable production in West Africa. Current research efforts were aimed mainly at the large-scale commercial producers of alien vegetables, whose problems are more likely to reach the attention of research workers and who are better organized to make use of the findings. The needs of the domestic rural producers should not be ignored because these people contribute so much to the maintenance of nutritional standards in large areas of developing countries.

7.4. Marketing and transport

7.4.1. *Economic aspects of production*

In discussing the economic aspects of vegetable production, Haswell (1971) has cited the situation in Kano, Nigeria as an example of vegetable growing in the urban fringe of a large city; the vegetable production here is badly organized, despite good road communication with urban markets. Vegetable production is generally a subsidiary occupation, undertaken mostly by women and children while most able-bodied men find work in the industrial labour force.

7.4.2. *Transport*

An economic survey of the movement of locally produced foodstuffs in Northern Nigeria was prepared by the Ministry of Agriculture, Kaduna, in 1951. Information on the transport of vegetables from production areas to collecting points on main highways in the Philippines, which had been discussed in some detail (Haswell 1969), was compared with the less developed situation in Nigeria (Haswell 1971).

7.4.3. *Marketing*

In many parts of Africa direct selling in the local market is practised by the grower or members of his family. The mobility of small amounts of produce is often uneconomic, if the time taken in transporting the produce to market is taken into account, but the social and recreational aspects of marketing are an important cultural factor.

Alternative marketing systems, in which urban wholesalers organize deliveries by primary buyers from producing areas, selling part of the produce direct to consumers and part on the retail market, or a system in which trading is in the hands of a co-operative or state trading organization, were discussed by Haswell (1971).

Studies of the marketing systems in many tropical areas are handicapped by the lack of reliable data on the real costs of production in rural areas. It is thus difficult to assess whether any revision of marketing systems would be likely to stimulate increased production. What is clear is that, as many urban areas become more densely populated, demand is progressively outstripping supply.

Ajeabu (1970) has outlined the transformation of a peasant system of extensive agriculture to an intensive system, in response to market influence in the environs of Lagos. This has involved a shift from traditional forms of communal labour to hired labour and a degree of specialization in the production of fruit and vegetables.

7.5. Vegetable production for export

A profitable and extensive vegetable export industry has developed in recent years in some francophone countries of West Africa, whereas in anglophone areas this trend has been relatively insignificant. Samba (1971) has outlined the situation in Senegal up to 1970. He estimated that the area under market-garden crops is about 6000 ha, of which 2800 ha are in the Cap-Vert area. The average estimated yield is 10 t/ha per annum and the value is 3·6 million CFA francs. The production for export arises from a season-by-season programme agreed between producers, exporters, and technical staff from the Direction des Services Agricoles.

The following crops have become the most widely grown: French beans, aubergines, melons, strawberries, sweet peppers, cucumber, and courgettes (zuccini squash).

Samba (1971) has drawn attention to a detailed work calendar, including dates of seed distribution, dates of sowing or planting in the nursery, and dates of harvest which has been proposed for Senegal. Seed is supplied by the exporter on interest-free credit, and cash payment is made by the producer when the crop is delivered for export. Crop protection is supervised by the technical agent of each production centre, assistance being given with materials and equipment by the Direction des Services Agricoles. At harvesting time, each co-operative or producer group appoints an agent to transport the vegetables, supervise weighing and grading, and collect and distribute the cash to individual producers.

Grading and packing, which are mechanized, are the responsibility of the exporter. Rejected produce is weighed and subtracted from the total weight. Payment is on the net weight. The airline company receives several days' notice of the space required for cargo, and this is confirmed at least 24 hours before the flight departs from Dakar. In 1966 legislation passed in Senegal specified standards for both produce and packaging, and rigorous application of these standards has ensured continuing high quality.

The production of vegetables for export by air to Europe in winter has enabled Senegal also to export to other African countries which were previously supplied by European producers. An increasing quantity of vegetables is now sent to Europe by sea at times of the year when air export ceases to be cost-competitive. Total vegetable exports during the period 1967-70 were as shown in Table 7.3.

Senegal is now one of the major exporters of French beans to France. The quantity of French beans exported was higher than that of any other vegetable, but potatoes, tomatoes, aubergines, salad crops, cabbage, peppers, melons, and onions were also exported in appreciable amounts during the period 1967-70. Quality-control specifications are being applied to an increasing number of other vegetable export crops and further market openings for produce are being continually investigated.

TABLE 7.3

Vegetable exports from Senegal, West Africa 1967-70 (t per annum) (from Nabos 1971)

	By air freight	By sea
1967-8	1003	163
1968-9	940	242
1969-70	1009	422

Two main constraints which influence further development of this programme are the availability of economic air-freight facilities for increasing quantities of produce, and investment in improved irrigation facilities in some of the production areas. The severe drought in the Sahel countries of West Africa in the early 1970s may have had a great effect on the proposed expansion of this programme. The escalating cost of air freight, due to fuel price increases, is also an important factor likely to influence future developments in the export of vegetables to Europe.

7.6. Seed storage

An investigation of the germination capacity of vegetable seeds in tropical climates and the significance of moisture content has been described by Soitout (1971a). Experiments which are currently being undertaken are centred on three kinds of vegetable: lettuce, onion, and tomato, one cultivar of each vegetable being selected for testing.

Two types of packaging were used: metal boxes such as are used for canning food and aluminized bags which have been heat-sealed. Preliminary results indicate that the seeds in the metal containers retained their viability for longer periods than the seeds in the aluminized bags; the differences in viability widened considerably with time. For periods up to 12 months, however, there was little difference in the germination percentage of all seeds, although the lettuce seed declined in viability after about 15 months in bag storage; onion seed also showed a lowered viability after being stored for 6–9 months in bags. The seeds of all three vegetables stored in metal boxes retained a high viability for 21 months; the lettuce seed showed the most rapid fall in germination after this period.

7.7. Research programmes related to tropical vegetable-crop production

7.7.1. *Crop physiology*

Some techniques of research which are in common use in temperate countries have been successfully applied to many tropical crops grown in Africa, but Gietema-Groenendijk and Flach (1971), in their review of the current situation, point out the need for basic research on many aspects of growth and development.

It has been found that appreciable differences exist between the net assimilation rate of many tropical and temperate plants. The photosynthetic pathways of the more efficient tropical plants, including *Amaranthus* and *Portulaca*, differ from those of temperate plants; there are also many other physiological and anatomical differences which have not yet been fully explored (Downton 1973). The growth rate of some tropical plants, for example, can be 2–3 times greater than that of some of the less efficient temperate plants.

The distribution of dry matter in tropical vegetables is another topic that has not been explored to any appreciable extent (Samson 1971). Growth-analysis studies on plants such as *Amaranthus* and *Basella*, along the lines described by de Vries, Ferwerda, and Flach (1967), could supply data which are not currently available. The improvement of the ratio of leaf to stem in leaf vegetables such as spinach could also be investigated, as it has already in crops such as rice and maize.

In referring to current work being undertaken at Wageningen, Samson (1971) also mentions that plants of Indian spinach and yard-long bean were being treated with CCC growth-retardant. Both plants have a climbing habit and it would be advantageous if their total length could be reduced by this chemical treatment without affecting the total yield of either crop.

Sinnadurai (1970) reviews the literature on the response of onion cultivars to day-length and temperature in relation to onion cultivation in West Africa. Long days and high temperatures favour production of bulbs, while short days and/or low temperatures stimulate flower initiation and inhibit bulb growth for many cultivars. Suitable cultivars for northern Ghana include some long-day types, although the relatively cool night temperatures in this area stimulate flower initiation in some cultivars.

7.7.2. *Plant breeding*

The selection and breeding of new cultivars of vegetables in West Africa has been very limited, although current work is in progress at IITA, Ibadan and other research centres. This neglect is largely due to the direction of limited resources towards the improvement of perennial crops, cereals, and other field crops (Ferwerda and Wit 1969); reference to this situation has also been made by Oomen (1964). One of the main continuing constraints is that the very limited number of workers on vegetable breeding are in research and experimental stations which often lack adequate facilities and resources.

7.7.3. *Plant protection*

Diseases of tropical vegetables. A review of the more important diseases of vegetable crops in the French West Indies has been given by Messiaen, Fournet, Beyries, and Quiot (1971). Many of these diseases occur widely in hot tropical areas, including Africa.

Pathogens which are not specific to any particular family of plants include *Pythium aphanidermatum* and *Sclerotium rolfsii*, which attack many seedlings. Most fungicides had been found to be relatively ineffective against *Pythium*, but systemic fungicides such as chloroneb were being used with success. Many cultivars of cowpeas and French beans appear to become resistant to *Sclerotium* after development of the first two leaves, but others, particularly of highly bred French beans, remain susceptible until flowering.

Sclerotium rolfsii can be a serious pathogen on older plants of various vegetables, including yams. This fungus is characterized by the white mycelium which grows in the soil and on the diseased parts of plants and by the common symptom of wilting. Control has been effected with pentachloronitrobenzene (PCNB) and systemic fungicides. Experimental work in Israel indicates that normal amounts of ammonium nitrogen in the soil inhibit *Sclerotium*. Leached tropical soils and the use of nitrate fertilizers as a nitrogen source may therefore favour the development of the fungus.

F

Investigations of fungicides for the control of *Sclerotium* wilt in Ghana have been carried out by Addison and Chona (1971). Several fungicides were phytotoxic at concentrations sufficient for controlling the pathogen. It was found that formaldehyde at 10 ml in 500–800 ml of water completely controlled the disease. Mercuric chloride at 1 g in 500 ml water was also effective, but hardly to be recommended for practical use on account of toxicity of residues.

Rhizoctonia solani can be a serious pathogen in seasons with more than 200 mm of rainfall, and crops of the Chenopodiaceae and Leguminosae appear particularly susceptible. No fungicide has yet come into general use against *Rhizoctonia* although difolatan may be partially effective. Lima bean and the Egyptian, or dolichos, bean appear to be more resistant than cowpeas (including yard-long beans), which are in turn more resistant than French beans.

1. *Pathogens of the Solanaceae.* At low elevations tomato late blight (*Phytophthora infestans*) rarely develops, owing to the high temperatures in these areas, but serious fungal pathogens of tomatoes include *Alternaria solani*, *Phoma destructiva*, and *Corynespora cassiicola*. Leaf mould (*Cladosporium fulvum*) also occurs in sheltered areas, as does *Stemphylium solani*. Cultivars of high-temperature-adapted tomato, which are resistant to *Cladosporium* and *Stemphylium*, and which have been tested in the French West Indies, include Indian River, Floralou, and Floradel; these are also resistant to at least one race of the Fusarium wilt pathogen *Fusarium oxysporum*. In Senegal, sprinkler irrigation applied to tomatoes in the middle of the day has been reported to reduce the dissemination by wind of spores of *Alternaria*.

Crops of egg plant and sweet pepper grown in the French West Indies are liable to attack by anthracnose caused by *Colletotrichum* spp., which severely damages the fruits. Eradication of wild species of *Solanum* in growing areas is recommended, also spraying with one of the dithiocarbamate fungicides. Benomyl, however, has been found to be ineffective. Several cultivars of egg plant are resistant to *Colletotrichum* and a programme has been initiated to produce new cultivars of F_1 hybrids of egg plant resistant to anthracnose and also to bacterial wilt.

2. *Bacterial wilt.* Bacterial wilt diseases caused by strains of *Pseudomonas solanacearum* are among the most serious diseases of many tropical crops, including the Solanaceae.

In outlining the studies undertaken in the French West Indies on the breeding of tomatoes for resistance to bacterial wilt, Kaan, Pecant, and Beramis (1971) also referred to the work done with egg plant and sweet pepper. A discussion of the reaction of bacterial wilt to environmental conditions included reference to the fact that this disease does not normally develop if the soil temperature is less than 21 °C and that irrigation during dry periods appears to encourage the incidence of wilt (Digat 1967). Crops grown on calcareous soils do not appear to be affected to any appreciable extent by the disease, and even inoculation does not result in a rapid spread (Messiaen 1971; unpublished data).

Strains of tomato which are normally tolerant of bacterial wilt become more sensitive in the presence of nematodes (Ben Halim 1967; Temiz 1968). Very young plants are tolerant of the disease (Winstead and Kelman 1952), and assessment of the degree of resistance of any cultivar can often only be made relatively late in the growing period. Introduction and testing of resistant cultivars from IRAT (West Indies), North Carolina, and Hawaii (USA) showed that selections CRA66 and OTB2 from North Carolina and 199.39.15, L91, and L11 from IRAT were most tolerant under the conditions prevailing in the French West Indies.

Trials with strains of species such as *Lycopersicon chilense*, *L. hirsutum*, *L. peruvianum*, and *L. pimpinellifolium* showed varying tolerance levels. CRA66 has been selected as the most useful parent for further work, although it has small fruits of about 50 g, a thin pericarp, and bitter green pulp. It has, however, some resistance to several fungi such as *Fusarium oxysporum*, *Stemphylium solani*, and *Cladosporium fulvum*. It has also vigorous growth characteristics and good fruit set.

The existence or possible development of different physiological races of bacterial wilt was discussed by Kaan *et al.* (1971), who also outlined the difficulties involved in identifying differences in virulence due to the complex interaction between host, pathogen, and environmental conditions. In field experiments, root inoculation was adopted in preference to stem inoculation, owing to the sensitivity of some tolerant tomato strains to this treatment (Adsuar 1961). The plants were treated when 6 weeks old, the roots being slightly wounded before 5 ml of a bacterial suspension was applied to each plant.

Research on the inheritance of resistance to bacterial wilt is still not very advanced (Digat and Derieux 1968), and further work is in progress.

Possible links between tolerance to bacterial wilt and sensitivity to nematodes, type of growth, low productivity, and fruit characteristics are being investigated. Resistance appears to be polygenic, which contributes to the difficulties involved in these investigations.

It appears that many strains of tomato exhibit varying degrees of tolerance in different growing areas; a tolerant cultivar could therefore have a varied range of adaptation. Results obtained in Hawaii (Mohanakumaran, Gilbert, and Buddenhagen 1968) suggest that the tomatin content of tomato roots may be directly connected with tolerance to wilt.

Kaan *et al.* (1971) report that many commercial cultivars of egg plant appear to be less sensitive to wilt than most tomato cultivars, infected plants often surviving until the first fruits are harvested.

Messiaen *et al.* (1971) have reported that grafting experiments using root stocks resistant to bacterial wilt have been successful with tomatoes; local selections of resistant cultivars have been used, together with *Solanum integrifolium* of Japanese origin. Resistant root stocks, such as *Solanum torvum*, have also given successful results when grafted with egg plant, and plants of this combination have survived for more than a year. When egg plants are grafted on to *Solanum integrifolium*, however, the plants survive for only 4–5 months. It appears possible that the superiority of *Solanum torvum* over *S. integrifolium* as a root stock for egg plant may be due to a higher degree of resistance to root-knot nematodes.

Evidence from Sierra Leone (Godfrey-Sam-Aggrey 1973) suggested that the resistance of tomato seedlings to wilt varied with the age of the seedlings at transplanting and that mechanical damage during this operation predisposes seedlings to the disease.

Nematode disorders of vegetable crops. The problems caused by nematodes in vegetable-crop production in West Africa have been referred to by Netscher (1971) and Luc and de Guiran (1960). Most of the research work described was concerned with root-knot nematodes and Netscher (1970) has indicated the severity of the problem in Senegal. Tomato yield could be reduced from 30–40 t/ha to 16 t/ha if *Meloidogyne* was present in the soil, and lettuce yields could similarly be reduced to 30 per cent of the number of plants normally harvested from uninfested soils. Seinhorst (1965) has reviewed crop losses associated with different nematode populations.

Control measures were discussed in some detail by Netscher (1971). Most of the current work is directed towards control through soil fumigation and crop rotation, but local cultural methods can sometimes be effective. The use of the three most commonly used nematicides, DD (dichloropropane-dichloropropylene), EDB (ethylene dibromide), and DBCP (dibromochloropropene), was reviewed, and the problems of their high cost and requirement of expertise for effective application discussed. DD was regarded as being overall the most suitable nematicide for use in West Africa since, although it is more phytotoxic than DBCP, it has a shorter residual effect. Brief reference was also made to Vapam (sodium *N*-methyl dithiocarbamate) and methyl bromide. These also function as bactericides, insecticides, fungicides, and weedicides and are useful for seedbed-treatments, although they are extremely expensive. The main obstacle to the use of methyl bromide is its high mammalian toxicity. Although some dithiocarbamates and phosphorylated esters are effective systemic nematicides, information on effects as residues in vegetables is required before they can be safely recommended for use in West Africa.

In discussing crop rotation as a means of control, Netscher (1971) referred to the use of susceptible trap crops during a fallow period as a means of reducing the *Meloidogyne* population. Several vegetables which are only slightly susceptible might also be used in this way (Town 1962).

Resistant cultivars of normally susceptible plants have already been produced (Kehr 1966; Fassuliotis 1968). The main problem which has been encountered, however, has been the considerable morphological and physiological variation which exists among species and strains of *Meloidogyne*, with the result that crops which are resistant in some areas prove to be susceptible in others. Work on such variation in the nematology laboratory of ORSTOM in Dakar has confirmed that *Meloidogyne incognita*, *M. javanica*, and *M. arenaria* are not only widespread in West Africa but have overlapping host ranges.

Crops particularly susceptible to root-knot nematodes include tomato, egg plant, beans, carrot, cucumber, and melon. Those which are slightly susceptible are sweet and chilli pepper, cabbage, and leek. Resistant crops include strawberry, onion, and groundnut. Netscher has emphasized the need to control weeds and other potential host plants of

Meloidogyne in areas where vegetables are being intensively cultivated and the importance of ensuring that infested seedlings are not transplanted to soils which are free from *Meloidogyne*.

The addition of organic material to infested soils appears to decrease losses due to *Meloidogyne*, although it is difficult to assess whether the increases in yield are due to improved plant vigour or a reduction in the parasitic nematode population. The possibility of the occurrence of polyphagous species of nematode which will also attack West African vegetables should be considered; the presence of *Trichodorus minor* in the vicinity of the roots of celery, leek, cauliflower, egg plant, tomato, potato, and water melon has already been reported from Senegal (Netscher 1970). This species is related to *T. christiei*, which is known to affect several crops grown in the USA.

Godfrey-Sam-Aggrey (1973) reported bacterial wilt (*Pseudomonas solanacearum*) and nematodes (*Meloidogyne* sp.) as the main factors limiting tomato production in Sierra Leone. He reviewed previous work there on control measures and detailed current experimental work with nematicides.

7.7.4. *Cultural practices*

General comments. Experimental work carried out at either a local or regional level and aimed at investigating means of improving traditional cultivation practices is still required in many areas of West Africa, in order to determine appropriate techniques for vegetable-crop production under different circumstances. Suggested improvements are normally published in the annual reports and extension leaflets of the relevant Ministries of various countries of West Africa, but data to support the advice are generally sparse or absent.

The traditional practice of mixed cropping has been investigated by Ufer (1971) who recommended that dolichos bean be sown between rows of a cereal crop. When the ears of the cereal have been removed, the stalks are left as supports for the climbing bean.

7.7.5. *IRAT vegetable research programmes*

IRAT vegetable programmes in francophone countries have been regularly reviewed by Soitout (1967, 1969, 1971a). Soitout (1969) included an outline of the ecological situation of the various research stations and referred to the vegetable research centres in Gabon, French Guinea, Upper Volta, Niger, Senegal, Reunion, Cameroun, Madagascar, and the

Central African Republic. Particular emphasis has been given in the programmes at these centres to tomato, lettuce, French beans, potatoes, cabbage, cauliflower, cucumber, melon, and asparagus, and Soitout (1971b) has provided a resumé of current investigations. Information from the IRAT programme will be found under each crop in the appropriate sections (pp. 111 *et seq.*).

7.7.6. *Vegetable-crop yields*

Yields of vegetables grown in the lowland humid tropics vary widely, but the factors responsible for this variation are numerous and difficult to define. The genetic constitution of cultivars of the same species which may have been given identical names may be responsible for observed variations in performance under similar conditions of soil and climate. The two main factors of soil and climate account, to a great extent, for variation in crop yield from season to season, even in ecologically similar areas of production. Variation in altitude, and temperature changes, particularly diurnal fluctuations, may also account for widely differing responses from vegetables which have the same genetic constitution and which are given similar cultural treatments. Day-length and light intensity are also factors which influence the growth and development of short-duration vegetable crops, particularly those which are derived from ancestral forms which originally evolved in temperate or sub-tropical regions.

Potential yields based on calculations are often highly misleading when applied to tropical conditions. This has been demonstrated by Vittum (1966), although high actual yields, which compare favourably with calculated yields, may be obtained under closely controlled experimental conditions (Table 7.4).

Samson (1971) reported yields at Wageningen from *Amaranthus* equivalent to 40 t/ha per annum of fresh vegetable matter. These plants were grown

TABLE 7.4

Vegetable crop yields (t fresh weight per ha) in tropical areas (from Vittum 1966)

	Average	Experimental	Potential
Pole beans	14·5	50·5	54·5
Lima beans	2·5	8·8	11·0
Cabbage	42·5	100·0	127·0
Spinach	23·3	34·3	54·5
Tomato	52·5	134·8	163·3

under greenhouse conditions, and plants of tanier which were given similar conditions produced leaf yields of only 30 t/ha per annum. Terra (1966) reported yields of 20 t/ha per annum for cassava leaves from plants established at high population density representing a possible leaf protein production of 2000 kg/ha per annum, based on a young-leaf protein content of 6–10 per cent.

Many vegetables, such as tomatoes, onions, and cucumbers, which take several months to mature, commonly produce lower yields when they are grown at low elevations in tropical climates than at higher elevations. The yields are also generally much lower than those obtained from the same cultivars grown in temperate countries. Associated with this are condensed life-cycles. To make valid comparisons between yields of vegetable crops grown in temperate and tropical areas, it is thus necessary to consider yield per unit time. In tropical countries it is possible to exceed temperate production by growing more crops per annum. This approach has been suggested as a basis for the comparison of the yields of tropical vegetable crops and cereal or root crops grown in tropical regions (Munger 1971).

Van Eijnatten (1971) stated that, in the drier parts of West Africa, where soils are often poor, the yield from a home garden may reach only $\frac{1}{2}$–1 t/ha per annum. This low yield was compared with market gardens in the same area which produced yields of 20–30 t/ha per annum. Intensive market gardens, on relatively fertile soils and with an adequate water supply, are capable of yielding up to 100 t/ha per annum.

An investigation of the yield and composition of eight tropical leaf-producing vegetables has recently been made by Schmidt (1971). Soils with two distinct levels of fertility were selected for the experiments. The vegetables compared were spinach (*Amaranthus cruentus*), Indian spinach, Chinese cabbage, *Brassica oleracea* var. *acephala*, *Corchorus olitorius*, kang kong, *Solanum melanocerasum* and water leaf (Appendix, p. 125). The average dry-matter yields of edible leaves of all the crops from soils of medium and high fertility were 1970 kg/ha and 3330 kg/ha respectively, and 2839 kg/ha and 5133 kg/ha in 2 consecutive years (after 70 days' and 120 days' growth, respectively). *Amaranthus cruentus* was the highest-yielding species, producing 3473 kg/ha per annum compared with an average of 2383 kg/ha per annum for Chinese cabbage, *Corchorus olitorius*, water leaf, and *Solanum melanocerasum*. *Amaranthus cruentus*, Indian spinach,

and *Solanum melanocerasum* responded most to high fertility.

The highest leaf total-nitrogen contents were found in *Solanum melanocerasum* and *Ipomoea aquatica*. Leaf calcium contents ranged between 2045 mg per 100 g and 412 mg per 100 g dry matter for *Amaranthus cruentus* and *Ipomoea aquatica* respectively. The highest leaf iron content was found in *Basella alba* (50 mg per 100 g dry matter), the average for the others being 34 mg per 100 g. Leaf iron contents were $22\frac{1}{2}$ per cent higher in plants grown on soils of medium fertility.

7.8. Research on specific crops

Published material on individual vegetable crops and their production potential in West Africa is not only limited in quantity but there are several important crops for which the data are completely inadequate. Indigenous African crops, such as *Talinum* and *Amaranthus*, are frequently mentioned but research information on their performance under the varying environmental conditions found in West Africa and their response to cultural treatments is not easily available in published form.

The following vegetable crops of temperate and sub-tropical origin were referred to by various contributors (see the following subsections) to the seminar held on vegetable crop research in West Africa in Ibadan in 1971: legumes, particularly French beans; tomatoes; egg plants; onions; lettuce; potatoes; cabbages; and various cucurbits. The information provided on that occasion is now reviewed, together with additional information where it is considered appropriate. Most of the data presented at the seminar do not appear to have been previously published in readily available journals or periodicals. The contributions to the seminar reviewed here can be obtained from IITA.

7.8.1. Legumes

French beans (Phaseolus vulgaris *L.*). A concise and detailed account of experimental work on vegetables, which has been conducted over a period of 10 years at the Farako-Ba station of Upper Volta, was presented by d'Arondel de Hayes (1971).

The value of legumes in rotational cropping is a factor which encourages the production of French beans in the Upper Volta area for either harvesting in the fresh state or as dry beans for export to other African countries or for local consumption. For the production of both fresh and dry beans, dry-season cultivation, with irrigation, is used for dwarf cultivars; climbing or pole cultivars are preferred for wet-season cultivation. The maturation period for dwarf cultivars for fresh pod production with irrigation is 60–70 days, while 80–90 days is required for most cultivars of climbing bean.

Dwarf beans for harvesting as green pods†

1. *Dry-season cultivation.* From trials with 32 cultivars the following have shown most promise: Contender, Tenderlong, Fin de Montclar, and Regalfin, but its crescent-shaped pod eliminated Contender from commercial production. Tenderlong has not become popular with local growers but other cultivars are being increasingly grown in many areas. Green-pod yields are about 7 t/ha.

2. *Wet-season cultivation.* Cultivar trials have shown that seeds sown early in July produce crops which are better than those sown earlier, but later-sown crops give the highest yields. Yields in any one year are found to vary considerably, owing to the intensity of the rainfall, disease incidence, and variation in harvesting conditions. Soil temperatures in May and June are very high, reaching 50–5 °C in the top 5 cm during the middle of the day; this results in poor germination. Rainfall is normally heavy from June to September. Owing to these factors, only the existence of a major market outlet could justify the risks involved in attempting to produce vegetables during the wet season.

In West Cameroun, at an elevation of 1400–1500 m, rust (*Uromyces phaseoli*) is prevalent throughout the year. Anthracnose is also serious in the wet season. Trials with 24 cultivars resulted in the selection of four dwarf cultivars which produced yields varying from 7 t/ha to 10 t/ha.

Climbing beans, for harvesting as green pods.

1. *Wet-season cultivation.* Trials in 1965 involving four cultivars with a variation in sowing date from early April to early October indicated that seeds sown in early July gave good results, approaching 25 t/ha with the cultivar Sossogbe. Sowings made in early August also produce acceptable yields.

The cultivar Sossogbe is found to give the highest production during either wet or dry seasons. It is similar to a traditional European cultivar Blanc de Juillet and has been established in the area for a considerable period. The cost of supplying and erecting

† The use of *Phaseolus vulgaris* for edible seed is considered in Chapter 4.

supports for climbing beans is a significant proportion of the total production cost, and the crop is also demanding in labour requirements.

The winged bean (Psophocarpus tetragonolobus) *as a vegetable crop.* The potential importance of the winged bean in Africa has recently been investigated by Karikari (1969, 1972) and Pospisil, Karikari, and Bomah-Mensa (1971). This legume, which produces edible green pods, as well as seeds which are high in protein and tubers which are used as a root vegetable, is recommended as a vegetable worthy of further consideration for growing in areas in West Africa where diets are at a low level. The botany and chemical analysis of the crop were first described, together with agronomic details obtained from field experiments, by Masefield (1961, 1967), who studied the root nodulation of the crop and discussed its agricultural potential. The role of this crop in the tropics has recently been reviewed thoroughly (NAS 1975) by a panel of scientists working on the crop.

7.8.2. *Tomatoes*

The importance of disease and nematode resistance in selecting cultivars of tomatoes for tropical cultivation has already been described (pp. 109–10).

In a review of work on the effects of various temperatures on tomato growth and development, Daly (1971) has drawn particular attention to the importance of night temperatures on stem elongation and the effects on growth of humidity and light. The generally accepted conclusion has been that the tomato plant requires a night temperature below 17 °C but above 13 °C, together with a day temperature of approximately 23 °C for successful flower formation.

After preliminary trials in the French West Indies in 1965–7, two tomato cultivars from Florida, Manalucie and Indian River, were selected for field experiments in Martinique, aimed at exploring the possibility of producing high-quality fruits for as long a period during the year as possible. These experiments were undertaken during the period 1968–70 under irrigated conditions. Individually staked plants were trained to a single stem, and side shoots were removed. Fertilizer was applied as needed according to foliar diagnosis. The plant population was 23 500 per ha.

Yields from plants established during periods of rising temperatures were low: e.g. 16 t/ha were harvested from Indian River and 15 t/ha from Manalucie. The yields obtained from plants which were developing during periods of low (or falling) temperature, however, were much higher: e.g. 40 t/ha from Indian River and 33–36 t/ha from Manalucie. Both the number of fruits per plant and the weights of individual fruits were found to contribute to the higher yield during the dry, cool season. Internodes were longer, the first-flower truss was produced higher on the stem, a lower number of flowers developed per truss, and the percentage of fertilized flowers was lower on plants grown during the wet season. The period between planting and first harvest is another variable for any given cultivar when planted at regular intervals throughout the year. This period is apparently extended by increased diurnal temperature variation. The average variation in harvesting date for the cultivars tested was approximately 12 days. Applications of a fruit-setting hormone spray at anthesis were not effective.

The commercial production of tomatoes in West Africa has been discussed by Quinn (1971, 1973), who defined the main factors limiting production as being diseases associated with the high rainfall and humidity.

In the northern Guinea savanna and Sudan savanna zones of northern Nigeria, however, leaf diseases are less prevalent, and nematode as well as bacterial infections are probably limited by the desiccation of the soils for up to 7 months of the year. Tomato cultivation is therefore successful, if irrigation is available, although yields are reduced when both day and night temperatures are high.

Mulching of both staked and unstaked crops with grass gave significant increases in yield of crops grown in the wet season but not in the dry season. The effects of mulching and staking on soil temperatures, sun-scorch, weed control, and the incidence of blossom-end rot were all discussed at the IITA, Ibadan seminar, and an economic assessment was made of the various treatments. Spraying with fungicides during the dry season was also discussed and found to be uneconomic.

Low soil temperatures occurring in valley bottoms have been found to retard growth, and some amelioration of this situation can be achieved by using black polyethylene as a mulching material.

At Samaru selection of suitable cultivars for processing showed that the following are suitable for this area: Cirio 56, Harvester, Marzanino, Piacenza 0164, and Ronita. The last-named has shown resistance to root-knot nematode throughout northern

Nigeria. For the fresh market Alicante, Enterpriser, Healani, and Ife No. 1 are recommended.

Jackson (1971) and Quinn (1973) have discussed the increasing interest in the production of tomatoes for processing into *purée* and associated products in northern Nigeria. According to Jackson (1971) the practical problems are primarily concerned with modifying the findings resulting from cultivar testing, fertilizer experiments, plant protection, and crop-density studies so as to suit the requirements and capabilities of the local growers. In particular, crop-protection practices are not being taken sufficiently seriously, and the practicality of irrigation is another major constraint on tomato production.

The establishment of two tomato-canning factories in Ghana in 1968 stimulated investigations into the selection of cultivars for processing; the results of trials, conducted during 1967, have been reported by Apte, Dirks, Eyeson, Ghansah, and Sundararajan (1969). Forty-two cultivars were tested at two sites for their commercial acceptability for processing. In the Guinea savanna zone site, the cultivar CPC2 out-yielded Roma. In the Sudan savanna zone, Red Top and MH/VF 145-21-4P gave the highest yields, but laboratory analyses for pulp yield, total soluble solids in pulp, content of reducing sugar, and acidity indicated that MH/VF 145 B, Pearson A-1, and VF Roma, although lower yielding, had the necessary processing qualities when grown under local conditions, and these were therefore selected for further trials.

In a later trial at Wenchi, described by Nsowah (1970*b*), 20 imported cultivars regarded elsewhere as suitable for processing all failed to produce fruits of suitable quality when grown under the local conditions.

The assessment of eight tomato cultivars considered suitable for growing for fresh market use in northern Ghana and the response of two of these cultivars to staking has been reported by Nsowah (1969*a*). Trials were at agricultural stations and on selected farms in various districts; the environmental conditions of these trials therefore varied appreciably. Highly significant differences in yield were observed between both cultivars and years and also a significant cultivar × locality interaction. However, cultivars Molokai (M44A) and Anahu (M90) out-yield the other cultivars tested in most of the trial areas. Improved Zuarungu shows promise for dry-season cultivation on the Accra plains (Amuti 1971).

In the francophone countries of West Africa the

incidence of bacterial wilt has promoted the adoption of more sophisticated cultural methods than simply growing plants in open ground. Hydroponic or soil-less cultivation techniques are being investigated. Selection of wilt-resistant cultivars is proceeding and trials with bactericides for use with field-grown crops are being conducted. Yields obtained during the wet season tend to be very variable, and the Roma cultivar has been found to be most likely to give adequate yields under wet, humid conditions. Trials in Upper Volta indicate that Moneymaker, Saint Pierre, Roma, and Kaki yielded from 52·1 t/ha to 61·6 t/ha. Dry-season irrigated production in Upper Volta had been encouraging but successful production depends on a fully adequate water supply and the use of cultivars at least partially resistant to nematodes and bacterial wilt.

In Niger, early maturity in tomatoes is regarded as of paramount importance, and Fireball and Marmande are recommended for use in the fresh state and Roma for processing. Recent introductions, including Pusa Early Dwarf and Geneva 11, are promising. The yields of most of the cultivars tested, under good conditions, lay between 60 t/ha and 80 t/ha. In experiments on 14 cultivars of tomato during the period 1967-8, the cultivars VFN8, Ronita, and Piernita showed the highest resistance to nematodes.

At a station in West Cameroun at an altitude of 1400 m, 22 cultivars were studied during different growing seasons; one treatment included growing plants under plastic shelters. Genotype × environment interactions have been studied by Praquin and Marchand (1970). Cultivars which yielded more than 60 t/ha when grown under plastic shelters included Marglobe, Homestead 24, Xokomo, Marion, Indian River, and Merveille des Marches.

7.8.3. *Egg plant* (*garden egg, brinjal, or aubergine*) (Solanum melongena)

Choudhury (1971b) has reviewed current research on this species at research stations and institutes in India. The centre of origin of the species has been ascribed by different authors to both the tropical African and the Indo-Burmese regions, and it is possible that evolution from related wild progenitors may have taken place in each area. The egg plant is widely cultivated in India and is among the highest-yielding vegetable crops. Many cultivars have been produced during the last 30 years for different ecological conditions and consumer preferences.

There is an incompatibility mechanism in this species which depends on heterostyly. Only flowers with long and medium-length styles set fruit by self-pollination, and the proportion of each type of flower that occurs on a plant depends on both genotype and environment (Prasad and Prakash 1968). Studies on the natural cross-pollination of egg plant have been undertaken by Sambandam (1965), and Rajasekaran (1970a, 1970b, 1970c) has described the cytological characteristics of the egg plant and related wild species.

There is hybrid vigour in the egg plant suggesting the possible practical value of producing F_1 hybrid cultivars (Sambandam 1962, 1964; Raman 1964; Mishra 1966; Thakur, Singh, and Singh 1968). Breeding for resistance to diseases, pests, and nematodes has recently been developed at the Indian Agricultural Research Institute, and particular attention has been given to the pathogen *Phomopsis vexans*, against which several related wild species of *Solanum* have resistance. Diseases with virus-like symptoms described as 'little leaf', caused by a mycoplasma, and mosaic virus are becoming important problems in some areas.

The most serious pest of egg plant in many parts of India is the fruit-borer (*Leucinodes orbonalis*). Cultivars vary in susceptibility to this pest. A cultivar resistant to aphids has recently been released. In trials to investigate resistance of cultivars and related species to root-knot nematodes (*Meloidogyne javanica*, *M. incognita*, and *M. arenaria*) *Solanum sisymbri-folium* was found to be highly resistant to all three species. Several cultivars also had high resistance.

Experiments on fruit set, using 2,4-D sprays were conducted by Krishnamurthi and Subramaniam (1954). The growth substance was applied in a water spray, in lanolin paste, and in talc dust to mature flowers, and a fruit set of 122 per cent of normal was recorded. Choudhury and Ray (1966) report that soaking the seeds for 24 hours in a solution of 2,4-D at a concentration of 2·5 p.p.m., and a whole-plant spray with the same growth substance at a concentration of 5 p.p.m., plus the addition of a 1 per cent urea spray at the time of flowering, increases fruit set and increases both early and total yield by 30-5 per cent over that of the untreated control.

A preliminary analysis at IARI of 15 egg-plant cultivars for ascorbic acid and protein content indicated variation in ascorbic acid from 14·0 mg per 100 g to 78·2 mg/100 g dry weight and protein content from 8·45 per cent to 12·70 per cent dry weight.

Nsowah (1969*b*) studied morphological variations and yield components (number, weight distribution, and average fruit weight) of 17 cultivars of egg plant. Fruits became smaller as plants aged.

Subsequently (Nsowah 1970*a*), using 16 cultivars of egg plant from various sources and working during the main wet season, found significant positive correlations between flower initiation and opening and fruit maturity. There was a negative correlation between fruit number and the average weight of the first fruit.

7.8.4. *Chilli pepper (or capsicum)*
Variety trial data have been reported from Upper Volta, the Central African Republic, and the West Indies (Soitout 1971*b*). Carré doux d'Amerique, Yolo Wonder, and California Wonder give yields varying from 9 t/ha to 13 t/ha in the West Indies.

7.8.5. *Onions*
Nabos (1971), in his review of the present state of research into the onion crop by IRAT at Niamey in the Niger region, states that the principal vegetable crop in Niger is the onion. Production for the period 1965–9 was approximately 38 000 t from 1850 ha, with an average yield of about 20·5 t/ha. The main growing period is from November to April, during the cool, dry season, which is characterized by minimal rainfall, a relative humidity of less than 35 per cent, a daily evapotranspiration rate of 6·0–7·5 mm, and a day-length of less than 12 hours. Crops are normally established at the bottom of valleys or beside streams or ponds, since irrigation is required during the growing period. The leaves are processed and used as a seasoning in sauces, and represent an economic return of about 30 per cent of the value of the mature bulbs.

Since 1962 work has been devoted to selecting cultivars for production potential for consumption as fresh and stored bulbs or, more recently, for dehydration. Galmi, Madaoua, and Soumarana, all with a violet skin colour, were selected in the initial screening. The first two cultivars have a maturation period of 140 days; Soumarana matures in 160 days. Madaoua and Galmi both yield in the region of 60–80 t/ha; they can be stored in a good condition for 6 months. Soumarana is lower yielding. The criteria for a 10-year selection programme (1965–75) include: yield, maturation period of less than 150

days, reduced flowering rate in the first year, uniformity of size, light violet to pale yellow colour of bulbs, and good storage qualities.

Most imported European, North American, and Australian cultivars fail to produce bulbs in the short-day conditions of West Africa. Some American hybrids, however, are adapted and yield up to 100 t/ha. These include Early Texas Yellow Grano, Yellow Granex, Red Star, Dessex, White Granex, and New Mexico White Grano. Most of these do not travel or store well under local conditions and their production is therefore currently limited to areas close to markets.

Since 1968, emphasis has been directed to research into cultivars which are adapted to the requirements for dehydration, for which a white colour, a high dry extract rate, and strong taste are essential. Current work is centred around the Blanc de Soumarana cultivar, and commercial tests show that this cultivar gives a dehydrated product with good appearance and colour and a dry-matter yield of 11–12 per cent, but only a moderately strong taste. Further work is being undertaken to improve the pungency of the taste.

Mean bulb weight decreases approximately linearly with plant density and densities of 400 000–500 000 plants per ha gave the most commercially acceptable bulb size, ranging from 176 to 138 g per bulb. Planting distance can be either 15 × 15 cm on level ground or on flat-topped ridges 30 cm apart, with two rows 10 cm apart on the ridge top and plants 15 cm apart in the row.

Experiments at IRAT on the preparation of bulbs for seed production have involved treatments of different planting material, i.e. whole bulbs, trimmed bulbs, or pieces of bulb which had already started growth after a dormant period, combined with two planting densities. Yields for whole bulbs planted in 1970 are shown in Table 7.5. It therefore appears that seed yield increases with planting density. In the experiments it was found that fragments of pre-germinated bulbs, planted at a density of 450 000

TABLE 7.5

Onion: plant density, bulb, and seed yield (from Nabos 1971)

Number of bulbs/ha	Weight of bulbs (t/ha)	Seed yield (kg/ha)
83 300	11	1208
166 600	22	2081

pieces per ha, gave a yield of only 786 kg/ha. Removal of the top half of the bulb before planting led to a loss in yield of 30 per cent compared with whole bulbs and was also associated with a reduced survival rate.

Bulb-storage experiments at IRAT have been in progress since 1965, with bulbs placed on slatted trays in a dark, well-ventilated store. Temperatures varied from 25 °C to 39 °C, and the relative humidity from 20 per cent to 85 per cent. The main cause of rot during storage was *Aspergillus*. The most important factor related to survival for the 6 months of the trials was the cultivar. Many local selections are found to be superior to exotic cultivars which do not normally store well for more than 3 months. Early maturing and strongly-coloured local onions have the best keeping qualities. Smaller bulbs store better than large ones of the same cultivar. Bulbs which are topped immediately after lifting when mature and dried for several days before storage gave the best results. Topping of immature bulbs after lifting increases storage loss. Layers of bulbs in the slatted storage cribs should not be more than 50 cm in depth.

Trials with both short- and long-day onion cultivars have been reported by Inyang (1966). Results of more recent cultivar trials in the northern states of Nigeria have been reported by Green (1972a). Few of the introductions tested have desirable combinations of characters for local needs. Selection of cultivars for the Nigerian market from local strains is strongly advocated and has recently been undertaken by the Institute for Agricultural Research, Samaru. If an export crop should prove to be economically viable (Green 1971), Texas Grano or similar cultivars, which are adapted to short day-lengths, could form the basis of such an industry. Although onion growing in the wet season is difficult, the market value of bulbs harvested from October to February can make even the low yields obtained during this period profitable. Compared with the other cash crops being grown currently in the northern states of Nigeria, onion growing is financially attractive.

In addition to the findings in Niger, no evidence was found in Nigeria either to support the local practice of tissue removal to stimulate flower production. Here also, cutting the bulbs and separating the shoots depressed seed yield by 35–62 per cent (Green 1972b).

The effectiveness of the weedicides chloroxuron, chlorprophan, chlorthal dimethyl, monuron, nitrofen, and fluorodifen have been assessed for an irrigated transplanted onion crop in northern Nigeria (Green 1972c). The cost of the effective chemical chlorthal dimethyl in the early 1970s did not allow economic use of the weedicide where sufficient cheap labour was available as an alternative.

7.8.6. *Cucurbit vegetables and fruits*

Choudhury (1971a) has listed 43 references to recent research on cucurbit vegetables in India. He points out that two-fifths of the known genera originated in Africa, two-fifths from America, and one-fifth from tropical Asia. The cucumber probably originated in Africa but also grows wild in northern India.

IARI have recently recommended several newly bred cultivars of cucumber, water melon, *Luffa*, bottle gourd, bitter gourd, winter squash, pumpkin, and musk melon for cultivation by vegetable growers in various parts of India. Reviewing genetic and breeding work undertaken in recent years in India, Choudhury (1971a) refers to the inheritance of sex forms (Thakur and Choudhury 1965, 1966, 1967), the inheritance pattern of flower and fruit characters in *Cucumis melo* (Kang and Bains 1963), and the monograph on the taxonomy and distribution of the Indian Cucurbitaceae by Chakravarti (1959).

F_1 hybrids of bottle gourds (Choudhury 1966; Choudhury and Singh 1971) give yields which are 80–100 per cent higher than those of the parental commercial cultivars. Srivastava and Nath (1971), testing 90 F_1 hybrid combinations of bitter gourd, recorded yields up to 86·9 per cent above the parental average yield.

Indian breeding for disease resistance is being directed towards the screening of cultivars of musk melon against powdery mildew (*Sphaerotheca fuliginea*), and some sources of resistance and field tolerance have been isolated (Choudhury, Sivakami, and Singh 1971). Khan, Khan, and Akram (1971) have reported that all cucurbits they have tested except *Coccinea* sp. are affected by *Sphaerotheca*.

Breeding for pest resistance has been mainly directed to finding resistance to the red pumpkin beetle (*Aulacophora foveicollis*) and the fruit-fly (*Dacus cucurbita*). Nath (1964) found resistance to red pumpkin beetle only in bottle gourd, but Choudhury and Vashistha (1971) report on selections and cultivars of musk melon, bottle gourd, and water melon which possessed varying degrees of tolerance to the beetle. Resistance is governed by a single dominant gene and associated with a low cucurbitacin content. Chelliah (1971) and Sambandam (1971) report that *Cucumis callosus* possesses a high degree of resistance to fruit-fly and that susceptibility appears to be controlled by two dominant complementary genes.

There has been continuous work in India on chemical promotion of fruiting in cucurbits since Maheswari (1957) reported that it was possible to modify sex in cucurbits by chemical sprays when plants were subjected to high temperatures and long-day conditions. Choudhury and Phatak (1959*a*, 1959*b*, 1960) report that maleic hydrazide, naphthalene acetic acid, indole-β-acetic acid, and gibberellic acid in concentrations varying from 5 p.p.m. to 50 p.p.m. increases the percentage of female flowers in cucumber; gibberellic acid also increases fruit yields. The two-leaf stage is the most critical period in the growth cycle for the application of chemicals (Choudhury and Singh 1968). Applications of boron, calcium, and iron can also affect sex expression in the bottle gourd (Choudhury and Babel 1969) and water melon (Choudhury and Elkholy 1970). Boron at 3 p.p.m., and calcium at 20 p.p.m., are most effective in stimulating the production of female flowers of the cultivar Sugar Baby of water melon. Boron promoted higher total soluble solids, fewer seeds, and thinner rind. Gibberellic acid at 10 p.p.m. and tri-iodobenzoic acid at 25 p.p.m. also induce a high number of female flowers and fruit set in the water-melon cultivar New Hampshire Midget (Choudhury and Gopalakrishna 1965).

d'Arondel de Hayes (1971) has given a review of trials with gherkins for processing and export at the IRAT station at Farako-Ba in the Sudan zone of Upper Volta. The high labour inputs for this crop probably account for the recent fall in production in Europe. The crop has to be hand-harvested every 2 days to maintain production, and fruits should not exceed 6 cm in length and 19 mm in diameter. Cultivar trials during the 1967-8 period indicated that the cultivar Epros outyielded the three other cultivars included in the trial, giving a yield of 13·6 t/ha. The optimum sowing time was found to be within the period 10-30 November.

Research on, and production of, melons has been undertaken in several francophone countries, and Soitout (1971*b*) has reviewed this. Serious pathogens of melons in the tropics are *Erysiphe cichoracearum*, *Pseudoperonospora cubensis*, and *Fusarium oxysporum* f.s. *melonis*. Fruit-flies such as *Dacus bivittatus* also generally require control. Cultivars such as Delicious 51 and SR59 are recorded as giving yields of 20 t/ha at the Upper Volta research station at Farako-Ba; Cantaloup Charentais and Smith's Perfect are less productive with yields of 10 t/ha. In the West Indies Cantaloup Charentais is the only important commercial cultivar, but for local market consumption the lower-yielding Hale's Best 45, PMR 45A, and Smith's Perfect are also grown. On local acid volcanic soils, liming at 3 t/ha gave 36·15 t/ha of melons with a control yield in unlimed land of 10 t/ha.

7.8.7. *Leaf crops, including brassicas*

Norman (1972) has provided information on cultural details from Ghana of crops such as African spinach (*Amaranthus hybridus*), Indian spinach, *Celosia argentea*, West African sorrel, *Corchorus acutangulus* and *C. tridens*, roselle, okra, and water leaf.

Sinnadurai (1970) has reported data from pot experiments on Indian spinach with nitrogenous fertilizers, including sulphate of ammonia, potassium nitrate, urea, and ammonium nitrate at levels of application of 3 g and 6 g per plant. Ammonium sulphate, applied at a rate of 6 g per plant, gave the highest yields.

Lettuce for selling to the markets of principal towns and cities is one of the most remunerative vegetable crops in Ghana. Crops grown in the forest zone at the Department of Horticulture, University of Science and Technology, Kumasi, although given fertilizer at the time of planting, normally became pale yellow in colour about 4 weeks before harvest. Norman (1969) showed that side-dressings of nitrogenous fertilizer, applied 18 days before the first harvest, were highly beneficial with the lettuce cultivar A1.

Soitout (1971*b*) has reviewed work on lettuce in francophone Africa. Great Lakes 659 has been generally the most successful cultivar throughout the inter-tropical zone. MR52, Reine des Glaces, and New York 515 Improved have also yielded well in lowland high-rainfall conditions. At higher elevations in the equatorial zone Kagraner Sommer, Gloire de Nantes, and Reine des Glaces all give yields between 25 t/ha and 30 t/ha.

Cabbage. Soitout (1971*b*) reports that in the evergreen forest region the cultivar Greengold Hybrid was selected for dry-season cultivation, with Comet Hybrid, Grey Green Hybrid, and Glory of Enkhuizen as possible alternatives. Cultivars which are successful during the dry season in the humid tropical region included Boston, Golden Acre, and Brunswick. Yields of 60 t/ha are obtained during the dry season in the Central African Republic from Brunswick and Quintal d'Alsace, which have maturation periods of 80–90 days. At 1400 m elevation in Cameroun, suitable cultivars are Copenhagen Market, Quintal d'Alsace, Boston, Milan Gloire d'Automne, and Milan Gros des Vertus. Recommendations for fertilizer application rates were given for Sierra Leone by Godfrey-Sam-Aggrey and Williams (1972) including the suggestion that 5·6 kg/ha of borax should be added in order to reduce the incidence of heart rot which had been observed to affect some cultivars of cabbage, Chinese cabbage, and mustard.

Cauliflower. Most cauliflowers are highly sensitive to photoperiod, and hence sowing date is found to be most important. The crop has been studied in areas of francophone Africa (Soitout 1971b). French cultivars Selandia, Nain hatif, Boule de Neige, Everest, and Avalanche, and Early Italian Giant, Early Patna, Kibo Giant, and Snowball Improved have been successfully grown in Upper Volta.

Although not widely grown in West Africa at the present time, there is an increasing interest in the cultivation of cauliflower in anglophone parts of West Africa, particularly in Ghana. Research by Apte (1968) in the equatorial forest zone of Ghana has been concerned with the selection of cultivars which are more suitable for local conditions than the cultivar Sutton's Tropical (syn. Early Patna). This has, until recently, been the most widely grown cultivar in the tropics, but it had been observed to exhibit considerable variability in maturity and curd quality. Earlier work by Town (1964) and White (1967) showed that the Snowball selections of European origin did not possess adequate resistance to the high temperatures typical in Ghana.

Apte (1968) has reviewed experiments with cauliflower in Ghana, as well as the classification of cauliflower cultivars and attempts to cultivate cauliflower under high-temperature conditions in other countries. Seeds of 37 cultivars have been imported from the Netherlands, Sweden, Denmark, the United Kingdom, India, Hawaii, and Japan, and plants of each cultivar were grown in observation trials. Cultivars of European origin either produce no curds or only malformed curds. Cultivars from India, Japan, and Hawaii produce normal curds, and the Japanese cultivar Snow Queen gives the best results, based on curd quality, uniform maturity,

and total yield. Other cultivars, including Tropical 45 Days, Tropical 50 Days, Tropical 55 Days, Snow King hybrid, and Snow Peak hybrid are considered suitable for growing on a limited scale. A classification was provided of the cultivars used in the trial, based on their physiological response to temperature.

Amaranthus spp. Recent research on species of *Amaranthus* has indicated that some hitherto neglected species of this genus have a high lysine content (Downton 1973). Species of *Amaranthus* photosynthesize by the C-4 pathway, the major initial product of photosynthesis being aspartic acid, which is necessary to the formation of the amino acid lysine. Analysis of the leaves and seeds has shown that the seeds of *Amaranthus edulis* are rich in protein, with a high lysine content. The lysine content of the leaves was found to be 5·9 g per 100 g of protein; and that of the seeds, which are also eaten with the leaves, was 6·2 g per 100 g of protein.

7.9. Suggestions for future research on vegetable crops in West Africa

7.9.1. *Research priorities*

The definition of research priorities with regard to vegetable crops must depend on a clear assessment of the factors affecting production and consumption in any specific area. Where the demand for vegetable crops is expanding rapidly, as in some of the francophone countries of West Africa, it is likely that an intensive research programme with clearly defined objectives related to local problems will be economically justified.

The demand for research on vegetable crops in less specialized situations than these commercial undertakings may be more difficult to justify on purely economic grounds. Selection of more productive cultivars, the investigation of management techniques, reducing field losses by plant protection measures, and post-harvest losses by the adoption of appropriate handling and storing techniques may, however, all be useful if scarce monetary resources permit. Marketing problems rather than technical problems are a major constraint to the increase of vegetable production in many African countries.

An increase in the use of mechanization for preparation of the soil and subsequent cultural operations may be justified in some areas where labour-intensive methods of horticultural crop production are no longer practicable.

Research aimed at solving the problems of commercial production could make a significant contribution towards an improvement in the variety and volume of vegetables available for local consumption in many developing countries. Whether government aid, in the form of research funds and facilities, should be devoted more towards assisting the small-scale producers or commercial growers is a political decision, depending largely on the government views on both overseas trade and domestic nutrition. Experience suggests clearly that for sophisticated export-orientated production, the main emphasis must be on the more advanced and better-capitalized growers. The limiting factor in most areas appears to be the availability of qualified and motivated manpower.

7.9.2. *Vegetables and nutrition*

Although urban populations are likely to become increasingly dependent on non-traditional foods, the nutritional value of many local vegetables is widely recognized by nutritionists and health workers. Changes in government policy in many countries have wisely discouraged the importation of exotic fruits and vegetables, and it appears likely that, as populations increase, there will be a continuing and increasing urban demand for locally produced food crops. Ensuring the maintenance of, and perhaps improving, the nutritional value of local vegetables, while attempting to increase their yield, therefore appears to be important.

The amino-acid spectrum as well as the protein content of vegetable crops needs further research. Investigations into the amounts of β-carotene and other vitamins present in leaf vegetables, which have already been carried out by some workers, could usefully be extended. The effect of various methods of preparation of the leaves for consumption on their nutritional value is a most important complementary activity to research on the crops themselves. Although recognized for a long time, the presence of poisonous substances in many vegetables has only recently begun to be quantitatively assessed. Research appears to be particularly necessary on the effects on the consumer of a high intake of oxalate, since this may prevent the normal absorption of minerals such as calcium, nitrogen, iron, and phosphorus. The selection and breeding of cultivars with low levels of even mildly poisonous substances in their tissues may be necessary where daily intake rates are high.

In the Solanaceae, alkaloids such as solanine and solanidin in the leaf and fruits are potentially hazardous to health. Investigations into the effect of environment and cultural treatment on the production of these and other harmful substances in plants as well as genetic variation for their production could be most valuable.

7.9.3. *Production economics and marketing*

The general lack of data on production, handling, and marketing costs and the great variation in yields and growth periods of vegetables in tropical countries make very difficult and uncertain the sort of economic analysis needed before specific plans can be prepared for development projects. The role of the small-scale producer in economic horticultural development is particularly difficult to evaluate, and as a result such people are rarely the recipients of aid, even when politically considered to be most in need of assistance. Investigations into the organization and costs of both existing and potential small-scale operations, so as to provide a basis from which to attract support, could be of benefit to the producing community. Such investigations should embrace social as well as technical aspects of production, since both have a direct bearing on production capability. Present rates of consumption of vegetables have been estimated for various areas, but more detailed information is generally required, in particular to assess the elasticity of supply and demand.

7.9.4. *Extension services*

The number of extension workers available and qualified to assist producers of horticultural crops in increasing and improving their production appears to be generally inadequate in West Africa (bearing in mind however that inappropriate or impractical advice can never be successfully extended even by a numerically strong corps). Without an efficient extension service, itself well briefed with the results of research, the messages will not get through. Extension workers should also provide feedback information to research workers. An effective extension service is needed for any viable attempt to improve production. The most effective way of organizing an extension service may not necessarily be the method currently being adopted in any particular area and, careful reconsideration of the organizational infrastructure may be required to ensure effective implementation of research findings. A different sort of infrastructure may be appropriate for different sorts of agricultural or horticultural enterprise.

7.9.5. *Crop physiology*

Seed-storage investigations towards the goal of making available seeds of improved viability will become increasingly important as specialized high-potential cultivars are selected and seed costs rise. Conditions for specialized seed production enterprises need also to be determined.

While fertilizer studies, including the response of crops to new proprietary compounds and formulations and responses to trace elements, are already being undertaken in many developing countries, the responses of traditional or indigenous crops have generally still to be investigated. For most vegetables there is a need to examine responses of different cultivars, as well as different species, since those which have not been selected for responsiveness are often unable to take full advantage of applied nutrients because of genetic limitations.

The use of other agrochemicals has not yet become widespread in many tropical areas, but exploration of the potential value of some of the more advanced techniques and chemicals used elsewhere may be appropriate in some situations. Studies on the use of plant-growth substances such as cycocel could be illuminating. Increased testing of weedicides for vegetable crops is likely to result in a reduction of labour requirements in areas where semi-commercial or large-scale production is being promoted. Such possibilities call for careful economic evaluation.

Basic research into photosynthetic pathways has promoted an interest in examining tropical crop performance more closely, and it appears possible that a more intensive programme to make use of crops of high photosynthetic efficiency (C-4 pathway) such as the Amaranthaceae, might produce economic results by the selection of species for leaf-protein production.

Responses to variation in day-length have been studied in detail only for a very few tropical crop species, such as rice and cowpeas. Many vegetable crops have not been adequately investigated, and many species may possess useful genetic variation for response to day-length. The response to temperature of flowering and fruiting is another subject which merits more extensive examination, particularly in the family Solanaceae. The water requirements of both exotic and indigenous vegetables in specified environments are not well known. Further information on this subject could be of practical use to the commercial-scale producer, particularly in areas which are dependent on irrigation, where water supplies are variable throughout the year; such information would enable growers to make the best use of the irrigation resources available.

7.9.6. *Crop protection*

Techniques which lead to a reduction in crop losses from diseases and pests have been studied extensively in agricultural crops in tropical areas, particularly for those crops which have an industrial or export value. Some investigations have been conducted on vegetable crops, but emphasis has naturally been placed on vegetables such as beans, tomatoes, and onions, which are important as large-scale commercial crops for processing or export.

Nematodes, and particularly root-knot nematodes, cause severe losses in vegetables of the family Solanaceae. Cultural methods and control, such as the use of fallow or alternate crops which are not susceptible to nematodes or soil-less cultivation, have been attempted with reasonable success in some areas. The treatment of soils with an increasingly wide range of nematicides is already practised in some commercial operations, but continuing work on the production of nematode-resistant or tolerant cultivars appears to be highly desirable.

Bacterial wilt (*Pseudomonas solanacearum*) of tomatoes and many other crops also causes severe losses in the humid lowland tropics. The development of resistant cultivars appears to offer the most economic long-term solution.

The need for further research on relatively inexpensive but effective control measures for pests, diseases, and weeds is obvious. An integrated approach to crop protection would be highly preferable to a piecemeal one.

7.9.7. *Plant breeding and germplasm resources*

Samson (1971) stated that an intensive programme of introduction of vegetable cultivars which appear likely to be climatically adapted to West Africa, should be accorded high priority. This aggressive introduction programme should be accompanied by screening, observation, and selection of suitable commercial cultivars and seed distribution to production areas. Plant breeding should then follow, aimed at improving characters such as high and early yields, high methionine and lysine contents, resistance to diseases, insects, and nematodes in this well adapted and high-quality germplasm. Samson suggests that some of the 'wild' vegetable species of high nutritional value could be modified to become useful vegetable crops. Much of the basic taxonomic work on many of these has still to be undertaken. Initial investigations could usefully be concentrated on the genera *Amaranthus*, *Basella*, and *Solanum*, of which some species are already well-known food crops.

The preservation of germplasm reservoirs by the establishment and maintenance of collections of indigenous forms of vegetables is highly desirable for the success of many future breeding programmes, but this may best be done at specialized centres elsewhere than in West Africa, rather than competing for the limited development resources which Africa can offer.

7.9.8. *Crop management*
Both Samson (1971) and Tindall (1971) have stressed the need for research on all the basic production factors for vegetables under West African conditions. In general, more information is required on the management practices necessary to obtain high yields, including plant-population studies and studies on the timeliness of operations such as crop establishment, fertilizer application, and irrigation.

Several research workers have investigated the patterns of existing mixed-cropping systems in both East and West Africa, and quantitative techniques are now being applied in order to evaluate the yields of various crop combinations under experimental conditions. Further information on appropriate crop combinations and treatments, including fertilizer application and the density of each component, would make possible economic assessment of inter-cropping techniques in comparison with conventional temperate practices. Research into the use of plastic materials as mulches and for the protection of seedlings from heavy rainfall will be appropriate if local cheap supplies of plastic sheet material become available.

7.9.9. *Post-harvest problems*
Investigations into methods of preventing post-harvest losses are urgently required in many areas. Although some work has been done on the preservation of fresh horticultural produce by the use of cool stores, many of the factors most likely to affect produce stored under these conditions have still to be examined quantitatively.

In recent years chemical residues in horticultural crops have come under critical scrutiny in many temperate countries. Although the application of chemicals in tropical areas has not yet reached a comparable level, a surveillance of residues in crops produced under sophisticated technology and destined for export should be routinely undertaken in the near future. This is necessary in order to protect the reputation of the producing country, as well as the interests of the ultimate consumers.

References
ADDISON, E. A. and CHONA, B. L. (1971). Evaluation of certain fungicides for the control of *Sclerotium* wilt of tomato (*Lycopersicon esculentium* Mill.) caused by *Sclerotium rolfsii* Sacc. *Ghana J. agr. Sci.* 4, 89-91.

ADSUAR, J. (1961). Use of the root-dipping inoculation method in testing tomatoes for wilt resistance. *J. agric. Univ. P. Rico* 45, 116-18.

AJEAGBU, H. I. (1970). Food crop farming in the coastal area of South Western Nigeria. *J. trop. Geogr.* 31, 1-9.

AMUTI, K. (1971). Tomato cultivars suitable for the dry season on the Accra plains, Ghana. *Ghana J. agr. Sci.* 4, 113-15.

ANDREWS, J. C. and VISER, E. T. (1951). The oxalic acid content of some common foods. *Fd Res.* 16, 306-12.

APTE, S. S. (1968). Cauliflower cultivar selection for curd formation at Kumasi, Ghana. *Ghana J. agr. Sci.* 1, 143-51.

—— DIRKS, R. F., EYESON, K. K., GHANSAH, A. K., and SUNDARARAJAN, A. R. (1969). Suitable tomato varieties for the canneries in Ghana. *Ghana J. agr. Sci.* 2, 73-80.

BEN HALIM, M. M. (1967). The role of root-knot nematodes in predisposing certain tomato varieties to southern bacterial wilt. *Diss. Abstr.* 26, 4161-2.

BUSSON, F. (1965). *Plantes alimentaires de l'Ouest africain.* Le Conte, Marseilles.

CHAKRAVARTI, H. L. (1959). Monograph on Indian Cucurbitaceae. *Rec. bot. Surv. India* 17, 1-234.

CHELLIAH, S. (1971). On the resistance in *Cucumis callosus* to the fruit fly. *Proceedings of the seminar on cucurbits, Udaipur.*

CHOUDHURY, B. (1966). Exploiting hybrid vigour in vegetables. *Indian Hort.* 10, 56-8.

—— (1971a). Research on improvement of cucurbits in India. *Ford Foundation/IITA/IRAT seminar on vegetable crop research, Ibadan.*

—— (1971b). Research on egg plants in India. *Ford Foundation/IITA/IRAT seminar on vegetable crop research, Ibadan.*

—— BABEL, Y. S. (1969). Sex modification by chemicals in bottlegourd. *Sci. Cult.* 35, 321-2.

—— ELKHOLY, E. (1970). Chemical sex modification in watermelon. *Int. hort. Congr.* 18, Abstr. 418.

—— GOPALAKRISHNA, P. K. (1965). Effect of plant regulators or certain physiological aspects on sex and fruiting behaviour in watermelon. Thesis, IARI, New Delhi.

CHOUDHURY, B. and PHATAK, S. C. (1959a). Sex expression and sex ratio in cucumber as affected by plant regulator sprays. *Indian J. Hort.* **16**, 162-9.

—— (1959b). Sex expression and fruit development in cucumber as affected by gibberellin. *Indian J. Hort.* **16**, 233-5.

—— (1960). Further studies on sex expression and sex ration in cucumber as affected by plant regulator sprays. *Indian J. Hort.* **17**, 210-16.

—— RAY, D. (1966). Studies on effects of plant regulator treatments on brinjal. Thesis, IARI, New Delhi.

—— SINGH, N. (1968). Effect of plant regulator sprays at different environments on sex and fruiting in cucumber. *Proc. Indian Sci. Congr.* **55**, 589-90.

—— —— (1971). Pusa Meghdoot and Pusa Manjari, two high yielding bottlegourd hybrids. *Indian Hort.* **16**, 15-32.

—— SIVAKAMI, N., and SINGH, B. (1971). Preliminary studies on breeding muskmelon resistant to powdery mildew. *Proceedings of the seminar on cucurbits, Udaipur.*

—— VASHISTHA, R. N. (1971). Breeding cucurbits resistant to red pumpkin beetle. *Proceedings of the seminar on cucurbits, Udaipur.*

DALY, J. (1971). The influence of climate on the development and fruiting of tomatoes. *Ford Foundation/IITA/IRAT seminar on vegetable crop research, Ibadan.*

D'ARONDEL DE HAYES, J. (1971). Complementary crops of possible interest in countries of the sudanian zone. *Ford Foundation/IITA/IRAT seminar on vegetable crop research, Ibadan.*

DE BRUIJN, G. H. (1971). Étude du caractère cyanogenetique du Manioc (*Manihot esculenta* Crantz) *Meded. Landb. Hogesh.* **71**, 13.

DE VRIES, C. A., FERWERDA, J. D., and FLACH, M. (1967). Choice of food crops in relation to actual and potential production in the tropics. *Neth. J. agric. Sci.* **15**, 241-8.

DIGAT, B. (1967). Survey of bacterial wilt of Solanaceous crops in French West Indies and in French Guiana. In *Proc. 5th ann. Meet. Carib. Fd Crop Soc. Paramaribo* Surinam. July 24-31, 1967, pp. 75-85.

—— DERIEUX, M. (1968. A study of the varietal resistance of tomato to bacterial wilt II. The practical value of F₁ hybrids and their contribution to the genetic basis of resistance. In *Proc. 6th ann. Meet. Carib. Fd Crop Soc.* St. Augustine, *Trinidad*, pp. 95-101.

DOWNTON, W. J. S. (1973). *Amaranthus edulis*: a high lysine grain amaranth. *Wld Crops* **25**, 20.

FASSULIOTIS, G. (1968). Resistance of *Cucumis* species to a root-knot nematode, *Meloidogyne incognita acrita*. *Nematologica*, **14**, 6 (Abst.).

FERWERDA, F. P. and WIT, F. (1969). *Outlines of perennial crop breeding in the tropics. Misc. Pap. LandbHoogesch. Wageningen* **4**, 511 pp.

FOOD AND AGRICULTURAL ORGANIZATION (1969b). *Production yearbook* **23**. FAO, Rome.

GIETEMA GROENENDIJK, E. and FLACH, M. (1971). Het

L-syndroomé, een oversicht van recent: onderzoek in fotosynthese. *Landbouwk. Tjdschr.*, 's-Grav. **83**, 276-85.

GODFREY-SAM-AGGREY, W. (1973). Tomato production in Sierra Leone: problems and future research. *Acta Hort.* **33**, 97-102.

—— WILLIAMS, B. J. (1972). Upland soils of Sierra Leone for selected vegetable crops. *Wld Crops* **24**, 30-4.

GREEN, J. H. (1971). The potential for the bulb onion crop in the Northern States of Nigeria. *Samaru agric. Newsl.* **13**, 54-61.

—— (1972a). Cultivar trials with onions (*Allium cepa* L.) in the Northern States of Nigeria. *Niger. agric. J.* **8**, 169-74.

—— (1972b). The influence of bulb size, bulb cutting and separation of axillary shoots on seed production on onion (*Allium cepa* L.). *J. hort. Sci.* **47**, 365-8.

—— (1972c). Preliminary trials with herbicides in irrigated onions at Samaru, Nigeria. *Hort. Res.* **12**, 119-25.

HASWELL, M. (1969). Economics of agricultural development in the Philippines. *Civilisation* **19**, 437-50.

—— (1971). The economics of the production, transport and marketing of vegetable crops in Africa. *Ford Foundation/IITA/IRAT seminar on vegetable crop research, Ibadan.*

INYANG, O. A. (1966). The potential for the bulb onion crop in the northern States of Nigeria. *Samaru agric. Newsl.* **13**, 54-60.

JACKSON, C. A. (1971). Problems associated with establishing a tomato processing industry in northern Nigeria. *Ford Foundation/IITA/IRAT seminar on vegetable crop research, Ibadan.*

KAAN, F., PECANT, P., and BERAMIS, M. (1971). Selection of *Solanaceae* for resistance to bacterial wilt. *Ford Foundation/IITA/IRAT seminar on vegetable crop research, Ibadan.*

KANG, V. S. and BAINS, M. S. (1963). Inheritance of some flower and fruit characters in muskmelon. *Indian J. Genet. Pl. Breed.* **23**, 101-6.

KARIKARI, S. K. (1972). Pollination requirements of winged bean (*Psophocarpus tetragonolobus*) in Ghana. *Ghana J. agr. Sci.* **5**, 235-9.

KEHR, A. E. (1966). Current status and opportunities for the control of nematodes by plant breeding. *Rep. U.S. Dep. Agric.* **ARS 33-110**, 126-38.

KHAN, A. M., KHAN, M. W., and AKRAM, M. (1971). Status of cucurbit powdery mildew in India. *Proc.* (*Abst.*) *3rd int. Symp. Pl. Path.* 144-5.

KRISHNAMURTHI, S. and SUBRAMANIAM, D. (1954). Some investigations on the types of flowers in brinjal based on style length and the fruit set under natural conditions and in response to 2,4-D as a plant growth regulator. *Indian J. Hort.* **11**, 62-6.

LUC, M. and DE GUIRAN, G. (1960). Les nématodes de l'Ouest Africain; liste préliminaire. *Agron. trop., Nogent* **15**, 497-512.

MAHESWARI, P. (1957). Hormones in reproduction. *Indian J. Genet. Pl. Breed.* **17**, 386-97.

MASEFIELD, G. B. (1961). Root nodulation and agricultural potential of the leguminous genus *Psophocarpus*. *Trop. Agric., Trin.* **38**, 225.

—— (1967). The intensive production of grain legumes in the tropics. *Proc. Soil Crop Sci. Soc. Fla.* **27**, 338-46.

MESSIAEN, C. M., FOURNET, J., BEYRIES, A., and QUIOT, J. B. (1971). Some important diseases in vegetable-growing under tropical climates. *Ford Foundation/IITA/IRAT seminar on vegetable crop research, Ibadan.*

MISHRA, G. M. (1966). Preliminary chemical studies on four varieties of brinjal and their F₁ hybrids. *Sci. Cult.* **28**, 439-40.

MOHANAKUMARAN, H. Y., GILBERT, J. C., and BUDDEN-HAGEN, I. (1968). Relationship between tomatin and bacterial wilt resistance in tomato. *Proc. Am. phytopath. Soc., Pacific Division.* Abst. in *Phytopathology* **59**, 14.

MUNGER, H. M. (1971). Vegetable production in the tropics. *Ford Foundation/IITA/IRAT seminar on vegetable crop research, Ibadan.*

MUNRO, A. and BASSIR, O. (1969). Oxalate in Nigerian vegetables. *W. Afr. J. biol. Chem.* **12**, 14-18.

NABOS, J. (1971). The onion in Niger. *Ford Foundation/IITA/IRAT seminar on vegetable crop research, Ibadan.*

NAS (1975). *The winged bean, a high protein crop for the tropics.* National Academy of Science, Washington, D.C.

NATH, P. (1964). Resistance of cucurbits to the red pumpkin beetle. *Indian J. Hort.* **21**, 77-8.

NETSCHER, C. (1970). Nematodes parasites des cultures maraîchères au Sénégal. *Cah. ORSTOM Sér. Biol.* **11**, 209-29.

—— (1971). Problems caused by nematodes in vegetable crop production in the inter-tropical zone. *Ford Foundation/IITA/IRAT seminar on vegetable crop research, Ibadan.*

NORMAN, J. C. (1969). Studies of the growth rate and refertilization of transplant lettuce in Ghana. *Ghana J. agr. Sci.* **2**, 39-43.

—— (1972). Tropical leafy vegetables in Ghana. *Wld Crops* **24**, 217-19.

NSOWAH, G. F. (1969a). Review of tomato variety and staking trials in Northern Ghana 1962-66. *Ghana J. agr. Sci.* **2**, 7-18.

—— (1969b). Genetic variation in local and exotic varieties of garden eggs (egg-plant). 1. Variation in morphological and physiological characters. *Ghana J. Sci.* **9**, 61-73.

—— (1970a). Effects of sowing date on flowering and yield in varieties of egg-plants. (*Solanum melongena* L.). *Ghana J. agr. Sci.* **3**, 99-108.

—— (1970b). Preliminary studies of tomato processing varieties at Wenchi, Ghana. *Ghana J. agr. Sci.* **3**, 199-201.

OKE, O. L. (1966). Chemical studies on the more commonly used leaf vegetables in Nigeria. *J. W. Afr. Sci. Ass.* **11**, 42-8.

OOMEN, H. A. P. C. (1964). Vegetable greens, a tropical undevelopment. *Chron. Hortic.* **4**, 3-5.

—— (1971). The significance of leaf vegetables for tropical diets. *Ford Foundation/IITA/IRAT seminar on vegetable crop research, Ibadan.*

PHANSALKER, S. V. (1960). Nutritive evaluation of vegetable proteins. *Proceedings of the Symposium on Proteins* (Central Food Technical Research Institute, Mysore, India), pp. 345-54.

PLATT, B. S. (1965). Tables of representative values of foods commonly used in tropical countries. *Commun. Med. Res. Coun.* 302.

POSPISIL, F., KARIKARI, S. K., and BOMAH-MENSA, E. (1971). Investigations of winged bean in Ghana. *Wld Crops* **23**, 260-4.

PRAQUIN, J. Y. and MARCHAND, D. (1970). Premiers résultats des récherches maraîchères dans les zones d'altitude de l'ouest-Caméroun. *Agron. trop., Nogent* **25**, 660-81.

PRASAD, D. N. and PRAKASH, R. (1968). Floral biology of brinjal. *Indian J. agric. Sci.* **38**, 1053-61.

QUINN, J. G. (1971). The prospects for the commercial production of tomatoes in West Africa. *Ford Foundation/IITA/IRAT seminar on vegetable crop research, Ibadan.*

—— (1973). Nigeria: prospects for a tomato paste industry. *Span,* **16**, 25-7.

RAJASEKARAN, S. (1970a). Cytology of the hybrid *S. indicum × S. melongena* var. *insanum. Curr. Sci.* **39**, 22.

—— (1970b). Cytogenetic studies on hybrid *S. indicum × S. melongena* and its amphidiploid. *Euphytica* **19**, 217-24.

—— (1970c). Sterility in an inter-varietal hybrid *Solanum melongena × S. melongena* var. *bulsearensis. Madras agric. J.* **57**, 194-5.

RAMAN, K. R. (1964). Further studies on hybrid vigour in brinjal. *Madras agric. J.* **51**, 79.

SADIK, S. (1971). Oxalate contents of some leafy vegetables. *Ford Foundation/IITA/IRAT seminar on vegetable crop research, Ibadan.*

SAMBA, B. (1971). Export of vegetables by air from Senegal; organization of production, collection, sorting, packing and despatch. *Ford Foundation/IITA/IRAT seminar on vegetable crop research, Ibadan.*

SAMBANDAM, C. N. (1962). Heterosis in egg plant. *Econ. Bot.* **16**, 71-6.

—— (1964). Early exhibition of heterosis in brinjal hybrids. *J. Annamalai Univ.* **258**, 12-17.

—— (1965). Natural cross pollination in brinjal. *Proc. Participant Seminar, Univ. Tennessee* **4**, 27-30.

—— (1971). Exploiting *Cucumis callosus* in breeding muskmelon varieties resistant to fruit fly. *Proceedings of the seminar on cucurbits, Udaipur.*

SAMSON, J. A. (1971). The improvement of tropical vegetables. *Ford Foundation/IITA/IRAT seminar on vegetable crop research, Ibadan.*

SCHMIDT, D. R. (1971). Comparative yields and composition of eight tropical leafy vegetables grown at two fertility levels. *Agron. J.* **63**, 546–50.

SEINHORST, J. W. (1965). The relationship between nematode density and damage to plants. *Nematologica* **11**, 137–54.

SINNADURAI, S. (1970). The effect of nitrogen fertilizers on Indian spinach (*Basella alba* L.). *Ghana J. agr. Sci.* **3**, 51–2.

SOITOUT, M. (1967). Le Gabon et les cultures légumières. *Agron. trop., Nogent* **22**, 66–9.

—— (1969). Les récherches légumières entreprises par l'IRAT. *Agron. trop., Nogent* **24**, 327–92.

—— (1971*a*). Some problems relative to the storage and marketing of vegetable seeds in the tropics. *Ford Foundation/IITA/IRAT seminar on vegetable crop research, Ibadan.*

—— (1971*b*). Vegetable crop research undertaken by IRAT in francophone West Africa. *Ford Foundation/IITA/IRAT seminar on vegetable crop research, Ibadan.*

SRIVASTAVA, D. N., and NATH, P. (1971). Hybrid vigour in bittergourd and its retention in F_2 generation. *Proceedings of the seminar on cucurbits, Udaipur.*

TEMIZ, K. (1968). Investigations on the role of plant parasitic nematodes in *Pseudomonas solanacearum* infection in some tomato varieties. *Rev. appl. Mycol.* **47**, Abstr. 3577.

TERRA, G. J. A. (1966). *Tropical vegetables.* Communication 54/e, p. 107. *Royal Tropical Institute, Amsterdam.*

—— (1967). *Rapport sur l'amélioration de la nutrition par la création de jardins familiaux.* Report TA/54/55. FAO, Rome.

THAKUR, M. R. and CHOUDHURY, B. (1965). Inheritance of some quantitative characters in *Luffa acutangula. Indian J. Hort.* **22**, 185–9.

—— —— (1966). Inheritance of some qualitative characters in *Luffa. Indian J. Genet. Pl. Breed.* **26**, 76–86.

—— —— (1967). Interspecific hybridization in *Luffa* species. *Indian J. Hort.* **24**, 87–94.

—— SINGH, K., and SINGH, J. (1968). Hybrid vigour studies in brinjal. *J. Res., Punjab. agric. Univ.* **5**, 492–5.

TINDALL, H. D. (1971). Vegetable crop research in anglophone West Africa. *Ford Foundation/IITA/IRAT seminar on vegetable crop research, Ibadan.*

TOWN, P. A. (1962). Intensive vegetable production in the forest zone of Ghana. *Int. hort. Congr. (Brussels),* **16**, 464–72.

—— (1964). *A summary of vegetable variety trials.* (mimeo). University of Science and Technology, Faculty of Agriculture, Department of Horticulture. Kumasi, Ghana.

UFER, M. (1971). *Manihot* and *Dolichos. Trop. Root Tuber Crops Soc. Newsl.* **4**, 45–50.

VAN EIJNATTEN, C. L. M. (1969). *Report on home gardens for improved human nutrition 1966–1968.* FAO, Rome.

—— (1971). Home gardens: principles and experience. *Ford Foundation/IITA/IRAT seminar on vegetable crop production, Ibadan.*

VITTUM, M. T. (1966). Maximum yields of processing vegetables. *Hort. Sci.* **1**, 50–2.

WHITE, R. A. (1967). *A summary of experience, observations and trials in vegetable production in Ghana.* USAID, Accra.

WINSTEAD, N. N. and KELMAN, A. (1952). Inoculation techniques for evaluating resistance to *Pseudomonas solanacearum. Phytopathology,* **42**, 628–34.

WU LEUNG, W. T., BUSSON, F., and JARDIN, C. (1969). *Food composition table for use in Africa.* FAO, Rome.

Appendix: common botanical names of vegetables

African spinach	*Amaranthus hybridus* L.
Aubergine	*Solanum melongena* L.
Bambara groundnut	*Voandzeia subterranea* (L.) Thou.
Bitter gourd	*Mormordica charentia* L.
Borecole, kale	*Brassica oleracea* var. *acephala* D.C.
Bottle gourd	*Lagenaria siceraria* (Mol.) Standl.
Brinjal	*Solanum melongena* L.
Cabbage	*Brassica oleracea* L. var. *capitata* L.
Cassava	*Manihot esculenta* Crantz
Cauliflower	*Brassica oleracea* L. var. *botrytis* L.
Chilli pepper	*Capsicum annum* L. var. *longum* Sentd.
Chinese cabbage	*Brassica chinensis* L.
Cocoyam	*Colocasia antiquorum* Schott
Cowpea	*Vigna unguiculata* (L.) Walp.
Cucumber	*Cucumis sativus* L.
Dolichos bean	*Lablab niger* Medik.
Egg plant	*Solanum melongena* L.
Egyptian bean	*Lablab niger* Medik.
Fluted pumpkin	*Telfairia pedata* Hook.
French bean	*Phaseolus vulgaris* L.
Garden egg	*Solanum melongena* L.
Gherkins	*Cucumis sativus* L.
Indian spinach	*Basella alba* L.
Kang kong	*Ipomoea aquatica* Forsk.
Lettuce	*Lactuca sativa* L.
Lima bean	*Phaseolus lunatus* L.
Melon	*Cucumis melo* L.
Musk melon	*Cucumis melo* L.
Okra, okro	*Hibiscus esculentus* L.
Onion	*Allium cepa* L.
Oyster nut	*Telfairia pedata* Hook.
Pumpkin	*Cucurbita maxima* Duch. ex Lan.
Roselle	*Hibiscus sabdariffa* L.
Sweet pepper	*Capsicum annuum* L.
Sweet potato	*Ipomoea batatas* Lam.
Tannia	*Xanthosoma sagittifolium* (L.) Schott
Tomato	*Lycopersicon esculentum* Mill.
Water leaf	*Talinum triangulare* (Jacq.) Willd.
Water melon	*Citrullus lanatus* (Thunb.) Mansf.
West African sorrel	*Corchorus olitorius* L.
Winter squash	*Cucurbita maxima* Duch. ex Lan. var. *maxima*
Yard-long bean	*Vigna urgenculata* (L.) Walp.
Yautia	*Xanthosoma sagittifolium* (L.) Schott

8. Forage and fodder crops

L. V. CROWDER AND H. R. CHHEDA

8.1. Introduction

Systematic research on forage and fodder crops in West Africa began expanding during the 1950s. Grass and legume research has been largely directed towards evaluation of introduced species suitable for intensely managed sown pastures rather than improvement of natural and semi-natural grazing lands (Adegbola 1971a; Borget 1971; Haggar 1969a; Oyenuga 1971a). Information gained has led to improved livestock nutrition on government farms, experiment stations, and some ranches but has little application to the traditional livestock production systems. Most grazing lands should not and cannot be destroyed and replaced by improved forage species at the present stage of livestock development. Under some conditions the introduced species could serve as temporary leys for supplemental grazing and dry season feed, as silage and hay, and in ways to enhance the utilization of natural grasslands.

Cattle keeping and production is concentrated in the northern regions of low tsetse-fly incidence, although types with resistance and tolerance, e.g. Ndama and West African Shorthorn, have been kept in the south for many years. Other breeds also prosper, despite the prevalence of trypanosomiasis, when provided with a high plane of nutrition throughout the year (Oyenuga 1967).

Areas suitable for ecological large-scale livestock production are characterized by periods of extreme drought, ranging from about 8 months in the Sudan zone to about 4 months in the southern Guinea and derived savannas. This imposes a direct influence on the amount and quality of herbage available during the year and an indirect effect on livestock performance. Animals find ample forage of fair quality during the wet period and gain weight, but lose much of it in the dry season, resulting in a 'stop-and-grow' type of production.

Lack of adequate year-round nutrition is probably the most important factor contributing to low animal output. Research work clearly indicates that the nutritive value of most native and natural grasses growing on low fertility soils is generally poor. During the dry season there is a further sharp decline in quality. It is largely the poor quality and low intake of forage that contribute to the slow and interrupted growth rate, haggard conformation, delayed maturity, diminished reproductive rate, and low production of meat and milk (Oyenuga 1958, 1967).

In the last two decades progress has been achieved towards increased productivity of several cash and human food crops. Inadequate funds, technical manpower shortages, abundance of natural grasslands, and low cattle and human population densities have led to a lack of high priority being given to the genetic improvement of pasture crops.

This chapter summarizes the results of forage and fodder crops research in West Africa. In this and other regions in the tropics and sub-tropics more attention has been given to the sub-tropics and less to humid lowland areas. Roseveare (1948) has reviewed literature concerning the grasslands of Latin America and Crowder has updated research findings in 1971. More detailed accounts were given for Colombia by Crowder (1967) and, with a listing of 180 publications, by Lotero (1970); Brazil by Grossman, Aronovich, and Campello (1965); Mexico by Guevarra-Calderón (1967); and Puerto Rico by Vicente-Chandler (1967). Rangeland research in East Africa was discussed by Heady (1960); Naveh (1966); Pratt, Greenway, and Gwynne (1966); and Bogdan and Pratt (1967). Rangeland research with references to legume breeding was dealt with by Bogdan (1971), to grass breeding by Boonman (1971), and to seed production by Combes and Verburgt (1971). Risopoulos (1966) and Jurion and Henry (1969) summarized the work with pasture and forage species in Zaire. Since the mid-1950s Australian researchers have added to our knowledge of introductions and evaluations, agronomic studies, breeding and selection, seed production, and grazing trials (Barnard 1966; Hutton 1970; Norman 1966; Norris 1971;

Wilson 1968). Information of a general nature has been cited by the Commonwealth Bureau of Pastures and Field Crops (1962), Davies and Skidmore (1966), McIlroy (1972), Semple (1970), Whyte (1968), Whyte, Moir, and Cooper (1959) and Whyte, Nilsson-Leissner, and Trumble (1966).

8.2. West African environment, vegetation, and grasslands

West Africa is bounded by the Atlantic Ocean on the south and west, the Sahara on the north and plateaux and peaks of the Cameroun and Adamawa highlands on the east, lying between about 5° and 20° N. latitude. It is divided into east-west belts by relief, climate, vegetation, soils, agriculture, and animal husbandry, and to some extent ethnic groups, societies, and religions.

8.2.1. Climate

Climate is a predominating factor in the way of life in West Africa and rainfall is the vital element (Richard Molard 1956). Its distribution distinguishes

the climatic zones arranged south to north (Aubréville 1949; Keay 1959; Monod 1957; Morgan and Pugh 1969; Rossetti 1965; Whyte 1968). They closely follow the vegetation zones (Boughey 1957; Keay 1959) and grass-woody (Rattray 1960) communities recognized in West Africa. Over 2500 mm of precipitation occur annually on parts of the coastal area but along the 15° latitudinal line may not exceed 250 mm (Fig. 8.1) (Thompson 1965). Data indicate high unreliability between years with regard to amount, regime, onset, and cessation (Hopkins 1965a).

Mean daily temperatures range from 20 °C to 32 °C and increase progressively from the coast, except during the harmattan season (Hopkins 1965a). The diurnal range fluctuates between 3 °C and 10 °C near the coast but between 8 °C and 20 °C in the north.

8.2.2. Vegetation

The vegetation consists of woody or grass-woody plant communities with some areas of open grass-

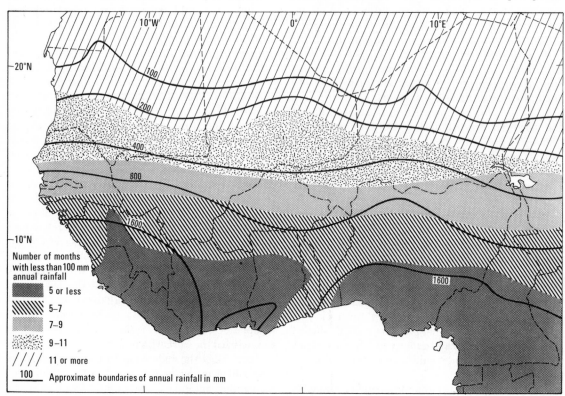

FIG. 8.1. Months with less than 100 mm rainfall and approximate boundaries of amounts (in millimetres). (Adapted from Thompson (1965).)

land. The extent of these areas is directly related to the number of months with less than 100 mm of rainfall and the minimum relative humidity at 1300–1500 hours (Fig. 8.1). Zones are recognized by the abundance of particular tree, shrub or bush, and grass species, which form mixed communities. The existing vegetation has been modified by burning, cultivating, tree-felling, and grazing. Zones described here (Fig. 8.2) have been simplified from various classifications (Aubréville 1949; Boughey 1957; Church 1960; Keay 1959; Lawson 1966; Roseveare 1953; Trochain 1955).

Lowland forest. The forest zone occurs in a 60–120 km wide belt along most of the coast from Sierra Leone eastward and includes the humid, mangrove, and swamp forests. A traveller might suspect a continuous forest cover, but much of the land is occupied by farmland under cultivation or bush fallow. Tall coarse grasses invade these areas and natural clearings, but grass cover is sparse or absent under the tree cover.

Derived savanna. This forest savanna occupies a narrow belt adjacent to and north of the lowland forest (Hopkins 1965a; Nye and Greenland 1960; Rose Innes 1971). The area is regularly swept by fires and many tall trees have been destroyed, leaving low growing types about 10 m high, gnarled shrubs, and bushes. Tall, coarse grasses cover abandoned farmlands but the forage, seldom grazed in the wet season, becomes parched and dry during the 3–4 month drought period, and provides the bulk of inflammable material. Reduction in frequency of burning causes an invasion of woody species and a rapid change back to forest as noted in the Olokemeji Forest Reserve in Nigeria (Charter and Keay 1960; Hopkins 1965b) and in the Kokondekro Forest Reserve in the Ivory Coast (Mensbruge and Bergeroo-Campagne 1958; Rose Innes 1971).

Montane forest. In the Guinea and Cameroun highlands a tall canopied montane rain forest exists with woodland at higher elevations where less precipitation occurs. It is easily destroyed by fire and gives rise to grassland (Church 1960).

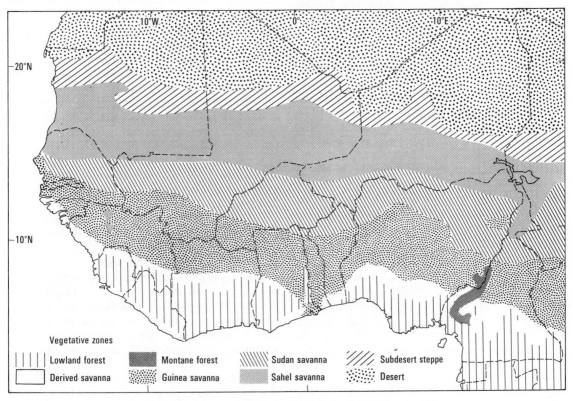

FIG. 8.2. Vegetative zones of West Africa. (Redrawn from Keay (1959).)

Guinea savanna. This is the largest vegetation zone, which lies between about 8° and 13° N latitude. It is largely woodland savanna with fire-resistant, broad-leaved, deciduous trees. Southern and northern sub-zones are recognized on the basis of moisture and vegetative cover (Church 1960; Hopkins 1965a). Tall, perennial tussocked grasses grow beneath the scattered trees and in open places. The cattle population is sparse, and a heavy growth of grass accumulates by the end of the wet season, then dries out and is readily burned during the dry season.

Sudan savanna. This is one of the most distinct zones in terms of climate and vegetation, being some 75–150 km wide and extending from Senegal to Nigeria and beyond. The southern region is wooded with many scattered single trees growing up to 10 to 15 m in height. They are shorter in the north and merge into shrub savanna with thorn bush. The grass cover is short, from 30 m to 150 cm high when mature, less heavily tufted, more feathery, and with finer leaves and stems than in the Guinea savanna. Much of the area is burned annually, but fires are less severe because of the smaller quantity of inflammable material. The zone is often densely populated with livestock since most regions are generally free of the tsetse fly.

Sahel savanna. The wooded steppe, originally thorn woodland, is open with scattered trees 5–10 m high. They are deciduous with deeply penetrating and spreading roots. There are many *Acacias* and thorn shrubs 2–3 m high with short conical bases and divided stems. Fleshy *Euphorbia* spp. appear among rocky sites. Grasses are short, discontinuous, wiry, and tufted, but heavily grazed by cattle and sheep. In some areas the land surface is bare. Fires are less serious than farther south, since there is not the density of tree cover and no great accumulation of grass for intensive burning.

Sub-Saharan steppe. The southern Sahara is fringed with a sub-desert steppe. In many places there is a dispersed permanent vegetation of small shrubby plants and bushes with *Acacias*, small trees and shrubs similar to those of the upper Sudan savanna, but more dwarfed. After summer rains annual grasses and herbs appear and soon mature. They grow as isolated tufts and much of the soil is bare.

8.2.3. Grasslands

The grasslands of West Africa comprise natural and semi-natural or indigenous components. Relatively few grassed savannas exist and most grazing lands occur in mixture with woody plant communities; i.e. woodland, tree, and shrub savanna.

The term savanna has been applied singly or with descriptive affixes to vegetation ranging from open grassland to woodland. Variations and modifications of grasslands, savannas, and steppes are encountered and considerable controversy has existed regarding terminology and descriptive nomenclature (Keay 1959; Phillips 1959; Pratt *et al.* 1966). The classification of savanna and steppe is based on the nature of the herbaceous layer and the density of woody vegetation (Cole 1963). At the Yangambi Conference (Boughey 1957), savanna was used for formations of grasses at least 80 cm high and forming a continuous, dominating layer. Steppe was described as open vegetation, sometimes with woody plants and perennial grasses widely spaced, usually less than 80 cm high and having interspersed annual plants.

Grassland distribution. Few climatic and ecological maps show the distribution of grasses but some include vegetation types with which grasses are associated (Keay 1959; Phillips 1959). One portrays the grass cover of Africa and describes the major grass communities on a floristic basis (Rattray 1960). A portion of the map redrawn and limited to West Africa is shown in Fig. 8.3. In a given region a grass genus emerged as the predominating type on the basis of percentage frequency and was selected to designate a grassland zone. Many other genera occur and in some localities no single one appears as the dominant grass. The most typical genera chosen as representatives of different communities in West Africa include *Pennisetum*, *Hyparrhenia*, *Andropogon*, *Cenchrus*, and *Aristida*.

The *Pennisetum* belt is closely associated with secondary tree growth following destruction of the forest where grasses characterize disturbed areas, commonly occurring in clearings.

The *Hyparrhenia* belt extends across the derived savanna zone and moves into the southern Guinea woodland savanna. It is considered to be a fire sub-climax. A number of communities with local dominants exist, the dominance depending on soil texture, degree of drainage, and extent of fire.

The *Andropogon* belt comprises the largest belt of tall grasses and spreads over the northern Guinea and most of the Sudan savannas. The most prevalent

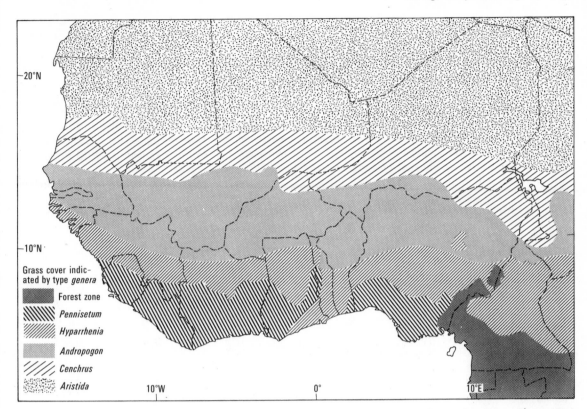

FIG. 8.3. Grass cover of West Africa as indicated by type genera. (Rattray (1960), by permission of the Food and Agriculture Organization.)

species is *Andropogon gayanus* associated with other species of the same genus and species of many other genera.

The *Cenchrus* belt lies astride the Sahel savanna but protrudes southward into the upper Sudan savanna. It is a region heavily grazed without recovery periods so that assessment of predominating species becomes difficult.

Grasses of the genus *Aristida* fringe the Sahara and occur over many parts of Africa. They frequently preponderate because of excess grazing and heavy trampling, which is the case on the subdesert steppe.

Most grasses have a wide range of adaptation and extend over several belts of the indicated type genera so that a comingling exists among plant communities. It would be impossible to list all of the prevalent species encountered but a few surveys are noteworthy (Adam 1958, 1966; Bille, Lamarque, Lebrun, and Rivière 1968; Boudet 1963; Peyre de Fabreques 1965; Rains 1968; Rattray 1960; Rose Innes 1962, 1967).

G

8.3. Evaluation and genetic improvement of forage species

A first step towards genetic improvement is the identification of important species for immediate utilization and for use as sources of genetic material. The following characteristics are generally considered for the selection of pasture and forage species: (1) seed supply or vegetative material, (2) ease of establishment, (3) rapid early growth for ground cover and weed competition, (4) production of palatable and nutritious herbage during the growing season, (5) seeding during the dry season, (6) drought tolerance, (7) rapid regrowth after grazing and the dry season, (8) ease of eradication, (9) persistence under grazing, and, for legumes, (10) nitrogen-fixing capability, implying efficient *Rhizobium* symbiosis, and (11) compatible association with grasses.

8.3.1. Search for adapted grasses and legumes
Over 300 grass and legume species belonging to more than 50 genera have been evaluated for use in pastures.

Those being used or showing potential are listed in Table 8.1 (Ahlgren, Adegbola, Eweje, and Salami 1959; Borget 1969; Cadot 1971*a*; Crowder 1973; Evans 1961; Foster and Mundy 1961; Rains 1963). These assessments indicate that:

(1) dry-matter yields of unselected genotypes of grasses are quite high;

(2) potentially useful grass species are of tropical African origin and potentially useful legumes mainly South American;

(3) grasses are cross-pollinated or apomictic and legumes self pollinated;

(4) most have been introduced rather than collected from their areas of origin;

(5) evaluations have been made on a limited number of strains with few data from long-term trials;

(6) some species and often the same genotypes have been repeatedly re-evaluated.

8.3.2. Search for suitable genotypes of adapted forages

Research has been confined mainly to (1) *Panicum maximum*, (2) *Cynodon* spp., (3) *Pennisetum purpureum*, (4) *Andropogon gayanus*, and (5) *Pennisetum pedicellatum*.

1. *Panicum maximum*. At Adiopodoumé in the Ivory Coast a clonal 'bank' consisting of 800 entries obtained from natural populations, research

TABLE 8.1

Forage species adapted to West Africa

	Origin (area of diversity)	2n, chromosome number	Breeding behaviour	Zone of adaptation	Average yields (t/ha)
GRASSES					
Andropogon gayanus	Tropical and S. Africa	20, 40	C.P.	G.S.; Su.	4–6
A. tectorum	Tropical Africa	20	C.P.	G.E.; G.S.	4–7
Brachiaria brizantha	E. Africa	36, 54	Apo.	G.S.	6–9
B. ruziziensis	E. Africa	18	C.P.	G.E.; G.S.	8–10
B. decumbens	Tropical Africa	36	Apo.	G.E.; G.S.	7–10
Cenchrus ciliaris	Central and S. Africa	36	Apo.	Su.; Sa.	5–6
Chloris gayana	E. and S. Africa	20, 40	C.P.	G.S.; Su.	10–14
Cynodon nlemfuensis	E. Africa	18, 36	C.P.	G.E.; G.S.	12–16
Cynodon spp.	E. and S. Africa	18, 36	C.P.	G.E.; G.S.	10–14
Digitaria decumbens	S. Africa	27	Sterile	G.E.	5–7
D. smutsii	S. Africa	18	C.P.	G.E.; G.S.	5–8
Hyparrhenia spp.	E. and S. Africa		C.P.+Apo.	G.S.; Su.	6–8
Melinis minutiflora	E. and W. Africa	36	Apo.	G.E.; G.S.	10–13
Panicum maximum	Tropical and subtropical Africa	16, 32, 48	Apo.	G.E.; G.S.	10–25
P. coloratum	E. and S. Africa	18, 36, 54	C.P.+Apo.	G.E.; G.S.	4–7
Pennisetum purpureum	Tropical Africa	28	C.P.	G.E.; G.S.	10–25
P. pedicellatum (annual)	N. and W. Africa	28	C.P.	G.S.; Su.	5–9
Setaria sphacelata	E. and S. Africa	18, 36, 54	C.P.	G.E.; G.S.	12–15
Tripsacum laxum	Central America	36, 72	C.P.	G.E.	10–15
LEGUMES					
Calopogonium mucunoides	S. America	36	S.P.	G.E.; G.S.	
Centrosema pubescens	S. America	20	S.P.	G.E.; G.S.	
Glycine wightii	E. Africa and S.E. Asia	22, 44	S.P.	G.E.; G.S.	
Macroptilium atropurpureum	S. America	22	S.P.	G.S.; Su.	Mostly grown in association with grasses
Pueraria phaseoloides	S.E. Asia	22	S.P.	G.E.	
Stizolobium deeringianum	N.E. Asia	22	S.P.	G.E.; G.S.	
Stylosanthes humilis	S. America	20	S.P.	Su.	
S. gracilis	S. America	20	S.P.	G.E.; G.S. Su.	

C.P. = Cross-pollinated
S.P. = Self-pollinated
Apo. = Apomictic

G.E. = Guinea equatorial (forest and derived savanna
G.S. = Guinea savanna

Su. = Sudan
Sa. = Sahel

stations, and selections from breeding lines has been assembled. Information on the mode of reproduction and variability of dry-matter yield, tillering, flowering and seed production, nutritive quality, and tolerance to diseases and pests has been gathered (Pernes 1971; Pernes, Combes, and R'n'e-Chaume 1970; Combes and Pernes 1970). *P. maximum* is largely apomictic but diploids ($2n = 16$) collected from East Africa are completely sexual. Most of the tetraploids are facultative apomicts with 2-3 per cent sexuality. Aneuploids and higher polyploids are also encountered. Chromosome doubling of diploids results in completely sexual tetraploids which could be used in a breeding programme with apomictic tetraploids. Lines having desirable agronomic characteristics and tolerance to diseases and pests have been identified. Sixteen promising clones gave average dry-matter yields of 30 t/ha per annum with top production up to 40 t/ha. The dry-matter digestibility ranged around 50 per cent and nitrogen content around 2·5 per cent.

P. maximum clones were assessed in Nigeria and S.112, a selection of Kenyan origin, was reportedly superior in terms of dry-matter and crude protein yields (Adegbola 1971a).

2. *Cynodon* spp. Biosystematic and breeding investigations utilizing over 200 collections of the 8 species within the genus, at the University of Ibadan (Chheda 1971) revealed wide genetic variability. Some species are only either diploid ($2n = 18$) or tetraploid ($2n = 36$) while others have both diploid and tetraploid forms. Under lowland humid conditions the non-rhizomatous *C. dactylon* var. *elegans* and var. *coursii*, *C. aethiopicus*, and *C. nlemfuensis* offer high pasture potential with several ecotypes yielding over 15 t/ha per annum of dry-matter. *C. plectostachyus*, though high-yielding in the year of establishment, shows poor persistence when cut regularly.

Cynodon IB.8, a selection of *C. nlemfuensis* var. *nlemfuensis* collected from the Lake Manyara area of Tanzania, has been released for general distribution. It is extremely vigorous and a highly self-sterile tetraploid ($2n = 36$), easy to establish vegetatively, drought-tolerant, and greatly superior to the local *Cynodon* ecotype. With fertilization and good management dry-matter yields of about 20 t/ha per annum and animal live-weight gains of about 650 kg/ha have been obtained. The *Cynodon* IB.8 pastures persist under heavy grazing for more than 5 years. Regional trials were conducted to identify *Cynodon* genotypes

suitable for different ecological areas of southern Nigeria. Wide differences in crude protein content (9-13 per cent), *in vitro* organic-matter digestibility (45-60 per cent), and utilization under grazing (28-70 per cent) have been noted (Chheda 1973).

3. *Pennisetum purpureum*. Promising collections of southern Nigeria origin evaluated under a 6-week cutting regime showed wide variation with respect to dry-matter yield (8-12 t/ha per annum under low-fertility conditions), crude protein (10-17 per cent), crude fibre (25-29 per cent), *in vitro* organic-matter digestibility (51-60 per cent), and agronomic characteristics such as leafiness, stem thickness, flowering behaviour, drought tolerance, and persistence (Chheda, Aken'Ova, and Crowder 1973b). Two promising clonal selections which yield over 20 t/ha per annum of dry matter have been recommended for distribution.

4. *A. gayanus*. Natural populations within latitude $7°-13°$ N and longitude $4°-14°$ E were sampled and 150 collections studied in northern Nigeria (Foster 1962). Variation in vegetative characters and breeding behaviour were reported. Diploid ($2n = 20$) and tetraploid ($2n = 40$) races were recognized. Inbreeding depression in this allogamous species is commonly encountered. Studies on the floral biology helped towards formulation of a breeding programme and indicated that the growing season could be extended by using selections from more southerly latitudes. Selection work was initiated at Samaru in 1956 and later discontinued.

5. *P. pedicellatum*. This is an annual forage species capable of growing in areas with 3-4 months of wet season. Regions of northern Cameroun, with 600-1200 mm per annum rainfall, were surveyed and 79 ecotypes assembled (Barrault 1971). They were classified into early-, medium-, and late-flowering groups. Selection within each group was based on leaf/stem ratios, single-plant forage production, and disease resistance. Materials are being combined to form composites which could be distributed commercially. Inbreeding work is in progress in order then to develop synthetics by panmixis among selected inbreds of proven combining ability. The results obtained in Cameroun could be of use in the northern Guinea savanna and Sudan zones of West Africa.

8.3.3. Production of superior intraspecific and interspecific hybrids

Herbage productivity and nutritive quality of superior hybrids can be preserved through vegetative production and/or the production of apomictic seed. The discovery of sexual types in *Panicum maximum* (Combes and Pernes 1970) allows recombination of existing variation of the apomictic ecotypes, a possible hybridization with *P. trichocladum* and *P. intestum*, and recovery of the apomictic mode of seed production.

Chheda (1971) reported interspecific hybridization among *Cynodon aethiopicus, C. dactylon, C. incompletus,* and *C. nlemfuensis* and observed that several *C. nlemfuensis* × *C. dactylon* hybrids, along with intraspecific hybrids of *C. nlemfuensis*, exhibited heterosis. These are being evaluated in different ecological regions of southern Nigeria.

Random hybridization of S$_2$ inbreds of *Pennisetum purpureum* (Chheda, Aken'Ova, and Crowder 1973) has resulted in a number of populations comparable in yield and quality to the superior clonal selections. Chheda, Aken'Ova, and Crowder (1974) produced 2831 F$_1$ hybrids between the tetraploid *P. purpureum* and the diploid *P. typhoides* cv. 'maiwa'. Twenty-two were selected on the basis of vigour, leafiness, drought-tolerance, and flowering behaviour and evaluated for their forage potential along with two promising elephant-grass selections. Some hybrid genotypes had similar dry-matter yield potential to that of elephant grass but a greater proportion of leaves, increased acceptability, and higher utilization by cattle and higher *in vitro* organic-matter digestibility. They were late in flowering and flowered for short periods of time. A grazing trial, in which unselected hybrids were used, revealed about 35 per cent higher liveweight gains when compared to those on elephant grass pasture (Chheda *et al.* 1974). It was proposed that production of commercial hybrid seed production using male sterile 'maiwa' millet and selected elephant-grass parents would be feasible.

8.3.4. *Seed production*

An inadequate supply of high-quality seed imposes a serious constraint on the use of adapted species in West Africa, whether indigenous or introduced, as no seed is yet produced commercially. Some species have been vegetatively propagated, but procurement of planting materials and labour, as well as the uncertainty of survival, constitute deterrents to their extensive use.

The seed yields of important pasture species, particularly of grasses, are low and of poor quality (Cadot 1971*b*). The seeds lose viability rapidly in the humid environment and, within a year of harvest,

germination falls below 20 per cent, unless stored in air-conditioned rooms. Genotypic variability, prolonged flowering, sparsity of seed-producing tillers, high degree of sterility, low seed retention due to shattering, spikelet diseases, and bird damage result in uneconomic seed returns. Experience from elsewhere in Africa (Whyte *et al.* 1959; Boonman 1971; Combes and Verburgt 1971) indicate that with proper selection procedures and cultural techniques seeds can be economically produced from grass species such as *Chloris gayana, Cynodon* spp., *Panicum maximum, P. coloratum, Pennisetum purpureum, Melinis minutiflora, Setaria sphacelata*, and *Andropogon gayanus*.

Okigbo and Chheda (1966) observed significant varietal and location differences in seed production of *Cynodon* spp. Differences in seed yields among ecotypes of *Panicum maximum* were observed in the Ivory Coast (Pernes 1971) and up to 120 kg/ha per annum of viable seed from three harvests were obtained. Haggar (1967) obtained two- or three-fold seed increase of *A. gayannus* with 222 kg/ha of nitrogen application and observed that management systems which promoted early tillering resulted in increased seed yields.

Seeds of legumes such as stylo, centro, puero, and calopo are hand-harvested for local use, e.g. on a monastery farm near Palimé in Togo and various government ranches in Nigeria.

8.4. Establishment and renovation of the grass cover

The improvement of natural grasslands usually includes some means of bush control and entails the introduction of adapted forage species. Any operation which requires clearing and cultivation is unlikely to give immediate economical returns and should only constitute one phase of a long-range programme if undertaken at all.

8.4.1. *Oversowing grazing lands*

Successful establishment of grasses and legumes sown into natural grazing lands depends upon the choice of appropriate and compatible species as determined primarily by the climatic potential of the area, the application of suitable techniques of land treatment, and the methods of sowing (Bogdan and Pratt 1967; Cadot 1971*a*). Under favourable conditions scratch ploughing or light discing will probably be sufficient but more intensive mechanical operations are required in harsher environments. In some instances aerial herbicide treatment and pelleted seeding treatments might have a place. Water management practices are needed to reduce soil erosion, to prevent washing away of seeds and to conserve moisture for seedling growth.

Stylosanthes gracilis (syn. *S. guyanensis*) has been successfully established in natural rangelands by overdrilling (Rains 1963), by broadcasting seed after the passage of heavy discs (Letenneur 1971) or into scratched grooves (Foster 1961*b*), or, more simply, by feeding seed to cattle. *S. humilis* and *S. mucronata* are other legumes which have shown promise in northern Nigeria (Haggar 1969*c*; Leeuw and Brinckman 1973). *Andropogon gayanus* is oversown by broadcasting seeds in cleared areas. With favourable soil moisture seeds germinate and seedlings grow to thicken the herbage cover (Rains 1963; Rose Innes 1967).

8.4.2. Seedbed preparation
The degree of successful establishment seems to be closely related to the extent of tillage prior to sowing (Bogdan and Pratt 1967; Rains 1963). A seedbed which has been well prepared, i.e. by land turning, harrowing, and firming, ensures soil and seed contact which favours germination, seedling emergence, and development. Vegetatively propagated species regenerate growth and become established on land less well prepared than that needed for sown species.

8.4.3. Method of sowing or transplanting
Seeds of herbage grasses and legumes are broadcast by hand, but mechanical equipment has been used on government farms and experiment stations. The semi-decumbent and spreading types, e.g. *Brachiaria* spp., *Cenchrus* spp., and *Chloris* spp., can be sown in rows, generally spaced about 1 m apart (Adegbola 1971*a*; Haggar 1969*c*; Nwosu 1960; Rains 1963). This allows lower seeding rates and eases hand weeding, when this is needed, to ensure establishment.

A number of species have been established vegetatively with stem cuttings, stolons, and crown splits, e.g. *Cynodon* spp. (Saleem 1972), *Pennisetum purpureum* (Adegbola 1971*a*), *Digitaria decumbens*, *Panicum maximum*, *Melinis minutiflora*, *Andropogon gayanus* (Borget 1969), and *Brachiaria* spp. and *Stylosanthes gracilis* (Nwosu 1961). These are usually transplanted by hand into rows marked at about 1·0 m and hills 0·50–1·0 m apart.

Hand-weeding is usually practised either by pulling or slashing, but weeds are controlled by mowing when they reach 20–30 cm in height, and by the use of weedicides (Adegbola 1971*a*).

Andropogon gayanus becomes well established when sown with maize and soya bean but *Mucuna* (*Stizolobium* sp.) and *Pennisetum pedicellatum* are too competitive (Haggar 1969*b*). Undersowing *Chloris gayana* into *Pennisetum typhoides*, planted in 1·0 m rows and 0·75 m hill spacings, gives adequate stands by the time of millet harvest (Rains 1963). Trials from other countries probably have application to West African conditions (Crowder 1971).

8.4.4. Time of sowing or transplanting
Sowing shortly after the rains begin takes advantage of nitrogen released by mineralization, but coincides with strong weed competition and severe rain storms. The first rains are unreliable and short periods of water stress may affect seedling survival. Early sowing in the Sudan zone is important because of the short growing season (Rains 1963). Further south sowing or transplanting could be delayed until rainfall is more dependable, or until the end of the short dry season where a bimodal pattern prevails (Ahlgren *et al.* 1959).

8.4.5. Seed rate
The quantity of seed used depends on viability and purity. Commercially grown and processed seeds have a high percentage of viable seeds. Those locally harvested contain excessive amounts of trash and plant parts, and germination may be 5 per cent or less. Seed rates range from 10 ka/ha to more than 40 kg/ha for grasses and from 5 kg/ha to 15 kg/ha for legumes (Adegbola 1965; Ahlgren *et al.* 1959; Bowden 1963; *Cahiers agriculture pratique en pays chauds* 1965–71; Rains 1963).

8.4.6. Depth of sowing
This is usually related to seed size, the seedling emergence, and survival of small-seeded species decreasing with deeper placement. The optimum depth of most grasses lies between 1 cm and 3 cm. In Nigeria, the seedling emergence of *Andropogon gayanus* declined markedly with sowings made deeper than 2 cm (Bowden 1963). Large-seed legumes such as *Centrosema*, *Leucaena*, *Clitoria*, and most *Crotalaria* spp. may be placed at 2·5–5·0 cm below the soil surface (Adegbola 1964) but small-seed types, such as stylo, glycine, puero, calopo, and *Desmodium* spp., must be placed at no more than 2·5 cm deep.

8.4.7. Seed treatment

Seeds of most tropical legumes need some form of scarification to improve germinability. The degree of seed-coat hardness varies among species, seed lots, and, to some extent, with the stage of maturity at harvest. Commercial harvesting and processing ruptures the seed coats of many species, but hand-harvested seeds require treatment (Adegbola 1971*a*; Rains 1963). Herbage legume species grown in West Africa generally produce an abundance of nodules, but on land not previously cropped with a legume there may be an advantage in inoculation (Rains 1963). In pot experiments stylo plants from seeds inoculated with selected strains of *Rhizobium* spp. produced twice as much root and top growth as those from a local strain and uninoculated seeds (Adegbola and Onayinka, 1966*a*). Studies in Australia have shown that most tropical legumes are highly specific in their requirements for effective symbiotic *Rhizobium* but such data are lacking for West Africa (Norris 1971).

The use of seed protectants and pelleting has not been investigated, but trials from other areas suggest they may have value under some conditions. Rock phosphate pelleting is more desirable for tropical legumes as lime may reduce inoculation (Norris 1971).

8.4.8. Grass–legume mixtures

Combinations of *Pennisetum purpureum*, *Panicum maximum*, *Cynodon* spp., *Brachiaria* spp., and *Melinis minutiflora* with centro are suitable for the forest zone (Ahlgren *et al.* 1959; Borget 1969; Evans 1961; McIlroy 1962; Moore 1962). Calopo and puero frequently appear spontaneously and provide short-term grazing when intersown. These same species generally thrive in the derived and southern Guinea savanna, but *Andropogon gayanus* with centro or stylo is likely to be more productive (Foster and Mundy 1961). Stylo with *A. gayanus* or *Chloris gayana* provides valuable mixtures in the northern Guinea and southern Sudan zones (Rains 1963). In practice it is unusual to find pastures of grass–legume mixtures except on government farms and experiment stations.

Complex mixtures are usually sown but after one or two seasons the components comprise one or two grasses and a single legume (Moore 1962; Oyenuga and Olubajo 1966). In many instances the legumes flourish sporadically or may disappear.

8.4.9. Fertilizers

Forage and fodder species can usually be established without applied fertilizers. With continuous cropping, removal of residues, and no fertilizer inputs, however, soil fertility declines sharply so that applied phosphate and potash are required for grasses and legumes, and nitrogen may be needed to boost early growth of grasses alone. Pot experiments provide a reliable guide for determining soil nutrient deficiencies and needed increments of nitrogen, phosphorus, and potassium (Borget 1969), but in general from 100 kg/ha to 200 kg/ha of phosphate and 50–100 kg of nitrogen are sufficient for the establishment and early growth (Adegbola 1964). Response to lime has been marginal or negligible (Ahlgren *et al.* 1959; Saleem 1972).

Vegetatively propagated grasses usually do not respond to applied nitrogen until they become established and several harvests are made (Crowder 1971; Saleem 1972). The need for potash depends on the inherent status of the soil (Rains 1963; Saleem 1972).

8.4.10. First-season management

Protection from grazing after seedling emergence or regeneration of growth of vegetatively propagated species is necessary to assure good stands. Usually 6–8 weeks after sowing, or until plants attain a height of 40–60 cm is ample time to allow for establishment; then grazing which is sufficiently light to avoid uprooting young plants and excessive trampling may be permitted for the next 10–12 weeks. It has been recommended that cutting or grazing of species such as *Andropogon gayanus* (Rains 1963) and *Stylosanthes gracilis* (Adegbola 1965) be deferred until the end of the first growing season.

8.4.11. Renovation of pastures

Rangeland dominated by *Andropogon gayanus* responded to applied nitrogen in Nigeria (Adegbola *et al.* 1968). The application of 110 kg/ha nitrogen increased dry-matter production and herbage intake by about 50 per cent and more than doubled animal gains for one season as compared to controls with no applied nitrogen.

Heavy stocking and overgrazing of natural rangeland leads to the invasion by weedy species. In Nigeria dry-season fallow (ploughing to 15 cm, with disc-harrowing in December and May), followed by seeding with *Andropogon gayanus* alone or in mixture with stylo reduced the stand of the weed species *Imperata cylindrica* from 95 per cent to 3 per cent within 6 months, as compared to overseeding without fallow. This treatment was also superior to discing and the application of dalapon followed by overseeding (Adegbola *et al.* 1970).

8.5. Agronomic management practices

Agronomic studies carried out on small plots under cutting managements provide an understanding of plant responses to selected practices. They also help in the evaluation of a large number of treatments and their interactions within a short period of time under different environmental conditions, at a much lower cost than would be possible with animal studies. The results provide guides in the planning and design of grazing and feeding experiments and give an indication of hay and silage yields.

8.5.1. Weed and pest control

Improper management, such as over- or undergrazing, can result in the invasion of brush and weeds. Fire, properly used, is an effective tool for reducing brush and tree invasion, maintaining desirable grass species, and stimulating rapid regrowth of grasses (Leeuw 1971; Rains 1963; Rose Innes 1971).

Clearing of savanna woodlands improves carrying capacity because of increased ground cover of grass (Rains 1963; Leeuw and Brinckman 1973). Slashing does not always have a significant long-term beneficial effect because of rapid regrowth of cut shrubs. Mechanical brush-cutting generally costs less than contract hand-slashing (Leeuw and Couper 1970). Neither appears to be economical in view of the low productivity of most rangelands, except at intervals of 3–5 years, as a measure of preventing tsetse-fly re-infestation (Leeuw 1971).

At Ibadan, cutting *Cynodon* spp. and *A. gayanus* at less than 6-week intervals, and at 5 cm height rather than 15 cm height above ground level, increased the percentage of weeds. Dressings of ample fertilizer for sustained plant growth reduces invasion by weeds (Ahlgren *et al.* 1959; Chheda and Akinola 1971a; Saleem 1972).

Biological control of weeds occurs with the use of *Cynodon* IB.8 (Chheda, unpublished). Root exudates from the actively growing grass inhibit the germination of some weed seeds and suppress the seedling growth of others.

Stylo and other competitive legumes such as centro and *Stizolobium* (*Mucuna*) spp. dominate *Imperata* when there is no interference by burning or by cattle (Risopoulos 1966). In Nigeria, *A. gayanus* reduced *Imperata* stands to a greater degree than *S. gracilis*, or a mixture of the two (Adegbola *et al.* 1970). The earlier seedling growth and heavier shade of the *Andropogon* may have accounted for the disparity.

Proper management of grazing animals provides another means of biological weed control. Well-regulated grazing maintains a desirable cover and prevents weed invasion (Ahlgren *et al.* 1959; Rains 1963). The inability of chemicals completely to eradicate brush and weedy grasses (e.g. *Imperata cylindrica*) and their high costs make them uneconomical (Adegbola *et al.* 1970; Little and Ivens 1965). A preventive control is effected by the use of high-quality seeds, i.e. those free of weeds in the first place. 'Ecological' protection can be achieved by choosing species adapted to the local soil and climatic conditions.

8.5.2. Cutting management

The interval between harvests and extent of defoliation profoundly affect herbage production, nutritive value, botanical composition, regrowth potential, and species survival. In general, an extended period between cuttings has the following results: an increase in the percentage contents of dry matter, crude fibre, lignin, and cell wall; an increase and then a decrease, or fluctuation, in total dry-matter production and nitrogen-free extract; a decrease in leaf/stem ratio, percentages of crude protein, mineral constituents (P, K, Ca, and Mg), ash, and soluble carbohydrates; an increase and then a decrease in the amount of nitrogen uptake by the plant and nitrogen recovery; a rapid decline in animal intake and in digestibility. With time, changes occur in the direction of juvenile conditions because of new tiller and leaf development, but never approach the immature stage because of the mass of aged material (Haggar 1970a; Haggar and Ahmed 1970; Lansbury, Rose Innes, and Mabey 1965; Oyenuga 1959a, 1959b, 1959c, 1960; Saleem 1972).

Increased dry-matter yields with extended cutting intervals are consequences of additional tiller and leaf formation, leaf elongation, and stem development. The period of maximum forage production varies with different grass species (Bowden 1963; Haggar 1970a; Nourrissat 1965). More nutritious forage is obtained with reduced cutting intervals so that an optimal harvest period must be chosen for a balance between quality and yield (Borget 1969, 1971; Cadot 1971a; Chheda and Akinola 1971a, 1971b).

Too-frequent cutting reduces total forage yields, depletes carbohydrate reserves, causes a decline in root development, and adversely affects regrowth potential (Akinola *et al.* 1971*a*; Rains 1963). At Ibadan, *Cynodon* IB.8 developed more new roots and achieved deeper soil penetration under a 6-week than a 4-week cutting scheme (Mackenzie and Chheda 1970). The higher energy reserves could be measured by the greater etiolated regrowth of sod plugs kept moistened in dark chambers (Akinola, Mackenzie, and Chheda 1971*b*).

Cutting near to ground level increases total and seasonal herbage yield as compared to more elevated cutting heights, but adversely affects plants in the same way as too-frequent harvesting. Over time the plants become weakened, stands thin out, weeds invade, and spots of bare ground between plants lead to soil erosion. A cutting height of 15–18 cm above ground level in Nigeria was needed to maintain productive stands of *A. gayanus* (Ahlgren *et al.* 1959; Rains 1963) and *C. nlemfuensis* (Saleem 1972), but a *Cynodon*-centro mixture performed well with 5, 10, and 15 cm treatments (Moore 1965). Grass–legume combinations are difficult to maintain under cutting management (Adegbola 1966) because their growth habits differ and patterns of physiological development do not coincide.

8.5.3. Fertilization

Response of established grass to applied nitrogen is dramatic in terms of increased dry-matter production. In northern Nigeria top-dressings of 50, 100, and 150 kg/ha of nitrogen boosted dry-matter yields of *A. gayanus* to 3550, 4140, and 4630 kg/ha per annum, respectively, as compared to 1330 kg/ha for the zero nitrogen treatment. Additional amounts of nitrogen up to 250 kg/ha, however, brought about a decrease in herbage yields. Yields of *Panicum maximum* were 3750, 5580, and 8500 kg/ha of dry-matter with 0, 50, and 200 kg/ha of nitrogen, respectively, and the use of 88 kg/ha of nitrogen versus nil doubled the quantity of forage of *Chloris gayana*, *Cenchrus ciliaris*, *Setaria sphacelata*, and *Hyparrhenia rufa* (Rains 1963). At Ibadan, *Cynodon nlemfuensis* produced averages of 10·2, 15·8, 17·8, 17·4 t/ha per annum of hay with 0, 84, 168, and 259 kg/ha of nitrogen top-dressing, respectively (Saleem 1972). Equally spectacular results have been obtained in the African francophone countries (Borget 1969). Split applications gave greater quantities of forage and more even distribution than a single application (Borget 1971; Saleem

1972). Additions of nitrogen after every second cutting or grazing provided as much total forage but greater fluctuations of individual harvests as additions than after every cutting. Nitrogen recoveries of 50–75 per cent were common with levels of 75–100 kg/ha per annum of applied nitrogen, and frequently the amount recovered exceeded 100 per cent because the deep-rooted grasses were able to use additional mineralized soil nitrogen (Chheda and Akinola 1971*b*; Moore 1965; Rains 1963; Saleem and Chheda 1972; and Chheda 1974).

Fertilizer nitrogen usually increased crude protein but depressed the P, K, Ca^{++}, and Mg^{++} content of herbage, due in part to a dilution effect by increased herbage production. It also decreased crude-fibre percentage, improved tiller density, and leaf area (Akinola *et al.* 1971*a*), and increased dry-matter digestibility (Saleem 1972). Root penetration, distribution, and bulk weight were improved by applied nitrogen (Mackenzie and Chheda 1970; Rains 1963). The regrowth potential of *Cynodon nlemfuensis* rose with increasing levels of nitrogen up to 164 kg/ha per annum because of increased nutrient reserves (Akinola *et al.* 1971*a*).

Single application of 164 kg/ha per annum of sulphur-coated urea nitrogen to an established cover of *C. nlemfuensis* at Ibadan gave an increased total dry-matter production over one wet season of 18·85 t/ha as compared to the same quantity of common urea, which increased total dry-matter production by only 17·45 t/ha, because sulphur-coated urea gave sustained production over a longer period. Split instalments of three or six equal doses favoured the common urea with increases of about 30 per cent over a single dressing as compared to 5 per cent increase for the sulphur-coated urea. Nitrogen recovery of the common form averaged 78 per cent while that of the sulphur-coated was 66 per cent. Total soil nitrogen at the end of the season was about 60 kg/ha greater for the sulphur-coated urea (Saleem 1972).

Application of phosphate fertilizers augmented forage yields on soils inherently low in phosphorus and not previously fertilized, particularly in combination with nitrogen (Borget 1971; Bowden 1963; Moore 1965; Rains 1963). Use of 225 kg/ha per annum of single superphosphate increased *A. gayanus* hay yields 30–50 per cent when combined with 100–200 kg/ha of nitrogen as compared to the non-phosphate treatment. Response to added phosphorus can be expected when the herbage phosphorus content drops to about 0·15 per cent, as shown by yield increases of *C. nlemfuensis* at Ibadan, where it was determined that the critical phosphorus concentration occurred at about 0·22 per cent (Saleem 1972).

Generally, response to added potassium does not appear until after several harvests and where herbage is removed (Ahlgren *et al.* 1959; Chheda and Saleem 1973*a*). In Ghana a mixed NPK fertilizer increased herbage and root growth of *Cenchrus ciliaris* (Asare 1969), but gave erratic results with *Panicum maximum* in eastern Nigeria (Tewari 1968).

A balance of grass to legume was maintained under grazing with an annual application of 125 kg/ha of single superphosphate to mixtures of centro with *Cynodon nlemfuensis* and *Pennisetum purpureum* at Ibadan (McIlroy 1962). Use of 0, 220, and 440 kg/ha per annum of superphosphate applied to the centro–*C. nlemfuensis–P. purpureum* mixture plus *Panicum maximum* and stylo gave 8·1, 10·0 and 12·0 t/ha of dry matter, respectively at Ibadan (Oyenuga and Hill 1966). Adding magnesium, copper, and molybdenum to the lower 220 kg/ha per annum of superphosphate increased forage production about 20 per cent but had no effect on the 440 kg/ha per annum treatment. Phosphorus–potassium and nitrogen–phosphorus–potassium fertilizers applied to *P. maximum* with centro and stylo showed variable results in eastern Nigeria (Tewari 1968). The optimal level of superphosphate as a single dressing at the time of sowing was 250 kg/ha for *S. gracilis* and 500 kg/ha for *S. humilis* in northern Nigeria (Leeuw 1969–71).

Grasses usually do not respond to application of limestone, but small amounts of it may enhance the growth and development of legumes. At Ibadan, roots of *C. nlemfuensis* growing in open-bottom boxes filled with 50 cm of acid soil, penetrated the 50 cm soil layer below the boxes, mobilizing calcium and transporting it to the upper 25 cm soil layer via the decaying root mass and residue of leaves and stems. As a consequence, the soil pH of the basement 15 cm soil layer dropped 0·7 units over the period of 1 year (Saleem 1972).

8.5.4. Irrigation

Few experiments with irrigation have been conducted. At Tarna (Niger) *Pennisetum purpureum* with 2132 mm per annum of water (rain+irrigation) and 600 kg/ha of nitrogen yielded 340 t/ha per annum of fresh material when cut 7 times (Borget 1971). *Phaseolus atropurpureus*, given the same treatment produced 100 t/ha per annum of fresh weight. Irrigation studies are under way in the Ivory Coast and are envisaged in Senegal (Cadot 1971*a*) and in Nigeria with the development of large irrigation projects (Haggar 1969*a*).

8.6. Grazing management and animal feeding

Herbage grasses in the tropics grow rapidly with an adequate and evenly distributed supply of moisture and nutrients. Unless cut or grazed at an appropriate time their nutritive value declines sharply with maturity and animal output is drastically curtailed (Oyenuga 1971*a*, 1971*b*).

8.6.1. Forage utilization

Pastures in the humid zones have been rotationally grazed and forage utilization determined by differences in forage before and after grazing. Short-growing grass–legume mixtures at Ibadan provided about 15 t/ha per annum of dry matter on offer as compared to about 25 t/ha per annum for tall-growing grasses and legumes (Olubajo and Oyenuga 1971; Oyenuga and Olubajo 1966). *Cynodon*-legume pastures gave 60 per cent herbage utilization as compared to 45–55 per cent for elephant (*Pennisetum purpureum*) and guinea grass (*Panicum maximum*)-legume pastures during the early wet season. Utilization of both dropped to about 25 per cent during the dry season.

At Ibadan, *Pennisetum purpureum* yielded 7·83 t/ha of dry matter on offer within a period of 132 days (May–September 1972) and unselected F_1 hybrids of *P. purpureum* and *P. typhoides* produced 6·32 t/ha. With 6-week rotational grazing periods, herbage utilization was 35 per cent and 46 per cent, respectively. During the full growing season (about 8 months) selected *P. purpureum* and F_1 hybrids each produced about 20·0 t/ha of dry matter, with 45 per cent and 57 per cent utilization, respectively (Aken'Ova 1974).

Mowing *Cynodon* IB.8 at 10 cm and 18 cm above ground level, after rotational grazing, gave substantial yields of forage, 10·0 t/ha and 11·1 t/ha dry-matter, 49 per cent and 38 per cent utilization, respectively, from mid-May 1970 to January 1971. Cutting at 25 cm and no mowing yielded 6·8 t/ha and 5·6 t/ha and 27 per cent and 22 per cent utilization, respectively (Chheda and Saleem 1973*b*).

8.6.2. Grazing systems

Continuous grazing in northern Nigeria produced slightly higher live-weight gain (15·8 kg/ha per annum) as compared to deferred and three-paddock rotational grazing (12·1 kg/ha and 11·2 kg/ha respectively) when stocked at 0·37 head per ha. In the southern Guinea zone, however, a rotational system was superior to continuous grazing (104·0 kg/ha and 85·9 kg/ha, 2 years total of 6 months grazing per year) when stocked at one head per ha, which was attributed to a greater degree of species selectivity and rest periods for recovery of the desirable grasses (Leeuw 1971).

Continuous grazing in humid areas permits the build-up of ticks, and internal parasites, in grazing cattle, and, possibly, nematode infestation of the soil. Grazing young stock on the same pasture as older cattle causes heavy infestation of helminth parasites and retarded growth. This problem is considerably reduced under rotational grazing (McIlroy 1962).

At the IEMVT Bouaké Station in the Ivory Coast Ndama bulls stocked at 2 ha per head showed average annual daily gains of 0·24 kg with daytime grazing of natural pastures oversown with stylo. When supplemented with 1·0 kg of rice bran per head daily gains increased to 0·33 kg and, with day and night grazing, they rose to 0·40 kg (Letenneur 1971).

In northern Nigeria, cattle grazing pure stands of stylo full time gained 0·17 kg per head per day; those grazing rangeland by day, and stylo every night gained 0·14 kg per head per day, those grazing rangeland by day and stylo every second night gained 0·02 kg per head per day; and those grazing stylo every fourth night lost 0·03 kg per head per day. Having nightly access to the stylo pasture was equal to feeding 0·68 kg *per caput* of undelinted cotton seed over a 10-week period (Leeuw 1969–71). For year-round use of stylo, animals grazed on separate wet- and dry-season pastures, but a four-paddock system would be more suitable for the growth pattern of herbage components (Leeuw and Brinckman 1973).

8.6.3. Sown versus natural pastures

Animal output from sown pastures in the derived savanna of Nigeria was 4 times that of natural rough grazing land (Ogor and Hedrick 1963). Productive pastures with various improved grasses (species of *Cynodon*, *Panicum*, *Pennisetum*, and *Brachiaria*) have been maintained at the University of Ibadan (Chheda

1971; Oyenuga 1971*a*), at the ORSTOM Research Station near Adiopodoumé and the IEMVT Station near Bouaké in the Ivory Coast (personal communications), in other African francophone countries (Borget 1969), and at Njala University in Sierra Leone (personal communication). In northern Nigeria sown pastures of *Chloris gayana*, *Digitaria smutsii*, *Brachiaria brizantha*, and *Stylosanthes gracilis* produce 150 kg/ha or more of live-weight gain during the wet season as compared with 18 kg/ha on unimproved shrub savanna. Increased production from sown pastures is related to a high level of management, which includes fertilization and eradication of undesirable plants (Leeuw 1969–71; Leeuw and Brinckman 1973).

8.6.4. Stocking rate and live-weight gains

White Fulani steers grazing at 0·42–1·25 head per ha on cleared shrub savanna in northern Nigeria gave average daily gains from nil to 0·18 kg during the dry season, the optimum stocking rate being 0·62 head per ha with a 0·17 kg daily gain per head. At the Mokwa Cattle Ranch, Gudali bulls grazed semi-natural *Andropogon gayanus* grassland during the wet season at 4 head per ha, 2 head per ha, and 1 head per ha. One head per ha gave daily and total live-weight gains of 0·49 kg per head and 84 kg/ha, respectively (three-year average of 171 grazing days per year); 2 head per ha gave daily and total *per capita* live-weight gains of 0·28 kg and 96 kg, respectively; 4 head per ha was discontinued because of over-stocking (Leeuw 1971).

In Senegal, grassland dominated by *Andropogon amplectens* and *Zornia glochidiata* permitted the grazing of 0·4 head per ha (200–230 kg animals) during the wet season (520 mm per annum rainfall) and 0·13 head per ha during the dry season (Valenza 1965).

At the IEMVT Bouaké Station carrying capacities have been determined as follows: (1) bush savanna—0·25 head per ha (Ndama weighing about 300 kg per head); (2) cleared savanna with natural grasses—0·5–1·0 head per ha (depending on season); (3) stylo oversown on disced natural grazing land—1·5 head per ha; (4) stylo sown on a well-prepared seedbed—3·0 head per ha; and (5) *Panicum maximum* transplanted in rows (25 × 50 cm), fertilized and irrigated —13·7 head per ha (Letenneur, personal communication).

Young Ndama heifers at the Fashola Stock Farm in Nigeria on sown *Cynodon* or *P. maximum* pastures gained up to 1·0 kg/day during the early and mid-wet season (April–June), 0·22 kg/day during July–October on *Cynodon*, and 0·33 kg/day on *Panicum*, with total gains of about 260 kg/ha per annum on either (Ogor and Hedrick 1963).

C. nlemfuensis rotationally grazed (1 week grazing and 6 weeks rest) with 5 steers per ha and fertilized with 82·5 kg/ha of nitrogen in six equal instalments, permitted 0·67 kg/day average gains per head during April–August at the University of Ibadan. This dropped to 0·32 kg/day in September–October, and rose slightly in November–December. Total live-weight gains approached 660 kg/ha during the 9-month period (Chheda 1974).

Steers on *Cynodon*–centro (about 66 per cent grass) showed daily gains of 0·32 kg per head and 0·056 kg per head for the wet and dry seasons, respectively, producing a live weight of about 260 kg/ha per annum at Ibadan (Okorie, Hill, and McIlroy 1965; Oyenuga 1971a). On *Pennisetum purpureum*–centro daily gains were 0·36 and 0·145 kg for the two seasons, with annual live-weight gains of 308 kg/ha per annum. Over an 8-year period live-weight gains sometimes exceeded 0·65 kg per head per day in the early wet season. Daily nutrient intakes by growing and fattening steers were comparable with National Academy of Sciences/National Research Council allowances. Sown pastures provided sufficient ingested dry matter, crude protein, total digestible nutrients, and digestible energy to permit gains commensurate with the accepted USA standards for beef cattle, but actual increases were about one-half of those expected (Oyenuga 1971a).

The indigenous West African cattle are essentially local races which developed under conditions of a low plane of nutrition, a scarcity of feedstuffs (including grasses of inferior quality), and a system of cattle hoarding. Compared with improved beef breeds they are poor converters. Thus a genetic restraint is considered to limit their output when grazed on high-quality forage.

8.6.5. *Fertilization of pastures*

Little commercial fertilizer is applied to pastures and few data are available to demonstrate their benefits to animal output. At the Fashola Stock Farm in Nigeria Ndama and Keteku steers grazing *Andropogon gayanus* fertilized with 110 kg/ha of nitrogen showed live-weight gains of 0·76 kg per head per day and of 245 kg/ha during June–December, in comparison to 0·38 kg per head per day and

110 kg/ha for nil nitrogen during the same period. Stocking rates and animal days per ha were able to be increased with applied nitrogen. The beneficial effects of applied nitrogen were noted in the grass silage made from excess herbage and fed during the dry season. Unfortunately, this trial did not continue longer than a year (Adegbola, Onayinka, and Eweje 1968).

Fertilized grasses at the Shika Station at Zaria, northern Nigeria showed differential responses among the species grown. *Chloris gayana* gave twice the live-weight gain (300 kg/ha during the wet season) as that obtained from local *Cynodon dactylon* given the same nitrogen treatment. Growth rates of young stock grazing *Chloris* and *Digitaria smutsii* fertilized with 50 kg/ha per annum of nitrogen and 25 kg/ha of phosphate declined over 3 years because the cover was invaded by *Sporobolus pyramidalis*. The *Digitaria* pastures fertilized with 100 kg/ha of nitrogen showed no such drop in quality (Leeuw and Brinckman 1973).

8.6.6. *Milk production*

Efforts toward improving milk production have mainly been by genetic upgrading of the cattle by selection and cross-breeding rather than by pasture management (Hill 1964). At Ibadan the milk production of White Fulani lactating cows did not vary significantly when grazed on different mixtures of *Cynodon*, *Pennisetum purpureum*, *Panicum maximum* and centro and when fed supplemental concentrates. Slight increases occurred when centro comprised 20 per cent or more of the cover. Production declined as plants aged, but daily yields and live-weight gains indicated that nutrient supplies were in excess of the genetic potential of the animals to produce milk (Olaloku 1972). Half-bred and three-quarter Friesians produced more than their White Fulani dams, but less than imported pure-bred Friesians at the Vom Station (Knudsen and Sohael 1970). White Fulani cows on *Digitaria decumbens* pastures averaged 5·8 kg fat-corrected milk per day and cross-bred Friesians 7·5 kg when fed 1·0 kg of concentrate and grazed at 2 cows per ha over a 3-month period (Leeuw 1969–71). This performance compared favourably with cows on rangeland when given 1·5 kg of concentrate per kg of milk produced.

Cross-bred Ndama × Jersey produced about 20 per cent more milk than both parents at the Bouaké Station (Mathon, Louvre, and Letenneur 1972) and had trypanosomiasis resistance. Production of the three-quarter Jersey dropped and susceptibility to the disease increased.

8.6.7. *Grazing behaviour*

Fulani cattle in northern Nigeria under the traditional system of unrestricted bush grazing during the day and kraaling at night showed peaks of grazing activity during the morning and afternoon (Haggar 1968). Grazing continued for a longer period during the wet than during the dry season because of supplementary feed and reduced quantity and quality of herbage. All herds walked for about $5\frac{1}{2}$ hours each day in the wet period and for about $3\frac{1}{2}$ hours in the dry period. Ruminating time was spread more uniformly during the wet season. Behaviour of free-grazing West African Shorthorns in Ghana showed four grazing peaks—three during the day and one at night (Rose Innes 1963). During 24 hours 43 per cent was spent grazing, 26 per cent ruminating, and 31 per cent idling.

8.7. Chemical composition and quality of herbage

Assessment of herbage quality involves the study of nutritive value (i.e. the chemical composition, digestibility, and nature of digested products), and the intake by and acceptability of forages to the grazing animals.

8.7.1. *Chemical composition*

Proximate chemical analyses, based primarily on the Weende method, are of limited value for estimating nutritive value because of changes in chemical composition with plant maturity and environmental conditions and of difficulties in separating chemical components. The close negative relationship between crude fibre and digestibility often reported in temperate grasses is not always applicable to tropical grasses (Butterworth 1965). Crude fibre is not always the less digestible fraction of forage and it was found to be more digestible in several West African herbages than a nitrogen-free extract (Oyenuga and Olubajo 1966; Okorie *et al.* 1965; Ademosun 1970*a*). Oyenuga (1968*a*) emphasized the need for further research both because of the rapid rise in crude-fibre content that commonly occurs with plant age and because of the possible ability of indigenous cattle to digest crude fibre in greater proportions than can improved breeds.

Chemical analysis, despite its limitations, is necessary to evaluate the critical nutrient levels for both the forage and animal. Nutrition studies have been undertaken for 40 years (Anderson 1933) but most data for estimates of crude protein, crude fibre, ether extract, nitrogen-free extract, ash, and major minerals such as P, Ca, K, and Mg have been accumulated in the last 15 years (Ahlgren *et al.* 1959; Ademosun and Baumgardt 1967; Ademosun 1970*a*, 1970*b*; Chheda and Akinola 1971*b*; Haggar 1970*a*; Lansbury 1958, 1960; Lansbury *et al.* 1965; Miller 1961; Miller, Rains, and Thorpe 1963, 1964; Okorie *et al.* 1965; Oyenuga 1957, 1959*a*, 1959*b*, 1960; Oyenuga and Hill 1966; Oyenuga and Olubajo 1966; Rose Innes and Mabey 1964; Rose Innes 1965, 1967; Sen and Mabey 1965; Saleem 1972; Wilson and Lansbury 1958). Summarized results by Oyenuga (1958), Ademosun and Baumgardt (1967), and Rains (1963) showed the interspecific variations among forages and differences due to stage of growth.

An appraisal of published work permits the following generalization to be made.

1. The main factors influencing chemical composition of herbage are (a) stage of growth at harvest, (b) leaf/stem ratio, (c) fertilizers and fertility level of the soil, and (d) climatic conditions.

2. The inverse relationship between herbage quality and yield is more pronounced in tropical than temperate grasses.

3. Dry-matter percentages vary from 15–20 per cent for tall-growing species and 25–30 per cent for low-growing types after 6–8 weeks' regrowth during the early wet season. Increasing the interval of cutting, the onset of the dry season, and the onset of flowering cause dry-matter increases ranging from 50 per cent to 70 per cent.

4. With 3–4 weeks of regrowth after the onset of rains, crude-protein content varies from about 7 per cent to 13 per cent of dry-matter. As plants mature it drops drastically, being 4–6 per cent after 3–5 months and falling below 3 per cent in the dry season. Nitrogen fertilization and associated legumes improve the crude protein content of pasture herbage.

5. Lignin ranges from 3 per cent to 6 per cent for grasses and from 6 per cent to 10 per cent for legumes.

6. Intra- and interspecific differences in crude fibre and nitrogen-free extract occur with forages containing 25–35 per cent and 35–50 per cent, respectively, at the grazing stage.

7. Phosphorus content of grasses ranges between 0·10 per cent and 0·50 per cent with lower values occurring on grasses grown on infertile soil. The

low phosphorus levels in herbage can be improved by superphosphate applications.

8. The calcium level in grasses is higher and less variable than the phosphorus level, ranging between 0·40 per cent and 0·90 per cent, but lower levels have been reported. Potassium is adequate in most herbages. Little information is available on trace elements.

9. Browse plants contain 2–4 times more crude protein (12–21 per cent) than grasses, have adequate calcium (0·28–2·36 per cent) and are similar in crude fibre and calorific values.

8.7.2. *Digestibility*

Proximate analyses give a quantitative measure of the nutrient content of forages but do not indicate their precise nutritive value. Raymond (1969) considered herbage quality to be a product of (1) intake, (2) digestibility, and (3) utilization of digested forage. These determine the nutrient intake of ruminants fed a given forage. Measurement of digestibility is highly important in evaluating herbage quality since it also influences the other two factors.

In temperate countries regression equations have been computed from chemical analyses and *in vitro* digestion trials, making it possible to predict forage digestibility from chemical-composition data. Use of these regression equations to predict nutritive value of tropical forages might be misleading (Duckworth 1946; Oyenuga 1968a).

Various techniques, as described below, have been used to estimate forage digestibility in West Africa.

Direct methods. The conventional *in vivo* technique, which determines the amount of dry matter eaten and voided in the faeces, has been used to study the nutritive value of various grass–legume mixtures (Okorie *et al.* 1965; Oyenuga and Olubajo 1966); *Stylosanthes gracilis, Pennisetum purpureum, Panicum maximum,* and *Cynodon nlemfuensis* (Ademosun 1970a, 1970b, 1973; Ademosun and Kolade 1973); *Andropogon gayanus* (Haggar 1970b); eight forage grasses and three legumes (Miller and Rains 1963), and Miller, Rains, and Thorpe (1963); and four browse plant species (Mabey and Rose Innes 1964a, 1964b, 1966a, 1966b; Rose Innes and Mabey 1964a, 1964b). The workers used local cattle, sheep, or goats in these trials.

Indirect methods (faecal index technique). Olubajo (1970) and Olubajo and Oyenuga (1970, 1971) estimated digestibility of grass–legume mixtures from faecal output of grazing animals using chromium trioxide. Herbage intake was determined from regression equations derived from conventional indoor digestibility and faecal nitrogen data.

In vitro digestion. The two-stage biological and enzymatic *in vitro* digestibility procedure of Tilley and Terry (1963) was employed by Ademosun and Baumgardt (1967) for 16 pasture grasses and 5 legumes; Saleem (1972) studied the decline in digestibility with age of *Cynodon* IB.8; Haggar (1970a) the seasonal changes in digestibility of the leaf and stem fractions of *A. gayanus*; and Olubajo, van Soest, and Oyenuga (1974) the digestibility of *Cynodon nlemfuensis, Panicum maximum,* and *Pennisetum purpureum*.

Mba, Oke, and Oyenuga (1971) observed a significant correlation ($r = 0.896$) between *in vivo* and *in vitro* organic matter digestibility for *Cynodon*-centro pastures.

Analytical methods. Recognizing the need for rapid laboratory methods, Ademosun and Baumgardt (1967) used various analytical techniques such as the availability index of van Soest and Marcus (1963); total apparent digestibility estimates, which comprise a summative equation proposed by van Soest and Wine (1966); digestible dry matter estimation by the regression equation of Bredon, Harker, and Marshall (1963), along with *in vitro* digestible dry matter estimations. They concluded that total apparent digestibility might be more useful since it considers the components of forages.

The above-mentioned studies allow the following generalizations to be made.

1. Digestibility of most West African forages is lower than temperate species, both at the grazing stage and maturity, ranging between 55 per cent and 70 per cent for regrowth of 5–9 weeks.

2. Reduced digestibility and poor nutritive value is attributable to high heat intensity which causes a rapid growth, enhanced maturity with decreasing leaf/stem ratio, increased crude fibre content, acid detergent fibre and cell wall contents, and higher lignification.

3. The rate of decline in digestibility varies from 0·2 per cent to 0·6 per cent per day, with a regrowth period of 4–9 weeks as compared to about 0·5 per cent per day in temperate grasses. Saleem (1972) observed up to 1 per cent per day decline in *in vitro* organic matter digestibility of *Cynodon* IB.8.

4. Digestibility data of stem and leaf fractions and horizontal layers of herbages are needed. Haggar and Ahmed (1970) showed *in vitro* digestibility decline of 0·5 per cent and 0·4 per cent per day, respectively, for leaves and stems in the vegetative phase but, after flowering, stems declined 0·75 per cent per day.

5. Seasonal changes in crude protein digestibility are generally greater than dry-matter digestibility.

6. The apparent digestibility of crude fibre is as high or higher than nitrogen-free extract.

7. Browse plants are highly digestible and provide a valuable dry-season supplement to grasses and legumes (Mabey and Rose Innes 1964a, 1964b, 1966a, 1966b).

8. Genotypes of *Cynodon* spp., *P. purpureum*, and *Pennisetum* hybrids show 20 per cent differences of *in vitro* organic matter digestibility at 6 weeks regrowth, which indicates the possibility of genetic improvement for nutritive quality (Chheda 1973; Chheda *et al.* 1974).

9. Nitrogen fertilization reduces the rate of decline in digestibility at early stages of growth and maintains slightly higher digestibility at later stages, possibly due to the development of a greater proportion of fresh tillers.

10. Dry-matter digestibility is improved when low-quality hay is supplemented with high-protein feeds, but not with non-protein nitrogen sources such as urea (Haggar 1970b).

8.7.3. Intake

Although dry-matter digestibility of over 60 per cent is obtained, intake usually ranges from 1·5 per cent to 2·25 per cent of body weight when animals are stall-fed (Miller *et al.* 1963; Ademosun 1973; Ademosun and Kolade 1973). Under grazing conditions, when measured by the conventional clipping technique, dry-matter intake ranges from 2 per cent to 3 per cent of body weight for grass–legume mixtures (Okorie *et al.* 1965; Oyenuga and Olubajo 1966; Olubajo and Oyenuga 1970) and browse plants (Rose Innes and Mabey 1964b). This method overestimated intake by 15–25 per cent as compared to the nitrogen–chromic oxide faecal index technique, and the magnitude of overestimation was much greater in the dry season (Olubajo 1970).

The correlation between herbage digestibility and voluntary intake of West African grasses is lower than for temperate species. Inter- and intraspecific differences in dry-matter intake have been observed. Values for legumes are higher than for grasses possibly due to higher cell contents and lower proportion of cell-wall constituents. Local sheep tend to consume more dry-matter than goats (Ademosun 1970a, 1970b, 1973; Ademosun and Kolade 1973). Stage of maturity exerts a greater influence on intake of digestible nutrients in grasses than on dry-matter intake (Ademosun 1973). Maximum intake values of pasture species occur in the early wet season with minima during peak dry season when the crude protein content of the herbage is at its lowest (Haggar and Ahmed 1970; Olubajo and Oyenuga 1970). Intake (up to 20 per cent) can be increased by supplementing the lower-quality forages with legume hay or cotton seed (Haggar 1970b).

8.7.4. Nutritive value index

Ademosun (1970a, 1970b, 1973) and Ademosun and Kolade (1973) computed the nutritive value index of the various forages using the formula:

$$\text{Nutritive value index} = \frac{\text{relative intake} \times \text{percentage digestible dry matter}}{100},$$

and concluded that the apparently low nutritive value of West African grasses is primarily due to low intake of feed. This is accentuated during the dry season when the nutrient content in the herbage is at its minimum. The high values of the nutritive index for *Stylosanthes gracilis* suggest that legumes can play a very important role in increasing animal productivity.

8.8. Range management

'Range' and 'rangeland' refer to land areas covered with natural or semi-natural vegetation which provide a suitable habitat and extensive grazing for domestic and wild animals (Naveh 1966; Pratt, Greenway, and Gwynne 1966; Rains 1963). They are frequently characterized as receiving less than 750 mm per annum average rainfall, but may include areas with higher amounts where soils and topography are unsuitable for cultivation.

Rangelands comprise most of the northern Guinea, Sudan, and Sahel savannas, but the derived and southern Guinea savannas should also be considered as range because of the imposed animal and grazing land management. Range management is the system of manipulation and utilization of rangeland soils, vegetation, and animals for the production of goods and services needed by man, i.e. meat, milk, hides, skins, and wool (Heady 1960; Rains 1963).

8.8.1. *Systems of grazing management*

Most of the cattle occur in the northern regions in which tsetse flies are absent or of low incidence. Cattle are largely controlled by a few ethnic groups (e.g. the Fulani) as owners and/or herders. Communal grazing around villages is practised during the wet season, with migrations in search of grass, browsing, and water occurring during the dry season (Malunfashi 1969). Stable systems of range management exist on government farms and some ranches. A three-camp, one-herd scheme was described for the northern Guinea savanna (Rains 1963) and four systems were compared at the Shika Research Station in Nigeria (Table 8.2; Leeuw (1971)).

TABLE 8.2

The effect of grazing systems on the performance of bulls on range at Shika Research Station, 1961–5, stocked at 2·7 ha per head (from Leeuw 1971)

Grazing system	Live-weight gain (kg)		Supplementary concentrates (kg)†
	Per head per annum	Per ha per annum	
Continuous	43	15·8	138
Two paddocks, deferred	39	14·4	136
Three paddocks, deferred	33	12·1	136
Three paddocks, rotational	30	11·2	144

† Whole cottonseed fed during dry season.

Optimal production from a given range unit depends on (1) maintaining a balance of animal numbers with the forage resources, (2) the distribution of cattle, and (3) having sufficient flexibility in the grazing scheme to allow for adjustments to be made during stress periods (Heady 1960; Ogor and Hedrick 1963). The potential carrying-capacity is of primary importance and should be determined on a year-round basis (Rains 1963). Arable crop residues as supplements in animal feeding must be given consideration but their value may be difficult to determine. The density of the tree canopy and the presence or absence of browse species influence the possible stocking rate, especially of uncleared grazing land. In areas of low rainfall browse shrubs and trees may be of equal or of greater value than the grasses and other herbaceous plants, but having to include them adds to the complexity of range management.

About 5·6 ha of uncleared land of the Guinea savanna supports one adult head of indigenous cattle, but prolonged drought increases the needed area upwards of 8–12 ha (Letenneur 1971; Ogor and Hedrick 1963; Rains 1963; Rose Innes 1967). When cleared of scrub the natural grass cover thickens and permits a stocking rate of about one beast per 2·7 ha. Studies of stocking rates from 1·25 head per ha to 0·42 head per ha in northern Nigeria gave optimal gains at 0·62 head per ha (1·6 ha per head) with high-level management (Leeuw 1971). In the more arid and humid regions the potential carrying capacities are lower and higher, respectively.

One factor limiting the year-round utilization of vast areas of grazing lands is the incidence of tsetse flies, the vectors of trypanosomiasis (MacLennan 1969). *Glossina morsitans submorsitans* is found in savanna vegetation under tree and scrub cover, while *G. longipalpis* is restricted to heavy woodland. *G. palpalis* and *G. tachinoides* require riparian fringing vegetation. Measures for meeting the trypanosomiasis problem include exploitation of highly tolerant and resistant livestock, chemotherapy, and proper animal husbandry. Methods directed against the vectors include a combination of vegetation clearing and the application of persistent insecticides in which prescribed spraying techniques are used (Davies 1964, 1971; MacLennan and Aitchison 1963).

8.8.2. *Brush control*

Large areas of the savanna grazing lands are repeatedly burned every year, either by design or mistake. Fire has a greater and more direct influence on brush encroachment and on the productivity of grass and herbage than any other method used for brush control. Burning every third or fourth year checks brush encroachment and provides optimal forage production (Rains 1963). Annual fires can be induced early in the growing season and this can effectively reduce brush to open grassland suitable for grazing within 4–5 years. The same result can be achieved over a longer period of time by burning and grazing in alternate years. It is only on government farms, agricultural research stations, and forest and game reserves that attention has so far been given to controlled burning (Egunjobi 1970; Guilloteau 1958; Ramsey and Rose Innes 1963; Rose Innes 1971; Traore 1958; West 1965).

A rest period before burning is needed to permit the accumulation of enough combustible material to allow an early burn (Rains 1963). At least 1000 kg/ha of dry matter is required for an effective fire to control woody scrub 1 m high, and 1500–2500 kg/ha to control shrubs and taller trees. Early burning at the end of the growing season weakens perennial grasses because food reserves have not been completely transported to the crowns and roots. Late burning just prior to the onset of the rains controls scrub and brush with minimal damage to dormant grasses, but may weaken browse plants which leaf-out well before the wet season. The optimal period of rest and recovery after burning depends on the amount and distribution of the rains and the vigour and growth of grasses, but usually should be no less than 4–6 weeks.

An intense but rapid fire is most desirable. Temperatures at soil level rise sharply and vary from 100 °C to over 500 °C depending on the season, the height and density of the combustible material, and the wind speed, but they usually return to ambient levels within a few minutes. At 2 cm of soil depth the temperature may fluctuate from 3 °C to 14 °C above normal (Hopkins, 1963, 1965*a*; Masson 1948; Pitot and Masson 1951).

Fire promotes regrowth with a fairly constant 'period of sprouting' which varies from 6 days to 10 days for grasses and from 18 days to 30 days for trees (Hopkins 1963). The rapid growth of grasses following burning is associated with a rapid increase in the population of nitrifying micro-organisms (Vine 1968).

Limited studies have been conducted on hand-slashing and mechanical brush-cutting (Leeuw 1971) and on the use of cultural and chemical treatments (Adegbola *et al.* 1970).

8.8.3. *Provision of water*

On arid and semi-arid rangelands animals rarely have ready access to water and are moved to watering sites at fixed times (Rains 1963). Animal output is greater with daily watering than with every second or third day, and the distance walked between grazing and the water site affects productivity. For daily watering the total distance walked should not be more than 8–10 km and, if the return journey exceeds 15 km, it should be programmed every second day. When watering sites are widely separated, overgrazing occurs near the points of supply. Thus the more distant herbage is not fully utilized and is generally of low quality.

Mature West African cattle of about 325 kg live weight require about 7·5 l of water per head daily during the wet season and double this amount in the dry period (Rains 1963; Rose Innes 1963). Cattle drink more when watered daily than they do when watered at less frequent intervals.

8.8.4. *Improvement by re-seeding and fertilizing*

Re-seeding is not a substitute for poor management. Before a decision is made to re-seed, the system of grazing should be evaluated to determine whether modifications will bring about an improvement in the natural vegetation. If clearing and cultivation are involved, range re-seeding probably cannot be justified in economic terms, unless as part of an intensive reclamation scheme (Rains 1963). In the Sudan savanna zone of Nigeria, herbage yields of *Andropogon gayanus* were increased by top-dressing sulphate ammonia, and annual legumes were increased with dressings of super-phosphate in the northern Guinea savanna. More favourable results, however, can be obtained by applying fertilizer to sown pastures where moisture is sufficient for their establishment and growth and where viable economic returns can be expected.

8.9. Forage conservation and supplementary feeding

During periods of scarcity, improvement of animal performance can be achieved by making use of:

(1) conserved forage, mainly as hay and silage (Foster and Mundy 1961; McKell and Adegbola 1966; Thorpe 1964);

(2) drought-tolerant species such as *Stylosanthes gracilis*, *S. humilis* (Haggar 1969*c*), and those which extend growth into the dry season;

(3) crop residues of groundnuts, guinea corn, cotton, cowpeas, etc. (Rains 1963; van Raay and de Leeuw 1971; Oyenuga 1968*a*);

(4) industrial by-products such as oil cakes, rice bran, molasses, citrus pulp, brewery by-products, and sugar-cane tops (Mba, Oke, and Oyenuga 1971; Oyenuga 1968*b*);

(5) appropriate range and pasture management practices (Ahlgren *et al.* 1959; Chheda and Akinola 1971*a*; Crowder 1971; Heady 1960);

(6) high-yielding, selected, fodder and pasture species and possibly of irrigation (Borget 1971; Cadot 1971*a*).

8.9.1. *Materials for conservation*

Herbage availability during the wet season often exceeds animal demands. If the herbage is left in the field it loses its nutritive value but could be conserved for supplementary dry-season feeding (Adegbola and Onayinka 1966; Chheda and Akinola 1971*b*; Oyenuga 1957, 1960; Saleem and Chheda 1973; Saleem 1972). Fodder crops such as *Panicum maximum, Pennisetum purpureum, Zea mays, Sorghum vulgare*, and *P. typhoides* cv. '*maiwa*' with or without legumes such as *Vigna unguiculata* and *Stizolobium deeringianum* are suitable for silage (Miller 1960; Miller and Rains 1963; Miller 1969; Rains 1963). Good-quality hay can be made from *Stylosanthes gracilis, Glycine max*, and *Vigna unguiculata* (*syn. siensis*) (Foster and Mundy 1961; Miller 1969).

Miller (1969) studied the nutritive value of fresh and conserved pasture grasses, forage crops, and crop residues and concluded that conservation of most natural and sown herbage grown under low-fertility conditions should be discouraged because of the low nutritive value and unfavourable weather conditions. Choice of species, good management, and adequate fertilization (Chheda 1971; Oyenuga 1960; Saleem 1972) which increase yields and improve chemical composition also maintain higher digestibility and improved quality of the conserved product.

8.9.2. *Conservation processes*

Forage conservation aims to obtain a stable product suitable for animal feeding with minimum loss of nutritive value.

Bush foggage. This consists of excess herbage and browse plants left standing in the grazing area. Tall perennial grasses in the Guinea savanna and annual grasses in the Sudan savanna are the major constituents of the standing hay (Miller, Rains, and Thorpe 1964). Animals selectively graze during the early dry season and show small increases in live-weight gains (Miller 1969). Later in the dry season the herbage loses most of its nutritive value and animal performance declines sharply. In the Sudan zone of Nigeria 6-month-old conserved hay and foggage of *P. pedicellatum* left for 7 months had 4·78 per cent and 1·91 per cent crude protein, 0·23 per cent and 0·11 per cent phosphorus, respectively (Rains 1963).

Hay. High rainfall and humid conditions during the growing season in most of West Africa are not conducive to haymaking. With adequate facilities, how-

ever, the authors found that good-quality hay could be obtained from well-managed *Cynodon* pastures when cuttings were made at the end of the dry season at Ibadan. For *Andropogon gayanus* haymaking during the time of maximum stem elongation in early October represents the best combination of bulk and quality (Haggar 1970*a*). In general, hay from tropical grasses is of poor quality, and legumes such as soya bean, mucuna, cowpeas, and stylo have been recommended for this method of conservation (Foster and Mundy 1961; Miller 1969; Nwosu 1960).

Silage. Silage made from tropical pasture species does not compare favourably with that of temperate species (Catchpoole and Henzell 1971; Miller 1969). Tropical grasses are generally coarse and stemmy, high in crude fibre, low in soluble carbohydrates, and high in ratio of structural to non-structural carbohydrates. These factors result in greater retention of cell contents, delayed fermentation, and proliferation of clostridial organisms which lead to the production of butyric acid, loss of energy, protein breakdown, and effluent losses.

For desirable silage about 8–12 per cent lactic acid is produced, giving a final pH around 3·8–4·2. Most tropical silages, except maize and cereal-legume mixtures, have a pH of 5·0 or above, indicating very little lactic acid production (Miller 1969). Lactobacilli inoculations of silage material have given promising results, as well as additions of molasses, mineral acids, antibiotics, sulphur dioxide, salts, and milled cereals. Recommendations for such additives depend on the cost and the farmers' knowledge of silage making. Low moisture suppresses clostridial activity (Murdock 1966), hence wilting of herbage before filling the silos is recommended (Miller 1969), even though this reduces the carotene content of the herbage. Rains (1963) reported that trench or pit silos with roofs make satisfactory structures. An earth covering would also be effective.

8.9.3. *Animal intake of conserved forages*

Voluntary intake of conserved forages by animals is lower than fresh material (Miller *et al.* 1964; Miller and Iduma 1967), primarily due to poor nutritive value and lower digestibility, resulting in greater retention time in the rumen. Low herbage protein may not provide enough nitrogen for rumen microorganisms resulting in reduced digestibility. Reduced silage intake has been attributed to organic acids which affect palatability and disturb pH in the rumen (Raymond 1969). Transformation of dried forages into wafers or pellets increases voluntary intake due to higher rate of passage through the digestive tract. Intake is further increased if the pelleted forage comes from a fertilized pasture (Minson and Milford 1967). Thorpe (1964) reported high digestible crude protein from sorghum-mucuna and maize-cowpea mixtures. Incorporating rice bran with maize silage (Mba *et al.* 1971) and cottonseed or groundnut cake with *Andropogon gayanus* hay (Haggar 1970*b*) increased protein supply and animal forage intake.

8.9.4. *Supplementary feeding*

Lansbury (1960) considered 3·5 per cent digestible crude protein in the herbage necessary for maintenance and 5·5 per cent for increased live weight for a 225 kg bullock consuming dry matter at 2 per cent of its weight. The unimproved grasslands and some standing hay may provide sufficient energy, but low crude protein limits animal performance during the dry season (Lansbury 1958, 1960; Miller 1969). Nwosu (1960) suggested conservation of *Stylosanthes gracilis* for dry-season protein supplementation and Haggar (1969*c*) the growing of *S. gracilis* as a pure crop with ration-grazing during the dry season. Feeding urea as a part of the protein supplement has been examined, but a decline in dry-matter digestibility was noted when *A. gayanus* hay having less than 3 per cent crude protein was supplemented with this non-protein source (Haggar 1970*b*). Alternative protein sources include cottonseeds, oil cakes, groundnut haulms, rice bran, and legume hays (Oyenuga 1968*a*; Rains 1963).

Vitamin A deficiency in grazing animals occurs due to loss of carotene in the bleached dry-season herbage; some is also lost when forage is cut and dehydrated (Lansbury 1960; Miller 1960). Rains (1963) maintained that cattle were capable of storing sufficient carotene during the wet season to meet their requirements during the dry period. Carotene supplement in the form of red palm oil increased live weights of animals provided with a ration of bleached grass hay and protein supplement (Miller and Iduma 1967). Animals browsing trees and shrubs during the dry season do not show vitamin A deficiency (Rains 1963).

Poor performance of animals subsisting on low-quality herbage is accentuated by mineral (particularly phosphate) deficiencies as reported by Rains (1963) in the northern Guinea savannas of Nigeria, Lansbury (1960) in the Nungua grasslands of Accra, Oyenuga (1957, 1968*a*) and Saleem (1972) on established pastures in southern Nigeria. In general, established pastures properly fertilized and managed are not mineral deficient but may induce nutritional disturbances such as hypocalcaemia and hypo-magnesaemia (Whyte *et al.* 1959; Adegbola 1971*b*). This can be corrected by mineral supplements in the feed or drinking water and use of salt licks.

8.9.5. *Economic considerations*

Rains (1963) compared the unit cost of crude protein and digestible protein for common feedstuffs in northern Nigeria and found more favourable cost-benefit ratios with concentrates such as groundnut cake and cottonseed. He concluded that conservation of low-quality herbage in the form of hay and silage would be highly uneconomical. Couper (1971) demonstrated that the cost of producing maize silage could be reduced by over 60 per cent in northern Nigeria by use of the herbicide Atrazine and by doubling the plant population.

8.10. Forage and fodder crops in farming systems

The growing of selected grasses and legumes in lieu of forest, bush, and natural fallow could reduce the interval between crops and regenerate or maintain the soil nutrient status and other characteristics (Vine 1968). Large quantities of herbage, however, must be accumulated and returned to the soil if the fertility and humus level are to be sustained.

8.10.1. *Vegetative cover and soil humus*

Under established forest fallow production of aerial parts amounts to about 10 t/ha per annum of dry matter and roots about 3 t/ha (Nye and Greenland 1960). In the savanna regions production of grasses depends on rainfall and varies with species. The high-grass Andropogoneae yield approximately 8 t/ha per annum of dry matter but much is lost by burning. Roots in the 0–30 cm soil layer contribute about one-third of this amount while leaf and stem litter from scattered trees and shrubs add another 2·5 t unless burned.

The increase or decrease in soil humus content follows a logarithmic curve and is governed by a decomposition constant applicable to the particular conditions of climate and soil (Laudelot 1962; Nye and Greenland 1960). In the lowland tropics this constant varies between 2 per cent and 5 per cent annually within the 0–30 cm soil layer and in the savanna soils from 0·5 per cent to 1·2 per cent

(Greenland and Nye 1959). In Nigeria a short-term grass fallow (4 years of *Cynodon* IB.8) increased the organic matter content of the 0–15 cm soil layer to 1·46 per cent as compared to 0·85 per cent for continuous cultivation (Babalola and Chheda 1972). Other studies have also shown significant increases of humus with a 3–5 year grass or legume fallow (Nye and Greenland 1960; Stephens 1960).

8.10.2. *Accumulation of nutrients*

An important feature of forest fallow is the mobilization of soil nutrients from deeper soil layers and their accumulation in plant organs. The accumulation of minerals by savanna regrowth is considerably less. *Andropogon-Hyparrhenia* grass fallow in Ghana, undisturbed for 20 years, mobilized slightly less than 25 per cent as much nitrogen, 40 per cent as much potassium, but 80 per cent as much phosphorus as forest fallow (Bartholomew, Meyer, and Laudelot 1953; Nye and Greenland 1960). The deep-rooted grasses gradually absorb subsoil nutrients and deposit them in the topsoil by leaf and stem litter and root residue (Nye and Greenland 1960; Nye and Hutton 1957; Saleem 1972). Various grasses and legumes accumulate amounts of 100–400 kg/ha of nitrogen, 40–60 kg/ha of phosphorus, and 90–500 kg/ha of potassium in 3–4 years (Laudelot 1962).

8.10.3. *Soil structure*

Land tillage operations involved in crop production tend to (1) destroy soil structure, (2) increase density, (3) decrease total porosity, (4) cause a decline in rainfall acceptance, infilltration, and percolation rate, and (5) increase soil compaction (Cunningham 1963; Fauck, Moureaux, and Thomann 1969). During the fallow period soil structure and other physical properties are restored, but the rate and degree of regeneration varies with the type of fallow and the composition of the vegetative cover.

In Nigeria the mean-weight diameter of soil aggregates in the upper 30 cm soil layer under long-term secondary forest averaged 1·0 mm, as compared with 0·8 mm in soil continuously cropped for 20 years (Babalola and Chheda 1972). A 4-year fallow of *Cynodon nlemfuensis* after 15 years of continuous cropping markedly improved soil structure, the aggregate size being 1·13 mm. This difference probably reflected a partial influence of the binding effect of grass roots rather than a lasting stabilized crumb structure.

The structure of subhumid savanna soils is less likely to be improved by fallow than that of the forest and humid savanna regions because of the inherent humus content (Nye and Greenland 1960). The effects of a rest period under bunch grasses are extremely transient once the land is cleared.

8.10.4. *Use of grasses and legumes*

Soil-conserving crops. A dense cover of grass affords almost as much protection against soil erosion as does forest with a closed canopy (Dugain and Fauck 1959). In Nigeria, on 3·0 per cent slope, soil losses from bare soil, maize, cowpea, and *Cynodon nlemfuensis* occurred at the ratio of 8:5:2:1, respectively, over the growing season (Babalola and Chheda, 1972).

Cover in perennial crops. Cover crops are sown for soil protection in plantations of perennial crops such as oil palm, coconut, rubber, coffee, cocoa, and banana. These are usually perennial leguminous creepers, e.g. centro, puero, calopo, or shrubby herbs such as stylo (Delegation of the Republic of Dahomey 1965; Gaudefroy-Demombynes 1957; Kannegieter 1967; Nye and Hutton 1957). A disadvantage is the competition for soil moisture and nutrients and climbing onto the plants of the perennial crops. Clean cultivation allowed more rapid growth of coconuts for the first 3 years on sandy soil in the Ivory Coast, after which centro was recommended as ground cover (Boyer 1965; Villeman 1963).

Contour strip cropping and buffer-strips. On highly erodible soils suitable rotations can be practised by establishing alternate strips of erosion-susceptible crops and close-growing protective crops (Roose and Bertrand 1971). Such buffers may be combined with contour terracing, the width of strips being adjusted to the land gradient.

Green manure crops. In order to incorporate residues into the soil, alternating leguminous plants with food or cash crops may sustain economic yields over some period of time. There is usually no net accumulation of soil nitrogen, but the green manure arrests its loss as compared to bare soil. It does not always forestall a decline of the organic matter content of the soil. Programming the green manure into the cropping system is important in respect to time of planting and disposal (Foster 1961a).

In Nigeria, yield data were taken from various row crops in rotations with *Stizolobium* sp. after newly cleared secondary forest (Doyne 1937; Faulkner 1934; Vine 1953). For about 10 years no difference occurred between drying and burning the herbage or incorporating it into the soil as fresh material which suggested that the benefit came largely from minerals. Digging in the green manure increased yields beyond this period, indicating the value of nitrogen and minerals.

Interplanting maize with leguminous crops usually results in a decline of grain yield as compared to maize alone, especially when fertilized (Agboola and Fayemi 1971). When sown prior to maize the legumes effectively reduce soil erosion but compete for nutrients and soil moisture. They also interfere with weeding and maize harvest, the creeping types climbing onto the maize stalks, causing them to break. If the intercrop legumes are sown after the last maize cultivation they may not become established or make only sparse growth because of competition for light, water, and nutrients.

8.10.5. *Mixed agriculture*

Mixed farming schemes have been attempted in northern Nigeria and Ghana since the 1930s (Alkali 1969). It was expected that cattle would be fed crop residues and grazed away from the smallholdings, which led to a major loss of the dung. Since about 1948 the aim has been to locate mixed farms on self-contained units and the numbers of these types have increased since 1960.

Regulated farming, where fields are deliberately sown with forage and fodder crops, then fertilized, fenced, and intensely grazed or managed for ley is not practised in West Africa because of the nomadic or semi-nomadic scheme of cattle raising. An un-regulated ley system is commonly found and refers to the secondary growth of grasses and herbs which are communally grazed without judicious management. The cotton holdings and groundnut–millet operations in West Africa may be classified as a form of unregulated ley (Ruthenberg 1971).

The value of the ley system has been shown by studies with crops grown after grass and legume fallows (Dennison 1959; Glele 1971; Nye 1958; Picard 1971; Stephens 1960; Watson and Goldsworthy 1964). In regions of adequate rainfall, rotations of about 3 years of fallow and 3 years of cropping maintain soil fertility at a moderate level if the soils are not initially infertile or exhausted by intensive cropping. *Pennisetum purpureum* has consistently shown advantages but other grasses have been used, e.g. *Panicum maximum*, *Brachiaria ruziziensis*, *Cynodon*, and *Andropogon* spp. In general, differences

in results have been small between management treatments such as leaving the grass to decompose on the plot, cutting and composting off the plot with return of the organic mater, and grazing the herbage.

A deficiency of available nitrogen often occurs in the first crop following a long-term grass fallow because of a wide carbon/nitrogen ratio of herbage and roots and low accumulation of nitrates in the soil (Berlier, Dabin, and Leneuf 1956; Greenland 1958; Nye and Greenland 1960; Rains 1963; Stephens 1960; Watson and Goldsworthy 1964). The condition may be caused by lack of nitrifying organisms and is often acute after a long-term *Andropogon* and *Hyparrhenia* fallow. Short-term fallows show less adverse effect, and the condition has not been noted with 3- and 4-year fallows of *Pennisetum purpureum*, *Panicum maximum*, and *Cynodon* spp.

An advantage of certain grass species is their influence on soil-borne pests. For example, *Cynodon* IB.8 and several other *Cynodon* species reduced the nematode population to zero after 12 months growth in the field at the University of Ibadan (Adeniji and Chheda 1971). Tomato yields after 2 years of *Cynodon* increased threefold as compared to 2 years of continuous tomato cropping.

Although yields of arable crops are increased after a sown-grass or grass–legume fallow, the total harvest does not usually compensate for the loss of cropping during the fallow period (Nye and Greenland 1960). Thus to be economically viable the system of alternate husbandry must considerably increase crop yields during the arable break and appreciably improve returns from animal production of the ley. This probably cannot be achieved without the use of fertilizers, improved managerial skills, and practices of crop and animal husbandry, as well as modifications of the land-tenure systems.

8.11. Research priorities

Technical and scientific knowledge of pasture and forage crops available in West Africa needs to be organized, synthesized, and interchanged. Several conferences and reports have drawn attention to this matter and made recommendations for the advance of research towards the solution of practical problems: (1) the Meeting on savanna development (1967); (2) the conference on livestock development in the dry and intermediate savanna zone (1969); (3) the seminar on forage crops research (1971); and (4) the report on the FAO conference on the establishment of cooperative agricultural research programmes—

Guinean zone (1971). The problems of grazing-land improvement are not only agricultural but also human and economic. Technical know-how is being obtained but emphasis must be given to the application of findings not only by the large-scale cattle man but also by people who have little or no capital and seldom have access to tools except the hoe and cutlass.

8.11.1. *Grazing resources*

An integrated grazing-land survey with an inventory of vegetation and numerical data on human and livestock populations would provide information on grassland and pasture resources. This necessitates the co-operation of West African Governments and the assistance of international organizations. In the arid, semi-arid, and subhumid regions a range-management approach is needed for herbage utilization. This requires a knowledge of rangeland condition and trend, whereby techniques and procedures are used to quantify the effects of factors such as stocking rates, seasons of use, density of cover, and burning on the stability or deterioration of grazing lands.

8.11.2. *Grazing and feeding systems*

Low-quality forage and inadequate feedstuffs during the dry season impose serious constraints on effective and efficient animal output. Research on year-round feeding systems is needed, with attention given to (1) planned grazing schemes to study the effects of controlled stocking rates and use of burning; (2) selection of herbage species to extend the grazing period; (3) the search for and assessment of legumes such as *Stylosanthes gracilis*, *S. humilis*, and *Glycine wightii* to provide supplementary sown pastures; (4) methods of their establishment and utilization; (5) additional studies of conserved herbage, concentrates, and industrial wastes as maintenance rations; (6) provision of water. These studies should be formulated and conducted by teams consisting of the pasture agronomist, plant breeder, animal scientist, ecologist, economist, and others. The calculation of cost-benefit ratios is highly essential to determine the economics of improved practices.

8.11.3. *Herbage quality and nutritive value*

Intensive research should be aimed towards understanding the nutritive value of tropical forages as well as the elucidation of suitable agronomic and pasture-management practices for maintaining herbage quality over a period of time. Precise knowledge of the nutrient requirements of tropical livestock is needed for optimum performance. Estimates of *in vitro* and *in vivo* digestibilities using forages of comparable physiological growth should receive attention. Satisfactory feeding standards which apply to tropical conditions are not available, making it necessary to utilize and revise those prepared for temperate zones. Research should be encouraged which leads to the development and formulation of feeding tables adequate for the tropics.

8.11.4. *Centres for the maintenance of strains and their improvement*

Procuring of planting stocks of forage grasses and legumes is very difficult, and frequently their origin and true identity are unknown. Regional centres should be established for the collection, characterization, maintenance, and distribution of seed stocks, where genetic improvement could be carried out: one in the Sudanian zone and the other in the Guinean zone. Such centres could provide leadership in organizing and standardizing regional testing of promising materials.

8.11.5. *Seed production*

Seeds of forage grasses and legumes are not produced commercially in West Africa. The demands for high-quality seeds will accompany an expanded and intensified animal industry. Basic information related to seed production must precede this development. The present selections, introductions, and cultivars must be assessed for uniformity of seed formation and maturity, or new cultivars may have to be developed. Research into cultural and management aspects, along with location of suitable environments for seed production, will be necessary before a viable seed industry can be established. Information from other places pertaining to harvesting, processing, storing, and handling should have application, but further studies are needed for tropical environments. Seed production is a specialized enterprise so that a cadre of knowledgeable growers would have to be organized.

8.11.6. *Mixed agriculture*

Integration of arable agriculture and animal production must go beyond the relocation of people on newly cleared lands with provision made for them to obtain farming equipment and livestock. It requires a modification of the socio-economic structure and a change in the way of life of those who by heritage are pastoralists or agriculturists. Only time and re-education can bring about such changes. The pasture agronomist and animal scientist can contribute by research applied to the use of forages in the development of alternative farming systems. Large quantities of herbage must be accumulated and plant nutrients incorporated into the soil if fertility is to be maintained. Studies are needed on the use of livestock for recycling of nutrients and on tillage equipment for handling the cover and root residues. Long-term arable cropping and animal rotational trials must be carried out using cash and food crops, pasture grasses, and legumes, and animals with the genetic potential to respond to improved practices. Improvement of indigenous livestock by cross-breeding is necessary to transform the existing marginally economical animal-keeping practice into a well-managed, profitable industry, capable not only of meeting the area animal protein requirements but also potentially able to generate surpluses for export and development of related industries.

Mixed farming is not the answer for improved agricultural and animal production in all regions. A co-ordinated effort by the plant and animal researcher, sociologist, ecologist, economist, and extensionist, with the financial support and backing of the State and Federal Governments, should be aimed towards problem solving in terms of efficient land utilization.

8.11.7. *Wildlife resources*

Although wild ungulate life has rapidly diminished in most of West Africa, the conservation and use of wildlife resources should be an integral part of savanna research. As arable agriculture and livestock enterprises expand, attention must be given to management plans for national parks and game reserves to ensure the preservation of wildlife habitats. The feasibility of game–stock management should be examined in regions where wild ungulates still exist.

References

ADAM, J. (1958). Principales graminées naturelles fourragères de l'Afrique occidentale ayant un intérêt économique pour l'alimentation du bétail. *Notes Afr. Bull. Inst. fr. Afr. noire* **80**, 98–102.

—— (1966). Les pâturages naturels et post-cultures du Sénégal. *Bull. Inst. fr. Afr. noire (A)* **28**, 450–537.

ADEGBOLA, A. A. (1964). Forage crop research and development in Nigeria. *Nigerian agric. J.* **1**, 34–9.

—— (1965). *Stylosanthes gracilis*: a tropical forage legume for Nigeria. *Proc. agric. Soc., Nigeria* **4**, 57–61.

—— (1971a). Research on forage crops in anglophone West Africa. *Ford Foundation/IRAT/IITA seminar on forage crops research in West Africa, Ibadan.*

—— (1971b). Livestock production problems associated with mineral deficiencies or excesses in West Africa with special emphasis on Nigeria. In *Mineral studies with isotopes in domestic animals*, pp. 165–75. Publication IAEA-PL-312-2/1. International Atomic Energy Authority, Vienna.

—— ONAYINKA, B. O. (1966a). Some observations on the responses of *Stylosanthes gracilis* to seed inoculation. *Nigerian agric. J.* **3**, 35–8.

—— —— (1966b). The production and management of grass/legume mixtures at Agege. *Nigerian agric. J.* **3**, 84–91.

—— —— EWEJE, J. K. (1968). The management and improvement of natural grasslands in Nigeria. *Nigerian agric. J.* **5**, 4–6.

—— —— —— (1970). The effect of cultural and chemical treatment on the control of spear grass (*Imperata cylindrica* (L.) Beauv. var. *africanus* (Anders.) Hubbard). *Nigerian agric. J.* **7**, 115–19.

ADEMOSUN, A. A. (1970a). Nutritive evaluation of Nigerian forages. 1. Digestibility of *Pennisetum purpureum* by sheep and goats. *Nigerian agric. J.* **7**, 19–26.

—— (1970b). Nutritive evaluation of Nigerian forages. 2. The effect of stage of maturity on the nutritive value of *Stylosanthes gracilis. Nigerian agric. J.* **7**, 164–73.

—— (1973). Nutritive evaluation of Nigerian forages. 4. The effect of stage of maturity on the nutritive value of *Panicum maximum* (Guinea grass). *Nigerian agric. J.* **10**, 170–7.

—— BAUMGARDT, B. R. (1967). Studies on the assessment of the nutritive value of some Nigerian forages by analytical methods. *Nigerian agric. J.* **4**, 1–6.

—— KOLADE, J. O. Y. (1973). Nutritive evaluation of Nigerian forages. 3. A comparison of chemical composition and nutritive value of two varieties of *Cynodon. Nigerian agric. J.* **10**, 160–9.

ADENIJI, M. O. and CHHEDA, H. R. (1971). Influence of six varieties of *Cynodon* on four *Meloidogyne* spp. *J. Nemat.* **3**, 251–4.

AGBOOLA, A. A. and FAYEMI, A. A. (1971). Preliminary trials on the inter-cropping of maize with different tropical legumes in Western Nigeria. *J. agric. Sci., Camb.* **77**, 219–25.

AHLGREN, G. H., ADEGBOLA, A. A., EWEJE, J. K., and SALAMI, A. (1959). *The development of grassland in the Western Region of Nigeria.* Ministry of Agriculture National Research ICA Project, Final Report 61.13.050. 133 pp.

AHMADU BELLO UNIVERSITY (1969). *Proceedings of the Conference on livestock development in the dry and intermediate savanna zones, Zaria.*

AKEN 'OVA, M. E. (1974). Improvement of *Pennisetum purpureum* Schum. for forage in the low altitude humid tropics. Ph.D. thesis, University of Ibadan, Nigeria.

AKINOLA, J. O., CHHEDA, H. R., and MACKENZIE, J. A. (1971*a*). Effects of cutting frequency and level of applied nitrogen on productivity, chemical composition, growth components and regrowth potential of three *Cynodon* strains. 2. Growth components: tillering and leaf area. *Nigerian agric. J.* **8**, 63–76.

—— MACKENZIE, J. A., and CHHEDA, H. R. (1971*b*). Effects of cutting frequency and level of applied nitrogen on productivity, chemical composition, growth components and regrowth potential of three *Cynodon* strains. 3. Regrowth potential. *W. Afr. J. Biol. appl. Chem.* **14**, 7–12.

ALKALI, M. M. (1969). Mixed farming in Northern Nigeria. *Proc. agric. Soc., Nigeria* **3**, 7–13.

ANDERSON, A. W. (1933). Problems of animal nutrition and animal husbandry in Northern Nigeria. Imperial (now Commonwealth) Bureau of Animal Nutrition, Technical Communication No. 4.

ASARE, E. O. (1969). Preliminary investigations to the response of buffel grass (*Cenchrus ciliaris* L.) to fertilization. *Ghana J. Sci.* **9**, 50–3.

AUBRÉVILLE, A. (1949). *Climats, forêts et désertification de l'Afrique tropicale.* Société d'Editions Géographiques, Paris.

AUDRU, J., LAMARQUE, D., LEBRUN, J. P., and RIVIÈRE, R. (1966). *Ensembles pastoraux du Logone et du Moyen Chari (République du Tchad).* Institut Élev. Méd. Vét. Pays Trop., Maisons-Alfort Val-de-Marne, France.

BABALOLA, O. and CHHEDA, H. R. (1972). Effects of crops and management systems on soil structure in a Western Nigerian soil. *Nigerian J. Sci.* **6**, 129–36.

BARNARD, C. (1966). *Grasses and grasslands.* Macmillan, London.

BARRAULT, J. (1971). Work to date in Northern Cameroon on *Pennisetum pedicellatum.* *Ford Foundation/IRAT/IITA seminar on forage crops research in West Africa, Ibadan.*

BARTHOLOMEW, W. V., MEYER, I., and LAUDELOT, H. (1953). Mineral nutrient immobilization under forest and grass fallow in Yangambi (Belgian Congo) region. *Inst. nat. Étude agron. Congo Ser. Sci.* **57**, 1–27.

BERLIER, Y., DABIN, B., and LENEUF, N. (1956). Physical, chemical and microbiological comparison of forest and savannah soils on the tertiary sands of the lower Ivory Coast. *Int. Congr. Soil Sci.* **5**, 499–502.

BILLE, J. C., LAMARQUE, D., LEBRUN, J., and RIVIÈRE, R. (1968). *Étude agrostologique des pâturages de la régions de savanes (République du Togo).* Institut Élev. Méd. Vét. Pays Tropical, Maisons-Alfort Val-de-Marne, France.

BOGDAN, A. V. (1971). Introduction and breeding of leguminous forage crops. *Ford Foundation/IRAT/IITA seminar on forage crops research in West Africa, Ibadan.*

—— PRATT, D. J. (1967). *Reseeding denuded pastoral land in Kenya.* Government Printer, Nairobi: (Report of the Kenya Ministry of Agriculture and Animal Husbandry).

BOONMAN, J. G. (1971). Experimental studies on seed production of tropical grasses in Kenya. 1. General introduction and analysis of problems. *Neth. J. agric. Sci.* **19**, 23–6.

BORGET, M. (1969). Résultats et tendances présentes des recherches fourragères de l'IRAT. *Agron. trop., Nogent* **24**, 103–55.

—— (1971). Recherches fourragères menées par l'IRAT en Afrique tropicale et à Madagascar. *Ford Foundation/IRAT/IITA seminar on forage crops research in West Africa, Ibadan.*

BOUDET, G. (1963). *Pâturages et plantes fourragères en République de Côte d'Ivoire.* Institut Élev. Méd. Vét. Pays Tropicales, Maisons-Alfort Val-de-Marne, France.

BOUGHEY, A. S. (1957). The physiognomic delimitation of West African vegetation types. *J. W. Afr. Sci. Ass.* **3**, 148–65.

BOWDEN, B. N. (1963). Studies on *Andropogon gayanus* Kunth. 1. The use of *Andropogon gayanus* in agriculture. *Emp. J. exp. Agric.* **31**, 267–73.

BOYER, J. (1965). Nature de la couverture du sol et influence sur le bilan hydrique d'une cocterale. *Oléagineux* **20**, 437–40.

BREDON, R. M., HARKER, K. W., and MARSHALL, B. (1963). The nutritive value of grasses grown in Uganda when fed to Zebu cattle 1. The relation between the percentage crude protein and other nutrients. *J. agric. Sci., Camb.* **61**, 101–4.

BUTTERWORTH, M. H. (1965). Some aspects of the utilization of tropical forages. *J. agric. Sci., Camb.* **65**, 233–9.

CADOT, R. (1971*a*). Forage research carried out by the IEMVT in Africa and Madagascar. *Ford Foundation/IRAT/IITA seminar on forage crops research in West Africa, Ibadan.*

—— (1971*b*). Seed quality of various forage species in the Ivory Coast. *Ford Foundation/IRAT/IITA seminar on forage crops research in West Africa, Ibadan.*

Cah. Agric. prat. Pays Chauds—Service Coop. Tech. Outre Mer (Supplement):

(1965). Le *Brachiaria*, Suppl. No. 2, 89–93; Siratro, Suppl. No. 4, 210–14 (*L'Agron. trop.* **20**).

(1966). Le *Cenchrus ciliaris*, Suppl. No. 2, 97–102; Le *Digitaria decumbens*, Suppl. No. 4, 189–94 (*L'Agron. trop.* **21**).

(1967). Le *Glycine javanica*, Suppl. No. 4, 203–9 (*L'Agron. trop.* **22**).

(1968). Le *Tripsacum laxum*, Suppl. No. 2, 79–90; Le *Melinis*, Suppl. No. 4, 197–203 (*L'Agron. trop.* **23**).

(1969). *Desmodium uncinatum* et *D. intortum*, Suppl. No. 1, 35–42 (*L'Agron. trop.* **24**).

(1970). L'herbe à l'éléphant, Suppl. No. 1, 37–46; Le Cynodon dactylon, Suppl. No. 3, 145–54 (*L'Agron. trop.* **25**).

(1971). Le kikuyu, Suppl. No. 1, 39–46 (*L'Agron. trop.* **26**).

CATCHPOOLE, V. R. and HENZELL, E. F. (1971). Silage making from tropical herbage species. *Herb. Abstr.* **41**, 213-21.

CHARTER, J. R. and KEAY, R. W. J. (1960). *Assessment of the Olokemeji fire-control experiment* (Investigation 254) *twenty-eight years after institution.* Niger. For. Inf. Bull. (new series), Vol. 3. Federal Government Printer, Lagos, Nigeria.

CHHEDA, H. R. (1971). *Cynodon* improvement in Nigeria. *Ford Foundation/IRAT/IITA seminar on forage crops research in West Africa, Ibadan.*

—— (1974). Forage crops research at Ibadan. 1. *Cynodon* spp. In *Animal production in the tropics* (eds J. K. Loosli, V. A. Oyenuga, and G. M. Babatunde), pp. 19-94. Heinemann Educational Books, Ibadan, Nigeria.

—— AKINOLA, J. O. (1971*a*). Effects of cutting frequency and levels of applied nitrogen on productivity, chemical composition, growth components and regrowth potential of three *Cynodon* strains. 1. Yield, chemical composition, and weed competition. *Nigerian agric. J.* **8**, 44-62.

—— —— (1971*b*). Effects of cutting frequency and level of applied nitrogen on crude protein production and nitrogen recovery by three *Cynodon* species. *W. Afr. J. Biol. appl. Chem.* **14**, 31-8.

—— SALEEM, M. A. M. (1973*a*). Effects of N and K fertilizers on *Cynodon* IB.8 and on soil in southern Nigeria. *Expl. Agric.* **9**, 249-55.

—— —— (1973*b*). Effect of height of cutting after grazing on yield, quality and utilization of *Cynodon* IB.8 pasture in southern Nigeria. *Trop. Agric., Trin.* **50**, 113-19.

—— AKEN'OVA, M. E., and CROWDER, L. V. (1973). Development of *Pennisetum typhoides* × *P. purpureum* hybrids for forage in the low altitude humid tropics. *Crop. Sci.* **13**, 122-3.

—— —— (1974). Forage crops research at Ibadan. 2. *Pennisetum purpureum.* In *Animal production in the tropics* (eds J. K. Loosli, V. A. Oyenuga, and G. M. Babatunde), pp. 95-101. Heinemann Educational Books, Ibadan, Nigeria.

CHURCH, R. J. W. (1960). *West Africa: a study of the environment and man's use of it* (6th edn), Chaps. 1, 2, and 3. Longmans, London.

COLE, M. M. (1963). Vegetation nomenclature and classification with particular reference to savannas. *S. Afr. geog. J.* **45**, 3-14.

COMBES, D. and PERNES, J. (1970). Variations dans les nombres chromosomiques du *Panicum maximum* Jacq. en relation avec le mode de reproduction. *C.r. hebd. Séanc. Acad. Sci., Paris,* **270**, 782-5.

—— VERBURGT, W. H. (1971). The commercial growing, cleaning and marketing of tropical and subtropical pasture seeds. *Ford Foundation/IRAT/IITA seminar on forage crops research in West Africa, Ibadan.*

COMMONWEALTH BUREAU OF PASTURES AND FIELD CROPS (1962). A review of nitrogen in the tropics with particular reference to pastures—a symposium. *Bull. Commonw. Bur. Past. Fld Crops,* No. 46, Farnham Royal, Bucks., England. 185 pp.

Conference on Livestock Development in the Dry and Intermediate Savanna Zones (1969). Proceedings. Ahmadu Bello University, Zaria.

COUPER, D. C. (1971). The cost of silage production at Shika Agricultural Research Station. *Nigerian agric. J.* **8**, 77-84.

CROWDER, L. V. (1967). Grasslands of Colombia. *Herb. Abstr.* **37**, 237-45.

—— (1971). Forage and pasture research in Tropical America. *Ford Foundation/IRAT/IITA seminar on forage crops research in West Africa, Ibadan.*

—— (1973). *Forage and fodder crops in farming systems.* IITA, Ibadan.

CUNNINGHAM LABORATORY STAFF (1964). Some concepts and methods in subtropical pasture research. *Bull. Commonw. Bur. Past. Fld Crops* **47**, Farnham Royal, Bucks., England, 242 pp.

CUNNINGHAM, R. K. (1963). The effect of clearing a tropical forest soil. *Soil Sci.* **14**, 334-45.

DAVIES, W. (1964). The eradication of tsetse in the Chad river system of northern Nigeria. *J. appl. Ecol.* **1**, 387-403.

—— (1971). Further eradication of tsetse in the Chad and Gongola river systems of north-eastern Nigeria. *J. appl. Ecol.* **8**, 563-78.

—— SKIDMORE, C. L. (1966). *Tropical pastures.* Faber, London.

DELEGATION OF THE REPUBLIC OF DAHOMEY (1965). Fodder crops and fallow in Dahomey. *Afr. Soils* **10**, 225-40.

DENNISON, E. B. (1959). The maintenance of fertility in the Southern Guinean Savannah zone of Northern Nigeria. *Trop. Agric., Trin.* **36**, 171-8.

DOYNE, H. C. (1937). Green manuring in southern Nigeria. *Emp. J. exp. Agric.* **5**, 248-53.

DUCKWORTH, J. (1946). A statistical comparison of the influence of crude fibre on the digestibility of roughage by *Bos indicus* (Zebu) and *Bos taurus* cattle. *Trop. Agric., Trin.* **23**, 4-9.

DUGAIN, F. and FAUCK, R. (1959). Erosion and run-off measurements in middle Guinea. *Inter-Afr. Soils Conf.* **3**. 597-600.

EGUNJOBI, J. K. (1970). Savannah burning, soil fertility and herbage production in the derived savannah zone of Nigeria. *Conf. W. Afr. Wildlife Cons.* **7**, pp. 52-8.

EVANS, D. C. P. (1961). A review of past and present work connected with grassland, pasture and fodder problems in Ghana. *Grassl. Symp.* **1**. Science Service Division, Ministry of Agriculture, Accra, Ghana.

FAUCK, R., MOUREAUX, C., and THOMANN, C. (1969). Bilans de l'évolution de sols de sefa (Casamance, Sénégal) après quinze années de culture continué. *Agron. trop., Nogent* **24**, 263-301.

FAULKNER, O. T. (1934). Some experiments with leguminous crops at Ibadan, Southern Nigeria, 1925-1933. *Emp. J. exp. Agric.* **2**, 93-102.

FOOD AND AGRICULTURE ORGANIZATION (1966). *Report of the meeting on savanna development, Khartoum, 1966.* FAO, Rome.

—— (1971). *Report of the Conference on the establishment of cooperative agricultural research programmes between countries with similar ecological conditions in Africa: Guinean zone.* FAO, Rome (ESR:LAR/71).

FOSTER, W. H. (1961a). Sowing dates and yields in Mucuna (*Stizolobium* spp.) in Northern Nigeria. *Newsl. No. 32 Min. Agric. N. Nigeria.*

—— (1961b). Note on the establishment of a legume in rangeland in Northern Nigeria. *Emp. J. exp. Agric.* **29**, 319-22.

—— (1962). Investigations preliminary to the production of cultivars of *Andropogon gayanus*. *Euphytica* **11**, 47-52.

—— MUNDY, E. J. (1961). Forage species in Northern Nigeria. *Trop. Agric., Trin.* **38**, 311-18.

GAUDEFROY-DEMOMBYNES, P. (1957). Observations sur la couverture du sol. *Bull. agron. Minist. fr. d'outre mer* **15**, 25-33.

GLELE, M. A. A. (1971). Sols fourragères et conservation de la fertilité: travaux réalisés au Dahomey. *Ford Foundation/IRAT/IITA seminar on forage crops research in West Africa, Ibadan.*

GREENLAND, D. J. (1958). Nitrate fluctuations in tropical soils. *J. agric. Sci.* **50**, 82-92.

—— NYE, P. H. (1959). Increases in the carbon and nitrogen contents of tropical soils under natural fallows. *J. Soil Sci.* **9**, 284-99.

GROSSMAN, J. (1956). Grazing experiments with beef cattle in Rio Grande de Sol, Brazil. *Int. Grassld Congr.* **7**, 528-37.

—— ARONOVICH, S., and CAMPELLO, E. DO C. B. (1966). Grasslands of Brazil. *Int. Grassld Congr.* **9**, 39-47.

GUEVARRA-CALDERÓN, J. (1967). Tropical forage plants. In *Development in tropical Latin America* (eds K. L. Turk and L. V. Crowder), pp. 256-71. Cornell University Press, Ithaca.

GUILLOTEAU, J. (1958). The problem of bush fires and burns in land development and soil conservation in Africa south of the Sahara. *Afr. Soils* **4**, 64-102.

HAGGAR, R. J. (1967). The production of seed from *Andropogon gayanus*. *Samaru Res. Bull.* **80**, 1-9.

—— (1968). Grazing behaviour of Fulani cattle at Shika, Nigeria. *Trop. Agric., Trin.* **45**, 179-85.

—— (1969a). Improved pastures in the northern states of Nigeria. In *Conference of livestock development in dry and intermediate savanna zones*, pp. 86-93. Ahmadu Bello University, Zaria.

—— (1969b). Use of companion crops in grassland establishment in Nigeria. *Expl Agric.* **5**, 47-52.

—— (1969c). A guide to the management and use of stylo (*Stylosanthes gracilis*). *Samaru agric. Newsl.* **11**, 63-6.

—— (1970a). Seasonal production of *Andropogon gayanus*. 1. Seasonal changes in yield components and chemical composition. *J. agric. Sci., Camb.* **74**, 487-94.

—— (1970b). The intake and digestibility of low quality *Andropogon gayanus* hay, supplemented with various nitrogenous feeds as recorded by sheep. *Nigerian agric. J.* **7**, 70-6.

—— AHMED, M. B. (1970). Seasonal production of *Andropogon gayanus*. 2. Seasonal changes in digestibility and feed intake. *J. agric. Sci., Camb.* **75**, 369-73.

HEADY, F. H. (1960). *Range management in East Africa.* Government Printer, Nairobi.

HILL, D. H. (1964). Animal breeding and improvement in Nigeria. *Outl. Agric.* **4**, 80-5.

HOPKINS, B. (1963). The role of fire in promoting sprouting of some savannah species. *J. W. Afr. Sci. Ass.* **7**, 154-62.

—— (1965a). *Forest and savannah: an introduction to West Africa.* Heinemann, London.

—— (1965b). Observations on savannah burning in the Olokemeji Forest Reserve, Nigeria. *J. appl. Ecol.* **2**, 367-81.

HUTTON, E. M. (1970). Tropical pastures. *Adv. Agron.* **22**, 1-73.

INSTITUTE FOR AGRICULTURAL RESEARCH, SAMARU (1971). *Shika Agricultural Research Station Biennial Report 1969-71.* The Institute, Samaru.

INTERNATIONAL INSTITUTE OF TROPICAL AGRICULTURE (1971). *Ford Foundation/IRAT/IITA seminar on forage crops research in West Africa, Ibadan.*

JURION, F. and HENRY, J. (1969). *Can primitive farming be mechanized?* Institut national pour l'étude agronomique du Congo, Brussels. (Trans. from French by Agra Europe, London.)

KANNEGIETER, A. (1967a). Zero cultivation and other methods of reclaiming *Pueraria* fallowed land for food crop cultivation in the forest zone of Ghana. *Trop. Agric. Mag. Ceylon agric. Soc.* **113**, 1-23.

—— (1967b). The cultivation of grasses and legumes in the forest zone of Ghana. *Int. Grassld Congr.* **9**, 313-18.

KEAY, R. W. J. (1959). *Vegetation map of Africa south of the Tropic of Cancer.* Oxford University Press, London.

KNUDSEN, P. B. and SOHAEL, A. A. (1970). The Vom herd: a study of performance of a mixed Friesian/Zebu herd in a tropical environment. *Trop. Agric., Trin.* **47**, 189-203.

LANSBURY, T. L. (1958). The composition and digestibility of some conserved fodder crops for dry season feeding in Ghana. *Trop. Agric., Trin.* **35**, 114-18.

—— (1960). A review of some limiting factors in the nutrition of cattle on the Accra Plains, Ghana. *Trop. Agric., Trin.* **39**, 185-92.

—— ROSE INNES, R., and MABEY, G. L. (1965). Studies on Ghana grasslands: yield and composition on the Accra plains. *Trop. Agric., Trin.* **42**, 1-18.

H

LAUDELOT, N. (1962). *Fallowing techniques of tropical soils.* United Nations, New York. *U.N. Conf. appl. Sci. Tech.* E/Confr. 39/C/26.

LAWSON, G. W. (1966). *Plant life in West Africa.* Clarendon Press, Oxford.

LEEUW, P. N. DE (Ed.) (1969–71). Shika Agricultural Research Station, Biennial Report. Institute for Agricultural Research, Ahmadu Bello University, Zaria, Nigeria.

—— (1971). The prospects of livestock production in the northern Guinea zone savannas. *Ford Foundation/IRAT/IITA seminar on forage crops research in West Africa, Ibadan.*

—— BRINCKMAN, W. L. (1973). Pasture and range development in the northern Guinea and Sudan Zone of Nigeria. *International symposium on livestock production in the tropics, Ibadan.*

—— COUPER, D. C. (1970). The cost of mechanical shrub control. *Samaru agric. Newsl.* 11, 53–6.

LETENNEUR, L. (1971). Le *Stylosanthes gracilis*: Synthèse des travaux effectués en Côte d'Ivoire et ses possibilités d'extension. *Ford Foundation/IRAT/IITA seminar on forage crops research in West Africa, Ibadan.*

LITTLE, E. C. S. and IVENS, G. W. (1965). The control of bush by herbicides in tropical and subtropical grassland. *Herb. Abstr.* 35, 1–12.

LOTERO, J. (1970). Publicaciones del programa pastos y forrajes del Instituto Colombiano Agropecuario. *Boln Inf.* 1. 16 pp.

MABEY, G. L. and ROSE INNES, R. (1964a). Studies on browse plants in Ghana. 2(a). Digestibility of *Griffonia simplicifolia*. *Emp. J. exp. Agric.* 32, 125–30.

—— —— (1964b). Studies on browse plants in Ghana. 2(b). Digestibility of *Baphia nitida*. *Emp. J. exp. Agric.* 32, 274–8.

—— —— (1966a). Studies on browse plants in Ghana. 2(c). Digestibility of *Antiaris africana*. *Expl Agric.* 2, 27–32.

—— —— (1966b). Studies on browse plants in Ghana. 2(d). Digestibility of *Grewia carpinifolia*. *Expl Agric.* 2, 113–17.

MCILROY, R. J. (1962). Grassland improvement and utilization in Nigeria. *Outl. Agric.* 3, 174–9.

—— (1972). *An introduction to tropical grassland husbandry* (2nd edn). Clarendon Press, Oxford.

MCKELL, C. M. and ADEGBOLA, A. A. (1966). Need for a range management approach for Nigerian grasslands. *J. Range Mgmt* 19, 330–3.

MACKENZIE, J. A. and CHHEDA, H. R. (1970). Comparative root growth studies of *Cynodon* IB.8: an improved variety of *Cynodon* forage grass suitable for southern Nigeria and two other *Cynodon* varieties. *Nigerian agric. J.* 7, 91–7.

MACLENNAN, K. J. R. and AITCHISON, P. J. (1963). Simultaneous control of three species of *Glossina* by the selective application of insecticide. *Bull. ent. Res.* 54, 199–212.

—— (1969). Trypanosomiasis. In *Conference on livestock development in dry and intermediate savanna zones,* pp. 145–55. Ahmadu Bello University, Zaria.

MALUNFASHI, A. T. (1969). Problems involved in settling the Fulani. In *Conference on livestock development in dry and intermediate savanna zones,* pp. 49–54. Ahmadu Bello University, Zaria.

MASSON, H. (1948). La température du sol au cours d'un feu de brousse au Sénégal. *Comm. no. 12, Conf. Afr. des Sols, Goma (Kivu), Congo Belge,* pp. 1933–40.

MATHON, J. C., LOUVE, J. L., and LETENNEUR, L. (1972). Interest and possibilities of cross-breeding Jersey-N'dama in the Ivory Coast. FAO, Rome. *Report FAO Conference on the Establishment of Cooperative Agricultural Research Programmes—Guinea Zone,* Document 34, pp. 201–5. (ESR:CAR/71.)

MBA, A. U., OKE, S. A., and OYENUGA, V. A. (1971). Studies on the *in vitro* OM digestibility and ME utilization in the small ruminant stock. Paper presented at the *7th Conf. Agric. Soc. Nigeria, Kano 7.*

MENSBRUGE, G. DE LA and BERGEROO-CAMPAGNE, B. (1958). Rapport sur les résultats obtenus dans les parcelles d'Ivoire. *Rep. deuxième Conf. forestière interafr.,* pp. 620–6. CSA/CCTA Pub. No. 43, London.

MILLER, T. B. (1960). The effect of feeding high protein and high carotene concentrates to young cattle during late dry season–early wet season in Northern Nigeria. *W. Afr. J. biol. Chem.* 4, 29–35.

—— (1961). Recent advances in studies of the chemical composition and digestibility of herbage. *Herb. Abstr.* 31, 1–9.

—— (1969). Forage conservation in the tropics. *Samaru Res. Bull.* 113, 159–62.

—— IDUMA, L. B. (1967). Nutrition of Zebu cattle in Northern Nigeria. 3, The importance of carotene supplements. *Expl Agric.* 3, 287–93.

—— RAINS, A. B. (1963). The nutritive value and agronomic aspects of some fodders in Northern Nigeria. 1. Fresh herbage. *J. Br. Grassld Soc.* 18, 158–67.

—— —— THORPE, R. J. (1963). The nutritive value and agronomic aspects of some fodders in Northern Nigeria. 2. Silages. *J. Br. Grassld Soc.* 18, 223–9.

—— —— —— (1964). The nutritive value and agronomic aspects of some fodders in Northern Nigeria. 3. Hay and dried crop residues. *J. Br. Grassld Soc.* 19, 77–90.

MINSON, D. J. and MILFORD, R. (1967). The voluntary intake and digestibility of diets containing different proportions of legume and mature pangola grass (*Digitaria decumbens*). *Aust. J. exp. Agric. Anim. Husb.* 7, 546–51.

MONOD, T. (1957). *Les grandes divisions chronologiques de l'Afrique.* CSA/CCTA, Publication No. 24, London.

MOORE, A. W. (1962). The influence of a legume on soil fertility under a grazed tropical pasture. *Emp. J. exp. Agric.* 30, 239–48.

—— (1965). The influence of fertilization and cutting on a tropical grass-legume pasture. *Expl Agric.* 1, 193–200.

MORGAN, W. B. and PUGH, J. C. (1969). *West Africa*, Introduction and Chaps. 5 and 6. Methuen, London.

MURDOCK, J. C. (1966). Grass silage. *Outl. Agric.* 5, 17-21.

NAVEH, Z. (1966). Range research and development in the dry tropics with special reference to East Africa. *Herb. Abstr.* 36, 77-85.

NORMAN, M. J. T. (1966). Katherine Research Station 1956-65. A review of published work. *Tech. Pap. Div. Ld Res. reg. Surv.* No. 28 C.S.I.R.O. Melbourne.

NORRIS, D. O. (1971). Leguminous plants in tropical pastures. *Ford Foundation/IRAT/IITA seminar on forage crops research in West Africa, Ibadan.* (See also *Trop. Grassld* 6, 159-69 (1972).)

NOURRISSAT, P. (1965). Influence de l'époque de fauche et de la hauteur de coupe sur la production d'une prairie naturelle au Sénégal. *Sols Afr.* 10, 365-77.

NWOSU, N. A. (1960). Conservation and utilization of *Stylosanthes gracilis*. *Trop. Agric., Trin.* 37, 61-6.

—— (1961). The propagation of *Stylosanthes gracilis* established from stem cuttings. *Nigerian Scient.* 1, 24-31.

NYE, P. H. (1958). The relative importance of fallows and soils in storing plant nutrients in Ghana. *J. W. Afr. Sci. Ass.* 4, 31-41.

—— GREENLAND, D. J. (1960). *The soil under shifting agriculture*. *Tech. Commun. Commonw. Bur. Soils* No. 51, C.A.B. Farnham Royal, Bucks., England.

—— HUTTON, R. G. (1957). Some preliminary analyses of fallows and cover crops at the West African Institute for Oil Palm Research, Benin. *Jl W. Afr. Inst. Oil Palm Res.* 2, 237-43.

OGOR, E. and HEDRICK, D. W. (1963). Management of natural pasturage (range grazing) in West Africa. *J. W. Afr. Sci. Ass.* 7, 145-53.

OKIGBO, B. N. and CHHEDA, H. R. (1966). Natural fertility and chromosome numbers in several strains of star grasses. *Nigerian agric. J.* 3, 72-5.

OKORIE, I. I., HILL, D. H., and McILROY, R. J. (1965). The productivity and nutritive value of tropical grass/legume pastures rotationally grazed by N'dama cattle at Ibadan, Nigeria. *J. agric. Sci., Camb.* 64, 235-45.

OLALOKU, E. A. (1972). The effect of the level of feed, nutrient intake and stage of lactation on the milk of White Fulani cows in Ibadan. Ph.D. Thesis, University of Ibadan, Nigeria.

OLUBAJO, F. O. (1970). Use of chomic oxide in indoor digestion experiments with tropical grass-legume mixture using white Fulani steers. *Nigerian agric. J.* 7, 105-10.

—— OYENUGA, V. A. (1970). Digestibility of tropical pasture mixtures using the indicator technique. *J. agric. Sci., Camb.* 75, 175-81.

—— —— (1971). The measurement of yield, voluntary intake and animal production of tropical pasture mixtures. *J. agric. Sci., Camb.* 77, 1-4.

—— VAN SOEST, P. J., and OYENUGA, V. A. (1974). Composition and digestibility of four tropical grasses in Nigeria. *J. Anim. Sci.* 38, 149-53.

OYENUGA, V. A. (1957). The composition and agricultural value of some grass species in Nigeria. *Emp. J. exp. Agric.* 23, 237-55.

—— (1958). Problems of livestock nutrition in Nigeria. *Nutr. Abstr. Rev.* 28, 985-1000.

—— (1959a). Effect of cutting on the yield and composition of some fodder grasses in Nigeria (*Pennisetum purpureum* Schum.). *J. agric. Sci., Camb.* 53, 25-33.

—— (1959b). Effect of stage of growth and frequency of cutting on the yield and composition of some Nigerian fodder grasses: *Andropogon tectorum* Schum: *W. Afr. J. biol. Chem.* 3, 43-58.

—— (1959c). Effect of frequency of cutting on the yield and chemical composition of some Nigerian fodder grasses: *Tripsacum laxum* Nash. *W. Afr. J. biol. Chem.* 4, 46-63.

—— (1960). Effect of stage of growth and frequency of cutting on the yield and chemical composition of some Nigerian fodder grasses: *Panicum maximum* Jacq. *J. agric. Sci., Camb.* 55, 339-50.

—— (1967). *Agriculture in Nigeria*, Chap. 9. FAO, Rome.

—— (1968a). *Nigeria's foods and feeding stuffs: their chemistry and nutritive value*. Ibadan University Press.

—— (1968b). Animal production in Africa meeting nutrient requirements of range cattle for optimum yield. *Rep. Conf. agric. Res. Priorities Econ. Dev. Afr., Abidjan* 3, 32-50.

—— (1971a). Biological productivity in West Africa. *J. W. Afr. Sci. Ass.* 16, 93-115.

—— (1971b). Methodology of grazing trials. *Ford Foundation/IRAT/IITA seminar on forage crops research in West Africa, Ibadan.*

—— HILL, D. H. (1966). Influence of fertilizer applications on the yield, efficiency and ash constituents of meadow hay. *Nigerian agric. J.* 3, 6-14.

—— OLUBAJO, F. O. (1966). Productivity and nutritive value of tropical pastures at Ibadan. *Int. Grassld Congr.* 10, 962-9.

PERNES, J. (1971). Problems posed by the improvement of the tropical forage species: *Panicum maximum* (Jacq.). *Ford Foundation/IRAT/IITA seminar on forage crops research in West Africa, Ibadan.*

—— COMBES, D., and R'N'E-CHAUME, R. (1970). Différenciation des populations naturelles du *Panicum maximum* Jacq. en Côte d'Ivoire par acquisition de modifications transmissibles, les unes par graines apomictiques, d'autres par multiplication végétative. *C.r. hebd. Séanc. Acad. Sci., Paris*, 270, 1992-8.

PEYRE DE FABREGUES, B. (1965). Études et principes d'exploitation de pâturage de steppe en République du Niger. *Revue Élev. Méd. vét. Pays trop.* 18, 329-32.

PHILLIPS, J. (1959). *Agriculture and ecology in Africa*. Faber, London.

PICARD, D. (1971). *Les plantes améliorantes et les prairies tempéraines*. Comité de Liaison des Organismes de Recherche Agricole Specialisés Outre-mer, Adiopodoumé, Ivory Coast.

PITOT, A. and MASSON, H. (1951). Données sur la température au cours des feux de brousse aux environs de Dakar. *Bull. Inst. fr. Afr. noire* 13, 711-32.

PRATT, D. J., GREENWAY, P. J., and GWYNNE, M. D. (1966). A classification of East African rangeland, with an appendix on terminology. *J. appl. Ecol.* 3, 369-82.

RAINS, A. B. (1963). *Grassland research in Northern Nigeria. Samaru Misc. Publ.* 1. Institute agric. Res., Ahmadu Bello University, Zaria. Headly Bros., London.

—— (1968). *A field key to the commoner genera of Nigerian grasses. Samaru Misc. Publ.* 7. Institute agric. Res. Ahmadu Bello University, Zaria. The Claxton Press (W. Afr.), Ibadan.

RAMSEY, J. M. and ROSE INNES, R. (1963). Some quantitative observations on the effect of fire on the Guinea savanna vegetation of Northern Ghana over a period of eleven years. *Afr. Soils* 8, 41-86.

RATTRAY, J. M. (1960). The grass cover of Africa. *FAO agric. Stud.* 49.

RAYMOND, W. F. (1969). The nutritive value of forage crops. *Adv. Agron.* 21, 1-108.

Report (1967) *Meeting on Savanna Development*, Khartoum, Sudan, Oct. 25-Nov. 6, 1966. UNDP/FAO MR/62338/8-67.

Report (1971) *FAO Conference on the Establishment of Cooperative Agricultural Research Programmes—Guinean Zone*, IITA, Ibadan, Nigeria, Aug. 23-8, 1971. (ESR:CAR/71).

RICHARD MOLARD, J. (1956). *Afrique occidentale française*. Berger Levrault, Paris.

RISOPOULOS, S. A. (1966). *Management and use of grasslands, Democratic Republic of the Congo*. Pasture and fodder crops studies No. 1. FAO, Rome.

ROOSE, E. J. and BERTRAND, R. (1971). Contribution à l'étude de la méthode des bandes d'arrêt pour contre l'érosion hydrique en Afrique de l'Ouest: résultats expérimentaux et observations sur le terrain. *Agron. trop.*, Nogent 26, 1270-83.

ROSE INNES, R. (1962). Grasslands and fodder resources. In *Agriculture and land use in Ghana* (ed. J. B. Wills), Chap. 23. Oxford University Press, London.

—— (1963). The behaviour of free-grazing cattle in the West African humid tropics: studies on a herd of West African Shorthorns on the Accra plains, Ghana. 1. Rainy season. *Emp. J. exp. Agric.* 31, 1-13.

—— (1965). The concept of the 'woody pasture' in the low-altitude tropical tree savanna environments. *Int. Grassld. Congr.* 9, 1419-23.

—— (1967). Grassland and livestock. In *Land and water survey in the Upper and Northern Regions, Ghana*, Vol. IV, pp. 109-88. FAO, Rome.

—— (1971). Fire in West African vegetation. *Tall. Timbers Fire Ecol. Conf.* 2, 147-73.

—— MABEY, G. L. (1964a). Studies on browse plants in Ghana. 1. Chemical composition. *Emp. J. exp. Agric.* 32, 114-24.

—— (1964b). Studies on browse plants in Ghana. 3. Browse/grass ratios. *Emp. J. exp. Agric.* 32, 180-90.

ROSEVEARE, D. R. (1953). Vegetation. In *The Nigerian handbook*, pp. 139-73. Government Printer, Lagos.

ROSEVEARE, G. M. (1948). *The grasslands of Latin America. Bull. imp. Bur. Past. Fld Crops* 36. Wm Lewis (Printers) Ltd., Cardiff.

ROSSETTI, C. (1965). *Ecological survey mission to West Africa. Studies on the vegetation (1959 and 1961)*. FAO, Rome. (UNSF/DL/ES/5.)

RUTHENBERG, H. (1971). *Farming systems in the tropics*. Clarendon Press, Oxford.

SALEEM, M. A. M. (1972). Productivity and chemical composition of *Cynodon* IB.8 as influenced by level of fertilization, soil pH and height of cutting. Ph.D. Thesis, University of Ibadan, Nigeria.

—— CHHEDA, H. R. (1972). Effects of level and time of nitrogen application on dry matter yield, crude protein and herbage utilization of a rotationally grazed *Cynodon* IB.8 pasture. *Nigerian agric. J.* 9, 93-9.

Seminar (1971). Papers presented to conference: forage crops research in West Africa. Ford Foundation/IRAT/IITA. Apr. 26-30. University of Ibadan, Nigeria (mimeo report).

SEMPLE, A. T. (1970). *Grassland improvement*. Chemical Rubber Company Press, Ohio.

SEN, K. M. and MABEY, G. L. (1965). The chemical composition of some indigenous grasses of coastal savanna of Ghana at different stages of growth. *Int. Grassld Congr.* 9, 763-71.

STEPHENS, D. (1960). Three rotation experiments with grass fallows and fertilizers. *Emp. J. exp. Agric.* 28, 165-78.

TEWARI, G. B. (1968). Response of grasses and legumes to fertilizer treatments in Nigeria. *Expl Agric.* 4, 87-91.

THOMPSON, B. W. (1965). *The climates of Africa*. Oxford University Press, Nairobi.

THORPE, R. J. (1964). Cereal-legume silage mixture for the Northern Guinea Zone, Nigeria. *Trop. Agric.*, Trin. 41, 41-5.

TILLEY, J. M. and TERRY, R. (1963). A two-stage technique for the *in vitro* digestion of forage crops. *J. Br. Grassld Soc.* 18, 104-11.

TRAORE, M. (1958). Mémorandum sur la protection des savannes en forêt classée de Niangoloko (Haute Volta). *Inter-Afr. Soils Conf.* 2, 677-83. CSA/CCTA, Publication No. 43. London.

TROCHAIN, J. (1955). Nomenclature et classification de types de végétation en Afrique Noire occidentale et centrale. In *Les divisions écologiques du monde. Moyens d'expresion, nomenclature, cartographie*, pp. 73-90.

Centr. Natl. Recherches Sci. (8th Botanical Congress), Paris, 1954.

VALENZA, J. (1965). Notes about stocking rate trials on grazinglands in the Republic of Senegal. *Int. Grassld Congr.* **9**, 9-12.

VAN RAAY, J. G. T. and DE LEEUW, P. (1970). The importance of crop residues as fodder. A resource analysis in Katsina Province, Nigeria. *Tijdschr. econ. soc. Geogr.* **61**, 137-47.

VAN SOEST, P. J. and MARCUS, W. C. (1963). A method for the determination of cell wall constituents in forages using detergent and the relationship between this fraction and voluntary intake and digestibility. *J. Dairy Sci.* **47**, 204.

—— WINE, R. H. and MOORE, L. A. (1966). Estimation of true digestibility of forages by the *in vitro* digestion of cell walls. *Proc. int. Grassland Congr.* **10**, 438-41.

VICENTE-CHANDLER, J. (1967). Intensive pasture production. In *Development in tropical Latin America* (ed. K. L. Turk and L. V. Crowder), pp. 272-95. Cornell University Press, Ithaca.

VILLEMAN, G. (1963). Entretien des cocoterais. *Oléagineux* **19**, 27-31.

VINE, H. (1953). Experiments on the maintenance of soil fertility at Ibadan, Nigeria, 1922-51. *Emp. J. exp. Agric.* **21**, 65-85.

—— (1968). Developments in the study of soils and shifting agriculture in tropical Africa. In *The soil resources of tropical Africa* (ed. R. P. Moss), Chap. 5. Cambridge University Press, London.

WATSON, K. A. and GOLDSWORTHY, P. R. (1964). Soil fertility investigations in the Middle Belt of Nigeria. *Emp. J. exp. Agric.* **31**, 290-302.

WEBSTER, C. C. and WILSON, P. N. (1966). *Agriculture in the tropics*. Longmans, London.

WEST, O. (1965). *Fire in vegetation and use in pasture management with special reference to tropical and subtropical Africa. Pub. Commonw. agric. Bur.* **1/1965**. Farnham Royal, Bucks., England.

WHYTE, R. O. (1968). *Grasslands of the monsoon*, Chap. 4. Faber, London.

—— MOIR, T. R. G., and COOPER, J. P. (1959). *Grasses in agriculture. FAO agric. Stud.* **42**.

—— NILSSON-LEISSNER, G. and TRUMBLE, H. C. (1966). *Legumes in agriculture. FAO agric. Stud.* **21**.

WILSON, A. S. B. and LANSBURY, T. J. (1958). *Centrosema pubescens*: ground cover and forage crops in cleared rainforest in Ghana. *Emp. J. exp. Agric.* **26**, 351-4.

WILSON, B. (1968). *Pasture improvement in Australia*. Murray, Sydney.

9. Irrigation

C. DES BOUVRIE AND J. R. RYDZEWSKI

9.1. Introduction

Irrigation can be defined broadly as the artificial control of soil moisture for agricultural purposes with the aim of increasing crop production. In physical terms this control is achieved by applying water on the land when soil moisture becomes depleted and, at the same time, by making provision for the removal of excess water from the soil surface or profile. The variety of engineering installations is therefore wide, ranging from sub-irrigation or trickle-irrigation systems in arid regions, where no drainage facilities are required, to swamp reclamation, where the engineering problem may be confined to drainage and flood protection.

The purpose of irrigation, so defined, will thus differ in different ecological areas, but can be conveniently summarized as follows:

(1) in arid areas to provide the opportunity for settled agriculture;

(2) in semi-arid areas to permit the intensification of cropping, to achieve higher yields and a greater diversification of cash crops;

(3) in areas where the amount and distribution of rainfall is unpredictable, to minimize the risk of crop failure or to maintain a high level of yield;

(4) in areas of excessive moisture, to reduce the soil moisture level to that required for optimum plant growth;

(5) to meet the special water requirements of cash crops (such as rice or sugar cane) for which the climate, but not the natural field-water regime, is suitable.

Since decisions on irrigation development are usually made at a national level and since many countries in West Africa extend from the forest zone into the savanna zones, it is felt that this ecological variation, taken together with the geography of the principal river basins, makes it essential for the purpose of this chapter to consider the problem in the context of all of West Africa south of the Sahara. The countries involved are: Mauritania, Senegal, Gambia, Guinea, Sierra Leone, Liberia, Mali, Ivory Coast, Upper Volta, Ghana, Togo, Dahomey, Niger, Nigeria, Chad, and Cameroun, the last two having been included because of the importance of developments in the Lake Chad basin.

Whether or not irrigation agriculture should be introduced depends on a complex combination of climatic, topographic, economic, technical, and socio-political factors. These are briefly discussed in the sections that follow.

9.2. Climate in relation to irrigation

As is well known, the over-all climatic pattern of Africa is dominated by movements of air caused by zones of high pressure situated roughly at the latitude of the tropics. At the earth's surface these zones often appear broken up into cells of high pressure from which air flows towards the equator in a path affected by the rotation of the earth. Within the two tropical zones of high pressure is a broad belt where ascending currents of air lead to the formation of rain clouds. The general effect over the whole of Africa is illustrated in Fig. 9.1. The topography of the eastern and southern parts of Africa considerably complicates the climatic picture. But the West Africa region, because of its position relative to the ocean mass and of its mainly flat topography, presents a much more uniform climatic pattern which can very largely be explained in terms of the seasonal movement of the inter-tropical convergence zone (ITCZ). In West Africa the ITCZ is formed by the meeting of the southerly moisture-bearing winds from the Atlantic Ocean and the dry winds from the north-east (the harmattan). As pointed out by Cochemé (1971) and Rijks (1972) the movement of the ITCZ, with its rain, is in an annual cycle following the vertical position of the sun at noon but with a time lag of 5–6 weeks. Its extreme positions vary from about 25° N

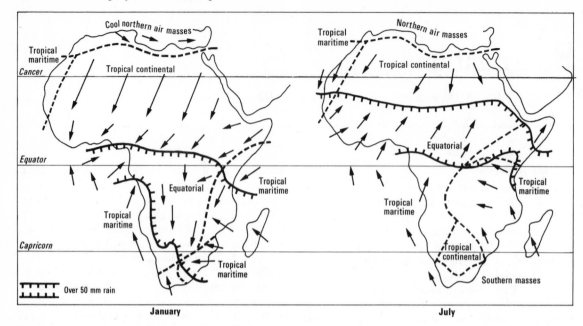

FIG. 9.1. Air masses and precipitation in Africa in January and July. The arrows indicate the directions of the prevailing winds at low altitudes. (Grove 1970.)

in July–August to about 5° N in January–February. This means that it rarely falls completely outside the West Africa region, considered for our agricultural purposes as lying between the latitudes of 4° N and about 17° N.

The northerly movement of the ITCZ in the northern hemisphere spring and its southerly retreat in the autumn accounts for the seasonality of the rains, which reach the northern parts of the region in late spring or early summer in a clearly marked single (tropical) wet season which becomes shorter as one moves north. In the south the ITCZ completely leaves the area during its northward movement so there is an interruption in the rains in late summer before the ITCZ returns on its southward path and the rains resume. This gives rise to a typical equatorial climate with two wet seasons (bimodal pattern of seasonal distribution), separated by a short dry spell, followed by a longer dry season in the winter.

The broad climate pattern is one of simple and regular isohyets running essentially parallel to the lines of latitude but it is, of course, locally modified by topography and the relative disposition of land and ocean.

From the agroclimatic standpoint, the region can be conveniently divided into four zones, which, as

Fig. 9.2 indicates, differ in distribution from the vegetation zones referred to elsewhere in this book.

Agroclimatic zone A, in which the annual rainfall is less than 300 mm and where settled agriculture is impossible without irrigation, corresponding roughly to the *Sahelian zone*.

Agroclimatic zone B, the *Sudano-Sahelian zone*, within which a single wet season permits some form of rain-fed agriculture. Its northern boundary is taken as the 300 mm per annum isohyet, while on the south it extends to the line indicating that the wet season (monomodal) does not last longer than 160 days. This boundary corresponds to a rainfall of between 700 mm per annum and 1000 mm per annum. Irrigation here is essential if double cropping is considered to be desirable. Once established, it would also serve to supplement any soil-water deficits during the wet season.

Agroclimatic zone C, the *Sudanian and Sudano-Guinean zone*, immediately south of zone B, extends down to the tropical rain forest zone in the coastal belt, defined by the 2000 mm per annum isohyet. Much of this area is subjected to the bimodal rainfall pattern, described above, which permits a larger variety of crops to be grown without irrigation, but at the beginning and end of

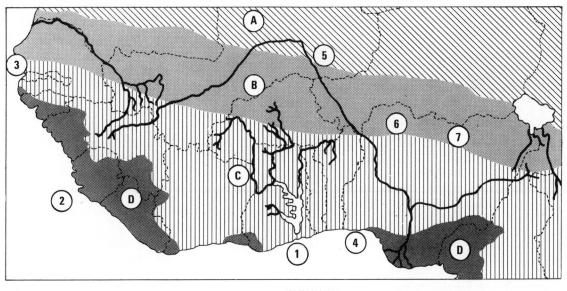

	ZONE A	Rainfall less than 300 mm per annum
	ZONE B	Rainfall more than 300 mm per annum but duration of availability of water less than 160 days. Maximum rainfall 700–100 mm per annum. Monomodal rainfall distribution
	ZONE C	Between Zone B and the 2000 mm per annum isohyet. Mostly bimodal rainfall distribution.
	ZONE D	Rainfall in excess of 2000 mm per annum. Includes coastal swamps

FIG. 9.2. Agroclimatic zones in West Africa.

both wet seasons there are periods of seasonally varied and unreliable rainfall. Therefore, for the attainment of reliable high yields, irrigation is advisable and for year-round cultivation it becomes essential.

Agroclimatic zone D, the *tropical rain forest* (*Guinean*) *zone*, which includes the coastal mangrove swamps, lies between the coast and the 2000 mm per annum isohyet. Controlled water management in this zone is primarily concerned with the removal of excess soil moisture by means of drainage systems, especially in areas where rainfall exceeds 3500 mm per annum.

The four zones described above correspond closely to the BWh, BSh, Aw, and Am categories of the Köppen classification of climate.

A further illustration of climatic conditions prevailing in the different zones is presented in Fig. 9.3 for the seven localities indicated in Fig. 9.2, as follows:

(1) Accra ($5°$ $33'$ N, $0°$ $15'$ W) in agroclimatic zone C;
(2) Freetown ($8°$ $30'$ N, $13°$ $17'$ W) in zone D;

(3) Bathurst ($13°$ $28'$ N, $16°$ $39'$ W) in zone D but towards zone C;
(4) Lagos ($6°$ $27'$ N, $3°$ $28'$ E) in zone C;
(5) Gao ($16°$ $19'$ N, $0°$ $09'$ W) in zone A but near boundary of zone B;
(6) Sokoto ($13°$ $02'$ N, $5°$ $15'$ E) in zone B;
(7) Kano ($12°$ $00'$ N, $8°$ $31'$ E) in zone B.

For each location diagrams are given for the monthly water balance, as defined in Fig. 9.3. The water balance curves are of particular interest since they vividly highlight the water deficits that would have to be made up to permit continuous cropping.

A more detailed presentation of the rainfall deficit problem in the region is made in Table 9.1, where the monthly figures for the difference between precipitation and potential evapotranspiration are given for 16 localities. Examination of this table reveals the danger of basing judgements on annual rainfall data. If Porto Novo is taken as an example it will be seen that, although the over-all water balance is strongly positive, the deficit in August acts as a severe constraint on successful crop production, even in the northern hemisphere summer months.

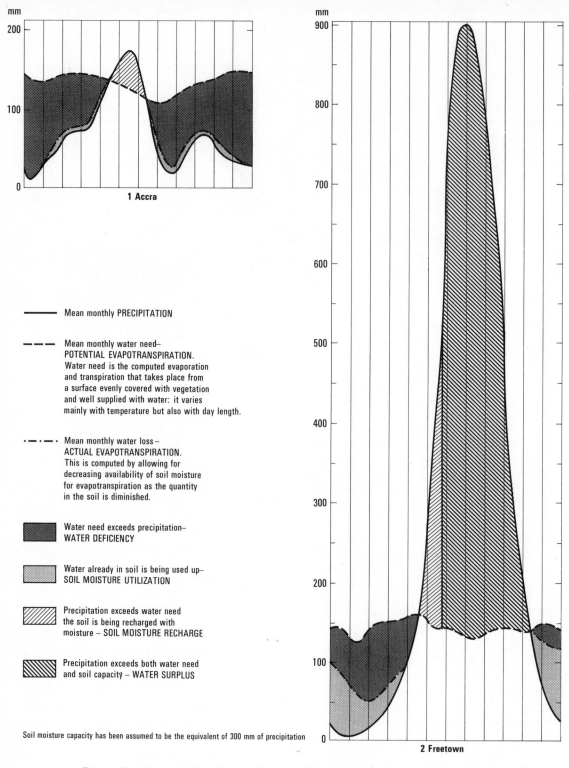

mm

200 ─

100 ─

0 ─

1 Accra

──────── Mean monthly PRECIPITATION

─ ─ ─ Mean monthly water need–
POTENTIAL EVAPOTRANSPIRATION.
Water need is the computed evaporation
and transpiration that takes place from
a surface evenly covered with vegetation
and well supplied with water: it varies
mainly with temperature but also with day length.

─ ∙ ─ ∙ ─ Mean monthly water loss –
ACTUAL EVAPOTRANSPIRATION.
This is computed by allowing for
decreasing availability of soil moisture
for evapotranspiration as the quantity
in the soil is diminished.

Water need exceeds precipitation–
WATER DEFICIENCY

Water already in soil is being used up–
SOIL MOISTURE UTILIZATION

Precipitation exceeds water need
the soil is being recharged with
moisture – SOIL MOISTURE RECHARGE

Precipitation exceeds both water need
and soil capacity – WATER SURPLUS

Soil moisture capacity has been assumed to be the equivalent of 300 mm of precipitation

mm
900 ─

800 ─

700 ─

600 ─

500 ─

400 ─

300 ─

200 ─

100 ─

0 ─

2 Freetown

FIG. 9.3. Monthly water balance diagrams (January to December) for seven locations in the region.

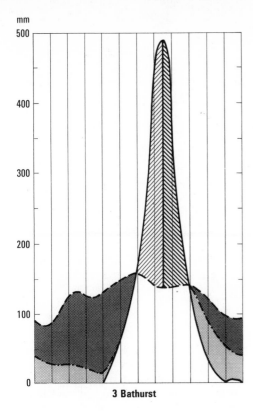

mm

3 Bathurst

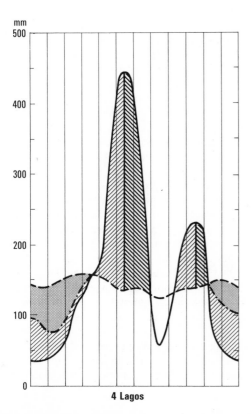

mm

4 Lagos

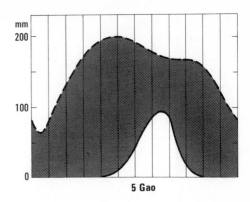

mm

5 Gao

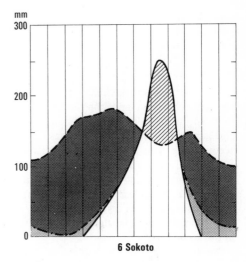

mm

6 Sokoto

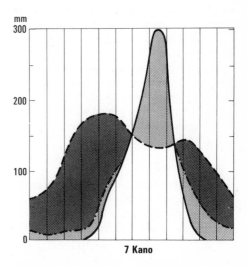

mm

7 Kano

TABLE 9.1

Rainfall deficit (precipitation minus potential evapotranspiration) for some ecological zones in West Africa (mm)

Country	Location	Jan.	Feb.	Mar.	Apr.	May	June	July	Aug.	Sept.	Oct.	Nov.	Dec.	Annual balance
Nigeria	Bida	−216	−236	−195	−109	+ 11	+ 82	+118	+199	+166	− 4	−174	−211	− 569
	Ibadan	−124	−132	− 44	+ 17	+ 51	+132	+ 64	− 24	+ 11	+ 74	− 55	−111	− 141
	Maidugri	−218	−264	−331	−340	−274	−126	+ 64	+130	− 13	−179	−252	−213	−2066
	Port Harcourt	− 82	− 53	+ 56	+ 80	+142	+201	+289	+200	+387	+177	− 2	− 82	+1317
Dahomey	Cotonou	− 38	− 36	+ 19	+ 34	+157	+290	+ 55	− 23	+ 26	+ 99	− 6	− 65	+ 512
	Porto Novo	− 70	− 60	− 4	+ 29	+150	+258	+150	− 21	+ 61	+113	+ 1	− 77	+ 560
Togo	Lome	− 89	− 68	− 45	+ 1	+ 63	+143	+ 1	− 50	− 35	+ 13	− 50	− 88	− 204
	Niamtougu	−224	−240	−169	− 74	− 2	+ 74	+152	+161	+233	+ 5	−144	−207	− 435
Ghana	Accra	− 91	− 68	− 44	− 13	+ 52	+ 94	− 12	− 43	− 37	− 13	− 57	− 78	− 310
	Akuse	−139	−126	− 45	− 15	+ 15	+ 83	− 28	− 63	− 19	+ 17	− 18	− 95	− 408
	Kumasi	−103	− 78	+ 10	+ 36	+ 85	+156	+ 63	+ 6	+107	+116	− 2	− 72	+ 324
	Tamale	−262	−262	−209	−117	− 19	+ 42	+ 56	+121	+141	− 17	−177	−239	− 942
Ivory Coast	Abidjan	− 45	− 33	+ 39	+ 74	+315	+537	+127	− 19	+ 1	+147	+116	+ 34	+1293
	Bouaké	−175	−142	− 76	+ 29	+ 22	+ 36	+ 9	+ 25	+126	+ 33	− 85	−131	− 329
Liberia	Cacaopa	− 9	− 8	− 1	− 3	− 2	+ 4	+ 4	+ 1	+ 9	+ 2	− 6	− 8	− 17
	Monrovia	− 1	− 1	+ 1	+ 3	+ 15	+ 35	+ 30	+ 14	+ 26	+ 22	+ 6	+ 1	+ 151

For certain crops the field-water regime required for their cultivation is not confined to evapotranspiration. In the case of rice, for instance, pre-watering and ponding on the paddy fields has to be taken into account. Fig. 9.4 shows field-water balance diagrams for rice cultivation in the inland swamps in Sierra Leone, situated on the boundary of zones C and D of Fig. 9.2. The clear indication is that controlled water management is essential and that, even in the case of single cropping, there is a need for irrigation in the second half of June.

Apart from the main problem of water balance, two other climatic factors which have an adverse effect on crop production have to be mentioned. First, most of the region can be subject to hot and dry harmattan winds during the winter. Secondly, in the northern parts, winter months can be cooler than some crops can tolerate without considerable reduction in yield.

With regard to the region as a whole it is apparent that, in the northern areas, irrigation can add one full cropping season and, in the south, supplementary irrigation, coupled with adequate drainage, could ensure year-round cropping at near-optimal soil-moisture conditions.

9.3. Land and water resources

Two-thirds of West African land surface south of the 12° N parallel consists of predominantly igneous and metamorphic rocks of a Pre-Cambrian shield, often referred to as the 'basement complex' (Ahn 1970).

Sedimentary formations deposited in lakes and shallow seas during successive geological periods from the Palaeozoic to the Quaternary and overlying the basement complex constitute the remainder. North of this parallel these proportions are approximately reversed.

Furthermore, the remains of at least two distinct peneplanation surfaces can be distinguished over much of the region, frequently covered, respectively, by bauxite and ironstone concretionary layers. The landscape as a consequence consists (in very broad terms) of upland plateaux and large open plains at 300 m above sea-level, which present monotonously little relief with a few prominent mountain ridges where the plains are interrupted by granite domes (*inselbergs*) and uplifts of the basement complex.

These geomorphological features are noteworthy since they have, in combination with differences in rainfall patterns, profoundly influenced the natural environment and have a significant bearing on its irrigation potential. Distinctly different drainage patterns have been cut in the old surface, with a system of flat uplands and river valley bottoms connected by colluvial slopes. Stream valleys are often more pronounced and numerous nearer the coast, but are gentler in slope and more widespread in the north.

Weathering processes are progressively slower as annual rainfall decreases. The uplands of the forest zone are associated with very uniform drift material

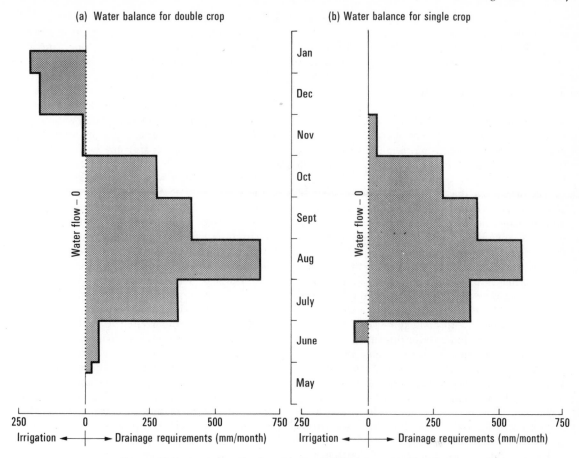

FIG. 9.4. Field water balance for rice cultivation on inland swamps in Sierra Leone.

rather than with broken-up remnants of the pene-plains' concretionary and ironstone layers, which are more commonly and extensively found in the savanna at shallow depth.

Furthermore, the rocks of the basement complex generally prove to be richer in soil parent materials than the younger quartz-rich sediments, mudstones and sandstones. The 'upland soils' of the savanna are thus found to be not only less fertile and lower in organic matter content but also more significantly of lighter texture and of a lower water-storage capacity than those of the forest zone. Crops grown on these soils under rain-fed agriculture prove to be most vulnerable when soil moisture deficits develop during the short, and not always reliable, wet season.

The 'colluvial' soils of the valley slopes, again as a result of the general topography, rarely extend over more than short distances in the forest zone but can

reach a width of several kilometres in the savanna. Here, at times, extensive areas of so-called 'ground-water laterites' appear at the bottom of the slopes, characterized by a hard iron-rich layer or a hardpan in the subsoils, causing serious impediment to drainage, thus making irrigation more difficult.

The more interesting soils from an irrigator's point of view are those which are distributed in relatively narrow strips of alluvial valley bottoms within command of the river. They are, however, not always easy to manage under cultivation. In the lower depressions they are often exposed to seasonal floods and are imperfectly drained. Due to their hydromorphic nature and high clay content they are difficult to cultivate by hand- or animal-drawn equipment and call for mechanized† cultivation, yet are well suited for irrigation of rice and sugar cane. The lighter

† Motorized in the sense of Chapter 13. Ed.

textured 'alluvial levée' soils are at times used for small-scale irrigated vegetable gardening.

A different situation is found in the large alluvial valleys of the lower Senegal River, the central Niger delta in Mali and along the shores of Lake Chad, where wide tracts of land are inundated seasonally. These inundations permit extensive flood-recession (or draw-down) crop cultivation on *fadamas* or seasonal grazing for the livestock of the Sahelian nomads. These locations comprise the area where the truly large potential for irrigation is to be found in the region.

Still further to the north lie extensive areas of gently sloping semi-arid savanna lands bordering on sand dune formations of more recent origin, on which extensive grazing is practised but which offer few opportunities for irrigation.

Particularly difficult to irrigate are the 'tropical black earths' (the 'black cotton soils') found under dry savanna-type climatic conditions and developed over base-rich rocks, covering extensive areas of the Accra plains and south of Lake Chad. They crack deeply on drying, have vertic properties,† and are excessively sticky and slowly permeable when wet.

Other soils with less favourable irrigation qualities are, for instance, those of the Senegal River delta and of the Guinea coastal swamps, which are heavy textured, slowly permeable, and saline. Yet, as is demonstrated below, they can be successfully reclaimed through deep drainage.

There is little doubt that the irrigation potential of West Africa's surface-water resources is very considerable indeed. Recent estimates of the average yearly run-off in the region's main river systems (FAO Regional Office, Accra) arrive at a total of 172 thousand million m³ per annum, distributed as follows (in thousand millions)

River Senegal at its delta	22
River Niger at Mopti	70 (but 30 at Niamey)
River Logone-Chari at Lake Chad	40 (but 30 at Niamey)
River Volta at Senchi	40 (but 30 at Niamey)

These figures correspond to only a portion, perhaps not more than 10–15 per cent, of the total precipitation falling on the basins. Yet most of the streams in

† Vertic properties implies the liability to vertical cracking on drying. This character is often associated with soils having a montmorillonitic clay fraction. Ed.

the upper catchments run dry for at least 6 months of the year outside the wet season. During much of this period large rivers and their major tributaries are reduced to almost a trickle, in sharp contrast to their peak discharges, which can, in fact, reach enormous proportions. An even distribution of streamflow in time and space is, therefore, singularly lacking, making an adequate year-round water supply very rare. Few major measures to regulate streamflow have been taken so far, apart from the diversion weir and the large reservoir on the Niger at Sansanding and at Kainji respectively, and the Akosombo dam on the Volta.

Irrigated agriculture has therefore to depend on run-of-the-river diversion or on pumping, unless storage reservoirs are provided, which in most instances is a costly operation, not facilitated by the generally flat topography. Numerous attempts have been made at small-scale storage in the upper reaches, with capacities from 100 000 m³ to 8 000 000 m³, but the use of these stores for agricultural purposes, as demonstrated below, has not always been undertaken with success (Food and Agriculture Organization 1967). Neither have investment costs for such structures been found to balance the returns. Instances of solid accomplishment and success, on the other hand, can be quoted where simple water-conservation works in the form of weirs or dykes have been constructed for flood-recession cultivation.

It can hardly be said that the region is richly endowed with sub-surface water resources (Secrétariat d'État 1973). Shallow ground-water less than 50 m deep is widespread (wells yielding 0·25–1·25 l/s) in the fractured and weathered upper strata of the crystalline pre-Cambrian shield. But although total annual replenishment to this formation is, without doubt, considerable, individual aquifers are rarely found to be sizeable in extent or thickness.

More abundant in water (well yields above 3 l/s), at times under artesian conditions, but generally at much greater depth (100–150 m), are the reservoirs of the three most important aquifer systems: the 'continental Intercalaire-Hamadan', the 'Maestrichtien', and the 'continental terminal', which have been estimated to hold several hundred thousand million cubic metres. Uncertainty, however, still exists as to their total annual recharge and consequently their safe yield, and much of their water is considered to be fossil or semi-fossil—'mining' of such water would have to be considered.

A third category of aquifer—and for irrigation

purposes perhaps the most interesting—is found in the alluvial deposits of the major river basins and of Lake Chad and to some extent of the coastal plains. At only a few locations, however, has tubewell pumping so far been attempted, and prospects for yields of the order of magnitude found in, for instance, the Indus valley are very slim indeed.

In only a few instances has quality, whether from surface or sub-surface sources, been found to be of below irrigation water standards. Notorious for high salinity risks are the waters of the lower Senegal caused by sea-water intrusion when flow in the river is low. Some aquifers, such as those of the Tchirezrine sandstone of the Irhazer in Niger, have been found to be polluted by rocks containing soluble salts rich in sodium chloride and calcium—or magnesium sulphate. The generally good chemical composition of the region's waters and soils constitutes a great asset for irrigation development without the need for intensive drainage provisions (compare the lower Indus basin).

In most river valleys, on the other hand, conditions prevail which predispose the transmission of water-borne diseases and the breeding of noxious insects. The effects of onchocerciasis (river blindness) (World Health Organization 1973), for instance, can best be illustrated by the 65 000 km² of deserted valley lands in the central West African region (mainly the Volta River system), a sizeable proportion of which could be developed for irrigated agriculture following the completion of the large-scale eradication campaign already initiated. Also bilharzia (schistosomiasis) (Organization for African Unity 1970) is highly endemic in the region. Even so, irrigation development can produce a dramatic increase in the prevalence of the 'disease of irrigated areas', particularly where reservoir storage is provided.

In conclusion, it may be said that the natural environment of the region is not particularly conducive to high-value crop production under irrigation. Yet a considerable potential has so far remained untapped and, for the region as a whole, no acute shortage of land and water resources exists. Costs of water-resources development therefore can be still kept within reasonable economic bounds, unlike the situation in other parts of the world where unfavourable land/man ratios have led to investments of over US $5000/ha. In certain areas, however, lands with an inherent supporting capacity of perhaps 20–30 persons per km² have been colonized up to a density of 100 persons per km². This is the case where the

retreat from disease-ridden river valleys places an insupportable burden of human occupation on the upland plateau lands. In other instances the failure of the rains, as in the period 1969–73, has locally exacerbated the inability of the region's savanna lands to support their ever-growing populations. These are the areas most urgently in need of supplementary irrigation, to relieve rainfall deficiencies, and of the development of ground-water reserves for use in periods of drought.

9.4. Human resources in relation to irrigation

Taking Africa as a whole, the only irrigated area which is historically important—and it is very important indeed—is the lower Nile valley. Here nature has provided a dependable source of water and suitable land immediately next to it. The world's other old-established systems of irrigation, such as those on the Indo-Gangetic plain, are equally blessed with good lands along the rivers, the flows of which do not suffer the great extremes known in Africa since they derive from a combination of direct run-off and snow-melt.

The topography of the continent of Africa has not encouraged the formation of alluvial valleys with river regimes easily adaptable to irrigation development. Historically, too, there has been no such population pressure as that which forced the peoples of southern Asia to struggle for as much food as possible from each piece of arable land. The peoples of Africa, grouped in numerous tribal units, tended to settle on steep terrain, more easily defended against hostile neighbours or wild animals and less affected by the diseases associated with river valleys, such as malaria, onchocerciasis, and bilharzia. In West Africa, this has at times resulted in overpopulated regions and underpopulated, disease-ridden, but more fertile, valley-bottom lands.

In some areas it has been the custom for women to do most of the agricultural work—an arrangement which could not easily be extended to intensive cropping. A further constraint has been, and still is, the presence of tsetse fly, making it impossible in some areas to use ox-power for ploughing and transport.

In addition, sections of the rural population have a 'target income' approach to increased earnings. This, coupled with a high leisure preference, may lead to a way of life the loss of which people in industrialized communities may secretly regret, but it does stand in the way of using the profit motive as a spur to higher agricultural production unless the profit increment clearly outweighs the traditional values forgone.

In short, the formula which combines man, land, and water to produce successful irrigated agriculture has been difficult to apply in tropical Africa. Besides, the available rainfall has usually been sufficient for subsistence agriculture or for a simple form of animal husbandry, so there appeared to be little reason for farming communities to become involved in solving the organizational problems arising in the establishment of successful irrigation schemes. The resulting lack of a tradition in the disciplines of irrigated agriculture, therefore, is a definite handicap to future development.

9.5. The need for irrigation

Since the provision of irrigation facilities should be considered as only one input, albeit an essential one, in the over-all process of intensive agriculture, the question posed here is where, given the above considerations on the environment and resources of the region, intensive agriculture should be a development objective.

From the over-all national point of view, in some of the countries, there may be a debate as to whether the attainment of given agricultural production targets should be achieved by intensive or extensive methods. It must be admitted that, in the short term, improvements in rain-fed agriculture can show a favourable benefit/cost aspect (Food and Agriculture Organization 1968).

The problem, however, can appear entirely different from the point of view of the rural community whose earning power, in comparison with those working in the emerging industrial and public sectors, begins to look increasingly unfavourable. The result of this disparity is the alarming population drift to urban areas (Food and Agriculture Organization 1968). Since for a long time to come the prosperity of most of the countries in the region will depend on the product of the rural sector, it is becoming a matter of some urgency to make the earning potential of the farmer or herdsman comparable with that of the industrial worker. Thus controlled water management, which brings with it an organizational structure capable of raising other aspects of agricultural technology, could become a useful instrument of government policy for regional development.

In the dry northern areas of West Africa such developments would result in the stabilization of currently nomadic communities, giving them a sense of civic consciousness and making it easier for the benefits of social progress to reach them. The herdsman was mentioned above because the authors feel that, in the West African regional context, he could become an important partner of the irrigation farmer. The world meat shortage, which is likely to become more acute in the future, suggests that it could be economically advantageous to grow fodder under irrigation and use it as an input for meat production. A measure which would have a more immediate effect, especially after the ravages of the Sahelian drought, is to establish focal centres for livestock fattening in the areas customarily covered by nomadic herdsmen.

The previous sections have demonstrated that, on the input side, water is an important constraint on the raising of cropping intensity in the region. The benefits on the output side are reported from many quarters where irrigated agriculture has been established alongside rain-fed cultivation. Increases in the total product have come here as a result of improved yields and from the introduction of double cropping. For instance, upland rice, grown under rain-fed conditions, may yield up to 2 t/ha, but in the same area full water control can raise the yield to 5 t/ha for each of two crops.

There are many more equally impressive statistics on the gross benefits of water control in agriculture, but their meaning cannot be evaluated without considering the other factors of production, such as fertilizers, improved seeds, and pesticides which, together with effective advisory services, marketing organizations, and credit facilities, form the 'package' of more advanced agricultural technology.

9.6. Irrigation in West Africa: general aspects

As has already been mentioned, the relative disposition of suitable agricultural lands and sources of water has, on the whole, not been favourable to irrigation development in Africa. This refers not only to sources of surface water but also to those of ground-water, where there is little likelihood of explorations revealing aquifers comparable to those of the Indo-Pakistan sub-continent.

Historically, therefore, the practice of irrigation with perennial water control is comparatively recent in tropical Africa, dating no further back than the 1930s. However, throughout the region there have been communities which have recognized the value of utilizing and partially controlling the soil moisture in valley bottoms. Where an annual cycle of floods is a marked feature of a river's regime, some cultivators relate their cropping calendar to the arrival and recession of the floods so as to raise one crop (in West Africa this is usually rice).

But hydrological phenomena are notoriously unpredictable in the short term, so some river-valley communities have introduced an element of water control by constructing a series of bunds which moderate the flooding of the lands and, what is more important, retain some of the water after the floods have receded. This extends the period of available soil moisture, thereby reducing the risk of complete crop failure (Comité Interafricain d'Études Hydrauliques 1972).

The contribution to agricultural production from these 'traditional' irrigated areas should not be under-estimated. Although the crop yields obtained are not high, the costs of providing such partial control are very low indeed (Rydzewski 1968). This type of development has, in addition, the advantage of achieving a definite increase in production without completely disturbing the local agrarian structure, as is likely to happen with the introduction of modern projects with complete water control.

We naturally wish to turn from statements based on general observation to some based on numerical data. Here we are confronted with the difficulty that accurate and up-to-date information is not available. What follows reflects the findings of the Survey of Irrigation Development (Rydzewski 1968, 1972a) for FAO's Indicative World Plan of Agricultural Development (the IWP). This study took the year 1965 as the 'base year' and made projections for the years 1975 and 1985. Information was collected, as

effectively as circumstances permitted, for most of tropical Africa.

With the qualification that the information here presented cannot be considered as accurate, especially when it relates to the estimates for 1975, a very general picture of the position of irrigated agriculture in tropical Africa emerges from Table 9.2 where the harvested irrigated area is compared with the total annually harvested area. The over-all proportion of just less than one-hundredth, compared with one-third for the Near East and North-West Africa, emphasizes the points made previously. We may go on and make comparisons with the vast developments in India and Pakistan, where a single canal command can be as large as the total irrigated area of West Africa, but little would be achieved by so doing.

Table 9.3 presents the best available record of the irrigated areas in various countries in tropical Africa as they were in 1965 and as the IWP saw their expansion by 1975. The general indications are that in the region as a whole the irrigated area is expected to increase by 64 per cent, in West Africa by 100 per cent, in Central Africa by 300 per cent, and in East and South Africa by 55 per cent.

By world standards the irrigation projects in tropical Africa are not very large. Since the Sudan is not in West Africa the Gezira-Managil Project of some 700 000 ha is not included. The largest existing projects in this region are of an order of magnitude smaller than that. Fig. 9.5 has been produced to give an over-all idea of the distribution of project size in tropical Africa for 1965, and as it could develop by 1975.

For 1965 it is interesting to observe the preponderance of small projects of less than 1000 ha. We must at once note that, at the other extreme, the

TABLE 9.2

The place of irrigation in total agricultural activity

	Arable land (millions of ha)	Area harvested annually (millions of ha)	Irrigated area (thousands of ha)	Crop intensity (per cent)	Harvested irrigated area as percentage of harvested area
West Africa	75	35	244	100	0·7
Central Africa	25	6	16	100	0·3
East and South Africa	50	22	240	100	1·1
Tropical Africa	150	63	600	100	0·95
Near East and North-West Africa	70	39	17 000	77	33·0

TABLE 9.3

Cultivated commanded areas (CCA) in tropical Africa in 1965, and estimate for 1975 (from Olivier 1972 and Cantor 1967)

Country	CCA (thousands of ha) 1965	1975	Country	CCA (thousands of ha) 1965	1975	Country	CCA (thousands of ha) 1965	1975
Mauretania	10·7	29·0	Chad	7·0	10·0	Ethiopia	94·7	153·9
Senegal	13·6	49·5	Cameroun	4·0	14·2	Uganda	3·0	27·5
Gambia	25·0	32·0	Congo	—	2·0	Kenya	15·1	27·1
Mali	117·0	150·3	Zaire	2·5	4·2	Tanzania	27·9	52·0
Upper Volta	0·1	6·0	Burundi	—	5·5	Mozambique	17·0	34·0
Niger	16·9	29·8	Rwanda	2·6	12·7	Malawi	0·6	11·0
Total for savanna zone	183·3	296·6	Total for Central Africa	16·1	48·6	Zambia	2·0	16·0
						Rhodesia	36·0	65·8
Guinea	41·0	64·0				Botswana	3·0	9·5
Sierra Leone	1·7	14·4				Swaziland	40·0	55·0
Liberia	0·1	10·0				Lesotho	—	3·0
Ivory Coast	4·8	21·3						
Ghana	0·2	16·1				Total for East and South Africa	293·3	454·8
Togo	0·2	15·2						
Dahomey	0·1	9·4				Total for tropical Africa	599·0	987·6
Total for coastal zone	48·1	150·4						
						(Republic of South Africa)	607·0	?
Nigeria	12·2	37·2						
Total for West Africa	243·6	484·2						

two large projects on the Niger River in Mali provide another peak in the figure. Within tropical Africa, medium-sized projects are much more in evidence in East and South Africa, where irrigation has had a slightly longer history. The projections for 1975 indicate a growing importance of small projects in West Africa and a shift towards medium-sized projects in the rest of the continent.

Figs. 9.6 and 9.7 attempt to show graphically how the unit cost of irrigation development is distributed in the region as a whole and in the two sub-regions. Again, use is made of data for 1965 and of estimates for 1975. The two 'humps' tend to separate the low-cost 'village project' from the fairly expensive so-called 'modern project'. Fig. 9.5 shows that small projects in West Africa tend to be low-cost 'village projects', while in the rest of the region they are quite likely to be plantations (possibly equipped for sprinkler irrigation) with a much higher unit cost.

A very rough comparison of the financial return to irrigation for 10 typical rice, sugar cane, and cotton irrigation projects in tropical Africa is presented in Table 9.4. The ratios between the estimated annual return R to irrigation and the approximate capital and recurrent costs expressed, by means of discounted-cash-flow methods, as an annual equivalent C, reveal an interesting relation to project unit cost. The average R/C for the five projects costing US $800 per ha or less is 3·90 while that for the five costing US $1000 per ha or over is 1·56. The low-cost projects are, in fact, the 'village projects' referred to previously, indicating that this type of development, though inferior in yield levels, appears to be economically more sound.

A further example of the impact of modest investment comes from northern Dahomey, where ox-ploughing and use of animal-traction to lift irrigation water from shallow wells has led to an almost fourfold increase in yearly farm revenue and has induced farmers to invest up to $400 each in these improvements (Food and Agriculture Organization 1971).

Obviously, a development programme on the scale envisaged for the region cannot be carried out

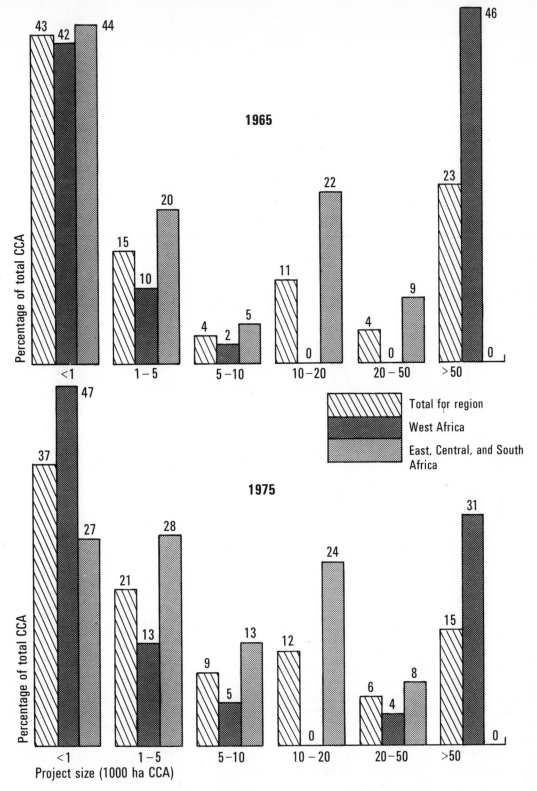

FIG. 9.5. Size of projects in tropical Africa (cultivable commanded areas).

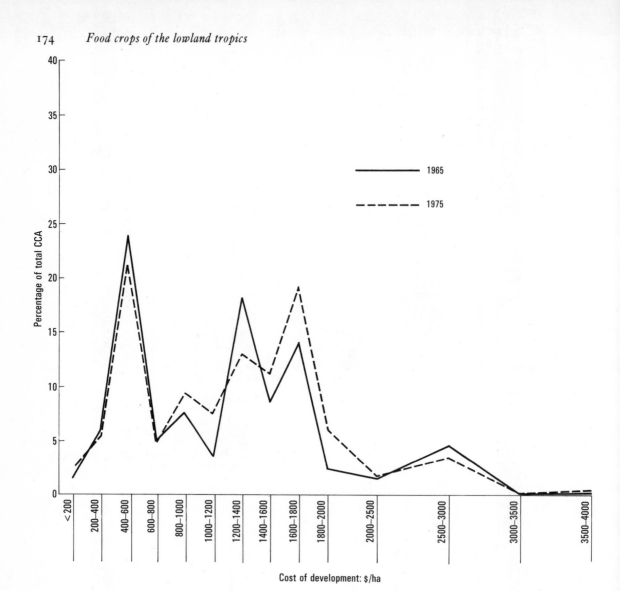

FIG. 9.6. Cost of development in tropical Africa, 1965 and 1975.

through small-scale projects alone. But these should not be neglected in favour of the admittedly more spectacular modern projects.

Decisions on development strategies should be based on data from projects already in operation but, as yet, such data are seldom available to planners. A step in this direction is a new survey of irrigation development in tropical Africa initiated in 1972 and now being carried out by FAO and Southampton University. When results of this survey become available it is hoped that comparisons, like those above, can be made with greater confidence.

9.7. Irrigation in West Africa: project development (Rydzewski 1968)

9.7.1. Introduction

In the previous section available figures were presented for the whole of tropical Africa in order to place the regional irrigation development achievement and programme in a broader perspective.

In this section attention is concentrated on the major features of irrigated agriculture in West Africa. For this purpose it is convenient to distinguish between the following groups: the large river valleys;

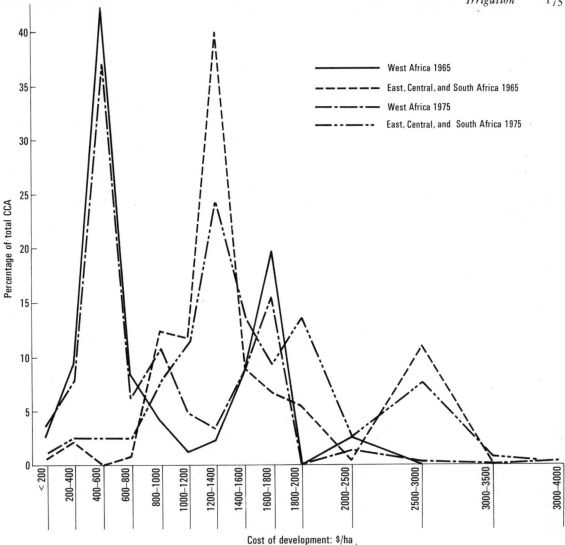

Fig. 9.7. Cost of development in West Africa and East, Central, and South Africa.

the small inland valleys and swamps; the coastal swamps; and ground-water development.

The major river basins and irrigation development areas in West Africa are shown in Fig. 9.8, which will be referred to in the following section.

9.7.2. *The Niger basin*

The River Niger rises in the Guinea highlands at an altitude of some 700 m and flows first in a north-easterly direction into Mali where it turns through approximately 90° to flow into Niger and Nigeria where it forms an extensive delta before

reaching the sea, having covered a distance of some 4000 km. This curved path means that the river begins its life in a zone of high rainfall (over 2000 mm per annum), flows into progressively drier climates, changes direction in agroclimatic zone A with rainfall less than 300 mm per annum, and then reverses the sequence of the climates it crosses, reaching the 2000 mm per annum rainfall zone at its coastal delta. The contribution of tributaries, the catchments of which receive less than 750 mm per annum of rain, is insignificant.

One of the more interesting features of the River

TABLE 9.4

Examples of returns to irrigated investment

	Costs (dollars/ha)			Returns to irrigation			
	Capital investment	Annual O.M.R.†	Approximate total annual equivalent (C)‡	Estimated increases in yield from irrigation (t/ha)	Producer price of crop (dollars/t)	Estimated annual return from irrigation (R)	(R)/(C)
RICE							
Niger flood plains (Mali)	400	6	38·7	1·2	51	61·2	1·58
Office du Niger (Mali)	1 700	26	165·0	3·0	51	153·0	0·93
Semry (Cameroun)	600	9	58·0	2·5	60	150·0	2·59
Bandama–Solomougou (Ivory Coast)	1 000	15	96·6	2·2	74	162·8	1·68
Small projects in north (Ivory Coast)	300	5	29·5	1·5	74	111·0	3·76
Mwea-Tebere (Kenya)	1 050	17	102·8	1·5	52	78·0	0·77
SUGAR CANE							
Marahue (Ivory Coast)	800	12	77·3	40	10	400	5·18
Asutsuare (Ghana)	1 700	60	199·0	60	10	600	3·00
COTTON							
Moggia project, on flood spreading (Niger)	400	6	38·7	2·0	124	248	6·40
Domet (Senegal)	2 000	30	193·2	2·0	136	272	1·41

† O.M.R.: Annual operation, maintenance, and replacement costs.

‡ Approximate total annual equivalent cost was obtained by converting the initial capital investment into an annual payment over 50 years having the same 'present value' with an 8 per cent discount rate, and adding the O.M.R. costs.

Niger is that, soon after passing Segou in Mali, it reaches a deltaic fan of Pleistocene sediments which form some 13 000 km² (1 300 000 ha) of lands, which, though not uniformly fertile, are considered to be suitable for irrigation development. This is shown as (1) in Fig. 9.8.

This inland delta has been the scene of two irrigation development programmes. The 'Office du Niger' Project was conceived in 1932 as the French equivalent of the Gezira project in the Sudan. However, unlike the Gezira, it did not overcome all the difficulties facing it in the early years. But, although it has not reached the intended 400 000 ha under irrigation, its present area of about 60 000 ha makes it by far the largest project in West Africa and an important source of rice and cotton production. Since the construction, in 1946, of the Markala barrage (Sansanding) which feeds the Macina and Sahel Canals on the left bank, the expansion of the irrigated area has been slow but steady. What is more important is the effort being made now to raise the cropping intensity (defined

in Rydzewski (1968)) from a figure of 70 per cent in 1965 (cf. Gezira, 50 per cent) to 100 per cent in 1985. Development costs here are estimated at US $1700 per ha.

The Office du Niger, which occupies the 'dead delta' area, finds itself in the middle of much less sophisticated irrigation activity than on the 'live delta'. Here some 80 000 ha of rice are irrigated using the flood-retention techniques previously described, which cost about US $400 per ha to install. It is expected that the area under this type of irrigation can be expanded to about 110 000 ha by 1985, accompanied by a very impressive rise in cropping intensity from 38 per cent in 1965 to 100 per cent in 1985. It is estimated that the ultimate potential cultivable commanded area (CCA) is 300 000 ha.

Throughout the length of the Niger valley outside the high rainfall areas (zone D) there is some flood plain (*fadama*) irrigation activity which makes a significant contribution to local agricultural production.

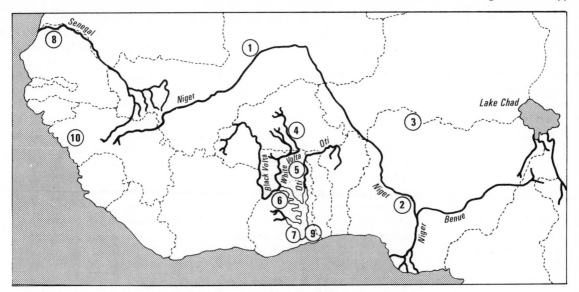

FIG. 9.8. West Africa: major river basins and irrigation development areas (see text).

Lower on the Niger, at Bacita, near Jebba in Nigeria (shown as (2) in Fig. 9.8), a commercial company has set up a modern sugar-cane irrigation project using water pumped from the river. Here about 2500 ha are irrigated, half by sprinklers and half by furrow. The overall cost was of the order of US $1500 per ha. In this area, which lies below the Kainji dam between Jebba and Lokoja, there is considerable scope for the development of pumped irrigation projects using the cheap hydroelectric power from Kainji.

At Lokoja the Niger is joined by its most important tributary, the Benue, which rises in the highlands of Cameroun and brings down with it a flow 1·4 times that of the Niger itself. A number of dam sites have been identified in the Benue basin and the land resources of some 280 000 ha of irrigable lands surveyed (by Nigerian Government staff and various consultants).

Another irrigation development in the Niger basin which has attracted considerable attention is that on the two tributaries, the Rima and the Sokoto, in north-western Nigeria (shown as (3) in Fig. 9.8). Here the ultimate potential is for about 145 000 ha of rice and sugar-cane irrigation at an over-all cropping intensity of 150 per cent. As yet no large-scale development has taken place there.

9.7.3. *The Volta basin*

The Black Volta has its source in the Banfora mountains of western Upper Volta, contributing 21 per cent of the total run-off of 38 thousand million m³ per annum, the White Volta and its tributary the Red Volta have a more direct southward course from the centre of Upper Volta, with 22 per cent of run-off, and the Oti rises in the Atakora mountains of Dahomey, with 25 per cent of run-off. All rivers join the Volta lake which stretches from the Akosombo dam over a maximum distance of 400 km. The entire basin of the Volta river system covers an area of 400 000 km², 85 per cent of which is evenly divided between Ghana and Upper Volta.

The irrigation development in the basin can be broadly classed under three categories, corresponding to the upper, the middle, and the lower basins. The upper basin, with a mean annual rainfall of between 600 mm and 1200 mm, north of Lake Volta (indicated by (4) and (5) in Fig. 9.8), is characterized by small- to medium-sized projects, often based on small dams across tributaries. A total of some 62 000 ha is now being considered for irrigation out of the estimated potential 400 000 ha. The middle basin, corresponding to Lake Volta itself, with a mean annual precipitation of between 1200 mm and 1500 mm (shown as (6) in Fig. 9.8), makes possible two distinct types of development. The first relies on the soil moisture retained when the seasonal draw-down in Lake Volta of about 3 m exposes about 120 000 ha of suitable land. At present only a very limited area of perhaps a few hundred hectares is cultivated on this basis. The second relies on modern pumping installations (costing about US $1700 per ha). At present only 6800 ha are considered to be suitable for irrigation in this manner, though the potential area is perhaps as large as 170 000 ha.

The lower basin is that commanded by the Akosombo dam (see (7) in Fig. 9.8). For the development of 4000 ha (3000 ha of sugar cane and 1000 ha of rice) in 1974 at Asutsuare it was found to be more economical to use Akosombo hydroelectric power to pump water from the river (Quartey-Papafio and Kemevor 1972). This type of water supply is likely to be favoured when the irrigation of the Accra plains is extended by a further 20 000 ha, although plans have also been prepared for a diversion barrage in the rapids at Kpong and gravity irrigation of about a third of the area.

9.7.4. *The Senegal basin* (Chaumeny 1972)

The Senegal River has its source 750 m up in the Fouta Djalon foothills in Guinea. The upper basin of the river is considered as that part upstream of the town of Bakel (about 50 km into Senegal from the Mali frontier) and is characterized by narrow valleys and rapids. The adjoining plains could be reached by pumped water but, on the whole, the conditions there do not encourage irrigation development. A small rice project is being considered at Kolimbine in Mali. After Bakel the river enters its lower basin, a sedimentary plain, where, in 800 km of its path to the sea, the river falls only 17 m. In this sector the tributaries, the most important of which is the Gorgol, contribute only intermittent storm run-off water. It enters the sea in a delta covering an area of some 400 000 ha.

In the lower basin the alluvial plain is 10–20 km wide. It is about 800 000 ha in extent, to which one should add the 100 000 ha of the Gorgol plain in Mauritania.

The entire area of the lower Senegal basin which has an irrigation potential is located within the agro-climatic zone B in Fig. 9.2.

The main problem in the development of the Senegal valley is the irregularity of river flow. Although the average flow is 765 m³/s this has little practical significance since whereas the minimum values can reach 1 m³/s in a wet year the flood peak

can be as high as 10 000 m³/s. This, and the low gradient, mean that, when the flow falls below 500 m³/s, saline water from the sea can be drawn as far as 200–250 km upstream. Thus it is easily seen that, though the annual discharge is between 8 and 40 thousand million m³ the river is of little value for irrigation development without the provision of storage, of flood protection, and of measures to prevent saline water intrusion.

As in the Niger basin, there are four types of irrigation practice in the lower basin of the Senegal. In the first, some 130 000 ha of rice are cultivated by a system which takes advantage of the annual flood but which involves no attempt at controlling it. The second, termed 'controlled submersion', covers about 12 000 ha. It involves the use of bunds to regulate the flooding of the fields. But since the vegetative period of rice is longer than the flood, the rice is planted with the first rains before the flood. This results in yield levels not much above those obtained in the uncontrolled state. The third method involves supplementing the controlled flood water with pumping. This immediately raises the costs considerably and is not a success without a full irrigation and drainage network. The fourth method comprises irrigation practice in the modern sense, implying ability to alter water level at will by either irrigation or drainage (see Chapter 2). Costs of development in the order of US $2000 per ha point to the need for double cropping to achieve economic viability. At present about 21 000 ha have been equipped for modern irrigation. Of these the 6000 ha of the Richard Toll Project in Senegal (indicated by (4) in Fig. 9.8) is the best known, especially for its experiments in mechanized rice production. These were not altogether successful and the project is now being converted to the growing of sugar cane.

9.7.5. *The Chad basin*

Climatic fluctuations and the consequent changes in the physical configuration of Lake Chad and its basin are of interest to geographers (Grove 1970). Although in the past the lake drained into the Atlantic through the Benue valley, today it is completely land-locked, receiving water primarily from the Lagone and Chari river systems at an average rate of about 50 thousand million m³ per annum. The area of the lake varies between 12 500 km² and 22 500 km² and the average depth between only 2 m and 6 m.

Four countries—namely, Nigeria, Niger, Chad, and Cameroun—have frontiers on the lake which, in the climatic divisions of Fig. 9.2, is horizontally cut by the line separating zone A and zone B.

Irrigation projects in the Chad basin can be broadly divided between those along the rivers draining into the lake and those which rely on the waters of the lake itself. Of the river-valley projects the most important have been on the Lagone. On the Cameroun side rice is successfully grown on some 6000 ha of the Yagona plains, while in Chad rice, sorghum, and cotton are grown experimentally on about 2000 ha in the Bongor Project. Both these project areas are capable of considerable expansion at a cost which should not exceed US $1000 per ha.

Of considerable interest are the traditional lake-side polders in Chad, where about 5000 ha of wheat are grown with water derived from shallow wells. There are plans to modernize this activity by investing about US $400 per ha on improved water control for cotton and wheat growing over an area which may ultimately reach 25 000 ha.

On the Nigerian side of the Chad basin small irrigation projects on the Yobe River in the north and on the Ebeji River in the south account for about 2500 ha under rice and wheat. But the most important event is that a modern irrigation development, involving the use of water pumped from the lake, is being actively studied for the growing of rice and cotton in the wet season and of wheat in the dry season at an over-all cropping intensity of 130 per cent. The area is shown as (9) in Fig. 9.8. Ultimately nearly 100 000 ha could be irrigated in this way, but designs are being prepared in 1974 for 24 000 ha at an estimated cost of US $2700 per ha, of which 70 per cent represents the cost of intakes, pumping stations, and conveyance to the project area.

9.7.6. *Small inland valleys*

The previous section has shown the wide range of possibilities for irrigation development in the large river valleys of West Africa, extending from most modern diversion and pumping installations down to traditional practices of imposing some degree of control on the natural river regime during the flood season. Low-cost partial-control projects can, in the short run, be economically attractive, so it is not surprising that many separate communities, away from the large river valleys, have also attempted to modify and use the farmable soil moisture conditions which exist where the valley bottoms widen sufficiently for agriculture to be feasible.

This localized widening of a small valley usually results in the formation of a seasonal swamp. A typical one would be 100 m wide and, say, 300 m long (i.e. 3 ha in area), with a longitudinal fall of 0.5 per cent. Such swamps, by definition, retain excessive moisture during the wet season, but they do tend to dry out completely soon after the rains have left. Hence water management here consists of removing the surplus water by open drains cut through any constriction at the downstream end of such a swamp and, if possible, of creating small reservoirs (or conservancies) upstream of the swamp for supplementary irrigation in the dry season.

Possibilities for this type of hydro-agricultural improvement exist in many areas where the annual rainfall exceeds about 1500 mm. In the section on climate in relation to irrigation above, the field water balance for rice cultivation on such inland valley swamps in Sierra Leone was presented in Fig. 9.4. It showed that, with minimal, but carefully timed, application of irrigation water and with adequate drainage, two crops of rice could be raised by a smallholder and his family. In Sierra Leone there is scope for developing some 8000 ha in this manner in the Northern Region alone, and possibly up to 20 000 ha in the country as a whole. Liberia offers a potential of nearly 250 000 ha for such development, while similar opportunities also exist throughout zone C (Fig. 9.2).

The cost of reclaiming inland swamps in the manner described can vary from as low as US $100 per ha to about US $400 per ha.

9.7.7. *The coastal zone*

In few areas of the region have local differences in hydrologic, pedologic, and topographic conditions created a greater variety of the natural environment than in the lowlands of the coastal zone. Here we can observe the interaction of the run-off from the coastal plateau, of the sea-water intrusion through tidal channels, of the interface between fresh and saline ground-water tables, and of the sharply varying precipitation patterns. A succession of coastal flats, lagoons, plateaux, and riverain lands are annually subjected to variable degrees of inundation by sea- or fresh water. As a result of all these factors the coastal-zone lands call for a much wider range of water-control measures than anywhere else in the region. In their natural state, such lands are usually considered intrinsically difficult to reclaim. Yet it is surprising how frequently the local farmers,

I

using traditional methods, have managed to bring them under successful cultivation. By the use of bunds they control the incoming and outgoing flood waters sufficiently to grow a good rice crop (e.g. yields of 1·5–2·0 t/ha) over the years. Such bunds are usually on the higher lands less exposed to regular and prolonged inundation. Nearer the coast successful cultivation of maize and vegetables has been centred around fresh-water pockets in the coastal dune lands, as, for instance, near Keta in Ghana (shown as (9) in Fig. 9.8).

To be successful such traditional practices have to take advantage of the most favourable conditions offered by the environment. Where such conditions do not prevail, modern reclamation measures have to be adopted to minimize the risk of crop failure. These measures must aim at maintaining an equilibrium between soil salinity, soil acidity, and soil fertility. This is not always easy to achieve since, for instance, the total elimination of saline conditions appears to accelerate the build-up of soil acidity (thioxic) to values as low as pH 1·9. But indications are (Food and Agriculture Organization 1970) that the development of thioxicy acidity can be kept within bounds by short-duration water fallowing with saline water in the dry season and freshwater inundation in the wet season and by keeping the water table close to the surface.

It is obvious that such a complex interaction between the soil, the water, and the crop requires a very high level of expertise, not only in reclamation engineering but also in soil chemistry and rice agronomy. It is therefore not surprising that, so far, not many areas have been developed in this manner, though an interesting case study has been made in the Kobak and Kawasa pilot projects in Guinea (Food and Agriculture Organization 1970), where about 3000 ha have been so reclaimed at a cost of nearly $2000 per ha. The rice yields obtained have been as high as 7 t/ha with an average of 5 t/ha. The ultimate potential in Guinea alone is estimated at 300 000 ha, and similar areas can be found in the delta of the Casamance in Senegal and Gambia as well as in the, as yet, essentially unexplored Niger delta.

As has been shown, the management of water is primarily concerned with flood protection and the removal of excess soil moisture. This means that attention has to be paid to various developments in modern drainage technology (FAO/UNESCO 1973) backed up by *in situ* trials.

Ground-water. While it is true that West Africa is not very richly endowed with ground-water resources, it is also true that a largely unrealized potential for increasing water supplies from this source has remained untapped so far. The reasons why are many, and vary from zone to zone, but the most common is the large expenditure involved in ground-water development, most notably the high pumping costs which are beyond the means of local users.

Yet ground-water has, in recent decades, gained recognition as an essential water resource and as an economically interesting development proposition, particularly in those areas away from large rivers and perennial streams where seasonally severe shortages of rainfall and surface water occur. This is equally true in those areas along perennial streams without proper surface-water storage sites and where ground-water development costs are easily competitive with those for surface water.

Certain other factors seem to favour sub-surface as against surface-water development. First, health hazards are considerably reduced. Secondly, development can be undertaken much more gradually as the need arises and, what is more, on the site of existing farmsteads, where it can be put to use with relatively little effort, an element which is of overriding importance if irrigation at even the smallest scale is to reach West African farmers. The need to start irrigation on a small scale, by which operation smallholders can become acquainted with the concept of growing a crop in the dry season, has been manifest (Rydzewski 1968; Food and Agriculture Organization 1971; des Bouvrie 1972). Thirdly, there is scope for providing training facilities in irrigation techniques and methods under self-help schemes, the benefits from which should eventually also assist in future large-scale surface-water irrigation development. Finally, projects consisting of such small elements can usually be constructed with local materials and within the capacity of locally available technology.

The hand-dug shallow open well (less than 50 m deep) is the type most widely used throughout West Africa (des Bouvrie 1972); it is also one of the cheapest methods of exploitation where a supply of a very few litres per second is required. It allows for water to be drawn by hand or animal-traction, or by wind-powered and engine-driven pumps. Its many advantages make it an almost invariable choice in the hard rock formations of zones B and C.

The second most common but more recently

introduced well is the borehole, drilled in coarse-grained sandstones (e.g. Maestrichtien), and selected to extract greater quantities of water (up to 10 l/s) at depths greater than 50 m. Little economic justification for irrigation from such sources can be found, however, in view of the large expenditures involved.

The third type, the large-diameter (up to 200 mm) tubewell is usually sunk in alluvium along, for instance, the Senegal River and around Lake Chad, but it is much less common.

A further source of water worth mentioning constitutes the largely unrealized potential of sub-surface flow in the Oueds and Koris of the Sahelian zone.

Although the overall impact of ground-water development for irrigation is not very impressive, it can be a source of considerable benefit at village level.

9.8. Irrigation development—general principles and practical problems

9.8.1. Introduction

Broad distinctions have been made between 'modern' and 'village-level' irrigation projects, and it has been pointed out that, in the short term, the latter can show a more immediate return on investment as well as serving as an introduction to the disciplines which irrigation agriculture impose on the farming community. But the demand for food products in West Africa and in the world as a whole will make it now necessary to plan and implement irrigation projects where both the water and the land resources are used more intensively and more efficiently.

This section is therefore devoted to the presentation, in outline, of the principles on which the successful planning of irrigation development depends. It is an interdisciplinary process which has to be approached simultaneously from many angles with considerable flexibility of mind but without losing sight of the declared objectives.

A good analysis of the sequence of events leading from the identification of a project to its full operation is given by Gosschalk (1972), who summarizes the problem by means of a diagram, here presented as Fig. 9.9. It would obviously be impossible, within the confines of this chapter, to consider separately each of the items listed thereon. The plan adopted here is to recognize irrigation development as an interaction of the water, the land, and the human resources and to examine their combined effect on intensifying agricultural production.

9.8.2. Estimation of water resources

The estimation of water available for a project is divided into two parts. The first consists of the task of observing the resource in its natural state and of predicting its variability with time. The second is concerned with the engineering manipulation of this water resource so that it may be apportioned to the project area in relation to the crop water requirements.

The study of the basic water resources lies in the province of the science of hydrology, usually separated into its surface-water and ground-water aspects. Both concern themselves in the first instance with observation of the physical phenomena with time and then, if, as is the case in West Africa, the time-base is rather short, turn to statistical predictions of likely patterns of future behaviour. It is obvious that the study of ground-water is more complex and costly but, since, in the region its role is likely to be confined to small projects, its precision need not be as refined as for major developments.

It is not appropriate in this book to summarize the principles on which modern hydrology is based. But interested readers will find relevant information in standard works on the subject (Ven Te Chow 1964; Nemec 1972; Wilson 1969; Todd 1958; Walton 1970; UNESCO 1972).

Since the accuracy of predictions is necessarily related to the quality of the data on which they are based, the provision of such data is of great importance in the region. Under the auspices of ORSTOM, yearly hydrological records have been collected since 1949 at 70 stations in the sub-Saharan francophone countries of the region, with particular reference to areas where rainfall is less than 1100 mm per annum (Secrétariat d'État 1973). Similar records have been obtained for the Volta basin in Ghana and for the Niger and Benue basins in Nigeria.

For such records to be of practical value one must ensure their completeness and continuity. The establishment and maintenance of hydrological networks over such a vast area is costly and calls both for the closest collaboration between riparian states and also for survey objectives to be given a problem-solving orientation. Since such networks would cross national boundaries it is essential to ensure that information from individual stations can easily be integrated so as to provide more useful figures for analysis and display on a regional scale. Furthermore, modern computer processing demands standardization in data-collection methodology and terminology. A step in the right direction has been taken by the

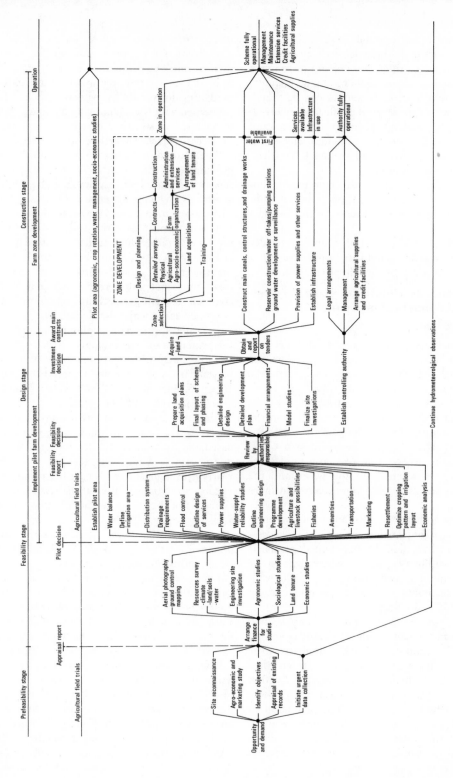

FIG. 9.9. Conceptual analysis for development of irrigation and drainage projects.

establishment of the three river basin commissions, namely for the Niger, the Senegal, and Lake Chad.

For the numerous small river valleys in the region it is obviously impossible to obtain other than sample information from carefully selected and instrumented typical catchments. In this respect reference should be made to the work of ORSTOM on 110 such basins, varying in area from 10 km² to 50 km² (Secrétariat d'État 1973).

As regards the inventories of ground-water resources the location and extent of the region's major aquifers have been identified on a small scale (Food and Agriculture Organization 1971; Greigert 1968; M.F.E.P. 1970; Archambault 1970; Gill 1969; Dijon 1971). Much remains to be done, however, with respect to assessing their safe yield and permissible rate of exploitation. This is achieved by the study of the over-all hydrological balance, together with field tests on instrumented wells. Once the basic parameters are known computer and analogue modelling techniques can be used to predict future behaviour (Thomas 1973).

We should not, however, neglect the considerable information that can be derived from the analysis of records (profiles, production, water quality) of the many thousands of wells already in use. In this connection the work of CIEH (1970) is noteworthy.

The second problem, that of managing the water resource for agricultural use, involves engineering works. In the case of surface water this means the provision of storage, the amount of which is dictated by the statistical security the designer wishes to impose on the water deliveries. Obviously the risk of reduced crop yields caused by deficits in water available to the crops at different growth periods must be weighed against the increase in costs needed to reduce such risks (Darley, Rydzewski, and Clark 1972). Such costs are considerably affected by the unit cost of storage (itself a function of the river flow regime and the valley topography) which in Africa has varied between such low figures as US $0·014 per m³ for Kariba and Kainji to US $0·024 per m³ for Aswan and US $0·250 per m³ for small dams (capacity under 1 million m³) in Upper Volta.

For ground-water, resource management should involve over-exploitation during dry months or even years when it has been statistically established that, during an unusually wet year, the aquifer can be replenished to its original level (e.g. experience in the Sous Valley in Morocco) (Armgroggi 1969).

As with surface-water storage the unit cost of

ground-water in the region can vary between wide limits. A typical shallow well in Niger could provide water at about US $0·063 per m³, while for a borehole in Dahomey it could be up to US $0·080 per m³. As is mentioned below great advantage can ensue from the conjunctive use of surface- and ground-water resources.

9.8.3. *Soil–water–plant relationships*

One of the greatest challenges yet to be fully met in the field of tropical irrigation science and practice remains the thorough understanding of the several parameters which in optimal combination constitute the formula for successful crop production. Among these, so-called soil–water–plant relationships determining what levels of soil moisture constitute an adequate supply to the plant at different periods during crop growth are of paramount significance.

Around the world, therefore, considerable attention is currently accorded to crop water-requirement studies. They have proved to be of interest to the planner to determine in theory, when to irrigate, how much at each stage of growth, at what rate or, taking into account prevailing rainfall patterns, whether to provide supplemental irrigation or no irrigation at all. This information has proved to be essential especially when the new high-yielding cultivars of rice and other crops which are known to be highly responsive to proper water management, are being introduced. Likewise, the aim is to gain insight into the water requirements and lengths of vegetative cycles of specific traditional crops to arrive at their optimal geographical distribution through the region (Dancette 1972).

Agroclimatologists are generally agreed (Franquin 1972) that an abundance of solar energy is available for photosynthesis, allowing for the production of two and even three crops per annum on condition that water and nutrient requirements are satisfied (and that socio-economic constraints do not prevent this). Clearly, conditions for irrigation are more favourable at higher latitudes, although during the dry season the benefits at times are to some degree offset by the effects of hot and strong midday winds and by low temperatures at night. Yet the taking of decisions with respect to irrigation dosage and schedule is relatively straightforward.

It is in the more humid south that the range of possible alternative solutions is increased by the occurrence of an occasional shower before or after the wet season, or by the irregularity of prevailing rains. Ideally it must be decided almost daily whether to irrigate or not, or even whether to drain the land. Here the need for knowledge of climate (in particular, rainfall probability, intensity, and duration), of soil depths, infiltration and water-retaining capacities, of the length and critical periods in the growth cycle of the region's crop species, and of other similar soil–water–plant parameters assumes particular relevance. As has been indicated above, water-resources development in the region is frequently found to be a costly proposition. It really pays, therefore, to ensure that the irrigation planner, who must in effect be a farmer, is provided with properly considered alternative solutions and has available appropriate basic data on which to base his decisions. He may, for instance, have to decide to change planting dates, or cultivars which in turn may have other consequences to the farming system, rather than supply supplemental irrigation at certain periods.

Numerous studies and experiments have been undertaken in this field in recent years. Not always, however, has due attention been accorded to the quality of the layout of research plots required to obtain meaningful data, or to the standardization of equipment and observation techniques. The results so far obtained do not, in consequence, allow for much correlation or extrapolation beyond the one or few seasons during which the observations were made.

Scope for improvement and further strengthening of such studies in the years ahead clearly exists. Gillet (1972), furthermore, emphasizes the importance of distinguishing between fundamental regional research, for which only the best expertise and equipment will do, applied research to establish the requirements of a particular area, and practical pilot experiments to observe the limitations of practical use of research information under actual farming conditions. The following paragraphs present a brief review of current research activity in this sphere.

In macroclimatology, the longest recordings have traditionally been taken at airfields throughout the region by aviation agencies such as ASECNA (Agence pour la Sécurité de la Navigation Aérienne en Afrique et à Madagascar). It is, however, now accepted as a general rule that data thus collected at widely spaced stations do not adequately represent those of the near-by agricultural environments. Even allowing for the necessary corrections, they do not allow for more than a broad identification of crop-production zones. For a particularly thorough analysis

of such climatological data, jointly with relevant data on soils and agronomy, reference should be made to the agroclimatological studies of the region's semi-arid area by Cochemé and Franquin (1967).

Research of a more detailed nature, however, is required for such studies to become meaningful at local levels. First in the order of the irrigator's needs ranks the long-term and specific knowledge of the components of the local water balance. In its simplest form $P = (ET) + S + R + I$, where P is precipitation, (ET) evapotranspiration, S soil moisture, R run-off, and I interflow (or acceptance).

Concerning precipitation P, summaries of rainfall data in the form of monthly means, including their standard deviation, can now be obtained from almost all national meteorological services. Yet this information is insufficient for irrigation purposes in view of the significant irregularities which frequently occur in rainfall patterns within periods of less than 1 month. An interesting new approach is now therefore being followed (Gigou 1972), following Manning (1956), in which daily rainfall figures are considered over shorter periods of, for example, $n = 12$ days' duration with $\frac{1}{2}n$ (i.e. 6) days overlapping. The averages, the mean (50 per cent probability), and the lower 10 per cent probability are subsequently presented in graphic form showing confidence limits simultaneously with monthly evapotranspiration figures. This method has the advantage of permitting the pin-pointing with precision of the most suitable cropping season as well as the periods in which supplemental irrigation is most likely to be beneficial.

Diverse ways and means have been developed to determine evapotranspiration (ET) ranging from the use of a formula based on parameters of climate, through recording the actual evaporation from the water surface of an open pan, to actual measurements of evapotranspiration from crops in the field.

Such methods are subjected to appraisal by an FAO *ad hoc* committee which has reviewed over 30 evapotranspiration formulae in use throughout the world, some based on meteorological parameters, others more of an empirical nature, and each having its own advantage or disadvantage (FAO, in preparation). In West Africa the Penman formula is often preferred (Rijks 1972) but the Turc equation has also been widely used (Lemoine and Prat 1972).

The same committee calls particular attention to the fact that these formulae are often used under conditions of climate and environment which differ widely from those for which they were developed originally. In particular the proportionality factor K, relating the potential evapotranspiration figures derived from climatic data to the actual measured rate of evapotranspiration from a short green crop, e.g. (ET) (grass), may vary considerably from station to station dependent on, for instance, the magnitude of deviation of maxima and minima from average daily temperatures, humidity, and wind velocity.

In spite of their limitations, the use of lysimeters to measure (ET) (grass) accurately *in situ* is therefore advocated by some researchers. These limitations are particularly pronounced where inadequate provisions have been made to surround the test sites with a 'buffer zone' of similar natural conditions.

Others have recommended the simpler and less expensive use of evaporation pans such as the Colorado or class 'A' open pan which, however, can be subject to the effects of advection. A 30 per cent increase in evaporation was, for instance, found in the desert area around N'jamena from a sunken tank in the bare soil as against one in a small irrigated plot of 400 m² (Riou 1972).

But in whichever manner maximum potential evapotranspiration has been determined as a standard reference (by formula, by lysimeter evapotranspiration pan, or field measurement) the need remains in all instances to relate these figures to the actual water requirements of a given crop by applying a crop coefficient k, where (ET) (crop) $= k$ $((ET)$ (grass)). This factor k reportedly can be as low as 0·4 in the early stages or 1·3 in the full stages of growth, when maximum crop foliage exposure occurs (Blaney and Criddle 1966).

In respect of the factors which influence moisture conditions in the soil, it may be recalled that West African soils, in particular those of the Sudano-Sahelian zone, are frequently found to be shallow and of light texture, to have a high water acceptance but a low available water capacity. These conditions, which are not normally favourable to irrigation may be aggravated by a permanent breakdown of soil structure and subsequent reduction in infiltration capacity, as well as loss of fertility through leaching once irrigation has been introduced. The importance of appropriate cultural practices, most notably stubble mulching to improve hydrophysical properties of such tropical soils and thus to increase water-use efficiency, is increasingly recognized as a research priority (Lal 1972).

As regards the run-off and interflow components of the water balance, finally, much useful information can be derived from the typical catchment studies referred to above.

The reader will readily appreciate that the overall accuracy of the water balance depends on the precision with which its individual components have been determined, resulting in either too optimistic or pessimistic water requirement predictions. The primordial need for reliable long-term records of comparable accuracy on the above-mentioned crop-production factors, therefore, can hardly be over-emphasized. These studies are by their very nature time-consuming and, indeed, quite expensive, not in the least due to the sophisticated equipment often needed (FAO/II/IAEA 1973). Their research objectives must be geared to meet the practical needs in the field and periodically be subjected to sensitivity tests to determine the most critical knowledge gaps. Regular consultations between specialists in the individual disciplines concerned are desirable to avoid costly duplication and to keep abreast of progress as well as of the limitations of current research.

The answer clearly must be sought in the setting up of regional programmes at well-equipped multi-disciplinary stations, where continuity can be ensured and which are organized either to deal with a certain crop, as is the case with the West African Rice Development Association (WARDA), or to cover a certain ecological zone, as applies to the International Institute for Tropical Agriculture (IITA). Another particularly imaginative new approach has been recently suggested by the FAO/UNESCO/WMO Interagency Group on Agricultural Biometeorology (1972). This proposal aims at establishing a dynamic production model permitting the forecasting of yield for selected cultivars of wheat in response to meteorological and management factors. Field experiments would be set up under rain-fed or irrigated conditions (or both) with an optimum supply of fertilizer at five strategic locations spread across the African continent. In the medium term the programme would aim at providing a running estimate of the chances that certain yields can be reached, even for environments very different from those of the original experimental conditions. Similar programmes for some of the region's other staple crops could subsequently be set up.

9.8.4. *Suitability of land*

Few elements are more important in the complex decision-making process inherent to an irrigation feasibility study than those involving land classification. The location, extent, and quality of the land, as a glance at Fig. 9.9 will confirm, have more than any other factor a direct bearing on the outline of the engineering design, on potential crop and livestock production, on flood protection and drainage requirements, and on the very choice of the irrigated area itself.

There is now little question (Storie 1964; Riquier, Bramão, and Cornet 1970) that the validity of the modern concept of land classification is generally accepted. This involves an integrated approach towards a final economic analysis and defines land classes in physical as well as in economic terms (down to the net potential income per unit area of the future farmer). The times are now, mercifully, gone when the results of soil surveys were evaluated not only in isolation from information on topography and availability of water but also outside the context of their economic significance to a particular development project.

It is useful to highlight the particular problems which arise when land classification is to be undertaken in terms of suitability for sustained irrigated agriculture. In particular, whenever water is more or less permanently added to or removed from a soil, the effect may be a significant modification of its physical, chemical and biological characteristics. A build-up of salinity may take place, changing the composition of the absorption complex, swamp lands may dry out and subside irreversibly or turn acid, and soil structure may change in such a way as to lower infiltration rates, rendering subsoil strata less permeable.

As has been stressed by several authors (Maletic 1966, 1967; Nielsen 1963), the responsibility therefore rests with the soil surveyor not only to collect information on the geographical distribution and intrinsic value of land resources but also to predict with reasonable exactitude the changes which may occur following irrigation development. He should keep in mind that in a feasibility study it is usual to expect that cost and return estimates should lie within 10-15 per cent of the final figures.

Instances of costly failures can be quoted which have resulted when land development costs at the project implementation stage, due, for instance, to unforeseen land-reclamation measures and deep drainage to control salinity, reached far beyond what had originally been found to be economically justifiable. The most common causes of such failures,

according to FAO experience, have been that the soil survey had not been properly planned as regards sampling density, depth, and soil mapping criteria and that inadequate attention had been accorded to investigations into specific soil characteristics, most notably soil-water relationships and other related factors.

In practice this means that a team approach should be adopted under which collaboration is ensured by soil scientists, agronomists, hydrologists, and engineers alike, to provide the over-all expertise beyond that which any individual specialist can offer. In this connection two schools of thought are brought to bear in land classification for feasibility studies.

1. The approach adopted by FAO (1969) which recommends 'a systematic soil survey, the findings of which are interpreted in the light of other physical, social and economic factors of the environment'.
2. The procedure followed by the US Bureau of Reclamation (1953, 1967) involves 'direct mapping of land units without a preliminary stage of soil survey, all pertinent factors of the environment being considered and mapped simultaneously'.

No general agreement, in spite of efforts in this direction (Food and Agriculture Organization 1969), has as yet been reached internationally as to the most suitable solution. The pragmatic solution for the irrigation planner in charge of assessing economic feasibility is therefore to familiarize himself with the standards considered acceptable by the international financing agency from whom development investment is to be solicited.

9.8.5. *The irrigation farm unit*

Within the region usually one can safely say that any new irrigation project would have a population density much higher than that which existed before the project was established. As a result people have to be attracted to the project from a wide ethnographic area. An extreme case arises in which resettlement of entire communities from distant areas is to be undertaken (Chambers 1969).

The point has already been made that irrigation should be considered as part of an overall heightening of agricultural technology. This means that the farmer and his family have to operate on a farm unit which, with the other inputs demanded by such activity, will give them an adequate return for all the additional effort and initiative that intensive agriculture entails. In designing any new irrigation project one is faced with the problem of achieving enough flexibility so that the project may function efficiently even if subsequent generations of farmers demand a higher income level than that which it is hoped to achieve initially. Also, as settlers may wish to farm on larger units, there is considerable urgency to ensure, in the first instance, that inheritance laws do not result in the fragmentation of holdings and if possible to allow for future consolidation. On settler projects one inheritor could, perhaps, be named as the recipient of cheap credit facilities, which would enable him to pay off the others. The possibility for enlarging of units in the future should not present many technical difficulties.

Planning the size of the optimum farm unit under present social and cropping conditions involves taking careful note of the farming calendar and its labour requirements. These show peaks at certain times (notably at harvest), so the planner must feel sure that, if additional labour is needed at such times, it will be available from settlements in the neighbourhood. With the raising of educational standards in rural areas one can no longer count on as many family helpers as one did before.

Physically the easiest unit to irrigate is one which is governmentally or privately managed, in which the staff are on a salary or wage. This is not typical of the region, so most planned systems have to cater for a number of individual units and to take into account the conflict between cost and operational convenience. This subject is discussed further below.

Due attention must also be paid to the laws governing the use of irrigation water. Water legislation, be it based on African customary law or on laws of Moslem, French, or British origin, is gradually being consolidated and rationalized on the basis of efficient resource utilization rather than on tradition, but much still remains to be done in this sphere (Food and Agriculture Organization 1973). Additionally, irrigation farmers should have easy access to credit at unexaggerated terms (a point which applies to all forms of intensive agriculture). It is axiomatic that the richer you are the easier and the cheaper are your credit terms, so part of the development package must consist of an equitable credit organization.

9.8.6. *Type of irrigation system and method of operation*

Since 'traditional' methods of utilizing surface-water and ground-water for irrigation have already been dealt with, this section is confined to a con-

sideration of modern, but not necessarily expensive, projects.

From the engineering point of view an irrigation project can be said to consist of (1) the regulation of a water resource to satisfy the project requirements, (2) the conveyance of the water to the project area, (3) the control of the soil water within the project area, and (4) the conveyance of excess water from the project area.

The supply of irrigation water to the project area presents the engineer with a great diversity of problems and with many possible solutions. In the exploitation of surface water it is very important to know at the outset whether the water is to be used solely for irrigation or whether its multi-purpose use is being considered. This will influence decisions on the need for and on the capacity of valley storage. Once the required amount of water has been made available in the river valley, various alternative means of delivering it to the project area have to be examined. Here a great deal depends on the position of the irrigable lands in relation to the source of water. Typical solutions to this problem are: (1) a gravity canal leading from a dam, (2) a gravity canal taking off from a barrage or weir with or without an upstream storage reservoir, and (3) pumping water from the river, lake, or reservoir. The point to note is that, with a gravity supply, the costs of providing the water are dominated by the initial capital cost of the engineering works while, with a pumped water supply, recurrent costs for power, maintenance, and replacement become relatively more important (Rydzewski 1974).

Ground-water development for irrigation presents its own problems. The water has obviously to be pumped to the surface but the range of possible pumping unit sizes is much narrower than is the case with surface water. The unit cost of the water is here closely related to well depth (Middleton, Hoffman, and Stoner 1966). Care must be taken to ensure that a reasonably accurate annual water balance for the exploited aquifer can be predicted, so as to avoid equipping for irrigation an area which could subsequently find itself with an inadequate water supply.

Useful development strategies have resulted from a combination of surface- and ground-water resources, allowing excessive (i.e. 'cheap') water during the wet season to provide a gravity-feed supply and to recharge an aquifer which is then over-pumped during the dry season (Johnson and Stoner 1974).

Concerning the basic capacity of the units which constitute the irrigation water supply, ground-water development (with its intrinsically small-capacity units) and small surface-water pumping stations give the planner a good opportunity to develop an area in stages. In regions where irrigated agriculture is new to the farming community this can be of great advantage, since 'pilot' projects can be established without excessive capital cost. This is patently not the case once a dam, barrage, or large pumping station has already been constructed.

Within the project area, i.e. the cultivable commanded area, the water has to be distributed by methods which vary according to the use of mechanization, to satisfy the crop-water requirements. In this context, crops may be divided into two groups: rice and other crops.

The cultivation of rice in paddy fields requires the maintenance of a level of water between set limits during specified periods of the crop cycle. Traditionally, the water level is controlled by manual labour, but scope exists for some degree of automation on larger projects.

With other crops there is greater variety in the methods of water application. Generalizing, we can say that these methods can be classed by the degree to which they substitute capital for labour. The same can be said of the manner in which a project is operated.

At field level, water is applied either by a variety of surface techniques (basin, border-check, furrow, corrugation, or trickle) or by overhead (sprinkler) systems. Removal of excess water is achieved by controlled drainage. Trickle and sprinkler irrigation systems can be very economical in the use of water. They can be fully mechanized, and their unit cost depends on the degree of mechanization specified. Progress in irrigation technology means that such systems need not be dismissed out of hand when different design alternatives for modern projects are being assessed. Even if it is considered socially undesirable to use capital-intensive irrigation methods, the cheapest form of mechanized systems (e.g. hand-moved sprinklers) could serve as a costing 'yardstick' for more conventional designs (Hay and Rydzewski, 1975). We would, for instance, be uneasy if a furrow irrigation system, with a lower field-application efficiency, began to be less economical than a sprinkler or trickle installation.

Decisions concerning what irrigation system to adopt depend not only on the economic cost of the irrigation water for each system but also on the value of the crop irrigated. It is not surprising to find that, when crops such as coffee, sugar cane, or tobacco are grown under irrigation, project managers are inclined to prefer a system which is less dependent on an unskilled labour force.

Given the climatic and soil conditions and the cropping pattern, the capacity of the irrigation system is related to the manner of its operation. One of the first questions posed is whether field irrigation can be carried out round the clock. With mechanized systems this presents no difficulties, but when manual labour is involved a great deal depends on the value the local community places on irrigation water. In southern Asia, where irrigation is first an insurance against hunger and only secondly an instrument of a cash economy, the farmer takes it as a matter of course that he will have to take his turn to receive water in the middle of the night. In West Africa this would normally not be a satisfactory arrangement, so a somewhat higher project cost would have to be accepted as an operational necessity. For instance, with a pumping scheme, delivery to the project area could be continuous but provision might be made for night storage so that irrigation takes place only during the hours of daylight. If the farms making up the project were of reasonable size, delivery to the farm boundary could be on a continuous basis, and each farmer could make his own decision on how to schedule his watering. But this would assume a community of experienced farmers and could be wasteful in the use of water if this were otherwise.

The operation of an irrigation system becomes more complex the more diverse is the cropping pattern. Obviously, with monocultures (e.g. of rice or sugar cane) considerable savings in project costs can be achieved by carefully drawing up schedules of water deliveries on a rota basis. But if farmers are given relative freedom on what to cultivate, they will at the same time show a preference for, and indeed may require, 'on-demand' water deliveries. This method of operation increases project costs but can be efficient in the hands of responsible irrigators.

Where there is little past experience of the response of farmers to irrigation, it is prudent to keep the design of a project as flexible as possible, not only because changes in cropping patterns may occur more frequently than expected, but also because a theoretically preferred method of operation may prove to be disappointing when tried out in practice. Whenever possible, therefore, it is recommended that a pilot project (not to be confused with an experimental station which could, though, be a part of it) should be established for testing agricultural and engineering design assumptions before heavy investment is incurred for full-scale development. Agricultural development planners are, after all, not so much interested in when a project is completed but in when it attains the designed full production, and that it shall not be a costly failure.

9.9. Development objectives and project appraisal

The need for irrigation development was discussed earlier. But, once such a need has been recognized in general terms, the project planners and designers need more specific objectives to which they are expected to direct their professional skills. The situation has often arisen in the past when the irrigation engineer, working to a 'traditional' brief, later finds his work evaluated by a set of rules or considerations which were not stated at the outset.

Irrigation development is, or should be, part of overall regional development achieved through a number of projects which should, at the same time, be closely coordinated and clearly defined. This is important since, when a project is evaluated, some form of comparison is made between the costs involved and the resulting benefits. The experienced irrigation engineer will be reasonably accurate in his estimates of the costs of delivering a dependable flow of water and of applying it to the land. The direct benefit of this input is an increased agricultural product, the quantity and value of which can be estimated, though not with the same degree of confidence. But it is not unusual to find that, to the construction of the works essential for providing the irrigation water, others are added, such as main access roads, housing, schools, hospitals, community leisure, and sports facilities, all of which, though highly desirable, are fundamentally part of a 'social advancement' project, the benefits of which are less easily quantifiable. The point being made is that such social programmes should be carried out but that their costs and their benefits should be presented as separate items.

The planner or politician at national or regional level might express the objectives of irrigation development as: maximizing the return per unit of capital invested, per unit of project area or per unit of water; maximizing the value of the agricultural output; maximizing the output of food crops (expressed in terms of bulk or of nutritional value) or reaching a target level of production; maximizing the output of crops for export or for import substitution so as to improve the country's balance of payments; minimizing government expenditure (i.e. by encouraging private sector participation); maximizing the farming family incomes; maximizing the number of families settled on the project; creating the maximum number of jobs, of a specified level; achieving a more equitable distribution of wealth within the country; etc.

It is obvious that some of the above aims are contradictory or mutually exclusive. So it is important to establish an order of priorities before planning and design begins. In practice, an important decision is that which specifies the intended cropping pattern. With this as a starting point one can begin to derive the optimum size of a farm so that a typical family can earn enough not to leave the land for superficially more attractive prospects in the cities. In this context it should not be forgotten that what may satisfy one generation may not necessarily satisfy the next.

Once the planner has ensured, albeit theoretically, that he can command all the required resources (including the human ones), he can embark on the economic appraisal of a number of technically alternative possibilities. At this stage it will usually have to be assumed that all the alternatives will be capable of providing the same soil moisture regime and hence the same benefits. The study can be regarded in such circumstances as being one of cost-effectiveness.

Recently there has come about a greater awareness of the influence of the factor of time in project development and appraisal (Mishan 1972; Overseas Development Administration 1972; Gittinger 1972). This is especially marked in the context of irrigation projects where there is a distinct time gap between the capital costs of the irrigation works and the benefits accruing from agricultural production (cf. a hydroelectric project where electricity can be sold very soon after construction is completed). If, therefore, comparisons of the values of costs and benefits have to be made they must be brought to a common time-base. The most convenient way of doing this is by means of discounted cash flow (DCF) techniques which recognize that in our society a commodity or service today is valued more highly than the same commodity or service tomorrow. The process of discounting so as to obtain the 'present value' of a sum realized at a future date is now well known (Wright 1967; Mishan 1972) and the simple calculation can be carried out with the help of tables (Gittinger 1973) or even with pocket electronic calculators specially designed for this purpose.

The skill of carrying out an economic appraisal of a project is largely one of making reasonable estimates of what the costs and the benefits are likely to be over the time horizon considered (usually 40 years). A convenient approach is to confine the cost stream to the capital investment and to the operation, maintenance, and replacement (OMR) costs, while including the on-farm production costs as a negative sum on the benefit stream. This makes it easier to arrive at net benefit estimates through typical farm budgets for a number of key years in the future. Unless no agriculture is possible without irrigation, it is reasonable to make forecasts for such key years only for the situation 'with project' and 'without project' and to consider the difference between the net benefits so obtained as attributable to irrigation.

Many texts have discussed in detail the differences between an economic and a financial analysis of a project (Overseas Development Administration 1972; Gittinger 1972). Put very simply, financial analysis deals in terms of market prices while in economic analysis values of the resources used or forgone are substituted for prices. There are various degrees of sophistication in arriving at these values, but meaningful results can be obtained by concentrating on three factors: (1) the realistic value of foreign currency, (2) the elimination of taxes and subsidies from the calculations, and (3) the 'shadow pricing' of unskilled labour when opportunities of alternative employment are few.

The cost and benefit streams are then discounted to the first year of construction, using a nationally agreed discount rate, and the result presented as (benefit/cost) or as (benefit–cost). The former is known as the benefit/cost ratio and should be greater than unity, while the latter is called the net present value and should be positive. Alternatively, discounting of both the cost and benefit streams is done using a range of discount rates to discover which rate makes the present value of the costs and of the benefits the same; this rate is known as the internal rate of return. Obviously, in all cases the over-all planner has to decide what criterion to apply for the acceptance or rejection of a project.

Discount rates used in 1974 are of the order of 9 per cent, so that £100 appearing in 'Year 8' of a cash stream has a value of only £50. This highlights the importance of planning an irrigation project so that benefits do not lag far behind the initial investment costs. In a region where farmers are as yet unused to the disciplines of irrigated agriculture, a project developed in stages (i.e. with costs able to be discounted as well as benefits) is likely to prove economically more attractive than one in which all the investment is concentrated in the first few years.

Here we have attempted to show that the various objectives set by planners should be subjected to a careful economic scrutiny. This does not mean that the economic objectives will always be paramount. In fact, it serves the useful purpose of revealing the real cost to the nation of satisfying various legitimate social and political objectives. It has to be left to the national planners to decide whether the attainment of such objectives through irrigation development is preferable to doing so by other means.

9.10. Project management

In the previous section attention centred on the identification of objectives and the relation between project costs and benefits. The job of minimizing costs within the set technical requirements and standards lies in the hands of the engineer, who will probably be on his next assignment before the project becomes productive. So it falls to the newly created management team to ensure that the economic benefits generated by the project reach and sustain the expected levels in their value and in their timing. In short, this means that if project performance is to be judged by reference to its design predictions then a very heavy responsibility rests on the shoulders of the management team.

Indications are that, in areas where irrigation is relatively new, project 'failure' can usually be attributed to the lack of experienced production agronomists who can take on the complex task of manipulating the various inputs to optimal advantage. But it is much easier to identify the problem than to suggest a solution. The finding of talent for leadership, especially in a basically conservative rural setting, is difficult enough, and to require it to be coupled with experience makes it a nearly impossible task in the short term.

Yet a beginning should be made somehow, and it does occur to us that, in countries where irrigation is now beginning but where a long-term development programme is envisaged, the problem could be tackled in a twofold manner. First, on medium and large projects management contracts could be given to foreign teams for a specified period, during which carefully selected indigenous talent would receive in-job training. Secondly, similar training could possibly be arranged on well-established and well-managed projects elsewhere in the region.

Here it must be emphasized that administrative ability and enterprise will be equally sought after by the industrial and the public sectors, so the incentives in agriculture should be commensurate with the difficulties of the job. If the typical cost and benefit streams of an irrigation project are examined, it is readily seen (Rydzewski 1972*b*) that the cost of management is not a very significant item. It is therefore of advantage to consider, under the particular conditions of each project, how a given increment in rewards to management is likely to be reflected in increased productivity.

9.11. Conclusions

If the development of agriculture in West Africa is considered as a whole, the impact of irrigation on the total agricultural product is not likely to be as marked as, say, in the Middle East. However, the variability of climate in the region clearly indicates that some countries will become increasingly dependent on irrigated agriculture.

Earlier in this chapter we listed a considerable number of possible and legitimate objectives of irrigation development. In the regional context the following appear to assume a dominant place:

(1) to make irrigation agriculture an instrument of economic and social advancement in the more arid areas;

(2) to take advantage of the export, or import-substitution, potential of certain cash crops;

(3) to combine animal fodder production under irrigation with the intensive production of meat for the export market;

(4) to create regional surpluses of basic foodstuffs and animal feeding stuffs as an emergency reserve against disasters such as the one which struck the Sahelian area in 1973.

The potential for irrigation in the region is, as has previously been pointed out, considerable, but its rate of development may be limited by the technical capacity for project implementation, and its rate of productivity by the present lack of operational experience.

References

AHN, P. M. (1970). *West African soils*. Oxford University Press, London.

ARCHAMBAULT, J. (1970). *Les eaux souterraines de l'Afrique occidentale*. CIEH, Ouagadougou.

ARMGROGGI, R. P. (1969). The management of water resources with reference to the role of groundwater. In *Proceedings of the ITC-UNESCO seminar on integrated surveys of river basin development, Delft*.

BLANEY, H. F. and CRIDDLE, W. D. (1966). Determining consumptive use for water developments. *Proceedings of the American Society of Civil Engineers Symposium on methods of estimating evapotranspiration, Las Vegas*.

CANTOR, L. M. (1967). *A world geography of irrigation*. Oliver and Boyd, Edinburgh.

CHAMBERS, R. (1969). *Settlement schemes in tropical Africa*. Routledge, London.

CHAUMENY, J. (1972). Le potentiel irrigable du bassin du fleuve Sénégal. *Ford Foundation/IRAT/IITA seminar; Prospects for irrigation in West Africa, Ibadan*.

COCHEMÉ, J. (1971). *Notes on the ecology of rice in West Africa*. FAO, AGS: WARDA/71/5, Rome.

—— FRANQUIN, P. (1967). *A study of the agroclimatology of the semi-arid area south of the Sahara in West Africa*. FAO/UNESCO/WHO, Rome.

COMITÉ INTERAFRICAIN D'ÉTUDES HYDRAULIQUES (1970). *Bulletins d'information*. CIEH, Ouagadougou.

—— (1972). *L'utilisation agricole des eaux de crue en Afrique*, CIEH, Ouagadougou.

DANCETTE, C. (1972). Research and main orientation in the field of plant-soil-water relationships. *Ford Foundation/ IRAT/IITA seminar; Prospects for irrigation in West Africa, Ibadan*.

DARLEY, P. R., RYDZEWSKI, J. R., and CLARK, W. M. (1972). A simulation model for the optimal design and operation of irrigation systems, based on yield-water deficit relationships. Paper presented at the *International Commission on irrigation and drainage symposium, Varna*.

DES BOUVRIE, C. (1972). Ground water: an important factor in West African development. *Ford Foundation/IRAT/ IITA seminar; Prospects for irrigation in West Africa, Ibadan*.

DIJON, R. (1971). The search for groundwater in the crystalline regions of Africa. *Nat. Resour. Forum* 1 (1).

FOOD AND AGRICULTURE ORGANIZATION (1967). *Étude critique des petits barrages en terre*. FAO. Rep. AT2429, Rome.

—— (1968). Indicative world plan for agricultural development to 1975 and 1985, in Regional study no. 3: *Africa south of the Sahara*. FAO, Rome.

—— (1969). Soil survey and land classification required for feasibility studies of water development projects. Paper presented at the second Session of the Regional Commission on land and water use in the Near East, Cairo.

FOOD AND AGRICULTURE ORGANIZATION (1970). *Étude d'un programme d'amenagement hydroagricole des terres rizicultivables de la basse Guinée.* FAO. Rep. FAO/ SF86/G UI: 2, Rome.
—— (1971). *Developpement de l'utilisation des eaux souter-raines.* FAO. Rap. tech. 1-3: AGL/SF/DAH/3, Rome.
—— (1973). *Water laws in Moslem countries.* FAO. Irrigation and Drainage Paper 20/1, Rome.
—— (1975). *Report of the ad-hoc consultation on crop water requirements.* FAO, Rome.
FAO/II/IAEA (1973). *Soil moisture and irrigation studies.* IAEA, Vienna.
FAO/UNESCO (1973). *Irrigation, drainage and salinity: an international source book.* Hutchinson, London.
FAO/UNESCO/WMO, Interagency Group on Agricultural Biometeorology (1972). *Global research project in agricultural biometeorology.* FAO, Rome.
FRANQUIN, P. (1972). L'analyse frequentielle du déficit hydrique et des pluies pour l'irrigation de complément. *Ford Foundation/IRAT/IITA seminar; Prospects for irrigation in West Africa, Ibadan.*
GIGOU, J. (1972). Frequency of rainfall and cropping cycle in the Ivory Coast. *Ford Foundation/IRAT/IITA seminar; Prospects for irrigation in West Africa, Ibadan.*
GILL, H. E. (1969). *A groundwater reconnaissance of the Republic of Ghana with a description of geohydrological provinces.* Water Supply Paper 1757, US Geological Service, Washington, D.C.
GILLET, N. (1972). Studies and experiments on agricultural use of water. *Ford Foundation/IRAT/IITA seminar; Prospects for irrigation in West Africa, Ibadan.*
GITTINGER, J. P. (1972). *Economic analysis of agricultural projects.* Johns Hopkins University Press, Baltimore.
—— (1973). *Compounding and discounting tables for project evaluation.* IRBD, Washington, D.C.
GOSSCHALK, E. M. (1972). The logic of engineering planning for efficient irrigation. *Ford Foundation/IRAT/IITA seminar; Prospects for irrigation in West Africa, Ibadan.*
GREIGERT, J. (1968). *Les eaux souterraines de la République du Niger.* BRGM, Paris.
GROVE, A. T. (1970). *Africa south of the Sahara.* Oxford University Press, London.
GUILLAUME, M. (1960). Les amenagements hydro-agricoles de riziculture et de culture de décrue dans la vallée du Niger. *Agron. trop., Nogent* 15, 73-91, 133-184, 273-319, 390-413.
HAY, G. T. and RYDZEWSKI, J. R. (1975). Comparative economics of sprinkler irrigation systems. Proceedings 10th Congress of the International Commission on Irrigation and Drainage, Moscow, 1975. (Preprint.)
JOHNSON, P. A. and STONER, R. F. (1974). Induced recharge using riverain wells. *J. Instn civ. Engrs* 57, 257-92.
LAL, R. (1972). Some hydrological characteristics of tropical soils in relation to irrigation. *Ford Foundation/IRAT/ IITA seminar; Prospects for irrigation in West Africa, Ibadan.*

LEMOINE, L. and PRAT, J. C. (1972). *Cartes d'evapo-transpiration potentielle.* CIEH, Ouagadougou.
MALETIC, J. T. (1966). Land classification survey as related to the selection of irrigable lands. Paper presented at the Pan-American soil conservation Congress, São Paulo.
—— (1967). Irrigation, a selective function: selection of project lands. Paper presented at the Water for Peace Conference, Washington, D.C.
MANNING, H. L. (1956). The statistical assessment of rainfall probability and its application in Uganda agriculture. *Proc. R. Soc.* B 144, 460-80.
M.F.E.P. (1970). *The occurrence of groundwater.* Ministry of Finance and Economic Planning, Accra.
MIDDLETON, A. A., HOFFMAN, G. R., and STONER, R. F. (1966). Planning the re-development of an irrigation system for control of salinity. Paper presented at the *Conference on civil engineering problems overseas, 1966.* Institution of Civil Engineers, London.
MISHAN, E. J. (1972). *Elements of cost-benefit analysis.* Allen and Unwin, London.
NEMEC, J. (1972). *Engineering hydrology.* McGraw-Hill, New York.
NIELSEN, D. R. (1963). Economics and soil science: copartners in land classification. Paper presented at the Land Classification Meeting, US Bureau of Reclamation Region 7.
OLIVIER, H. (1972). *Irrigation and water resources engineering.* Edward Arnold, London.
ORGANIZATION FOR AFRICAN UNITY (1970). *Symposium on schistosomiasis.* OAU, Addis Ababa.
OVERSEAS DEVELOPMENT ADMINISTRATION (1972). *A guide to project appraisal in developing countries.* HMSO, London.
QUARTEY-PAPAFIO, H. K. and KEMEVOR, E. D. (1972). Irrigation in the Volta basin. *Ford Foundation/IRAT/ IITA seminar; Prospects for irrigation in West Africa, Ibadan.*
RIJKS, D. (1972). Climate of West Africa in relation to irrigation. *Ford Foundation/IRAT/IITA seminar; Prospects for irrigation in West Africa, Ibadan.*
RIOU, C. (1972). Le bac d'eau et l'évaluation des consommation d'eau des couverts végétaux. *Ford Foundation/ IRAT/IITA seminar; Prospects for irrigation in West Africa, Ibadan.*
RIQUIER, J., BRAMÃO, D. L., and CORNET, J. P. (1970). *A new system of soil appraisal in terms of actual and potential productivity.* FAO, Rome.
RYDZEWSKI, J. R. (1968). *Irrigation development in Africa south of the Sahara: potentials and possibilities.* FAO. Rep. WS/73702, Rome.
—— (1972a). Irrigation development and potential in tropical Africa. *Ford Foundation/IRAT/IITA seminar; Prospects for irrigation in West Africa, Ibadan.*
—— (1972b). The significance of operation and maintenance costs in irrigation development planning. *Bull. int. Commis. Irrig. Drain.* (July 1972), 18 pp.

—— (1974). Determination of the operation and maintenance costs of irrigation projects. *Bull. int. Commis. Irrig. Drain.* (January 1974).

SECRÉTARIAT D'ÉTAT AUX AFFAIRES ETRANGÈRES (1973). *Recherches françaises au service de l'Afrique tropicale sèche.* The Secretariat, Paris.

STORIE, R. E. (1964). Soil and land classification for irrigation development. *Int. Congr. Soil Sci., Bucharest* 8, 873–82.

THOMAS, R. G. (1973). *Groundwater models.* Irrigation and Drainage Paper 21. FAO, Rome.

TODD, D. K. (1958). *Groundwater hydrology.* Wiley, New York.

UNESCO (1972). *Groundwater studies.* UNESCO, Paris.

US BUREAU OF RECLAMATION (1953). *Land classification: irrigated land use.* Bureau of Reclamation Manual, Vol. 5, Part 2. US Department of the Interior, Washington, D.C.

—— (1967). *Instructions for the conduct of feasibility grade land classification surveys of the Lam Nam Oon Project, Thailand.* US Department of the Interior, Washington, D.C.

VEN TE CHOW (1964). *Handbook of applied hydrology.* McGraw-Hill, New York.

WALTON, W. (1970). *Groundwater resource evaluation.* McGraw-Hill, New York.

WILSON, E. W. (1969). *Hydrology for engineers.* Macmillan, London.

WORLD HEALTH ORGANIZATION (1973). *Onchocerciasis control in the Volta river basin area.* WHO, Geneva.

WRIGHT, M. C. (1967). *Discounted cash flow.* McGraw-Hill, New York.

10. Insect and mite pests and their control

W. K. WHITNEY

10.1. Introduction

Arthropods, especially insects, constitute a major constraint on food production in the tropics. Not only do they cause great losses in quantity, but also seriously lower the quality of food. In addition to directly damaging the crops, many insect species are also notorious for their ability to spread plant pathogens, especially viruses and mycoplasmas. Furthermore, owing to high costs, most small farmers feel that they cannot afford to use chemical control measures.

Those who do use insecticides without heed to the necessary precautions run the risk of poisoning themselves or others and of destroying beneficial insects, and they may endanger the health of consumers of their products. Therefore it is essential that research workers have a clear understanding of the pests themselves and the interactions among the pests, their host plants, and the natural enemies of the pests, and that extension workers are well-informed of all possible hazards.

Wise, as well as economic, plant protection requires an appreciation of the concept of 'pest management' in which selective chemical control is integrated with other pest-suppressing operations such as use of insect-resistant crop cultivars, correct date and method of planting, weed control, encouragement of the pests' parasites and predators, timely harvesting and drying, appropriate crop rotations, and other suitable agronomic practices.

It is within the framework of this philosophy of insect control as a component of over-all plant protection that the following examples of insect problems and suggested solutions are presented. The scope of this discussion is limited to insect and mite pests of the major staple food crops of tropical Africa. This chapter deals with only the crop pests in the field since grain storage is the subject of another chapter.

10.2. Polyphagous insects

Three groups of insects—(1) grasshoppers and locusts, (2) armyworms and leafworms, and (3) termites—attack many different crops. These pests will be discussed before turning to the major pests of specific crops.

10.2.1. Grasshoppers and locusts

The director of the British Centre for Overseas Pest Research (formerly the Anti-Locust Research Centre) recently calculated that the annual worldwide potential damage by these pests was equivalent to a monetary value of about £75 million (Haskell 1972). A conservative estimate of the annual actual crop loss is 1 million t or more (Haskell 1970). Some examples of loss estimates in Africa are given in Table 10.1.

TABLE 10.1

A few examples of crop losses and control expenditures in Africa due to locusts

Locations	Periods	Losses and expenditures	References†
Nigeria	1925–34	£300 000 (mostly cereals) plus over £28 000 for famine relief	2
Northern Nigeria	1959	£14 000 spent on control	2
Kenya	1928–9	£300 000 per year	3
Morocco	1954–5	£4 500 000 in a season	3
Senegal	1957	16 000 tons of pearl millet and 2000 tons of other crops	3
Tanzania	1963	£18 700 (cereals; other crop losses not estimated)	4
Kenya	1963	£56 200 (cereals; other crop losses not estimated)	4
Ethiopia	1958	50 000–150 000 tons of cereals	1

† 1 = MacCuaig (1963), 2 = Onazi (1968), 3 = Taylor (1967a), 4 = Walker (1967).

The major grasshoppers and locusts which attack food crops in tropical Africa are listed in Table 10.2. However, there are more than 500 genera and approximately 5000 species in the superfamily Acridoidea (Dirsch 1965), many of which are potentially serious pests.

TABLE 10.2

Some locust and grasshopper pests of food crops in tropical Africa

Pests	Distribution	Crops	References†
LOCUSTS			
Anacridium spp.	General	Polyphagous	4, 5, 6, 8, 11
Locusta migratoria migratorioides Reish & Fairmaire	South of the Sahara	Many, esp. cereals	2, 4, 6, 8, 11
Nomadacris septemfaciata Serville	East, Central, South	Many, esp. cereals	2, 4, 6, 8, 11
Schistocerca gregaria Forskal	North of the equator	Polyphagous	4, 6, 8, 11, 12
GRASSHOPPERS			
Acrotylus spp.	General	Many, esp. cereals	8, 11
Anacatantops spp.	Central, East, South, West	Many, esp. cereals	11
Atractamorpha spp.	Central, East, South, West	Rice and others	5, 7
Caloptenopsis spp.	General	Cereals and others	8, 11
Cataloipus spp.	General	Rice and other cereals	5, 8
Catantops spp.	General	Polyphagous	4, 5, 8, 11
Chrotogonus spp.	General	Polyphagous	2, 11
Cyrtacanthacris spp.	Central, East, South, West	Many, esp. cereals	5, 8, 11
Eyprepocnemis spp.	General	Many, esp. cereals	5, 8, 11
Gastrimargus spp.	Central, East, South, West	Many, esp. cereals	5, 8
Hieroglyphus spp.	Central, East, West	Cereals and others	4, 11
Homoxyrrhepes spp.	Central, East, West	Rice and other cereals	5
Kraussaria spp.	Central, East, South, West	Many, esp. cereals	5, 11
Mesopsis spp.	Central, East, South, West	Cereals and others	11
Oedaleus spp.	General	General	1, 4, 5, 8, 11
Ornithacris spp.	General	General	4, 11
Oxya spp.	General	Rice and others	3, 7, 10
Phyllocercus spp.	West	Rice and others	5
Phymateus spp.	Central, East, South, West	General	4, 5, 8, 11
Pnorisa spp.	Central, East, South, West	Cereals	8, 11
Pyrgomorpha spp.	General	General	11
Spathosternum spp.	Central, East, South, West	Rice and others	5
Sudanacris spp.	East, North, West	Cereals and others	11
Zacompsa spp.	Central, East, West	General	5, 8
Zonocerus elegans Thunberg	East, Central, South	Polyphagous	8
Zonocerus variegatus (L.)	Central, East, West	Polyphagous	2–5, 8, 9, 11

† All distribution records checked against Johnston (1956, 1968). 1 = Batten (1969), 2 = Buyckx (1962), 3 = Feakin (1970), 4 = Forsyth (1966), 5 = Grist and Lever (1969), 6 = Haskell (1970), 7 = Jerath (1965a), 8 = Le Pelley (1959), 9 = Libby (1968), 10 = Phipps (1970), 11 = Schmutterer (1969), 12 = Wyniger (1962).

Control. Although locusts and grasshoppers have a number of natural enemies and a few diseases, Haskell (1970) considered that 'control by insecticides is the only feasible technique and is likely to remain so for many years'. Because locusts breed in localized areas and spread in swarms to various countries to do their damage, control operations need to be co-ordinated between national governments and organizations such as the United Nations which are well placed to achieve this (Onazi 1968).

Research on locusts and grasshoppers and techniques for their control has been led by the British Centre for Overseas Pest Research, the French Institut de Recherches Agronomiques Tropicales et des Cultures Vivrières (IRAT), and the Office de la Recherche Scientifique et Technique Outre-Mer (ORSTOM), in co-operation with many countries. An illustration of the large measure of economic improvement in control methods was indicated by Haskell (1970) who wrote that '... over 25 years the cost has been reduced from about £100 to about £2 per ton of locusts killed'. Haskell (1972) pointed out that, if the techniques now available for locust control are fully utilized, crop losses can be kept to economically insignificant levels. He explained that this success is due to an ecologically based control policy consisting of the following elements: (1) taxonomy and faunistics, (2) anatomy and morphology, (3) biology, physiology, and behaviour, (4) ecology, (5) control techniques, and (6) control strategy.

Control of non-swarming grasshoppers also depends primarily on insecticides. However, non-chemical control measures can assist in reducing crop damage. Several species of grasshopper multiply in weeds and natural vegetation around cultivated fields and become crop pests when the grasshopper populations reach a high level and when their native habitat becomes less attractive to them than the nearby crops. Thus elimination of such breeding areas can help reduce crop damage. Examples of suggested cultural control practices in rice farming include early flooding to float the eggs of *Oxya* spp. to the surface of the soil, where they fail to hatch, and ploughing soon after harvest to bring *Hieroglyphus* spp. eggs to the surface where they will also be destroyed (Feakin 1970).

As elsewhere biological control of grasshoppers in tropical Africa has not so far been very effective. Taylor (1964a) in Nigeria, for example, reported only 2·6 per cent parasitism of *Zonocerus variegatus* by the sarcophagid *Blaesoxipha filipjevi* Rohd.

Search for cultivar resistance to grasshoppers has been largely neglected in crops in Africa, but in America such resistance has been found, in maize, for example. Painter (1951) used, for the frontispiece in his classical book on host-plant resistance, a photograph showing an example of resistance to *Melanoplus* spp. in a maize cultivar. In tropical Africa many cassava plantings suffer severe defoliation by *Zonocerus variegatus* during the dry season almost every year (Kaufmann 1965). A few cultivars, however, are hardly touched by this omnivorous grasshopper (Hoare 1968; Singh, personal communication).

Chemical control may be profitable for African farmers at times of severe infestation. At such times insecticides are essential for protection of certain field research plots such as in plant breeding and agronomic experiments. Some suggested insecticidal sprays and dusts for grasshopper control are presented in Table 10.3.

Poisonous baits prepared by mixing a powdered insecticide with cereal products and sawdust, lightly moistened after mixing, are effective when scattered at the rate of 25–50 kg of the mixture per ha on grasshopper hatching grounds or infested fields, especially if the natural food supply is not abundant. Some effective insecticides and mixing rates (percentage

TABLE 10.3

Some insecticides used for grasshopper control by spraying or dusting

Insecticides	Formulations[1]	Rates[2]	Remarks
Aldrin	WP, EC, or D	0·3 kg	[3]
BHC (10 per cent gamma)	WP, EC, or D	2–2·25 kg	[3]
Carbaryl	WP or D	1–2 kg	7 days[4] Very toxic to bees
Chlordane	WP, EC, or D	1·5–2 kg	[3]
DDT	WP, EC, or D	1·5 kg	14 days[4]
Diazinon	WP or EC	0·5 kg	7 days[4]
Dieldrin	WP, EC, or D	0·2 kg	[3]
Lindane	WP, EC, or D	0·3–0·5 kg	7 days[4]
Malathion	WP, EC, D, or ULV	1–2 kg	2 days[4]

1 WP = wettable powder, EC = emulsifiable concentrate, D = dust, ULV = non-aqueous ultra-low volume.
2 Active ingredient per hectare.
3 Treat only vegetation surrounding crops. Do not apply to crops. Do not feed treated vegetation to livestock.
4 Suggested minimum waiting period between last application and harvest to allow dissipation of toxic residues.

active ingredient based on total dry weight of bait) are BHC (0·5–1·0%), lindane (0·1%), dieldrin (0·1–0·2%), aldrin (0·1–0·2%), or chlordane (0·5–1·0%).

10.2.2. *Armyworms*

The larvae of certain noctuid moths, especially *Spodoptera* spp. and *Mythimna* spp., are commonly called armyworms, because of their gregarious marching behaviour. They are widely spread and are serious defoliators of many crops in tropical Africa (Anon. 1972; Brown 1962, 1972; Buyckx 1962; Grist and Lever 1969; Libby 1968; Schmutterer 1969; Wyniger 1962).

The most notorious species of this group, the African armyworm (*Spodoptera exempta* Walker), is almost exclusively a pest of graminaceous plants and, although largely unreported, outbreaks occur somewhere in Africa nearly every year (Brown 1962, 1972). Outbreaks tend to occur when there is an unusually late onset of rains after a drought (Brown 1962; Lever 1969). Estimates of losses are scarce but Walker (1967) suggested annual cereal losses due to armyworms in Kenya and Tanzania at £2 280 000 and £937 400 respectively. Since infestations are usually localized, cereal crops on individual farms may suffer complete devastation. Brown and Mohamed (1972) simulated armyworm damage to maize and sorghum by cutting off leaves and stems. When leaves were removed from young plants little or no grain yield loss occurred, but when stems of young plants were cut or older plants were defoliated drastic reductions in the yields of grain resulted. In addition to crop damage, armyworms have been known to poison livestock grazing on heavily infested grasses (Brown and Mohamed 1972).

Control. Certain cultural practices such as removal of weeds in and around fields, ploughing to expose larvae and pupae in the soil, burning of stubble, flooding the soil, and early planting (late-planted, succulent plants are preferred by armyworms) have been suggested (Brown 1962; Feakin 1970; Grist and Lever 1969; Schmutterer 1969). Armyworms have many natural enemies, such as parasitic and predaceous insects, birds, toads, frogs and other vertebrate predators, and diseases. Armyworm outbreaks occur when these organisms are failing to provide sufficient control, and in these circumstances prompt application of a suitable insecticide is essential and economic (Anon. 1972; Appert 1964, 1970; Brown 1962, 1972).

Since the key to success is promptness, the East African Agriculture and Forestry Research Organization, in co-operation with the British Centre for Overseas Pest Research, is now monitoring the movements of, and forecasting outbreaks of, armyworms, so that farmers and insecticide suppliers can quickly meet the challenge of these pests (Anon. 1972; Brown 1972; Haskell 1972).

Effective but potentially dangerous insecticides for control of *Spodoptera* spp. fall into two main groups: (1) chlorinated hydrocarbons such as endrin, dieldrin, isodrin, and aldrin, which are generally considered too hazardous for the non-specialist to handle and may leave undesirable residues; and (2) less persistent but acutely toxic organophosphates such as parathion, methyl parathion, monocrotophos, dicrotophos, Cyolane(R) (diethyl 1,3-dithiolan-2-ylidenephosphoroamidate), and Cytrolane(R) (diethyl 4-methyl-1,3-dithiolan-2-ylidenephosphoroamidate) (Appert 1964; Brown 1962; Feakin 1970; Grist and Lever 1969; Serghiou 1971; Wyniger 1968; Zeid, Saad, Ayad, Tantawi, and Eldefawi 1968). A third group of safe and effective insecticides available for armyworm control, however, exists, as indicated in Table 10.4.

Taylor (1968a) showed that the African armyworm was effectively controlled on rice with the bacterial insect pathogen *Bacillus thuringiensis* Berliner sprayed at a rate of 5.35×10^{14} spores per ha, but this was not at that time, economically competitive with conventional insecticides.

TABLE 10.4

Some insecticides currently used for control of armyworms and related pests

Insecticides	Formulation†	Rate (kg a.i. per ha)	Suggested minimum time from last application to harvest
BHC	D, EC, WP	1–2·5	1 week
Carbaryl	WP	1–1·5	1 week
DDT	D, EC, WP, ULV	0·5–2	2 weeks
Diazinon	EC	0·5–1	2 weeks
Dichlorvos	EC	0·5–1	1 week
Endosulfan	EC	0·5–0·75	2 weeks
Fenitrothion	ULV	0·75	2 weeks
Fenthion	EC	0·25–0·5	2 weeks
Gardona(R)	EC, ULV	0·6	1 week
Malathion	EC, ULV	1–1·5	2 days
Toxaphene	EC	2–4	2 weeks
Trichlorfon	EC	0·25–0·38	1 week
DDT+lindane	D, EC	1–1·8+0·3–0·5	2 weeks
DDT+toxaphene	D, EC	0·5–1·25+1–2·5	2 weeks

† D = dust, EC = emulsifiable concentrate, WP = wettable powder, ULV = non-aqueous ultra-low volume.

10.2.3. *Termites (Isoptera)*

These social insects, often erroneously called 'white ants', are found everywhere in the tropics. While they damage many crops, and some build mounds which interfere with mechanized farming, termites nevertheless play a generally beneficial role in the breakdown and incorporation of organic matter in the soil. Furthermore, their subterranean mining activities may improve the soil texture (Wyniger 1962). In East Africa, Hesse (1955) showed that termite mounds were built of subsoil and that termites did not directly alter or affect the chemical characteristics of the soil, except possibly the base exchange capacity and total exchangeable bases. He showed that neither the presence of termite bodies nor termite-collected vegetable matter had any lasting effect on fertility of mound soils. He concluded that the improved growth of certain plant species on some mounds was related mainly to the improved soil structure and drainage. Since mounds are built of subsoil, termite-disturbed soil in some locations is less productive than surrounding soil. For example, the author found maize grown on plots previously occupied by mounds of *Macrotermes* spp. in Nigeria yielded 10 per cent less grain than adjacent plots with no obvious history of mounds (IITA 1971). Some termite species often cause serious crop damage for two or three years in newly cleared land and eventually disappear (Wyniger 1962). These termites and the

TABLE 10.5

Major termite pests of food crops in the African humid tropics

Termite genera	Crops	References†
Allodontermes	Root crops, cereals, grain legumes	5
	Miscellaneous food crops	3
Amitermes	Yams, cassava	5
	Miscellaneous food crops	3
Ancistrotermes	Root crops, cereals, grain legumes	5
	Miscellaneous food crops	3
Hodotermes	Miscellaneous food crops	3
Macrotermes	Root crops, cereals, grain legumes	5
	Rice	3
	Maize	1
	Maize	2
	Maize	4
Microtermes	Root crops, cereals, grain legumes	5
	Cereals, legumes, vegetables	6
	Maize	1
	Maize, miscellaneous food crops	3
	Maize	2
Nasutitermes	Maize	2
Odontotermes	Root crops, cereals, grain legumes	5
	Miscellaneous food crops	3
	Maize	1
Pseudacanthotermes	Miscellaneous food crops	3
Trinervitermes	Rice	3

† 1 = Appert (1970), 2 = Buahin (1971), 3 = Harris, W. V. (1969), 4 = Le Pelley (1959), 5 = Sands (1971), 6 = Schmutterer (1969).

food crops which they attack in the African humid tropics are listed in Table 10.5.

Sands (1971) considered that because the order Isoptera included insects of such a wide range of behaviour and feeding habits, it was likely that practically every crop species used by man is attacked by at least one termite species somewhere in the tropics. The most destructive termite attacks occur at times when plants are least resistant. In single-seasonal herbaceous crops (often called annuals) this is usually near the time of maturity and during dry periods. Some of the vascular tissues of a living plant may be cut by termites during periods of high moisture availability but associated wilting may not appear until a slight water stress occurs. While vigorously growing plants are less frequently attacked than drought-stressed or senescent plants, termite damage is often apparent on lush young maize plants as well as mature maize. Untreated maize plots averaged 29 per cent loss attributed to termites (IITA 1971).

In southern Tanzania, over a five-year period, Bigger (1966) found an average stand loss of 27 per cent and 33 per cent (90 per cent and 82 per cent maxima) in maize and soya beans respectively. He

showed that the use of small amounts of seed-treating† insecticides resulted in 336–560 kg/ha and 112–202 kg/ha seed yield increases in maize and soya beans respectively. Sands (1962) showed that cassava stands in Nigeria suffered as much as 40 per cent loss due to termite damage to cuttings and that attacks on yam sets and tubers sometimes resulted in 70 per cent yield loss. In Nigeria *Macrotermes natalensis* eats seedling roots of rain-fed rice, while *Trinervitermes ebenerianus* removes the leaves which in some localities results in complete crop loss (Harris, W. V. 1969). *Macrotermes* spp. attack the ripening rice crop, cutting stems at ground level and then feeding on the fallen heads, with losses of 15–20 per cent of the crop observed (Harris, W. V. 1969; IITA 1971).

Control. Although no successful method of biological control has yet been developed (Sands 1971), termites have natural enemies. According to Schmutterer (1969), the predatory animals which eat termites include certain species of ants, dragonflies, lizards, frogs, toads, birds, bats, and man. These depradations and the few known termite diseases (Bouillon, in Krishna and Weesner 1970) exercise too little control over termite populations to prevent crop losses in tropical Africa. The use of insecticides is therefore sometimes necessary.

Aldrin and dieldrin are the insecticides still mostly used for termite control. W. V. Harris (1969) gave details of useful insecticides and application methods. Sands (1971) summarized the current methods of chemical control for mound-building and subterranean termites and presented useful comments on the economics and limitations of several approaches to the use of pesticides against species which attack food crops. He dealt with the subject under the following headings:

1. Mound-poisoning.
2. General applications.
3. Local applications.

1. *Mound-poisoning.* Crops are frequently attacked by several species of termites and only some of them are mound builders. Therefore, mound-poisoning often fails to control damage. It is essential to have an accurate identification of the pest species before control is attempted.

2. *General application of insecticides to the soil.* Sands included ridge, furrow, and broadcast

† The term 'seed-dressing' is now usually kept for cleaning.

TABLE 10.6

Termite control: summary of some methods used successfully in the tropics

Methods	Insecticides and rates	Termites	References[†]
Mound poisoning (inject into the centre of the hive)	Aldrin or dieldrin, 200 ml of 6% emulsion per small mound	Harvesters (*Trinervitermes* spp.)	4
	Aldrin or dieldrin, about 20 l of 1% emulsion per large mound	*Macrotermes* spp.	4
General soil treatment (shallow incorporation 10–20 cm, into soil surface)	Dieldrin 2·5% dust or chlordane 10% dust, 30–70 kg formulation per ha gives 2–3 years protection	*Ancistrotermes* spp.	2
	BHC dust (0·65% gamma) 4·5 kg mixed with 200 l sawdust applied per ha	*Microtermes* spp.	3
	Gamma BHC dust, 1–2 kg a.i. or aldrin dust, 0·5 kg a.i./ha drilled with seed	Various spp.	3
	Dieldrin dust, 1·2–3·6 kg a.i./ha	*Microtermes* spp.	5
	Dieldrin 5% granules, 5 kg a.i./ha[‡]	*Odontotermes* spp.	6
	Aldrin or heptachlor, 3–5 kg a.i./ha[‡]	*Hodotermes* spp.	6
	Aldrin, 0·5 kg a.i./ha in planting rows	Various spp.	1
Local applications (seed-treatment)	Dieldrin 75% WP, 0·1 oz/lb seed (about 5 g a.i./kg seed). Dieldrin WP, 4 g a.i./kg seed	Various spp.	1
	Aldrin or heptachlor WP, 1·25–1·875 g a.i./kg seed	Various spp.	1, 6
	Dieldrin WP, 4–5 g a.i./kg seed	Various spp.	6
Dip cuttings or sets before planting	Aldrin WP, 0·25% a.i. in water	Various spp.	6
	BHC WP, 0·375% a.i. in water	Various spp.	6
	Dieldrin WP, 0·125% a.i. in water	Various spp.	6

† 1 = Bigger (1966), 2 = Buyckx (1962), 3 = Harris, W. V. (1969), 4 = Sands (1971), 5 = Schmutterer (1969), 6 = Wyniger (1968).
‡ These dosage rates appear excessive in view of other reports.

treatments in this category. These are the most widely used methods for termite control in all parts of the tropics. Effective control by this method depends upon uniform dispersion and shallow incorporation in the soil.

3. *Local applications of insecticides.* This category of treatments is economically and environmentally attractive. The category includes protection of seeds, seedlings, yam sets, cassava cuttings, etc. by dipping them in dust or water suspensions of appropriate insecticides before planting. Small amounts of insecticide used in this way have yielded handsome returns (e.g. Bigger 1966).

A summary of some successful chemical control methods for termite pests of food crops in Africa is given in Table 10.6.

10.3. Pests of cereals (maize, pearl millet, rice, and grain sorghum)

10.3.1. The major insect pests of cereals

Brenière (1971) suggested that borers are the most important insects attacking graminaceous crops in tropical Africa. Table 10.7 lists the major borer species, their distribution, and the cereal grain crops attacked. Most of these insects also attack other graminaceous plants such as sugar cane and several species of wild grasses.

Losses due to stemborers are difficult to assess precisely because of the number of insect species, the various types of damage, the stages of plant development when attacked, and often the associated presence of other insects and micro-organisms. For example, maize attacked by newly hatched *Busseola fusca* shows a characteristic pattern of holes and 'window panes' in the upper leaves resulting from the leaf-feeding activities of the young larvae before they enter the stalk via the funnel (Harris 1962). *Sesamia calamistis* larvae, on the other hand, bore directly into the stalk beneath the leaf sheath where the eggs are laid. Attack on young maize plants frequently results in stand loss. Older plants may be weakened by borers and lodge, subjecting the ears to rotting and depredation by termites and rodents. Some examples of the magnitude of assessed losses due to borers are given in Table 10.8.

TABLE 10.7

Major lepidopterous stem-borer pests of graminaceous food crops in tropical Africa

Insects	Plants†	Locations reported in Africa	References‡
NOCTUIDAE			
Busseola fusca (Fuller)	M, S, Mil	Widely distributed	1, 2, 4–6, 9–14, 16–24
Manga basilinea Bowden	Mil	Ghana, Nigeria	4, 9, 11, 17
Sesamia botanephaga Tams & Bowden	M, S, R	Ivory Coast, Ghana, Togo	2, 4–11, 13–15, 17, 20, 22
Sesamia calamistis Hampson	M, S, R, Mil	Widely distributed	1–18, 20–5
Sesamia cretica Lederer	S, M, Mil	Widely distributed	2, 10, 14, 16, 18, 20
Sesamia penniseti Tams & Bowden	Mil, M, S	East and West (esp. in wet forest zones)	5, 7, 9, 10, 11, 15, 17, 20, 21, 22
Sesamia poephaga Tams & Bowden	M, S, Mil	Widely distributed	2, 5, 6, 7, 9–11, 13, 15–17, 20
PYRALIDAE			
Acigona (=*Coniesta*) *ignefusalis* (Hampson)	Mil, S, M, R	West	1, 4, 5, 7, 9–12, 17, 19, 22, 24
Adelpherupa flavescens Hampson	R	Cameroun, Nigeria	3, 10, 15
Chilo partellus (=*C. zonellus*) Swinhoe	S, M, Mil	East	3, 4, 13, 14, 16, 18, 19, 23
Chilo spp.	R	Widely distributed	1, 3, 8, 10–17, 19, 25
Chilo zacconius Bleszynski (=*Proceras africana* Risbec)	R, S	West	3, 10
Chilotraea argyrolepia (Hampson) and other spp.	M, S, Mil	Widely distributed	1, 5, 9–11, 14, 16, 22, 23
Eldana saccharina Walker	M, S, Mil, R	Widely distributed	1, 2, 4–6, 9–17, 19, 24
Hypsotropa subcostella Hampson	R	Nigeria, Ghana, Uganda	1, 10, 13, 15, 16
Maliarpha separatella Ragnot	R	Widely distributed	3, 4, 8–13, 15, 16, 19
Scirpophaga spp.	R	Cameroun, Nigeria	1, 3, 4, 10, 12, 15, 17

† Listed in order of importance for each insect species. M = maize, Mil = millets, R = rice, S = sorghum.
‡ 1 = Adeyemi (1967), 2 = Appert (1970), 3 = Brenière (1969), 4 = Brenière (1971), 5 = Buahin (1971), 6 = Buyckx (1969), 7 = Endrody-Younga (1968), 8 = Feakin (1970), 9 = Forsyth (1966), 10 = Grist and Lever (1969), 11 = Harris (1962), 12 = IITA (1971), 13 = Ingram (1958), 14 = Jepson (1954), 15 = Jerath (1965*b*), 16 = Le Pelley (1959), 17 = Libby (1968), 18 = Schmutterer (1969), 19 = Starks (1969), 20 = Tams and Bowden (1953), 21 = Usua (1968), 22 = van Eijnatten (1965), 23 = Wheatley (1961), 24 = Whitney (1970), 25 = Will and Jones (1970).

Some of the other pests of maize, pearl millet, rice, and sorghum are listed in Table 10.9. There is some information available on the magnitude of losses associated with some of these other insects. A few examples follow:

Rhopalosiphum maidis on maize in Indiana: 10 per cent less grain in lightly infested fields, mainly due to smaller seeds and ears. Forty per cent loss in many fields (Everly 1960).

Paliwal (1971) found maize mosaic virus infected plants (aphid transmitted) in India were more susceptible to *Chilo partellus*.

Geromyia penniseti in pearl-millet heads in Senegal (Coutin and Harris 1969).

midges per head	37	204	285
grain yield per head (g)	12·1	6·6	0·7

Pachydiplosis oryzae on rice in northern Cameroun: 75 per cent of the crop was destroyed in the Lagone Valley in 1954 (Descamps 1956). Losses of 30–50 per cent are common (Feakin 1970).

Contarinia sorghicola on sorghum in Ghana and Nigeria: 1·34–3·48 per cent grain yield loss for each 1 per cent infested spikelets (Food and Agriculture Organization 1970). 5–10 per cent annual loss in Africa (Harris, K. M. 1969). Losses in Sudan often reach 25 per cent (Schmutterer 1969).

Diopsis thoracica on rice in West Africa: up to 80 per cent damage of tillers (Brenière 1969).

Atherigona varia soccata on sorghum in Uganda: 17 per cent grain yield increase by use of soil systemic insecticide even when infestation of untreated plots was exceptionally low (20 per cent, usually 40–80 per cent) (Barry 1972). In Thailand and Sudan, infestation is as high as 90 per cent (Sepsawadi, Meksongsee, and Knapp 1971; Schmutterer 1969).

10.3.2. Control of cereal insects

Cultural control of insects is the first method that should be considered because it is generally compatible with other means of control and it adds little

TABLE 10.8

Some examples of losses due to cereal stemborers in the tropics

Borer species	Location	Damage and losses (by crop)	References†
		MAIZE	
Busseola fusca	Northern Nigeria	1 larva per plant = 28 per cent grain loss	6
	Southern Nigeria	1–2 larvae per plant = 25 per cent grain loss	12
		4–5 larvae per plant = 75 per cent grain loss	
	East Africa	1 per cent decrease in number of infested plants = 35 lb more grain per acre (40 kg/ha)	13
	Tanzania	Total loss frequent in some areas. Regular loss of 5 per cent	8
Chilo partellus	Uganda	37 per cent grain loss when infested at 20 days after plant emergence	11
Sesamia botanephaga	Ghana	50–75 per cent of plants attacked in 1966–7	3
B. fusca and *S. calamistis*	Nigeria	Generally more than 20 per cent stand loss; Endrin sprays increased grain yield 26 per cent	6
Various	Ghana	Generally 20–35 per cent yield loss; entire crop may be lost	4
Various	Nigeria	Grain yield reduced 1 per cent for each 1 per cent of stem-length bored (measured at harvest); sprays gave 0·55–2·04 t more grain per ha	7
Various	Nigeria	4–60 per cent infestation, 4–33 per cent stand loss, up to 34 per cent grain yield loss	1
		SORGHUM	
Chilo partellus	Uganda	56 per cent grain yield loss when infested at 20 days after plant emergence	11
B. fusca	Tanzania	Usually 40–100 per cent infestation	8
	Northern Nigeria	Average of 33 borers per 100 stems, 1957–60	6
Acigona ignefusalis	Northern Nigeria	Average of 17 borers per 100 stems, 1957–60	6
Various	General	Up to 80 per cent of stems attacked	3
		RICE	
Maliarpha separatella	Madagascar	More than 80 per cent of stems generally attacked	2
	Swaziland	24·8 kg grain loss per ha for each 1 per cent infestation	5
Various	Tropical Africa	20 per cent average loss	9
Various	General	Each 1 per cent white head = 1–3 per cent yield loss	10
Various	Nigeria	23–37 per cent yield increase from sprays	7
S. calamistis	Nigeria	69 per cent infestation (73 borers per 100 stems)	7
		PEARL MILLET	
Acigona ignefusalis	Northern Nigeria	Average of 100 borers per 100 stems, 1957–60	6
B. fusca	Northern Nigeria	Average 6 borers per 100 stems, 1957–60	6
Various	Northern Nigeria	61 per cent bored stems = 15 per cent yield loss	6

† 1 = Adeyemi, Donnelly, and Odetoyinbo (1966), 2 = Brenière (1969), 3 = Brenière (1970b), 4 = Buahin (1971), 5 = Grist and Lever (1969), 6 = Harris (1962), 7 = IITA (1971), 8 = Jepson (1954), 9 = Lever (1971), 10 = Pathak (1968), 11 = Starks (1969), 12 = Usua (1968), 13 = Walker (1960a).

if any to production costs. Some important elements of cultural control of cereal pests are listed below:

Ploughing and soil tillage some days before planting to expose soil-inhabiting larvae and pupae to predators and the sun (Buyckx 1962; Schmutterer 1969).

Observance of suitable crop rotations and adjustment of planting dates to avoid peak pest populations in the highly susceptible plant growth stages (Brenière 1969; Buyckx 1962; de Pury 1968; Endrody-Younga 1968; Feakin 1970; Grist and Lever 1969; Harris 1962; Khan 1967; Pathak

1968; Schmutterer 1969; Walker 1963). Jepson (1954), however, correctly emphasized that without the availability of irrigation, agronomic factors rather than pests may dictate planting dates.

Control of weeds and volunteer host plants in and around fields (Appert 1970; Brenière 1969, 1970a, 1970b; Buyckx 1962; de Pury 1968; Feakin 1970; Grist and Lever 1969; Harris, K. M. 1969; Reddy 1967; Schmutterer 1969).

Soil amendments: (1) High fertility helps plants overcome insect damage (Appert 1970; Brenière 1969, 1970a, 1970b, 1971; Grist and Lever 1969;

TABLE 10.9

Pests of maize, pearl millet, rice, and sorghum in tropical Africa (except locusts, termites, and stemborers)

Insects	Crops† and plant parts damaged	References‡
COLEOPTERA		
Chrysomelidae		
Chrysispa viridicyanea (Kraatz)	R leaves	8, 11, 14
Lema spp.	Mil, R leaves	2, 16, 20
Sesselia pusilla Gerstacker	R rice yellow mottle virus vector	4
Trichispa sericea (Guerin)	R leaves	10, 13, 14, 16
Coccinellidae		
Epilachna spp.	M, Mil, R, S leaves	3, 5, 14–16, 20, 22, 23
Curculionidae		
Nematocerus sp.	M leaves, stems	7
Sitophilus spp.	M, R, S ripening grain, field infestations	1, 8, 15, 20, 22
Meloidae		
Cylindrothorax spp.	R flowers, immature grain	16, 18
Decapotoma affinis Olivier	R flowers, immature grain	16
Scarabaeidae		
Adoretus spp.	M stems, roots	1
Anomala spp.	M silks, immature grain	21
Heteroligus spp.	M, R stems, roots	8
Heteronychus spp.	M, R stems, roots	1, 3, 8, 13, 14, 16, 17, 23
Pachnoda spp.	M, Mil, R silks, immature grain	20
Silvanidae		
Cathartus quadricollis Guerin	M ripening grain, field infestation	1, 22
DIPTERA		
Anthomyiidae		
Atherigona varia soccata Rondani	Mil, S shoots	1, 11, 20, 21, 23
Atherigona spp.	M, Mil, R, S shoots	14
Cecidomyiidae		
Contarinia sorghicola (Coq.)	S flowers, immature grain	1, 12, 17, 20, 21
Geromyia penniseti (Felt)	S flowers, immature grain	9
Pachydiplosis oryzae (Wood-Mason)	R stems	5, 13, 14, 19, 20
Diopsidae		
Diopsis thoracica Westw.	R stems	5, 13–17
Diopsis apicalis Dalm.	R stems	5, 13–17
Diopsis spp.	Mil, R, S stems	11
HEMIPTERA		
Alydidae		
Mirperus jaculus (Thnb.)	Mil, S immature grain	19
Riptortus dentipes F.	R immature grain	15, 17
Stenocoris (*Erbula*) *southwoodi* Ahmad (*Leptocorisa apicalis* Auct.)	R immature grain	5, 8, 13–17, 23
Coreidae		
Acanthomia spp.	R immature grain	5, 15
Anoplocnemis spp.	M, R shoot tips, immature grain	3, 14, 15, 20, 22
Miridae		
Creontiades pallidus (Rambur)	M, Mil, S leaf buds	3, 20
Taylorilygus vosseleri (Poppius)	M, Mil, S leaf buds	3, 8, 11
Pentatomidae		
Agonoscelis spp.	M, S shoot tips, immature grain	3, 20
Aspavia spp.	R, S immature grain	5, 15–17
Nezara viridula L.	M, R, S shoot tips, immature grain	3, 5, 8, 13–17, 20, 22, 23
Pyrrhocoridae		
Dysdercus spp.	M, Mil, R, S shoot tips, immature grain	5, 11, 14, 16, 17, 20

K

TABLE 10.9 (*cont.*):

Insects	Crops† and plant parts damaged	References†
HOMOPTERA		
Aphididae		
Melanaphis (*Longiunguis*) *sacchari* (Zehnt.)	S young leaves, flowers	8, 20
Rhopalosiphum maidis (Fitch)	M, S young leaves, flowers	3, 8, 15, 17, 20, 23
Cercopidae		
Locris spp. and *Poophilus* spp.	M, Mil, R, S stems, leaves	5, 6, 13–17, 20, 22
Cicadellidae		
Tettigella spectra Distant (Complex)	R stems, leaf sheaths	5, 15
Cicadulina spp.	M, R maize streak virus vector, leaves	1, 8, 15, 20, 22, 23
Hecalus spp.	R leaves	13, 14
Nephotettix spp.	R leaves, potential virus vectors	3, 8, 13–15
Delphacidae		
Pereginis maidis (Ashmead)	M, R, S young stems, leaves	14, 20, 23
Sogatella spp.	M, Mil, R young stems, leaves	13, 15
Sogatodes cubanus (Crawf.)	R potential vector of hoja blanca disease	14, 15
Fulgoridae		
Zanna (=*Pyrops*) *tenebrosa* (F.)	R leaves	13, 14
Meenoplidae		
Nisia atrovenosa (Leth.)	M, R leaves	13, 14
LEPIDOPTERA		
Gelechiidae		
Sitotroga cerealella (Oliv.)	R, S ripening grain, field infestation	1, 8, 14, 15
Heliodinidae		
Stathmopoda auriferella (Wlk.)	S ripening grain, field infestation	1
Hesperidae		
Borbo sp. and *Parnara* sp.	R leaves	5
Pelopidius mathias (F.)	R leaves	5, 8, 13, 14
Noctuidae		
Chloridea (=*Heliothis Auct.*) *armigera* (Hubn.)	M, Mil, R, S immature grain	3, 8, 11, 17, 20, 21
Pyralidae		
Hymenia recurvalis (F.)	M leaves	3
Hyphilare loreyi Dup.	M leaves, stems	3
Marasmia trapezalis (Guenee)	M, R, S leaves	3, 14, 15, 22
Mussidia nigrivenella Rag.	M ears	1, 8, 15, 22
Nymphula spp.	R leaves	3, 8, 14, 15
Prodotuala grisealis Guenne	M leaves	15
Susumia exigua (Butl.)	R leaves	15
Tortricidae		
Cryptophlebia (=*Argyroploce Auct.*) *leucotreta* (Meyr.)	M immature grain	1, 8, 15, 17
ORTHOPTERA		
Gryllidae		
Brachytrypes membraneaceous (Dru.)	M seedlings	6
Gryllotalpidae		
Gryllotalpa africana (P. de B.)	R seedlings	13, 14, 16, 17

† M = maize, Mil = millet, R = rice, S = sorghum.
‡ 1 = Adeyemi (1967), 2 = Anon. (1965), 3 = Appert (1970), 4 = Bakker (1970), 5 = Brenière (1969), 6 = Buahin (1971), 7 = Bullock (1961), 8 = Buyckx (1962), 9 = Coutin and Harris (1969), 10 = Deeming (1971), 11 = de Pury (1968), 12 = Descamps (1956), 13 = Feakin (1970), 14 = Grist and Lever (1969), 15 = IITA (1971), 16 = Jerath (1965b), 17 = Libby (1968), 18 = Okwakpam (1971), 19 = Reddy (1967), 20 = Schmutterer (1969), 21 = Starks (1969), 22 = Whitney (1970), 23 = Wyniger (1962).

Reddy 1967) but may also make some crops more attractive to certain insects (Grist and Lever 1969; Khan 1967; Pathak 1968; Starks, Schumaker, and Eberhart 1971). (2) More borer damage has occurred in acidic (pH 4·48) soils than in neutral to slightly alkaline soils of pH 7–8 (Pathak 1968). (3) Addition of silica to silica-deficient soils gave reductions in rice stemborer damage and disease (Pathak 1968).

Water management (Feakin 1970; Grist and Lever 1969; Khan 1967; Pathak 1968).

Physical removal of insects by hand-picking, netting, light traps, roguing badly infested plants, and, most important, destruction of crop residues which carry pests over from one season to another (Adeyemi 1969; Appert 1970; Brenière 1969, 1970a, 1970b, 1971; Buyckx 1962; de Pury 1968; Feakin 1970; Grist and Lever 1969; Harris 1962, 1969; Jepson 1954; Khan 1967; Pathak 1968; Reddy 1967; Schmutterer 1969; Usua 1967).

Optimizing all cultural practices yields greatest value for each input (Appert 1970; Brenière 1970a, 1970b, 1971; IITA 1971; Whitney 1970).

Host-plant resistance has been considered an ideal means of insect control (Horber 1971). Van Emden (1972) pointed out many advantages of plant resistance over other methods of insect control, emphasized the need for more intensive research in resistance, and proposed a 'risk-rating' method for use in developing improved plant cultivars. Pathak (1970) presented an excellent summary of the genetics of host-plant resistance to insect pests. Although research and development on host-plant resistance has been badly neglected in Africa, some progress has been made.

Maize stemborers

The West African Maize Research Unit started screening and breeding work and identified one hybrid, bred from American material with European cornborer resistance, which had statistically significant resistance to *Busseola fusca* and *Sesamia* sp. (van Eijnatten 1965). Unfortunately, the programme was discontinued.

Preliminary screening work at IITA has shown some interesting differences in susceptibility among open pollinated cultivars and S_I lines. Mass-rearing of *Sesamia calamistis* and further screening for resistance is in progress (IITA 1971, 1972a; Whitney 1970).

Starks and Doggett (1970) reported material resistant to *Chilo zonellus* in East Africa.

Sorghum pests

In East Africa sorghum cultivar CK-60 has good resistance to *Chilo zonellus*, cultivar Serena has good recovery resistance to *Atherigona* spp., and cultivars Namatare and 5DX 142/9 are resistant to both of these pests (Starks 1969). Working with the shoot-fly-resistant cultivar Maldani 35-1 and a range of susceptible cultivars of sorghum in northern Nigeria, Langham (1968) suggested that resistance was due either to (1) a single recessive gene with heterosis in the F_I plants or (2) several genes displaying partial dominance and epistasis plus a heterotic effect. Resistance in sorghum cultivar Maldani 35-1 was thought to be due to the presence of prickle-hairs instead of the usual soft micro-hairs on leaf sheaths.

'Nunaba' strains of *Sorghum membranaceum* which are cultivated in West Africa have natural resistance to *Contarinia sorghicola* due to the glume characteristics (Harris, K. M. 1969).

Rice stemborers

There are many references to certain rice cultivars being more susceptible than others, e.g. OS-6 is more susceptible than TKM-1 to *Sesamia calamistis* (IITA 1972a). Indica types are more susceptible than japonicas (Grist and Lever 1969). Soto found *Oryza punctata*, a wild rice species of Nigeria, to be highly resistant to *S. calamistis* (IITA 1972a).

A considerable amount of host-plant resistance work has been done elsewhere and some of the resulting information might be fruitfully utilized in breeding programmes to improve the pest resistance of cereals in tropical Africa. Some examples are the following:

1. American maize cultivars with resistance to:
   (a) European cornborer, *Ostrinia nubilalis*, which is associated with 6-methoxybenzoxazolinone and related compounds (Beck 1965);
   (b) *Heliothis* earworms (Bennett, Josephson, and Burgess 1967; Douglas and Eckhardt 1957; Painter 1951);
   (c) other pests such as aphids, scarab larvae, *Sitophilus* spp., nitidulids, etc. (Painter 1951, 1958).

2. Pearl millet cultivars in India with resistance to lepidopterous leaf-folders, planthoppers, and lepidopterous larvae which attack developing grain heads (IARI 1971).

3. Asian rice cultivars with resistance to:

   (a) gall midge (Kovitvadhi and Leaumsang 1971; Shastry and Seshu 1971; Wickramasinghe 1971);

   (b) leafhoppers (Cheng and Pathak 1972; Pathak 1969);

   (c) planthoppers (Jennings and Pineda 1970a, 1970b);

   (d) stemborers (Djamin and Pathak 1967; Patanakamjorn and Pathak 1967; Pathak 1969; Pathak, Andres, Galacgac, and Raros 1971).

4. American sorghum cultivars resistant to aphids, stemborers, *Heliothis*, sorghum midge, lygaeids, and pentatomids (Chada, Atkins, Gardenhire,

and Weibel 1961; Painter 1951, 1958; Wiseman and McMillian 1970) and Indian cultivars resistant to borers and shoot-flies (IARI 1971; Singh *et al.* 1968).

Biological control, especially by the indigenous parasites and predators of cereal pests, is a very important factor in tropical Africa. Since Greathead (1971) has recently presented a comprehensive review of applied work in the Ethiopian region, this chapter will point out only some of the indigenous parasites and predators of cereal pests which undoubtedly exercise a significant measure of control (Table 10.10). When considering the use of insecticides and other control measures one must consider these naturally occurring beneficial insects and do

TABLE 10.10

Some insect parasites and predators which attack cereal pests in tropical Africa†

Parasites and predators	Host insects	References‡
COLEOPTERA		
Carabidae		
Scarites madagascarensis Dejean	*Heteronychus plebejus* (Klug)	10
Coccinellidae		
Various species	Aphids and leafhoppers	1, 9
DIPTERA		
Bombyliidae		
Geron sp.	Various lepidopterous borers	17
Chloropidae		
Mepachymerus baculus Speiser (=*Steleocerus lepidopus* Beck.)	*Diopsis thoracica* Westw.	2
Muscidae		
Atherigona sp.	*Sesamia* sp.	18
Phoridae		
Diploneura sp.	*Busseola fusca* (Fuller)	10
Megaselia scalaris (Löw)	*Chloridea armigera* (Hbn.)	17
Plethysnochaeta sp.	*Busseola fusca* (Fuller)	10
Pipunculidae		
Pipunculus risbeci Seguy	*Chilo zacconius* Blesz.	2, 17
Sarcophagidae		
Sarcophaga dispessa Vill.	*Dysdercus* spp.	16
S. dysderci Vill.	*Dysdercus* spp.	16
Syrphidae		
Various species	Aphids, cercopids, leafhoppers	1, 9, 21
Tachinidae		
Actia comitata Vill.	*Dysdercus* spp.	16
Actia spp.	*Busseola fusca* (Fuller)	1
Alophora nasalis Bez.	*Dysdercus* spp. and various pentatomids	16
Bogosia antinorii Rond.	Various pentatomids	9, 16
Bogosiella fasciata (F.)	*Dysdercus* spp.	8, 16

Table 10.10 (*cont.*):

Parasites and predators	Host insects	References†
DIPTERA (*cont.*):		
Tachinidae (*cont.*):		
Ceromyia sp. nr. *cuthbertsoni* Curran	*Chilo* spp.	10
Decampsina sesamiae Mesnil	*Busseola fusca* (Fuller)	18, 20
Decampsina sesamiae Mesnil	*Eldana saccharina* Walker	20
Decampsina sesamiae Mesnil	*Sesamia calamistis* Hampson	1, 20
Decampsina sesamiae Mesnil	*Sesamia* spp.	1, 20
Drino inconspicuus (Mg.)	*Chloridea armigera* (Hbn.)	8
Linnaemyia longirostris Macq.	*Chloridea armigera* (Hbn.)	16
Gen. nr. *Metopsisyrops*	*Hypsotropa* sp. probably *subcostella* Hamps.	10
Nemoraea bequarti Emden	*Sesamia* sp.	1
Nemoraea discoidalis Vill.	*Sesamia* sp.	1
Nemoraea discoidalis Vill.	*Busseola fusca* (Fuller)	1
Ocyptera xiphias Bez.	Various pentatomids	16
Siphona murina Mesnil	*Busseola fusca* (Fuller)	1
Siphona murina Mesnil	*Chilo partellus* Swinhoe	1
Siphona murina Mesnil	*Eldana saccharina* Walker	1
Sturmia imberbis Wied.	*Chloridea armigera* (Hbn.)	16
Sturmiopsis parasitica Curran	*Acigona ignefusalis* (Hamps.)	10, 11
Sturmiopsis parasitica Curran	*Sesamia* spp.	1, 2, 10, 11, 20
HEMIPTERA		
Anthocoridae, Miridae, Reduviidae	Various species	1, 6, 8, 9, 16
HYMENOPTERA		
Bethylidae		
Goniozus procerus Risbec	*Adelpherupa* sp.	10
Goniozus procerus Risbec	*Chilo zacconius* Blesz.	2, 10
Goniozus procerus Risbec	*Scirpophaga* sp.	10
Goniozus procerus Risbec	Various lepidopterous borers	17
Goniozus sp.	*Eldana saccharina* Walker	21
Goniozus sp.	*Maliarpha separatella* Ragnot	2, 10
Braconidae		
Apanteles belliger Wilk.	*Mythimna pseudo loreyi* (Rungs)	10
A. flavipes (Cam.) (introduced)	Various lepidopterous borers	9
A. procerae Risbec	*Chilo zacconius* Blesz.	2, 10
A. ruficrus (Hal.)	*Chilo zacconius* Blesz.	2, 10
A. sesamiae Cam.	*Busseola fusca* (Fuller)	1, 10, 11, 13
A. sesamiae Cam.	*Chilo partellus* Swinhoe	1, 10
A. sesamiae Cam.	*Sesamia* spp.	1, 4, 9-12
Apanteles sp.	*Busseola fusca* (Fuller)	4
Apanteles sp.	*Chilo* spp.	4, 10
Apanteles sp.	*Sesamia* spp.	4, 13
A. syleptae Ferriere	*Chilo zacconius* Blesz.	2, 10
Aulosalphes antennatus Granger	*Chilo zacconius* Blesz.	17
Bracon sp. nr. *annulicornis* (Brues)	*Pachydiplosis oryzae* (Wood-Mason)	19
Bracon brevicornis Wesm.	*Sesamia* spp.	1, 13
Bracon hebetor Say	*Busseola fusca* (Fuller)	1
Bracon hebetor Say	*Sesamia* spp.	1
Bracon quadratinotatus Granger	*Sesamia* spp.	17
Bracon sp.	*Chilo* spp.	20
Bracon sp.	*Chloridea armigera* (Hbn.)	16
Bracon sp.	*Maliarpha separatella* Ragnot	2
Braunsia occidentalis Enderlein	*Sesamia calamistis* Hamps.	5
Braunsia sp.	*Chloridea armigera* (Hbn.)	16
Chelonus curvimaculatus (Cam.)	*Busseola fusca* (Fuller)	1

TABLE 10.10 (*cont.*):

Parasites and predators	Host insects	References‡
HYMENOPTERA (*cont.*):		
Braconidae (*cont.*):		
Chelonus curvimaculatus (Cam.)	*Chilo* spp.	1
Chelonus curvimaculatus (Cam.)	*Chilo zacconius* Blesz.	17
Euvipio rufa Szepl.	*Busseola fusca* (Fuller)	1
Euvipio rufa Szepl.	*Chilo partellus* Swinhoe	1
Euvipio sp.	*Eldana saccharina* Walker	20
Euvipio sp.	*Sesamia* sp.	20
Glyptomorpha deesae Cam.	*Chilo partellus* Swinhoe	1
Habrobracon sp. nr. *triangularis* Szepl.	*Chilo zacconius* Blesz.	2
Perilitus sp.	*Chilo zacconius* Blesz.	2
Phanerotemella saussurei Kohl	*Maliarpha separatella* Ragnot	1, 2, 10
Phanerotemella saussurei Kohl	*Sesamia* spp.	1
Phanerotoma sp.	*Maliarpha separatella* Ragnot	2
Rhaconotus spp.	*Maliarpha separatella* Ragnot	2, 8, 10
Rhaconotus sudanensis Wilk.	*Chilo* spp.	14
Rhaconotus sudanensis Wilk.	Various lepidopterous borers	17
Rogas sp.	*Chilo zacconius* Blesz.	17
Ceraphronidae		
Trichosteresis sp. nr. *forsteri* Kieff.	*Atherigona varia soccata* Rondani	15
Chalcididae		
Brachymeria spp.	*Chilo* spp. and *Sesamia* spp.	1, 11
Hyperchalcidia soudanensis Stef.	*Acigona ignefusalis* (Hamps.)	10, 11
Hyperchalcidia soudanensis Stef.	*Busseola fusca* (Fuller)	1
Hyperchalcidia soudanensis Stef.	*Chilo* spp.	1, 2, 10, 12
Diapriidae		
Trichopria sp.	*Diopsis thoracica* Westw.	5
Encyrtidae		
Coccidencyrtus sp.	*Epilachna* sp.	5
Oencyrtus sp.	*Epilachna* sp.	5
Eulophidae		
Aprostocetus sp.	*Diopsis thoracica* Westw.	2, 5, 10, 21
Enicospilus sp.	Various lepidopterous borers	5
Pedobius amaurocoelus (Wtstn.)	*Epilachna* sp.	5
Pedobius furvus (Gah.)	*Acigona ignefusalis* (Hamps.)	11
Pedobius furvus (Gah.)	*Busseola fusca* (Fuller)	11
Pedobius furvus (Gah.)	*Eldana saccharina* Walker	20
Pedobius furvus (Gah.)	*Sesamia* spp.	1, 4, 8–12, 20
Pedobius sp.	*Busseola fusca* (Fuller)	10, 12
Pedobius sp.	*Chilo partellus* Swinhoe	10, 12
Pedobius sp.	*Sesamia* spp.	12
Tetrastichus atriclavus Wtstn.	*Acigona ignefusalis* (Hamps.)	10, 11
Tetrastichus atriclavus Wtstn.	*Busseola fusca* (Fuller)	1, 5, 10, 11
Tetrastichus atriclavus Wtstn.	*Sesamia calamistis* Hamps.	1, 10
Tetrastichus diplosidis (Crawford)	*Contarinia sorghicola* (Coq.)	3
Tetrastichus diopsisi Risbec	*Diopsis thoracica* Westw.	2
Tetrastichus pachydiplosisae Risbec	*Pachydiplosis oryzae* (Wood-Mason)	2, 19
Tetrastichus procerae Risbec	*Chilo zacconius* Blesz.	2
Tetrastichus sesamiae Risbec	*Chilo zacconius* Blesz.	10
Tetrastichus soudanensis Stef.	*Chilo zacconius* Blesz.	2
Tetrastichus spp.	*Contarinia sorghicola* (Coq.)	3
Tetrastichus spp.	*Geromyia penniseti* (Felt)	6

T<small>ABLE</small> 10.10 (*cont.*):

Parasites and predators	Host insects	References‡
HYMENOPTERA (*cont.*):		
Eupelmidae		
Eupelmus popa Gir.	*Contarinia sorghicola* (Coq.)	3
Eupelmus spp.	*Contarinia sorghicola* (Coq.)	3
Eupelmus spp.	*Geromyia penniseti* (Felt)	6
Eurytomidae		
Eurytoma lepidopterae Risbec	*Adelpherupa flavescens* Hamps.	10
Eurytoma lepidopterae Risbec	*Chilo zacconius* Blesz.	17
Eurytoma lepidopterae Risbec	Various lepidopterous borers	17
Eurytoma sp.	*Chilo partellus* Swinhoe	1
Eurytoma sp.	*Eldana saccharina* Walker	1
Eurytoma sp.	*Maliarpha separatella* Ragnot	2
Formicidae		
Dorylus spp.	Various species	8, 11
Pheidole megacephala F.	Various lepidopterous borers	1
Ichneumonidae		
Amesospilus sp.	*Sesamia calamistis* Hamps.	1
Charops sp.	*Busseola fusca* (Fuller)	1
Charops sp. nr. *spinitarsus* Cam.	*Sesamia* spp.	14
Chasmias sp.	*Acigona ignefusalis* (Hamps.)	10, 11
Chasmias sp.	*Busseola fusca* (Fuller)	1, 13
Chasmias sp.	*Chilo partellus* Swinhoe	10, 12
Chasmias sp.	*Sesamia calamistis* Hamps.	10
Coelocentrus sp.	*Chilo zacconius* Blesz.	10
Dentichasmias busseolae	*Chilo* spp.	20, 21
Enicospilus sp.	*Busseola fusca* (Fuller)	1
Enicospilus sp.	*Sesamia calamistis* Hamps.	1, 8, 20
Homocidus bizonarinis	*Atherigona varia soccata* Rondani	15
Ichneumon tananarive Hein.	*Sesamia calamistis* Hamps.	1
Isotima sp.	*Maliarpha separatella* Ragnot	2
Lathromeris ovicida	*Scirpophaga* spp.	5
Mesostenus basimaculata Cam.	*Dysdercus* spp.	8
Metopius discolor Tosq.	*Sesamia* spp.	14
Netelia opaculus Thoms.	*Chloridea armigera* (Hubn.)	16
Phygadeuon sp.	*Atherigona varia soccata* Rondani	15
Pimpla hova Seyrig	*Sesamia calamistis* Hamps.	1
Pristoderus sp.	*Maliarpha separatella* Ragnot	2
Procerochasmia glaucopterus Morl.	*Busseola fusca* (Fuller)	1
Procerochasmia glaucopterus Morl.	*Sesamia calamistis* Hamps.	1
Scenocharops sp.	*Maliarpha separatella* Ragnot	2, 4, 10, 20
Scenocharops sp.	*Sesamia* spp.	1, 10
Syzeuctus sp.	*Acigona ignefusalis* (Hamps.)	10
Syzeuctus sp.	*Chilo* spp.	10
Syzeuctus sp.	*Eldana saccharina* Walker	20
Syzeuctus sp.	Various lepidopterous borers	5
Xanthopimpla citrina (Holmgr.)	Various lepidopterous borers	9
Xanthopimpla stemmator (Thnb.)	*Chilo partellus* Swinhoe	1
Xanthopimpla stemmator (Thnb.)	*Sesamia calamistis* Hamps.	8
Mymaridae		
Gonatocerus spp.	Leafhoppers	5
Platygasteridae		
Platygaster diplosisae Risbec	*Pachydiplosis oryzae* (Wood-Mason)	2, 19
Platygaster sp.	*Geromyia penniseti* (Felt)	6

TABLE 10.10 (*cont.*):

Pests	Distribution	Crops	References†
HYMENOPTERA (*cont.*):			
Pteromalidae			
Dinarmus (=*Bruchiobius*) spp.	*Adelpherupa flavescens* Hamps.		10, 17
Norbanus sp.	*Sesamia* spp.		11
Trigonasta sp.	*Busseola fusca* (Fuller)		10
Scelionidae			
Asolcus basalis (Woll.)	Various pentatomids		9
Gryon (=*Hadronotus*) *gnidus* Nixon	*Acanthomia tomentosicollis* Stal.		16
Trissolcus (=*Microphanurus*) spp.	Various pentatomids		16
Platytelenomus busseolae (Gah.)	*Busseola fusca* (Fuller)		1, 4, 11
Platytelenomus busseolae (Gah.)	Various lepidopterous borers		5, 20
Platytelenomus hylas Nixon	*Sesamia* spp.		1, 8, 13
Telenomus spp.	*Chilo* spp.		5
Telenomus spp.	*Maliarpha separatella* Ragnot		2, 4, 5, 10
Telenomus spp.	*Scirpophaga* spp.		6
Telenomus spp.	*Sesamia* spp.		2, 8, 17
Scoliidae			
Campsomeris coelebs Sichel	*Heteronychus plebejus* (Klug)		10
Campsomeris pilosella Sauss.	*Heteronychus plebejus* (Klug)		10
Trichogrammatidae			
Gen. nr. *Bloodiella* sp.	*Scirpophaga* spp.		10
Trichogramma spp.	*Chilo* spp.		5
Trichogramma spp.	*Diopsis* spp.		5
Trichogramma spp.	*Scirpophaga* spp.		5
Trichogrammatoidea luteum (Gir.)	*Diopsis thoracica* Westw.		5
Trichogrammatoidea minutum (Riley)	*Cryptophlebia leucotreta* (Meyr.)		13
Trichogrammatoidea simmondsi Nagaraga	*Sesamia calamistis* Hamps.		10
NEUROPTERA			
Chrysopidae	Aphids and others		1
ORTHOPTERA			
Mantidae	Various pests		9
STRESPSIPTERA			
Corioxenidae			
Corioxenos antestiae Blair	Pentatomids		9

† See Tables 10.14 and 10.17.

‡ 1 = Appert (1970), 2 = Brenière (1969), 3 = Brenière (1970*a*), 4 = Brenière (1971), 5 = Unpublished reports of the Commonwealth Institute of Biological Control, West African Sub-Station, 6 = Coutin and Harris (1969), 7 = Deeming (1971), 8 = Forsyth (1966), 9 = Greathead (1971), 10 = Grist and Lever (1969), 11 = Harris (1962), 12 = Ingram (1958), 13 = Jepson (1954), 14 = Jerath (1965*b*), 15 = Langham (1968), 16 = Le Pelley (1959), 17 = Nickel (1964), 18 = Oboite (1968), 19 = Reddy (1967), 20 = Scheibelreiter (1971), 21 = Simmonds (1971).

all that is feasible to prevent their unnecessary destruction and, if possible, encourage their activities.

Chemical control of cereal pests is often necessary and profitable, especially in modernized, intensive agriculture. For example, Brenière (1969) indicated that when rice yields in West Africa reach 3 t/ha on a given farm, the judicious use of insecticides could increase yields by a further 1 t/ha. Whitney and Kang (IITA 1971; Whitney 1970) demonstrated the importance of optimizing all agronomic factors (e.g. plant population density, soil fertility, weed control) if maximum benefit is to be gained from insecticides and fungicides on maize. There is great need for reliable information on economic threshold levels of insect infestation and resultant crop losses before realistic decisions can be made regarding the use of insecticides (Chiarappa and Chiang 1971). A step in this direction was taken with the publication of the FAO manual on the evaluation and prevention of losses by pests, diseases, and weeds (FAO 1970).

Although many insecticide trials have been conducted in tropical Africa against cereal-crop pests, very few workers have reported data on cost/benefit ratios. Furthermore, many experiments included highly toxic insecticides which would be too hazardous to recommend for use by most farmers. A list of reasonably safe and reportedly effective chemical control measures for insect pests of tropical cereal crops is presented in Table 10.11.

TABLE 10.11

Some suggested chemical control measures for insect pests of tropical cereal crops

Pests, insecticides, formulations† and rates‡	Crops§	References‖
LEPIDOPTEROUS STEMBORERS		
Carbaryl D, WP, GR: 1–1·5 kg/ha, 2–4 applic.	M, R, S	1, 2, 4, 10, 13, 15, 17, 19, 21, 29
Carbaryl+lindane GR: 3+3 kg/ha, 2–3 applic.	R (flooded)	15, 25
Carbofuran GR: 0·5–1 kg/ha in seed furrow	R	7, 15
DDT D, EC, GR, WP: 1–2 kg/ha, 2–4 applic.	M, R, S	4, 8–11, 13, 17, 20, 23, 27–9
Diazinon D, EC, GR: 1–2 kg/ha, 2–4 applic.	M, S, R	4, 8, 9, 14, 15, 17, 18, 25
Endosulfan EC, GR, WP: 1–1·5 kg/ha, 2–4 applic.	M, S, R	13, 17, 21
Lindane EC, GR, WP: 1·25–3 kg/ha, after transplanting and at 30 days	R (flooded)	9, 13–15, 17, 18, 25
Lindane GR: 0·2–0·3 kg/ha, 2–4 applic.	M, S	8, 9, 21
Trichlorfon EC, GR: 0·2–0·4 kg/ha, 2–4 applic.	R	13, 29
DEFOLIATORS (Coleoptera and Lepidoptera)		
BHC D, EC: 1 kg/ha	R	14
Carbaryl WP: 1 kg/ha	R	14, 15, 24
Carbaryl D, GR: 1 kg/ha	M	4
Carbofuran WP (seed treatment): 2·5–5 g per kg seed (3 weeks control)	R	7
Carbofuran GR: 0·5–1 kg/ha in seed furrow	R	7
DDT D, EC, GR, WP: 1–2 kg/ha	M, Mil, R, S	3, 4, 11, 14
Diazinon D, GR: 1 kg/ha	M	4
SUCKING INSECTS (Hemiptera and Homoptera)		
BHC D: 2–3 kg/ha	M, R, S	29
Carbaryl D, WP: 1 kg/ha	M, R, S	29
Carbaryl GR: 2–3 kg/ha, 20-day intervals	R (flooded)	14, 15, 25
Carbaryl WP: 0·05–0·1% spray, 7-day intervals	R	14, 15
Carbofuran WP (seed treatment): 2·5–5 g per kg seed (4 weeks control)	R	7
Carbofuran GR: 0·5–1 kg/ha in seed furrow	R	7
DDT D, EC, WP: 1–1·5 kg/ha	M, R, S	14, 29
Diazinon D, EC: 0·25–0·5 kg/ha	M, R, S	29
Diazinon GR: 2 kg/ha, 10 and 30 days after transplanting	R (flooded)	5, 14, 25
Dimethoate EC: 0·05% spray, 0·25–0·5 kg/ha	M, R, S	4, 14, 29
Endosulfan EC, WP: 0·05% spray, 0·25–0·5 kg/ha	R	14, 29
Lindane GR: 2–3 kg/ha	R (flooded)	25
Lindane+carbaryl GR: 2+2 kg/ha	R (flooded)	25
Malathion EC, ULV: 0·5–1 kg/ha	M, R, S	14, 15, 25, 29
DIPTERA		
Atherigona spp.		
Carbofuran WP (seed treatment): 30–50 g per kg seed	S	6, 21, 26
Carbofuran GR: 1–3 kg/ha in seed furrow	S	6, 26
Chlorfenvinphos GR: 1–3 kg/ha in seed furrow	S	6, 26
Disulfoton GR: 1–3 kg/ha in seed furrow	S	21
Phorate GR: 1–3 kg/ha in seed furrow	S	21
Contarinia sorghicola		
Carbaryl D, WP: 1·5 kg/ha ⎫	S	16, 21
DDT D, WP: 1–1·5 kg/ha ⎪ Apply when 90 per cent of heads are out.	S	16
Diazinon EC: 0·6 kg/ha ⎬ Repeat in 3–5 days if needed	S	29
Endosulfan EC: 0·35 kg/ha ⎭	S	21

TABLE 10.11 *(cont.)*:

Pests, insecticides, formulations,† and rates‡	Crops§	References‖
DIPTERA *(cont.)*:		
Diopsis spp.		
BHC GR: 2·75–3·75 kg/ha	R (flooded)	29
BHC WP: 2·75–3·75 kg/ha	R	29
Diazinon GR: 2 kg/ha	R (flooded)	29
Diazinon EC: 2 kg/ha	R	29
Lindane GR: 1–2 kg/ha	R (flooded)	13, 17
Lindane EC: 1–2 kg/ha	R	13, 17
Pachydiplosis oryzae		
Diazinon GR: 2 kg/ha, 30 and 60 days after transplanting	R (flooded)	22
Phorate GR: 3–4 kg/ha, 2 weeks after transplanting	R (flooded)	12, 15
SOIL INSECTS (e.g. white grubs and cutworms)		
Bait: 0·3% heptachlor at 50 kg/ha		4
Bait: 1% BHC at 30–40 kg/ha		29
Seed treatment: aldrin, BHC, or heptachlor, 4 g per kg seed		4, 14
Soil treatment: Broadcast and incorporate aldrin, dieldrin or heptachlor, 1·5–2 kg/ha		4, 11, 14
Spray seedlings: DDT, 1·5 kg or dieldrin, 1 kg/ha		29

† D = dust, EC = emulsifiable concentrate, GR = granules, ULV = non-aqueous ultra-low volume, WP = wettable powder.
‡ Rates expressed in terms of active ingredient, except for baits.
§ M = maize, Mil = millet, R = rice, S = sorghum.
‖ 1 = Adeyemi (1967), 2 = Ahmed and Young (1969), 3 = Anon. (1965), 4 = Appert (1970), 5 = Bae and Pathak (1969), 6 = Barry (1972), 7 = Bowling (1969), 8 = Brenière (1970b), 9 = Brenière (1971), 10 = Buahin (1971), 11 = Buyckx (1962), 12 = Cantelo and Kovitvadhi (1967), 13 = Damotte (1967), 14 = Feakin (1970), 15 = Grist and Lever (1969), 16 = Harris, K. M. (1969), 17 = IITA (1971), 18 = IRRI (1967), 19 = Isa and Bishara (1966), 20 = Jepson (1954), 21 = Jotwani and Young (1972), 22 = Kovitvadhi and Leaumsang (1971), 23 = Nyasaland Department of Agriculture (1961), 24 = Okwakpam (1971), 25 = Pathak (1967), 26 = Sepsawadi, Meksongsee and Knapp (1971), 27 = Walker (1959), 28 = Walker (1960b), 29 = Wyniger (1962).

10.4. Grain-legume pests (cowpeas, grams, lima beans, pigeon peas, soya beans, etc.)

10.4.1. *Insect pests on cowpeas*

Insects constitute a major constraint on tropical grain legume production (Table 10.12). This problem is especially acute on cowpeas, probably because this is the only species of this group of long-standing cultivation in tropical West Africa (Taylor 1971b); therefore the following discussion will emphasize the pests of cowpeas to illustrate the magnitude of the problem.

Grain yields of cowpea on West African farms range from 150 kg/ha to 600 kg/ha (Delassus 1970; Taylor 1964b, 1967b, 1971b). Yields using the same cultivars under modest insecticidal spray regimes are easily increased by two- to tenfold (Appert 1964; Booker 1965b; Delassus 1970; IITA 1971, 1972b; Koehler and Mehta 1972; Taylor 1964b, 1967b, 1971b; Whitney 1972). Insect attack on some cultivars of cowpeas grown in monoculture without plant protection is so severe that no marketable seeds are produced (IITA 1971, 1972b; Whitney 1972).

10.4.2. *Insect damage in grain legumes*

Insect damage to grain legumes occurs throughout the life of the plant and continues in storage. Seedlings are subject to attack by wireworms, white grubs, crickets, cutworms, leafworms, leaf beetles, leafhoppers, bean-flies, aphids, and thrips (mainly *Sericothrips occipitalis*), resulting in stand loss, stunting, and distortion. Plants which escape or survive attack in the seedling stage may be damaged in the pre-flowering stage by the same complex of insects plus others such as stemborers, pentatomids, alydids, and coreids. At flowering time the larger Hemiptera, thrips (mainly *Taeniothrips sjostedti*), and *Maruca testulalis* cause extensive damage to flower buds and flowers, resulting in premature abscission of flowers, reduced pod set, and delayed maturity. The larger Hemiptera and *M. testulalis* seriously damage immature pods and developing

TABLE 10.12

Some pests of grain legumes in tropical Africa

Pests†		Plant parts damaged	References‡
ACARINA			
Tetranychidae	*Tetranychus* spp.	Leaves	24, 36, 37
COLEOPTERA			
*Bruchidae	*Acanthoscelides obtectus* (Say)	Seed. Field infestation	19
	Bruchidius atrolineatus (Pic)	Seed. Field infestation	1, 2, 4, 21, 35
	Callosobruchus maculatus (F.)	Seed. Field infestation	1, 2, 21, 24, 32, 35
Buprestidae	*Sphenoptera* spp.	Stems	1, 16, 24, 32
Cerambycidae	*Sophronica* spp. and other genera	Stems	9, 16
*Chrysomelidae	*Altica pyrilosa* (Erichs.)	Leaves	16
	Aphthona latipennis Pic	Leaves	24
	Asbecesta spp.	Leaves	33
	Aulacophora spp.	Leaves	16, 24
	Barombiella (=*Barombia*) *humeralis* Lab.	Leaves	24, 27, 29, 32, 36
	Luperodes (? = *Monolepta*) *lineata* Kars.§	Leaves. Virus vector	9, 11, 32, 33, 35, 36
	Luperodes quaternus Fairm.§	Leaves	2, 9, 16, 24
	Ootheca mutabilis (Sahlb.)	Leaves. Virus vector	2, 3, 9, 16, 17, 27, 29, 32, 35
	Ootheca spp.	Leaves	2, 16
	Pagria spp.	Leaves	5, 16
	Podagrixena (=*Podagrica*) spp.	Leaves	9, 24, 33
	Prosmidia conifera Fairm.	Leaves	16
	Sagra amethystina Guer.	Leaves	9
Coccinellidae	*Cheilomenes* spp.	Leaves	2, 9
	Epilachna spp.	Leaves	2, 16, 27, 33
*Curculionidae	*Alcidodes* spp.	Stems	2, 9, 16, 29, 32, 37
	Apion spp.	Flowers, seeds	27
	Nematocerus acerbus Faust	Leaves. Virus vector	2, 16, 36
	Piezotrachelus varius Wagn.	Seeds	1–3, 16, 17, 21, 25, 26, 32
Elateridae	Various genera and spp.	Seedlings, roots	1
Lagriidae	*Chrysolagria* spp.	Leaves, flowers, pods	29, 32
	Lagria villosa F.	Leaves, flowers, pods	2, 9, 16, 27, 29, 33, 36
	Lagria cuprina Thoms.	Leaves, flowers, pods	9, 32, 33
Lycidae	*Lycus* (*Acantholycus*) *corniger* Dalm.	Leaves, flowers	2, 36
Meloidae	*Coryna* spp.	Flowers	2, 16, 24, 32
	Cylindrothorax spurcaticolis Fairm.	Flowers	24
	Decapotoma affinis Oliv.	Flowers	2, 16, 27, 29
	Epicauta spp.	Flowers	2, 24
	Mylabris spp.	Flowers	2, 8, 9, 16, 24, 32, 33, 36
Scarabaeidae	*Pachnoda* spp. and other genera	Roots, flowers	1, 2, 32
Scolytidae	*Hypothenemus* (? = *Stephanoderes*) *hampei* (Ferr.)	Seeds	24, 36
DIPTERA			
*Agromyzidae	*Agromyza* sp.	Leaves, stems	16
	Melanagromyza candidipennis (Lamb.)	Leaves, stems	36
	M. chalcosoma Spencer	Maturing seeds	13, 14, 16
	M. phaseoli (Tryon)	Leaves, stems	1, 7, 10, 16, 24, 25, 34, 35, 37
	M. sojae Zhnt.	Leaves, stems	16
	M. spencerella Greathead	Leaves, stems	10, 35
	M. vignalis Spencer	Maturing seeds	2, 35
	Ophiomyia centrosematis (de Meij.)	Leaves, stems	10
HEMIPTERA			
*Alydidae	*Mirperus jaculus* (Thunb.)	Shoots, immature seeds	2, 17, 23, 29, 32, 33, 35
	Riptortus dentipes (F.)	Shoots, immature seeds	16, 17, 27, 29, 33, 35

TABLE 10.12 (*cont.*):

Pests†		Plant parts damaged	References‡
HEMIPTERA (*cont.*):			
Alydidae	*Riptortus* spp.	Shoots, immature seeds	9, 16, 23, 32
	Stenocoris southwoodi Ahmad	Shoots, immature seeds	9
*Coreidae	*Acanthomyia brevirostris* Stål	Shoots, immature seeds	2, 3, 17, 24, 29
	A. horrida Germ.	Shoots, immature seeds	2, 3, 16–18, 23, 28, 29, 33, 36
	A. tomentosicollis Stål	Shoots, immature seeds	16, 18, 21, 23, 32, 33, 36
	Anoplocnemis curvipes (F.)	Shoots, immature seeds	2, 3, 17, 24, 27, 29, 33, 35
Lygaeidae	*Spilostethus* (=*Lygaeus Auct.*) spp.	Shoots, leaves	2
Miridae	*Adelphocoris apicalis* Reut.	Shoots, leaves	2, 16, 33
	Campylomma spp.	Shoots, leaves	17, 24
	Crentides pallidus (Ramb.)	Shoots, leaves	16
	Halticus tibialis Reut.	Shoots, leaves	17
	Helopeltis bergrothi Reut.	Shoots, leaves	9
	H. schoutedeni Reut.	Shoots, leaves	24
	Lygus spp.	Shoots, leaves	16
	Tayloriligus vasseleri (Popp.)	Shoots, leaves	24
*Pentatomidae	*Agonoscelis versicolor* F.	Shoots, immature seeds	23
	Aspavia armigera (F.)	Shoots, immature seeds	2, 17, 32, 33, 35
	Chinapia acuta (Dallas)	Shoots, immature seeds	2, 23, 32, 33
	Cyclopelta tristis Stål	Shoots, immature seeds	16
	Dalsira costalis Germ.	Shoots, immature seeds	23
	Glypsus conspicuus Westw.	Shoots, immature seeds	2
	Halydicoris scoruba Dallas	Shoots, immature seeds	23
	Macrina juvenca Burm.	Shoots, immature seeds	2
	Nezara viridula L.	Shoots, immature seeds	2, 9, 16, 17, 23, 24, 32, 33, 35, 37
	Piezodorus spp.	Shoots, immature seeds	2, 16, 23, 32
Plataspidae	*Brachyplatys testudonigra* Deg.	Shoots, immature seeds	9, 16, 23, 33
	Coptosoma spp.	Shoots, immature seeds	16, 23, 33
	Libyasphis (=*Plataspis*) *vermicellaris* Stål	Shoots, immature seeds	16
Pyrrhocoridae	*Dysdercus superstitious* F.	Shoots, immature seeds	2, 17, 32
Tingidae	*Urentius enonymus* Dist.	Shoots, leaves	16
HOMOPTERA			
Aleyrodidae	*Bemisia* spp.	Leaves	16, 24, 36
*Aphididae	*Acyrtosiphon sesbaniae* Kalt.	Leaves, shoots	24
	Aphis craccivora Koch	Leaves, shoots, flowers, pods	2, 9, 24, 36
	Aphis spp.	Leaves, shoots	16, 24, 32, 35, 37
	Macrosiphum nigrinectaria Theo.	Leaves, shoots	16
Cercopidae	*Locris* spp.	Stems, leaves	2, 36
	Poophilus spp.	Stems, leaves	35
	Ptyelus grossus F.	Stems	16, 36
*Cicadellidae	*Empoasca dolichi* Paoli	Leaves	16, 27, 33, 35
	Empoasca spp.	Leaves	16, 17, 24
	Erythroneura spp.	Leaves	24, 35
	Nehela sp.	Leaves	16
Coccidae	Many genera and spp.	Leaves, stems	9, 16
Delphacidae	Many genera and spp.	Leaves	35
LEPIDOPTERA			
Arctiidae	*Amsacta* spp.	Leaves	1, 2
	Diacrisia spp.	Leaves	2, 9, 32
	Utethesia lotrix Cram.	Leaves	2, 24

TABLE 10.12 (*cont.*):

Pests†		Plant parts damaged	References‡
LEPIDOPTERA (*cont.*):			
Gelechiidae	*Lecithocera palpigera* Wlsm.	Seeds	16
Geometridae	Various spp.	Leaves	16
Gracilariidae	*Acrocercops coerulea* Meyr.	Leaves	16
	Acrocercops sp.	Leaves	24, 33, 35
*Lycaenidae	*Cupidopsis cissus* (Godart)	Pods, flowers	2
	Euchrysops malathana (Boisd.)	Pods, flowers	9, 16, 29, 32
	Lampides boeticus (L.)	Pods, flowers	1, 16
	Virachola (=*Deudorix*) *antalus* Hopf.	Pods, flowers	1, 2, 9, 16, 33
Lymatriidae	Various spp.	Leaves	16
Noctuidae	*Achaea catocaloides* Guen.	Leaves	16
	Agrotis ipsilon (Hufn.)	Leaves, seedlings	24
	Anomis erosa (Hubn.)	Leaves	1
	Anticarsia irrorata F.	Leaves	2, 16, 23
	Autographa gamma L.	Leaves	1
	Chloridea armigera (Hubn.)	Pods, leaves	6, 15, 16, 22–24, 36
	Mocis undata F.	Leaves	2, 9
	Pardasena virgulana Mab.	Leaves	16
	Peridroma saucia Hubn.	Leaves	37
	Plusia spp.	Leaves	16, 33, 35
	Spodoptera spp.	Leaves, pods	1, 2, 9, 16, 17, 23, 24, 32, 33, 35, 37
Pterophoridae	*Sphenarches caffer* Zell.	Pods	16, 32
*Pyralidae	*Etiella zinckenella* (Treit.)	Pods, seeds	9, 16, 22, 24, 37
	Lamprosema indicata F.	Leaf folder	16, 37
	Maruca testulalis Geyer	Flowers, pods, stems	1–3, 14, 16, 17, 24, 30, 33, 35
	Sylepta derogata (F.)	Leaf folder	23
*Tortricidae	*Cryptophlebia* spp.	Stemborer	16, 23
	Laspeyresia campestris Meyr.	Seeds in pods	16
	L. ptychora Meyr.	Seeds in pods	16, 17, 28, 33, 35
	Laspeyresia sp.	Leaf buds	33, 36
ORTHOPTERA			
Gryllidae	*Brachytrupes membranaceus* Dru.	Seedlings	16, 24, 36
	Gryllus bimaculatus Deg.	Seedlings	24
Tridactylidae	*Tridactylus* sp.	Seedlings	24, 36
THYSANOPTERA			
Phlaeothripidae	*Haplothrips* sp.	Flowers	20, 36
*Thripidae	*Caliothrips* spp.	Leaf buds	24
	Frankliniella schultzei Trybom	Flowers, leaf buds	20, 24, 36
	Sericothrips occipitalis Hood	Leaf buds	14, 35
	Taeniothrips sjostedti (Trybom)	Flowers, leaf buds	2, 6, 14, 20, 22, 31, 35
	Thrips tabaci Lind.	Leaf buds	24

† Major pests indicated by *.

‡ 1 = Appert (1964), 2 = Booker (1965*a*), 3 = Booker (1965*b*), 4 = Booker (1967), 5 = Bryant (1942), 6 = Delassus (1970), 7 = Emery (1957), 8 = Farrell and Adams (1966), 9 = Forsyth (1966), 10 = Greathead (1969), 11 = IITA (1971), 12 = IITA (1972*b*), 13 = Koehler and Mehta (1970), 14 = Koehler and Mehta (1972), 15 = Koehler and Rachie (1971), 16 = Le Pelley (1959), 17 = Libby (1968), 18 = Materu (1970), 19 = Nyiira (1970), 20 = Okwakpam (1967), 21 = Prevett (1961), 22 = Rachie and Wurster (1971), 23 = Robertson (1969), 24 = Schmutterer (1969), 25 = Sellschop (1962), 26 = Tardieu (1961), 27 = Taylor (1964*b*), 28 = Taylor (1965), 29 = Taylor (1967*b*), 30 = Taylor (1967*c*), 31 = Taylor (1969*b*), 32 = Taylor (1971*b*), 33 = van Halteren (1971), 34 = Walker (1960*c*), 35 = Whitney (1972), 36 = Whitney (unpublished), 37 = Wyniger (1962).

§ Probably *Paraluperodes quaternus* (Fairmaire).

seeds. Mature but unripe seeds are often heavily damaged by *Laspeyresia ptychora* and *Piezotrachelus varius* before harvest. Some species of bruchids, which are mainly storage pests, infest ripening pods in the field.

Although considerable defoliation by chewing insects can be tolerated by a well-established plant without significant reductions in seed yield (Begum and Eden 1965; Mehta 1971; Todd and Morgan 1972; Turnipseed 1972), many species of chewing insects and at least two species of thrips are capable of spreading cowpea mosaic virus (Bock 1971; Chant 1959; Dale 1949, 1953; Janzen and Staples 1970, 1971; Smith 1924; van Hoof 1962; Walters and Henry 1970; Whitney and Gilmer 1974).

Damage to soya-bean pods by chewing insects has been studied by a few workers. Smith and Bass (1972*a*) showed that one larva of *Heliothis zea* (Boddie) is capable of partially consuming at least six green pods. They indicated that ten larvae per metre of row introduced once, or three larvae per metre of row introduced twice, were sufficient to justify treatment. In another study Smith and Bass (1972*b*) showed that while up to 80 per cent pod-removal prior to the start of pod-filling did not reduce seed yields, as little as 10 per cent removal after pods were filled did significantly reduce yields. Kincade, Laster, and Hartwig (1971), using a mechanical technique to simulate podborer damage, showed that, on fully-extended but unfilled soya bean pods, yields were significantly reduced when damage reached the 20 per cent level. The larger Hemiptera (alydids, coreids, and pentatomids) not only cause direct damage to developing seeds (Daugherty, Neustadt, Gehrke, Cavanah, Williams, and Green 1964) but also may introduce pathogens such as *Nematospora coryli* Peglion (Clark and Wilde 1970*a*, 1970*b*, 1971; Daugherty 1967; Foster and Daugherty 1969; Materu 1970).

Table 10.13 shows some examples of damage and losses associated with certain pests of grain-legume crops.

Studies on the relative susceptibility of various grain-legume genera and species to insect pests are under way in tropical Africa (IITA 1971, 1972*b*; Whitney 1972). Preliminary results show that cowpeas are generally most heavily damaged. Pigeon peas have few insect problems until flowering, when thrips and *M. testulalis* cause serious flower damage. *M. testulalis* and other podborers and large Hemiptera cause heavy damage to developing pigeon-pea

pods. Soya beans and lima beans in tropical Africa also appear to have few insect problems except for Hemiptera which damage developing pods and seeds of all legume species evaluated. Workers in the USA have also shown significant differences in susceptibility to insects among various genera and species of grain legumes (Campbell and Brett 1966; Wolfenbarger and Sleesman 1961*a*, 1961*b*).

10.4.3. Control of grain legume insect pests

Some examples of cultural and management practices in relation to insect problems are listed below:

1. Bean-fly damage can be reduced by creation of conditions favourable for plants to grow rapidly and gain sufficient robustness to overcome attack and by covering the base of plants with soil during weeding and cultivation to promote adventitious rooting on the hypocotyls (Appert 1964).

2. Weeds in fields promote increased insect damage to cowpeas and soya beans (IITA 1971, 1972*b*; Whitney 1972). Other authors (e.g. Schmutterer 1969) recommend clean weeding and prompt destruction of crop residues.

3. Intercropping of cowpeas with maize, sorghum, millet, etc., as practised by African farmers, is thought to reduce insect problems. This theory has been verified experimentally in northern Nigeria (Raheja, personal communication) but tests in the more humid southern part of Nigeria were inconclusive (IITA 1971; Whitney 1972).

4. Early planting of quick-maturing cowpea cultivars helps to avoid snout-weevil attack, since these insects become more numerous only near the end of the rainy season (Appert 1964). Daugherty *et al.* (1964) have also shown that soya bean cultivars which require longer development periods between flowering and maturity suffer greater damage from pentatomids than cultivars which mature quickly.

5. Prompt and regular harvesting as pods mature minimizes damage by *Laspeyresia ptychora*, snout weevils, and bruchids (Appert 1964; Prevett 1961; IITA 1971; Whitney 1972).

6. Since bruchids oviposit on pods which the newly hatched larvae must penetrate to attack the seeds, prompt threshing and disposal of hulls should reduce seed damage (Appert 1964).

7. Legume pests in tropical Africa have a number of natural enemies (Table 10.14) which undoubtedly exercise a significant level of control. This is suggested by the fact that cowpea seed

TABLE 10.13

Some examples of damage and losses associated with grain-legume plant pests

Crops and pests	Damage and loss estimates	Location	References†
COWPEAS			
Bruchids	Field infestation: 2 per cent of seeds; 1 insect/seed = 3–5 per cent wt. loss; 3 or more insects/seed = reduced germination and viability	Uganda, Nigeria	3
Complex	10–60 per cent yield loss	West Africa	1
Complex	Yields increased 24–47 per cent by insect control	Uganda	11
Complex	Farmers' yields = 400–500 kg/ha. Improved varieties with insect control = 1200–2000 kg/ha	West Africa	5
Complex	Six- to ninefold increase in yield with insect control	Nigeria	14, 15, 16
Complex	Yields increased by as much as twenty-sevenfold with insect control	Ibadan, Nigeria	7, 8, 17
Complex	50–90 per cent of seeds from unprotected plants unfit for local market	Ibadan, Nigeria	7
Hemiptera	Pod damage 11–93 per cent	Northern Nigeria	2
Hemiptera	15–50 per cent damaged seeds	Ibadan, Nigeria	7, 8, 17
Laspeyresia ptychora	25 per cent weight loss in 100 per cent damaged seeds; 2–67 per cent damaged seeds	Ibadan, Nigeria	7, 8, 14–17
Maruca testulalis	Pod damage 5–94 per cent. Average of 3·2 seeds destroyed per pod	Northern Nigeria	2
Maruca testulalis	Pod damage 36 per cent	Uganda	11
Maruca testulalis	Damage: flowers 19–100 per cent; pods 15–31 per cent; seeds 5 per cent	Ibadan, Nigeria	7, 8, 17
Maruca testulalis	One larva can destroy 3–4 flowers; up to 30 per cent pod damage	Ibadan, Nigeria	14, 15, 16
Melanagromyza chalcosoma	13 per cent of seeds damaged. If damaged seeds germinate, stunted plants result	Uganda	10
Melanagromyza spencerella	Average of 5 per cent of seeds damaged	Ibadan, Nigeria	17
Piezotrachelus varius	Pod damage 13–50 per cent. Average of 3·8 seeds destroyed per pod	Northern Nigeria	2
Thrips	50 per cent yield loss in tests (in greenhouse)	Ibadan, Nigeria	7, 17, 18
LIMA BEANS			
Hemiptera	52 per cent seeds damaged	Ibadan, Nigeria	8
Laspeyresia ptychora	5 per cent seeds damaged	Ibadan, Nigeria	8
Leafhoppers	Yields reduced by 64–326 kg/ha	USA	6
PIGEON PEAS			
Complex	Yields increased fivefold with insect control	Ibadan, Nigeria	14–16
Hemiptera	31 per cent seeds damaged. Fewer, smaller and lower quality seeds	Ibadan, Nigeria	8, 12
Laspeyresia ptychora	6 per cent seeds damaged.	Ibadan, Nigeria	8
Maruca testulalis	22 per cent flowers damaged; 8 per cent pods damaged	Ibadan, Nigeria	7, 8
SOYA BEANS			
Hemiptera	36–47 per cent seeds damaged	Ibadan, Nigeria	8
Laspeyresia ptychora	5–9 per cent seeds damaged	Ibadan, Nigeria	8
Pentatomids	Fewer good seeds, smaller seeds with lower viability and 7 per cent less oil	USA	4, 9
Pentatomids	9 per cent yield loss. Price discounted for > 2 per cent damage. One bug per metre row justifies treatment	USA	13

† 1 = Appert (1964), 2 = Booker (1965*b*), 3 = Booker (1967), 4 = Daugherty *et al.* (1964), 5 = Delassus (1970), 6 = Eckenrode and Ditman (1963), 7 = IITA (1971), 8 = IITA (1972*b*), 9 = Jensen and Newsom (1972), 10 = Koehler and Mehta (1970), 11 = Koehler and Mehta (1972), 12 = Materu (1970), 13 = Miner (1966), 14 = Taylor (1964*b*), 15 = Taylor (1967*b*), 16 = Taylor (1971*b*), 17 = Whitney (1972), 18 = Whitney and Sadik (unpublished).

TABLE 10.14

Some insect parasites and predators which attack pests of grain legumes in tropical Africa†

Parasites and predators	Host insects	References‡
COLEOPTERA		
Coccinellidae	Aphids, scale insects, thrips, etc.	2, 5, 9
DIPTERA		
Syrphidae	Aleyrodids, aphids scale insects, thrips, etc.	2, 5, 9
Tachinidae		
Cadurcia auratocauda (Curr.)	*Sylepta derogata* (F.)	2, 6
Carcelia evolans Wied.	*Diacrisia* sp. and *Spodoptera* sp.	1, 6
Chlorolydella spp.	*Plusia* sp.	5
Drino sp.	*Diacrisia* sp.	6
Exorista (=*Tachina*) *fallax* Mg.	*Amsacta flavizonata* Hamp.	1
Hemiwinthemia sp.	*Maruca testulalis* Geyer	1
Linnaemyia spp.	*Plusia* sp.	5
Neoplectops mudinerva (Mesnil)	*Maruca testulalis* Geyer	1
Sturmia sp.	*Plusia* sp., *Diacrisia* sp., etc.	6
Thelairia spp.	*Diacrisia* spp.	6
Thelairosoma sp.	*Maruca testulalis* Geyer	1, 6
Voria capensis Vill.	*Plusia* sp.	1, 6
HEMIPTERA		
Anthocoridae		
Orius albidipennis (Reut.)	*Caliothrips* spp.	9
Lygaeidae		
Geocoris amalbis Stål	*Lygus* spp.	5
Miridae		
Dreaeocoris sp.	Aphids, mirids, scale insects	5
Trichophthalmocapsus jamesi China	Scale insects	5
Pentatomidae		
Macrorhaphis spurcata Walk.	*Chloridea armigera* (Hubn.), *Spodoptera* spp.	8
Reduviidae		
Various genera and spp.	Various genera and spp.	2, 5
HYMENOPTERA		
Aphelinidae		
Eretmocerus sp.	Aleyrodids	2
Prospaltella sp.	*Bemisia* spp.	2
Bethylidae		
Goniozus sp.	*Cryptophlebia williamsi* Bradley	4
Proprops nasuta Wtstn.	*Hypothenemus hampei* (Ferr.)	5
Braconidae		
Agathis sp.	*Maruca testulalis* Geyer	1
Apanteles spp.	*Diacrisia* sp., *Sylepta derogata* (F.)	2, 5, 6
Bracon hebetor Say	*Etiella zinckenella* (Treit.), *Maruca testulalis* Geyer, *Lampides* sp.	2, 4
Braunsia analis Kreichb.	*Maruca testulalis* Geyer	1
Chelonus sp.	*Cryptophlebia williamsi* Bradley, *Etiella zinckenella* (Treit.)	4
Cremnops sp.	*Sylepta derogata* (F.)	6
Euphorus spp.	Various mirids	2, 5
Heterospilus cuffeicola Schmied.	*Hypothenemus hampei* (Ferr.)	5
Meteoridea sp.	*Lamprosema indicata* F.	6
Meteoridea sp.	*Sylepta derogata* (F.)	6
Meteorus sp.	Noctuids, pyralids, tortricids	2, 6
Microgaster sp.	Lymatriids, noctuids	2, 6
Opius melanagromyzae Fischer	*Melanagromyza* spp.	3, 4, 11
Opius sp.	*Melanagromyza phaseoli* (Tryon)	3, 5
Phanaerotoma sp.	*Maruca testulalis* Geyer	1

TABLE 10.14 (*cont.*):

Parasites and predators	Host insects	References‡
HYMENOPTERA (*cont.*):		
Ceraphronidae		
Ceraphron dictynna Wtstn.	*Hypothenemus hampei* (Ferr.)	5
Chalcididae		
Brachymeria spp.	*Diacrisia* sp., *Plusia* sp., *Sylepta* sp.	6
Encyrtidae		
Leptomastix sp.	*Plusia* spp.	6
Eucoilidae		
Eucoilididea sp.	*Melanagromyza spencerella* Greathead, *M. vignalis* Spencer, *Ophiomyia centrosematis* (de Meij.)	3, 10
Gronotoma sp.	*M. vignalis* Spencer	1, 5
Eulophidae		
Achrysocharis sp.	*Melanagromyza* spp.	5
Chrysocharis sp.	*Melanagromyza* spp.	5
Closterocerus africanus Wtstn.	*Melanagromyza* spp.	5
Entedon spp.	*Piezotrachelus varius* Wagn.	1, 7
Euderus sp.	*Melanagromyza* spp.	10
Euplectrus nigroclaypeatus Ferr.	Various noctuids	6
Pedobius sp.	*Epilachna* sp.	6
Pedobius sp. nr. *vignae* Risb.	*Melanagromyza vignalis* Spencer	10
Teleopterus violaceus Ferr.	*Melanagromyza* spp.	5
Tetrastichus ovulorum Ferr.	*Epilachna* sp.	5
Trichospilus pupivora Ferr.	*Cryptophlebia williamsi* Bradley	4
Eupelmidae		
Bruchocida vuilleti Crawf.	Bruchids and apion weevils	7
Eupelmus sp.	Bruchids and apion weevils	7
Eurytomidae		
Eurytoma sp.	Bruchids and apion weevils	7
Gen. nr. *Eurytoma*	*Melanagromyza* spp.	10
Formicidae		
Dorylus sp.	*Maruca testulalis* Geyer	11
Ichneumonidae		
Charops sp.	*Lamprosema indicata* (F.)	2
Enicospilus spp.	Various arctiids and noctuids	6
Horogenes sp.	*Cupido cissus* Gdt.	1
Netelia rufescens Tosq.	*Plusia* sp.	5
Phaenolobus alcides Wlkn.	*Alcidodes* sp.	5
Pimpla spp.	*Plusia* sp.	5
Theronia melanocera Hlmg.	*Plusia* sp.	5
Xanthopimpla stemmator (Thunb.)	*Cryptophlebia williamsi* Bradley	4
Pteromalidae		
Anisopteromalus calandrae (How.)	Bruchids and apion weevils	1, 7
Dinarmus laticeps Ashm.	*Callosobruchus maculatus* (F.)	1
Habrocytus sp.	*Melanagromyza vignalis* Spencer	10
Mesopolobus (=*Xenocrepis*) sp.	*Piezotrachelus varius* Wagn.	1
Sphegigaster sp.	*Melanagromyza* sp.	5
Trigonastra sp.	*Melanagromyza* sp.	5
Trichogrammatidae		
Trichogramma sp.	Bruchids	7
NEUROPTERA		
Chrysopidae, Hemerobiidae	Aphids, scale insects, thrips, etc.	5, 9

† See also Tables 10.10 and 10.17.
‡ 1 = Booker (1965*a*), 2 = Forsyth (1966), 3 = Greathead (1969), 4 = Greathead (1971), 5 = Le Pelley (1959), 6 = Oboite (1968), 7 = Prevett (1961), 8 = Robertson (1969), 9 = Schmutterer (1969), 10 = Whitney (1972), 11 = Whitney (unpublished).

produced in plots treated with DDT or DDT-plus-lindane had a higher percentage *L. ptychora* damage than the smaller amount of seed produced in unsprayed check plots (IITA 1971; Whitney 1972) and by the increased level of podborer (*M. testulalis*) damage in plots sprayed with Dupont 1410, an insecticide known to be effective against Diptera and Hymenoptera but not against Lepidoptera (IITA 1972*b*).

As indicated by Taylor (1964*b*), it is doubtful whether cultural and biological control alone can solve the grain-legume pest problems. Therefore, considerable research emphasis has been placed on control by chemicals and more recently on host-plant resistance.

An example of the potential practical application of host-plant resistance in grain legumes was given by Painter (1958), who reported that as long as 20 years ago 50 per cent of the bush snap beans grown in North Carolina were cultivars resistant to the Mexican bean beetle, *Epilachna varivestis* Mulsant.

Resistance to leafhoppers in soya beans is associated with pubescence (Wolfenbarger and Sleesman 1963), the inheritance of which has been studied by Bernard and Singh (1969). Broersma, Bernard, and Luckmann (1972) have shown that the orientation of plant hairs on soya beans is more important than the number of hairs. They found that normal pubescence had a greater deterrent effect on leafhoppers than curly hairs. Insects other than leafhoppers, however, were not affected by pubescence (Kogan 1972). Beck (1965) reported that soya beans with hairy pods were more heavily damaged by the pod borer, *Grapholitha glicinivorella* Matsumura, which preferred to oviposit on hairy pods than on glabrous pods. Differential susceptibility among soya bean cultivars to other insects and spider mites has been demonstrated. (Clark, Harris, Maxwell, and Hartwig 1972; van Duyn 1972; van Duyn, Turnipseed and Maxwell 1971; Jensen 1972; Wolfenbarger and Sleesman 1961*a*).

Until recently most of the work on insect resistance in cowpeas was concerned with the circulionid, *Chalcodermus aeneus* Boheman, in the southern part of the USA. Significant differences in levels of susceptibility to circulionids among cowpea lines and cultivars have been shown for *C. aeneus* in the USA (Brantley and Dempsey 1955; Chalfant, Suber, and Canerday 1972; Cuthbert and Chambliss 1972; Todd and Canerday 1968; Wolfenbarger and Correa

1963), and for the related curculio *Piezotrachelus varius* in West Africa by Tardieu (1961), who found that coloured seeds were damaged less than white seeds. Factors reported to be associated with resistance include different levels of unidentified chemicals encouraging or discouraging feeding on both pods and seedlings (Canerday and Chalfant 1969; Cuthbert and Davis 1972), toxic factors in hulls (Chalfant, Suber, and Canerday 1972), and differences in pod-wall thickness—thicker hulls being less successfully penetrated (Chalfant, Suber, and Canerday 1972; Cuthbert and Davis 1972). Sources of resistance to curculionids were identified in wild species of *Vigna* and in *V. unguiculata* var. *sesquipedalis* (L.) Fruw. (Cuthbert and Chambliss 1972; Tardieu 1961). Other chemical factors possibly associated with resistance in legumes include saponins (Applebaum, Marco, and Birk 1969; Su, Speirs, and Mahany 1972) and the amino acid L-DOPA (L-dihydroxyphenyl-alanine) (Bell and Janzen 1971; Rehr, Janzen, and Feeny 1973). A cowpea cultivar called 'Polon Me' was reported as being resistant to aphids and drought in Ceylon (Anon. 1962). Bean-fly resistance in cowpeas was reported by Fernando (1941). In India strains of various pulse crops are being screened for insect resistance. Several cowpea lines appear promising against a chrysomelid leaf beetle and the leafhopper, *Empoasca kerri* Pruthi (Saxena, Kumar, and Prasad 1969).

Screening is under way in West Africa to detect insect resistance in a world collection of cowpeas which contains more than 4000 accessions (IITA 1972*b*). Good resistance to the foliage-damaging thrips, *Sericothrips occipitalis*, has been discovered, and prospects of finding resistance to pod pests are encouraging (IITA 1971, 1972*b*).

10.4.4. Chemical control of cowpea pests

The use of insecticides is still the only practical means of ensuring a profitable return from intensive cultivation of cowpeas and other grain legumes in tropical Africa. Research has been concentrated on conventional high-volume aqueous sprays. The most promising compounds and suggested dosage rates for sprays, ULV formulations, and dusts are listed in Table 10.15.

Control of insect pests of legumes must be based on preventive rather than curative treatments, because the timing of treatments in relation to the pests observed in any individual field is still not

TABLE 10.15

Some insecticides found useful as sprays and dusts for control of grain-legume pests

Insecticide	Formulations†	Dosage rates (kg a.i./ha)	Remarks‡	References§
Abate(R)	EC	1·0	Broad spectrum. Experimental	11
Azinphos-M (Gusathion(R))	WP	0·6	Broad spectrum. Experimental	11
Carbaryl (Sevin(R))	WP, D, ULV	0·6–2·2	Variable results, phytotoxic. 1 day*	3, 11–14, 16, 17, 20, 22–24, 27, 28
Carbofuran (Furadan(R))	WP	0·6	Broad spectrum. Experimental. See text re seed-treatment and soil systemics	11
Carbophenothion (Trithion(R))	EC	0·8	Controls spidermites and some insects. 7 days*	16
Chlorpyrifos (Dursban(R))	EC	0·6	Broad spectrum. Experimental	11
DDT	WP, EC, D	0·6–3·3	Does not control aphids or *Laspeyresia ptychora*. 7 days*	2, 7–9, 11, 13, 14, 16, 17, 22–24, 29
DDT+lindane	EC	1·1+0·1	Does not control aphids or *Laspeyresia ptychora*	3, 10, 11, 22, 23
Diazinon	EC, WP	0·6–1·1	Variable results. Does not control larger Hemiptera or *Maruca* at lower dosage. 7 days*	2, 11, 24, 29
Dicofol (Kelthane(R))	EC, WP	0·7	Controls mites only. 7 days*	24, 29
Dimethoate	EC, ULV	0·3–1·2	Controls only sucking insects at low rate	1, 11, 17, 21, 25, 27
Dupont 1410(R)	EC	0·6	May increase Lepidoptera damage. Experimental. Very effective against sucking insects	11
Endosulfan (Thiodan(R))	EC, WP	0·5–1·0	Broad spectrum	4, 10, 11, 17, 19, 25
Endrin	EC, WP	0·09	Bean-fly control only at this rate. Too hazardous to use at higher rates for control of other pests	15
Fenitrothion (Sumithion(R))	EC	1·1	Failed to control *Chloridea armigera*	4, 11, 13, 14
Gardona(R)	EC, WP, ULV	0·3–1·2	Failed to control aphids and Hemiptera at 0·6 kg or less	10, 11
Iodofenfos	ULV	?	Good control of curculio *Chalcodermus aeneus* Boh.	6
Lindane	WP, EC	0·3–0·6	Variable results	11, 22, 23
Lindane	ULV	0·6		11
Lindane	D 0·65%	?	Dust cotyledons for bean fly control	7
Lindane+dimethoate	EC	1·0+0·15	Broad spectrum	10, 26
Lindane+dimethoate	EC, ULV	0·6+0·3	Broad spectrum	11
Malathion	WP, EC, D	1·0–1·8	Variable results vs *Maruca*. WP better than EC. 3 days*	2, 16, 24, 29
Malathion	ULV	1·0	Variable results vs *Maruca*. Good curculio control	6, 11
Methomyl (Lannate(R))	SP	0·6	Broad spectrum. Experimental	11
Methoxychlor	WP, EC, D	1·0–1·8	Variable results at low rate. 7 days*	11, 16, 24, 27
Orthene(R)	SP	0·6	Broad spectrum. Experimental	11
Pirimiphos-M (Actellic(R))	EC	0·6	Experimental	11
Toxaphene	WP, EC, D	2·0–4·5	1 day (if only dry seeds are used)	5, 16, 24, 29
Toxaphene+DDT	EC, D, WP	2·8+1·4	Broad spectrum. 7 days*	16, 18, 24

† D = dust, EC = emulsifiable liquid concentrate, SP = soluble powder, ULV = non-aqueous ultra-low volume, WP = wettable powder.

‡ * = officially recommended in the United States; 1 day and 7 days = suggested minimum time between last treatment and harvest.

§ 1 = Appert (1964), 2 = Ayoade (1969), 3 = Booker (1965b), 4 = Delassus (1970), 5 = Dupree (1963), 6 = Dupree (1970a), 7 = Emery (1957), 8 = Farrell and Adams (1966), 9 = Hetrick (1946), 10 = IITA (1971), 11 = IITA (1972b), 12 = Jerath (1968), 13 = Koehler and Mehta (1972), 14 = Koehler and Rachie (1971), 15 = Lee (1962), 16 = North Carolina State University (1968), 17 = Ratcliffe, Ditman, and Young (1960), 18 = Riherd (1949), 19 = Robertson (1969), 20 = Schwartz, Osgood, and Ditman (1961), 21 = Shorey, Deal, and Snyder (1965), 22 = Taylor (1967b), 23 = Taylor (1968b), 24 = US Department of Agriculture (1968), 25 = Vaishampayan and Singh (1969), 26 = Whitney (1972), 27 = Wolfenbarger (1963), 28 = Wolfenbarger (1964), 29 = Wyniger (1968).

practical in tropical Africa (Appert 1964; Ayoade 1969; Delassus 1970). Since flower and pod damage by various insects invariably occurs in the absence of control measures, scheduled treatments should start at the time of blooming and be repeated every 5–7 days until the approach of maturity. Pre-flowering treatments may also be profitable if defoliating insects or thrips become numerous.

Seed-treatments with non-systemic, contact insecticides have been successfully used for many years. Some examples are listed below:

1. Diazinon 5 g a.i. per 100 g seed for control of seed-corn maggot, *Hylemya cilicrura* (Rondani) (McEwen and Davis 1965).
2. Aldrin, dieldrin, endrin, or heptachlor, 0·02–0·2 g a.i. per 100 g seed for protection of germinating seeds and control of bean flies (Appert 1964; Delassus 1970; Emery 1957; Walker 1960c; Wyniger 1968).

Recent research with systemic insecticides has shown that seedlings and young plants can be protected for 4–6 weeks by planting seeds treated with carbofuran or methomyl (Bowling 1968; Broersma and Luckmann 1967; IITA 1972b; McEwen and Davis 1965). Granular formulations applied to the soil also show promise against pre-flowering pests, as indicated below:

1. Aldicarb, 4·5 kg/ha, or carbofuran, 9 kg/ha, applied in seed furrow for nematode control gave, respectively, good control of thrips and *Spodoptera exigua* (Hubn.) on cowpeas (Chalfant and Johnson 1972).
2. Dimethoate, disulfoton, phorate, thionazin plus phorate, or carbofuran applied at sowing or as side-dressing to seedlings controlled thrips on lima beans and increased seed yields (Dupree 1970b).
3. Phorate or disulfoton, 2 kg a.i./ha, applied at sowing controlled chrysomelid leaf beetles on cowpeas and Asian *Vigna* in India for 6 weeks (Saxena, Kumar, and Prasad 1969) and gave good control of *Empoasca fabae* (Harris) and *Epilachna varivestis* Muls. on lima beans in the USA (Schwartz, Osgood, and Ditman 1961).
4. Carbofuran, 2·2 kg a.i./ha, or methomyl, 6·7 kg a.i./ha, applied as a side-dressing to soya beans in the USA gave good control of lepidopterous leafworms (Todd and Canerday 1972) and other soya bean pests (Turnipseed 1967).

5. Methomyl, Cytrolane[(R)], Cyolane[(R)], carbofuran, disulfoton, or phorate, 1 kg a.i./ha, applied in seed furrow gave 6 weeks protection of cowpea plants against a complex of chewing and sucking insects. Side-dressings at rates up to 2 kg a.i./ha at flowering time failed to protect flowers and pods (IITA 1972b).
6. Disulfoton granules, 1·1–2·2 kg a.i./ha, applied in bean-seed furrow is officially recommended in the USA (North Carolina State University 1968).

The choice of insecticide and method of application must be based on several factors, such as toxicity and hazards to man and other non-target animals, phytotoxicity, insect species involved, application equipment available, history of treatments, and cost and availability of the chemicals in the local market. Some insecticides which are recommended for use on grain legumes in temperate regions are not suitable for use in the tropics. For example, carbaryl wettable powder applied at recommended rates has produced severe phytotoxic reactions on cowpeas, pigeon peas, and soya beans (IITA 1972b; Koehler and Rachie 1971; Taylor 1968b). Emulsifiable concentrate formulations of many different insecticides caused slight to severe distortion of trifoliate cowpea leaves when young plants with developing leaf buds were sprayed (IITA 1972b). Hexachlorocyclohexane (BHC or HCH) imparted an undesirable 'off' flavour which persisted in processed cowpeas, but lindane (γ-BHC) left no taint (Wene and Griffiths 1954). The length of time that an effective insecticide can be used in a specific area before resistance develops to the product is also a matter of concern. Resistance of the cowpea curculionid *Chalcodermus aeneus* Boheman to the chlorinated hydrocarbon insecticides was reported in the USA more than 10 years ago (Wolfenbarger and Schuster 1963). The farmers' ultimate choice of an insecticide will probably be most heavily influenced by the immediate cost in relation to what he thinks he will gain by using it. Preliminary economic evaluations of cowpea spray trials made by Whitney and Hedley at IITA have shown excellent returns on the costs and high probabilities of making a profit (IITA 1971; Whitney 1972). The work of Taylor (1968c, 1969a) in integrated control of cowpea pests using *Bacillus thuringiensis* Berliner is noteworthy, but, as he pointed out, the immediate economics of using this selective microbial insecticide are unfavourable.

Finally, on the subject of the economics of using insecticides on tropical grain legumes, especially cowpeas, several workers including Booker (1965*b*), Koehler and Mehta (1972), Taylor (1964*b*), and Whitney (1972) have shown that insect control not only gives spectacular yield increases but also allows the crop to mature faster and in a more synchronous manner. This is advantageous both in terms of efficient land-use and harvesting.

10.5. Pests of root and tuber crops (cassava, sweet potatoes, yams)

10.5.1. Assessment of susceptibility and crop losses

Although roots and tubers are very important sources of carbohydrate foods in the humid tropics of Africa and suffer from numerous pests (Table 10.16),

TABLE 10.16

Some pests (other than armyworms, locusts, grasshoppers, and termites) of cassava, sweet potatoes, and yams in tropical Africa

Pest*	Host plants† and parts damaged	References‡
ACARINA		
Eriophyidae		
Aceria sp.	SP foliage	2
Tetranychidae		
Tetranychus spp.	*C, SP foliage (dry season)	5, 6, 18, 19
Mononychellus tanajoa Bondar	C foliage	3
COLEOPTERA		
Chrysomelidae		
Aspidomorpha spp.	SP, Y foliage	4, 10, 11, 13, 19
Chirida baumanni Spaeth.	SP foliage	10, 11
Crioceris livida Dalm.	Y foliage	4, 10
Laccoptera spp.	SP foliage	10, 11, 13
Coccinellidae		
Epilachna spp.	SP, Y foliage	4, 10
Curculionidae		
Alcidodes spp.	SP roots, leaves	11, 18
Blosyrus spp.	SP roots	11, 18
Cylas brunneus (F.)	SP roots, stems, leaves	2, 4, 10, 12
C. compressus Hartm.	SP roots, stems, leaves	11
C. formicarius (F.)	SP roots, stems, leaves	2, 4, 5, 6, 8, 13, 19
C. puncticollis Boh.	SP roots, stems, leaves	2, 4, 5, 6, 8, 10, 12, 18
Lagriidae		
Chrysolagria sp.	SP foliage	10
Lagria spp.	SP foliage	4, 10
*Scarabaeidae		
Heteroligus appius Burm.	Y tubers	9, 15, 16, 17
H. meles (Billb.)	Y tubers	4, 10, 15, 16, 17
Prionorcytes canaliculus Arrow	Y tubers	8, 10, 11, 12, 15-17
P. rufopiceus Arrow	Y tubers	10, 12, 15, 16, 17
DIPTERA		
Cecidomyidae (gall midges)	Y leaves, stems	1, 18
HEMIPTERA		
Coreidae		
Anoplocnemis tristator (F.)	C shoots	4
Pseudotheraptus devastans (Dist.)	C shoots	7
Miridae		
Halticus tibialis Reut.	SP foliage	10, 12, 19
Helopeltis spp.	SP, Y foliage	4, 11, 13

TABLE 10.16 (*cont.*):

Pest*	Host plants† and parts damaged	References‡
HOMOPTERA		
Aleyrodidae		
Bemisia spp.	*C, SP leaves (virus vectors)	2, 4–6, 8, 10–13, 19
Aphididae		
Macrosiphum euphorbiae Thomas	SP foliage (possibly virus vectors)	11
Cercopidae		
Ptyelus grossus F.	Y foliage, stems	4, 9, 18
Cicadellidae		
Empoasca facialis (Jac.)	SP foliage	13
Erythroneura spp.	SP foliage	11, 13, 19
Coccidae		
Coccus spp.	C foliage	11
Saissetia coffeae (Walk.)	C foliage	19
Diaspididae		
Aspidiella hartii (Ckll.)	Y tubers	12
Aspidiotus destructor Sign.	Y tubers	10, 11, 12
Hemiberlesia tectonae Lindinger	Y tubers	11
Pseudococcidae		
Aonidomytilus albus (Ckll.)	C foliage	4, 8, 11, 19
Ferrisiana virgata (Ckll.)	C foliage	4, 11, 12
Planococcus spp.	C, Y foliage	4, 11, 12, 19
Pseudococcus brevipes Ckll.	Y tubers	10, 11
LEPIDOPTERA		
Ageriidae		
Aegeria sp.	SP stemborer	11, 19
Tipulamima pyrostoma Meyr.	SP stemborer	2, 11
Hesperidae		
Tagiades flesus (F.)	Y foliage	11
Noctuidae		
Achasa sp.	SP foliage	11
Agrotis ipsilon Hubn.	SP seedlings	19
Nymphalidae		
Acraea acerata Hew.	SP foliage	2, 8, 11, 13
Pyralidae		
Eldana saccharina Walk.	C stemborer	14
Hymenia recurvalis (F.)	SP foliage	13
Saturniidae		
Bunea alcinoe Stoll	C leaves	10
Sphingidae		
Celerio lineata livornica (Esp.)	SP foliage	13
Hippotion celerio (L.)	SP foliage	11, 13
Herse (=*Agrius*) *convolvuli* (L.)	SP foliage	2, 4, 8, 10, 12, 13
Tortricidae		
Cryptophlebia (=*Argyroploce*) *batrachopa* Meyr.	Y	11
C. peltastica Meyr.	Y	11

* Major pests and plants attacked.
† C = cassava, SP = sweet potato, Y = yams.
‡ 1 = Barnes (1946), 2 = Buyckx (1962), 3 = Coursey and Booth (1976), 4 = Forsyth (1966), 5 = IITA (1971), 6 = IITA (1972c), 7 = INERA (1971), 8 = Jameson (1970), 9 = Jerath (1965c), 10 = Jerath (1967), 11 = Le Pelley (1959), 12 = Libby (1968), 13 = Schmutterer (1969), 14 = Simmonds (1971), 15 = Taylor (1964c), 16 = Taylor (1964d), 17 = Taylor (1971a), 18 = Whitney (unpublished), 19 = Wyniger (1962).

relatively little research has been done on the arthropod pests of these crops. Few quantitative damage and loss assessments have been made, but the following examples illustrate the magnitude of the pest problems:

1. White-fly-transmitted cassava mosaic infections may reduce yields by as much as 80 per cent (Beck 1971).
2. Sweet-potato weevil damage approaches 100 per cent in high-yielding clones, especially if harvest is delayed. The positive correlation between damage and yield appears to be related to cracking of soil and greater exposure of the larger roots (IITA 1972c; Pillai and Magoon 1969).
3. Yam-beetle damage to tubers ranges from 5 per cent to 70 per cent (always more than 20 per cent in the main production areas); insecticidal treatment of seed pieces or planting holes gives 20–77 per cent yield increases (yields often doubled) (Coursey 1967; Taylor 1971a).

Pests of root and tuber crops in tropical Africa have a number of natural enemies (Table 10.17), but

TABLE 10.17

Some insect parasites and predators which attack pests of cassava, sweet potatoes, and yams in tropical Africa†

Parasites and predators	Host insects	References‡
DIPTERA		
Cecidomyidae		
Schizobremia coffeae Barnes	*Planococcus kenyae* Le Pelley	3
Chamaemyiidae		
Leucopis (Leucopella) africana Mall.	*Planococcus* spp., *Ferrisiana virgata* (Ckll.)	3
Chloropidae		
Elachiptera vulgaris Adams	*Planococcus kenyae* Le Pelley	3
Dolichopodidae		
Medeterus normalis Curr.	*Planococcus kenyae* Le Pelley	3
Drosophilidae		
Gitonides perspicax Knab	*Planococcus kenyae* Le Pelley	3
Sarcophagidae		
Sarcophaga spp.	*Heteroligus* spp.	2, 4
Tachinidae		
Carcelia normula Curr.	*Acraea acerata* Hew.	3
Congochrysosoma fuscicata (Curr.)	*Achaea* sp.	1
Drino (Zygobothria) dilabida (Vill.)	*Herse convolvuli* (L.)	1
Drino (Zygobothria) dilabida (Vill.)	*Hippotion celerio* (L.)	3
Microthalma flavipes Mesnil	*Heteroligus* spp.	2, 4
Paratamiclea pallida Vill.	*Heteroligus* spp.	2, 4
Sturmia spp.	*Herse convolvuli* (L.)	1, 3
HYMENOPTERA		
Aphelinidae		
Aphytis chrysomphali (Merc.)	*Aspidiotus destructor* Sign.	1
Coccophagus spp.	*Coccus* spp., *Planococcus* spp., *Pseudococcus* spp., *Saissetia* spp.	1, 3
Encarsia lutea Masi	Aleyrodids	2
Eretmocerus mundus Merc.	Aleyrodids	2
Prospaltella sp.	*Bemisia* spp.	1, 2
Braconidae		
Apanteles acraea Wilkn.	*Acraea acerata* Hew.	3
A. coffea Wilkn.	*Cryptophlebia batrachopa* Meyr.	3
Metebrus lipsus Nixon	*Achasa* sp.	1
Chalcididae		
Brachymeria sp.	*Achaea* spp., *Aspidomorpha* spp.	3

TABLE 10.17 *(cont.)*:

Parasites and predators	Host insects	References[‡]
HYMENOPTERA *(cont.)*:		
Encyrtidae		
Acerophagus pallidus Timb.	*Ferrisiana virgata* (Ckll.)	1
Achrysopophagus aegyptiacus Merc.	*Planococcus kenyae* Le Pelley	3
Achrysopophagus aegyptiacus Merc.	*Pseudococcus* spp.	1, 3
Alphycus spp.	*Coccus* spp., *Saissetia* sp.	3
Anagyrus spp.	*Ferrisiana virgata* (Ckll.), *Pseudococcus* spp.	1, 3
Baeonusia oleae Silv.	*Saissetia* sp.	3
Cheiloneurus spp.	*Pseudococcus* spp., *Saissetia* sp.	3
Coccidoxenus spp.	*Saissetia* spp.	1, 3
Diversineurus spp.	*Saissetia* spp.	3
Encyrtus fulginosis Comp.	*Saissetia* spp.	3
Euaphycus spp.	*Coccus* spp.	3
Exicomps spp.	*Coccus* spp., *Saissetia* sp.	3
Gyranusa citrina Comp.	*Ferrisiana virgata* (Ckll.), *Planococcus* spp.	3
Leptomastidea spp.	*Ferrisiana virgata* (Ckll.), *Planococcus* spp.	3
Leptomastix spp.	*Ferrisiana virgata* (Ckll.), *Planococcus* spp., *Pseudococcus* spp.	1, 3
Mesopeltis truncatipennis (Wtstn.)	*Saissetia* spp.	1
Microterys spp.	*Saissetia* spp.	3
Neodiscodes martinii Comp.	*Ferrisiana virgata* (Ckll.), *Pseudococcus* spp.	1, 3
Pauridia perigrina Timb.	*Ferrisiana virgata* (Ckll.), *Pseudococcus* spp.	1, 3
Prochiloneurus sp.	*Planococcus kenyae* Le Pelley	3
Pseudoapycus spp.	*Ferrisiana virgata* (Ckll.), *Planococcus* spp.	1, 3
Tetracnemus sp.	*Ferrisiana virgata* (Ckll.), *Planococcus* spp.	1, 3
Thysanus sp.	*Planococcus kenyae* Le Pelley	3
Tropidiophryne melvillei Comp.	*Pseudococcus* spp.	3
Eulophidae		
Aprostocetus aspidomorphae Ferr.	*Aspidomorpha* sp.	3
Cassidocida africana Ferr.	*Aspidomorpha* sp.	3
Tetrastichus ovivorum Crawf.	*Aspidomorpha* sp.	3
Tetrastichus spp.	*Coccus* spp., *Pseudococcus* spp., *Saissetia* sp.	3
Eupelmidae		
Eupelmus saissetia Silv.	*Coccus africanus* Newst.	3
Ichneumonidae		
Mesochoris exploitus Wilkn.	*Aceraea acerata* Hew.	3
Pteromalidae		
Pachyneuron sp.	*Planococcus kenyae* Le Pelley	3
Scutellista cyanea Motsch.	*Coccus subhemisphaericus* Newst., *Saissetia* spp.	3
Scelionidae		
Telenomus brimo Nixon	*Herse convolvuli* (L.)	3
LEPIDOPTERA		
Lycaenidae		
Aslauga purpurascens Holl.	*Planococcus kenyae* Le Pelley	3
Spalgis lemolea Druce	*Planococcus kenyae* Le Pelley	3
Noctuidae		
Eublemma scitula Ramb.	*Coccus* spp., *Planococcus kenyae* Le Pelley	3
Ozopteryx basalis Saalm.	*Coccus* spp.	3

† See also Tables 10.10 and 10.14.
‡ 1 = Forsyth (1966), 2 = Gameel (1969), 3 = Jerath (1967), 4 = Le Pelley (1959), 5 = Taylor (1964*d*).

these beneficial insects appear to have little economical impact on the pests (Jerath 1967; Taylor 1971*a*). Likewise, attempts at cultural control have not met with much success except perhaps for sweet-potato weevils, where deep planting of non-infested cuttings, weed control in and around fields, early harvesting, prompt destruction of crop residues, and crop rotation is claimed to reduce weevil damage (Schmutterer 1969).

10.5.2. Host-plant resistance

Research on host-plant resistance of root and tuber crops to insects in the tropics is in its infancy. Differential susceptibility of certain cassava cultivars

USA (Cuthbert and Davis 1970, 1971); Cuthbert and Jones 1972). So far the most practical means of pest control on these crops appears to be the discriminate use of insecticides as indicated in Table 10.18.

10.6. Desirable future trends

The entomology of tropical food crops is so complex and extensive that the interdisciplinary-team research approach must be used to identify and remove the major constraints which all aspects of production, including pests, impose on food production. When the major constraints are removed, entomologists must be vigilant to changes in the 'pest status' of

TABLE 10.18

Some suggested insecticides and acaricides for control of pests on cassava, sweet potato, and yam plants

Pests	Crops†	Treatments and remarks	References‡
INSECTS			
COLEOPTERA			
Leaf beetles (Chrysomelidae)	SP, Y	Carbaryl WP: 0·25% a.i.	7
Weevils (*Cylas* spp.)	SP	DDT EC, WP: 0·1–2·5% a.i. dip slips before plantings	1, 8
	SP	Dieldrin D, WP: 0·8–1·0 kg a.i. per ha on soil at base of plants, repeat in 3–4 weeks at least 3 weeks before harvest	6, 8
	SP	Fenitrothion EC, WP: 0·1% a.i. spray monthly	2
Yam beetles (Scarabaeidae)	Y	Aldrin D: 0·1–0·2 g per seed piece before planting	1, 4, 5
		Aldrin D: 1 kg a.i. per ha in planting holes	5
HEMIPTERA			
Coreidae, Miridae, etc.	C, SP	Malathion D: 1·2–1·4 kg a.i. per ha	8
HOMOPTERA			
Leafhoppers (Cicadellidae)	C	Diazinon EC: 0·02% a.i.	8
Whiteflies (Aleyrodidae)	C, SP	Diazinon EC: 0·02% a.i.	8
		Dimethoate EC: 0·3–0·4 kg a.i. per ha	3
		Malathion EC: 0·04–0·05% a.i.	8
LEPIDOPTERA			
Leafworms	SP	Carbaryl WP: 0·4–0·5 kg a.i. per ha	8
		DDT EC, WP, D: 1–2 kg a.i. per ha	1, 8
ACARINA (mites)	C	Dicofol EC, WP: 0·7 kg a.i. per ha	8

† C = cassava, SP = sweet potato, Y = yams.

‡ 1 = Jameson (1970), 2 = Pillai and Magoon (1969), 3 = Schmutterer (1969), 4 = Taylor (1964*d*), 5 = Taylor (1971*a*), 6 = US Department of Agriculture (1968), 7 = Whitney (unpublished), 8 = Wyniger (1968).

to *Zonocerus variegatus* was mentioned in the section on grasshoppers. Observations in Nigeria indicated that some lines of cassava from Latin America may be resistant to tetranychid mites (IITA 1972*c*). While preliminary results of screening for resistance in sweet-potato lines to *Cylas* spp. in Nigeria have given promising leads (IITA 1971, 1972*c*), considerably more work on sweet-potato resistance to weevils and other pests has been carried out in the

those species which, although currently are of secondary or minor importance, may quickly become major pests under new systems of production.

A few of the most urgent entomological research and development needs in tropical Africa include:

1. Establishment of specific insect-related crop-loss factors and economic infestation thresholds (Chiarappa and Chiang 1971).

L

2. Elucidation of the interrelationships among arthropod vectors, host plants, and plant disease, with special emphasis on viruses and mycoplasmas (Maramorosch 1971).
3. Assessment of pest population dynamics and crop damage in relation to various farming practices, especially in relation to intercropping, multiple cropping, crop rotations, weed control, time of planting, etc.
4. Identification of the indigenous natural enemies of food-crop pests, measurement of their impact on the pests, and utilization of pest-control practices which will optimize the benefits of these parasites, pathogens, and predators (Koehler 1971).
5. In view of the apparent absence of hymenopterous parasites of aphids in West Africa, introduction of certain species of the Aphelinidae and Aphidiidae should be considered (Hagen and van den Bosch 1968).
6. Screening, breeding, selection, multiplication, and distribution of plant germplasm with resistance to major, widely distributed pests such as dipterous and lepidopterous stemborers of cereals, grain-legume pests (*Piezotrachelus aeneus*, *Maruca testulalis*, *Laspeyresia*, *ptychora*, bruchids, and several species of Hemiptera), sweet-potato weevils, and yam beetles. Concurrent with the more empirical development of resistant cultivars and lines, the plant factors associated with resistance should be identified so that further and more rapid progress can be made (Horber 1971; Pathak 1970).
7. Development of insecticidal programmes with a good margin of safety to man and other non-target organisms. These chemicals, formulations, dosage rates, methods, and timing of application must be compatible with the concepts of integrated control and pest management. Particular attention should be given to use of systemic compounds which are less affected by frequent rainfall, which quickly removes conventional pesticides from plant surfaces. Increased work with combinations of systemic seed-treatment insecticides and fungicides merits high priority. All insecticidal research must involve economic evaluations.

Finally, there is a great need for more intensive and more effective extension of improved pest-management technology to the farmer. Without this critical final step all the expensive entomological research and development efforts outlined above will be wasted.

Acknowledgements

Professor John Medler, University of Ife, Nigeria, kindly assisted in checking the scientific names of the insects.

References

ADEYEMI, S. A. O. (1967). In *Fifty years applied entomology in Nigeria*, pp. 13–19. Proc. Entomol. Soc. Nigeria. University of Ife, Ibadan, 57 pp.
—— (1969). Survival of stem borer populations in maize stubble. *Bull. entomol. Soc. Nigeria* 2, 16–22.
—— DONNELLY, J., and ODETOYINBO, J. A. (1966). Studies on chemical control of the stemborers of maize. *Nigerian agric. J.* 3, 61–6.
AHMED, S. M. and YOUNG, W. R. (1969). Field studies on the chemical control of the stem borer *Chilo partellus* on hybrid sorghum in India. *J. Econ. Entomol.* 62, 478–82.
ANON. (1962). New variety of cowpea, Polon Me, resistant to aphids and drought. Ceylon Dept. of Agriculture Report 1960. Cited in *Pl. Breed. Abstr.* 33, 101.
—— (1965). Les crioceres, insectes du mil en Afrique. Moyens de lutte. *Agron. Trop.*, Nogent 20 (1 suppl.), 53–4.
—— (1972). *The African armyworm*. East African Literature Bureau, Nairobi.
APPELBAUM, S. W., MARCO, S., and BIRK, Y. (1969). Saponins as possible factors of resistance of legume seeds to the attack of insects. *J. Agr. Food Chem.* 17, 618–22.
APPERT, J. (1964). Faune Parasitaire du Niebe (*Vigna unguiculata* (L.) Walp. = *V. catjang* (Burm.) Walp.) en Republique du Senegal. *Agron. Trop.*, Nogent 19, 788–94.
—— (1970). Les Insectes Nuisibles au Mais en Afrique et a Madagascar. Institut de Recherches Agronomiques a Madagascar, Division d'Entomologie Agricole. Doc. No. 223, Jan. 1970. 43 pp.
AYOADE, K. A. (1969). Insecticide control of the pod borer *Maruca testulalis* Gey. (Lepidoptera: Pyralidae), on Westbred cowpea (*Vigna* sp.). *Bull. entomol. Soc. Nigeria* 2, 23–33.
BAE, S. H. and PATHAK, M. D. (1969). Common leafhopper-planthopper populations and incidence of tungro virus in diazinon-treated and untreated rice plots. *J. Econ. Entomol.* 62, 772–5.
BAKKER, W. (1970). Rice yellow mottle, a mechanically transmissable virus disease of rice in Kenya. *Neth. J. Plant. Path.* 76, 53–63.
BARNES, H. F. (1946). *Gall midges of economic importance*, Vol. 1, p. 82. Crosby Lockwood and Son Ltd, London.
BARRY, D. (1972). Chemical control of sorghum shoot fly on a susceptible variety of sorghum in Uganda. *J. Econ. Entomol.* 65, 1123–5.

BATTEN, A. (1969). The Senegalese grasshopper *Oedaleus senegalensis* Krauss. *J. appl. Ecol.* **6**, 27-45.

BECK, B. D. A. (1971). Cassava production in West Africa. *Ford Foundation/IRAT/IITA seminar on root and tuber crops in West Africa, Ibadan.*

BECK, S. D. (1965). Resistance of plants to insects. *A. Rev. Entomol.* **10**, 207-32.

BEGUM, A. and EDEN, W. G. (1965). Influence of defoliation on yield and quality of soybeans. *J. Econ. Entomol.* **58**, 591-2.

BELL, E. A. and JANZEN, D. H. (1971). Medical and ecological considerations of L-Dopa and 5-HTP in seeds. *Nature, Lond.* **229**, 136-7.

BENNETT, S. E., JOSEPHSON, L. M., and BURGESS, E. E. (1967). Field and laboratory studies on resistance of corn to the corn earworm. *J. Econ. Entomol.* **60**, 171-3.

BERNARD, R. L. and SINGH, B. B. (1969). Inheritance of pubescence type in soybeans: glabrous, curly, dense, sparse, puberulent. *Crop. Sci.* **9**, 192-7.

BIGGER, H. (1966). The biology and control of termites damaging field crops in Tanganyika. *Bull. ent. Res.* **56**, 417-44.

BOCK, K. R. (1971). Notes on East African plant virus diseases. I. Cowpea mosaic virus. *E. Afr. agric. for. J.* **37**, 60-2.

BOOKER, R. H. (1965a). List of insect species found in association with cowpeas at Samaru. Samaru Misc. Paper No. 9, Institute of Agriculture Research, Ahmadu Bello University, Zaria, Nigeria.

—— (1965b). Pests of cowpea and their control in Northern Nigeria. *Bull. ent. Res.* **55**, 663-72.

—— (1967). Observations on three bruchids associated with cowpea in Northern Nigeria. *J. Stored Prod. Res.* **3**, 1-15.

BOWLING, C. C. (1968). Systemic insecticide seed treatment tests on soybeans. *J. Econ. Entomol.* **61**, 1224-7.

—— (1969). Tests with carbofuran to control insects on rice. *FAO int. Rice Comm. Newslett.* **18**, 29-35.

BRANTLEY, B. B. and DEMPSEY, A. H. (1955). *Southern pea varieties for middle Georgia.* Georgia Agric. Exp. St. Mimeo Ser. **7**.

BRENIÈRE, J. (1969). Importance des problems entomologiques dans le developpement de la riziculture de l'Afrique de l'Ouest. *Agron. Trop., Nogent* (Ser. Riz et Riziculture), **24**, 906-27.

—— (1970a). Les problemes des Lepidopteres foreurs des graminees en Afrique de l'Ouest. *Summaries of Papers, 7th International Congress on Plant Protection, 1970, Paris*, pp. 506-8.

—— (1970b). Economic problems caused by insects which infest sorghum in West Africa. *Ford Foundation/IRAT/IITA seminar on sorghum and millet, Bambey.*

—— (1971). Stalk borers of tropical cereal crops. *Ford Foundation/IRAT/IITA seminar on plant protection of tropical food crops, Ibadan.*

BROERSMA, D. B., BERNARD, R. L., and LUCKMANN, W. H. (1972). Some effects of soybean pubescence on populations of the potato leafhopper. *J. Econ. Entomol.* **65**, 78-82.

—— LUCKMANN, W. H. (1967). Seed treatment techniques and phytotoxicity studies on some grain and vegetable crops. *J. Econ. Entomol.* **60**, 821-3.

BROWN, E. S. (1962). *The African armyworm* Spodoptera exempta (*Walker*) (*Lepodoptera: Noctuidae*): *A review of the literature.* Commonwealth Institute of Entomology, London.

—— (1972). Armyworm control. *Pest Artic. News Summ.* (*PANS*) **B18**, 197-204.

—— MOHAMED, A. K. A. (1972). The relation between simulated armyworm damage and crop-loss in maize and sorghum. *E. Afr. agric. for. J.* **37**, 237-57.

BRYANT, G. E. (1942). Notes on the genus *Pagria* (Eumolpinae, Coleopt.). *Bull. ent. Res.* **33**, 31-4.

BUAHIN, G. K. A. (1971). Insect pests of maize and their control. Paper presented in a symposium *Maize, the wonder crop,* 1971, Faculty of Agriculture, University of Science and Technology, Kumasi, Ghana. (Mimeo)

BULLOCK, J. A. (1961). *Nematocerus* sp. (nr. *brevicornis* Hust.). A pest of cereals in Kenya. I. The bionomics and control of the adult and larva. *E. Afr. agric. for. J.* **27**, 24-32.

BUYCKX, E. J. E. (1962). *Precis des maladies et des insectes nuisibles recontres sur les plantes cultivees au Congo, au Rwanda et au Burundi.* INEAC, Bruxelles.

CAMPBELL, W. V. and BRETT, C. H. (1966). Varietal resistance of beans to the Mexican bean beetle. *J. Econ. Entomol.* **59**, 899-902.

CANERDAY, T. D. and CHALFANT, R. B. (1969). An arrestant and feeding stimulant for the cowpea curculio, *Chalcodermus aeneus* (Coleoptera: Curculionidae). *J. Georgia entomol. Soc.* **4**, 59-64.

CANTELO, W. W. and KOVITVADHI, K. (1967). Effectiveness of insecticides applied to root area of the rice plant in controlling the rice gall midge. *J. Econ. Entomol.* **60**, 109-11.

CHADA, H. L., ATKINS, I. M., GARDENHIRE, J. H., and WEIBEL, D. E. (1961). *Greenbug-resistance studies in small grains.* Texas Agric. Exp. Stn Bull. 982.

CHALFANT, R. B. and JOHNSON, A. W. (1972). Field evaluation of pesticides applied to soil for control of insects and nematodes affecting southern peas in Georgia. *J. Econ. Entomol.* **65**, 1711-13.

—— SUBER, E. F., and CANERDAY, T. D. (1972). Resistance of southern peas to the cowpea curculio in the field. *J. Econ. Entomol.* **65**, 1679-82.

CHANT, S. R. (1959). Viruses of cowpea, *Vigna unguiculata*, in Nigeria. *Ann. appl. Biol.* **47**, 565-72.

CHENG, C. H. and PATHAK, M. D. (1972). Resistance to *Nephotettix virescens* in rice varieties. *J. Econ. Entomol.* **65**, 1148-53.

CHIARAPPA, L. and CHIANG, H. C. (1971). The need, value and procurement of reliable crop loss information. *Ford Foundation/IRAT/IITA seminar on plant protection of tropical food crops, Ibadan.*

CLARK, W. J., HARRIS, F. A., MAXWELL, F. G., and HARTWIG, E. E. (1972). Resistance of certain soybean cultivars to bean leaf beetle, striped blister beetle, and bollworm. *J. Econ. Entomol.* **65**, 1669–72.

CLARKE, R. G. and WILDE, G. E. (1970a). Association of the green stink bug and the yeast-spot disease organism of soybeans. I. Length of retention, effect of molting, isolation from feces and saliva. *J. Econ. Entomol.* **63**, 200–4.

—— (1970b). Association of the green stink bug and the yeast-spot disease organism of soybeans. II. Frequency of transmission to soybeans, transmission from insect to insect, isolation from field population. *J. Econ. Entomol.* **63**, 355–7.

—— —— (1971). Association of the green stink bug and the yeast-spot disease organism of soybeans. III. Effect on soybean quality. *J. Econ. Entomol.* **64**, 222–3.

COURSEY, D. G. (1967). *Yams. An account of the nature, origins, cultivation and utilization of the useful members of the Dioscoreaceae.* Longmans, London.

—— BOOTH, R. H. (1976). Chapter 5 of present volume.

COUTIN, R. and HARRIS, K. M. (1969). The taxonomy, distribution, biology and economic importance of the millet grain midge, *Geromyia penniseti* (Felt), gen.n., comb.n. (Dipt., Cecidomyiidae). *Bull. ent. Res.* **59**, 259–73.

CUTHBERT, F. P. and CHAMBLISS, O. L. (1972). Sources of resistance to cowpea curculio in *Vigna sinensis* and related species. *J. Econ. Entomol.* **65**, 542–5.

—— DAVIS, B. W. (1970). Resistance in sweet potatoes to damage by soil insects. *J. Econ. Entomol.* **63**, 360–3.

—— —— (1971). Factors associated with insect resistance in sweet potatoes. *J. Econ. Entomol.* **64**, 713–17.

—— —— (1972). Factors contributing to cowpea curculio resistance in southern pea. *J. Econ. Entomol.* **65**, 778–81.

—— JONES, A. (1972). Resistance in sweet potatoes to Coleoptera increased by recurrent selection. *J. Econ. Entomol.* **65**, 1655–8.

DALE, W. T. (1949). Observations on a virus disease of cowpea in Trinidad. *Ann. appl. Biol.* **36**, 327–33.

—— (1953). Transmission of plant viruses by biting insects with particular reference to cowpea mosaic. *Ann. appl. Biol.* **40**, 73–81.

DAMOTTE, P. (1967). Service des avertissements Riz-Coton. Direction de la Protection des Vegetaux, Min. Agric., *Rep. Cote d'Ivoire, Rapport Biennal, 1966–1967 (Bouaké),* pp. 32–50.

DAUGHERTY, D. M. (1967). Pentatomidae as vectors of yeast-spot disease of soybeans. *J. Econ. Entomol.* **60**, 147–52.

—— NEUSTADT, M. H., GEHRKE, C. W., CAVANAH, L. E., WILLIAMS, L. F., and GREEN, D. E. (1964). An evalua-

tion of damage to soybeans by brown and green stink bugs. *J. Econ. Entomol.* **57**, 719–22.

DEEMING, J. C. (1971). Some species of *Atherigona* Rondani (Diptera, Muscidae) from Northern Nigeria, with special reference to those injurious to cereal crops. *Bull. ent. Res.* **61**, 133–90.

DELASSUS, M. (1970). Crop protection: Major problems observed and studied on edible legumes in various French speaking countries. *Ford Foundation/IRAT/IITA seminar on grain legume research, Ibadan.*

DE PURY, J. M. S. (1968). *Crop pests of East Africa.* Oxford University Press, Nairobi.

DESCAMPS, M. (1956). Deux dipters nuisibles au riz dans le Nord Cameroun—*Pachydiplosis oryzae* Wood-Mason, *Pachylophus* sp. aff. *lugens* Loew. *Phytiatrie-Phytopharm.* **5**, 109–16.

DIRSCH, V. M. (1965). *The African genera of Acridoidea.* Cambridge University Press and the Anti-Locust Research Centre, London.

DJAMIN, A. and PATHAK, M. D. (1967). Role of silica in resistance to Asiatic rice borer, *Chilo suppressalis* (Walker), in rice varieties. *J. Econ. Entomol.* **60**, 347–51.

DOUGLAS, W. A. and ECKHARDT, R. C. (1957). *Dent corn inbreds and hybrids resistant to corn earworm in the South.* US Department of Agriculture Technical Bulletin 1160.

DUPREE, M. (1963). *Studies on control of the cowpea curculio* (Chalcodermus aeneus *Boh.*). Georgia Agric. Exp. Stn Mimeo Ser. 167.

—— (1970a). Ultra-low-volume insecticide sprays for control of the cowpea curculio. *J. Georgia Entomol. Soc.* **5**, 39–41.

—— (1970b). Control of thrips and the bean leaf beetle on lima beans with systemic insecticides. *J. Georgia Entomol. Soc.* **5**, 48–52.

ECKENRODE, C. J. and DITMAN, L. P. (1963). An evaluation of potato leafhopper damage to lima beans. *J. Econ. Entomol.* **56**, 551–3.

EMERY, G. A. (1957). Bean fly. *E. Afr. Fmr Planter* **2**, 37.

ENDRODY-YOUNGA, S. (1968). The stem borer *Sesamia botanephaga* Tams & Bowden (Lep., Noctuidae) and the maize crop in central Ashanti, Ghana. *Ghana J. agric. Sci.* **1**, 103–31.

EVERLY, R. T. (1960). Loss in corn yield associated with the abundance of the corn leaf aphid, *Rhopalosiphum maidis* in Indiana. *J. Econ. Entomol.* **53**, 924–32.

FARRELL, J. A. K. and ADAMS, A. N. (1966). Cowpea and bean yield responses to insecticide. *Agric. Res. Council Central Africa, Ann. Rept. 1965,* p. 57.

FEAKIN, S. D. (ed.). (1970). *Pest control in rice. PANS Manual,* No. 3. Ministry of Overseas Development, London.

FERNANDO, H. (1941). The relative resistance of some cowpea varieties to *Agromyza phaseoli* Coq. *Trop. Agriculturist* **96**, 221–4.

FOOD AND AGRICULTURE ORGANIZATION (1970). *Crop loss assessment methods.* FAO, Rome.

FORSYTH, J. (1966). *Agricultural insects of Ghana.* Ghana University Press, Accra.

FOSTER, J. E. and DAUGHERTY, D. M. (1969). Isolation of the organism causing yeast-spot disease from the salivary system of the green stink bug. *J. Econ. Entomol.* **62**, 424-7.

GAMEEL, O. I. (1969). Studies on whitefly parasites *Encarsia lutea* Masi and *Eretmocerus mundus* Mercet (Hymenoptera: Aphelinidae). *Rev. Zool. Bot. Afr.* **79**, 65-77.

GREATHEAD, D. J. (1969). A study in East Africa of the bean flies (Dipt., Agromyzidae) affecting *Phaseolus vulgaris* and their natural enemies, with the description of a new species of *Melanagromyza* Hend. *Bull. ent. Res.* **59**, 541-61.

—— (1971). *A review of biological control in the Ethiopian region.* Commonwealth Inst. Biol. Contr. Tech. Communication No. 5, Slough.

GRIST, D. H. and LEVER, R. J. A. W. (1969). *Pests of rice.* Longmans, London.

HAGEN, K. S. and VAN DEN BOSCH, R. (1968). Impact of pathogens, parasites, and predators on aphids. *A. Rev. Entomol.* **13**, 325-84.

HARRIS, K. M. (1962). Lepidopterous stem borers of cereals in Nigeria. *Bull. ent. Res.* **53**, 139-71.

—— (1969). The sorghum midge. *Wld Crops* **21**, 176-9.

HARRIS, W. V. (1969). *Termites as pests of crops and trees.* Commonwealth Agricultural Bureau, London.

HASKELL, P. T. (1970). The future of locust and grasshopper control. *Outlook on agriculture* (Imp. Chem. Ind.) **6**, 166-74.

—— (1972). Recent progress in acridological research and its implications in the field of plant protection and agricultural development. *Acrida* **1**, 1-14.

HESSE, P. R. (1955). A chemical and physical study of the soils of termite mounds in East Africa. *J. Ecol.* **43**, 449-61.

HETRICK, L. A. (1946). Control of the cowpea curculio. *J. Econ. Entomol.* **39**, 268-9.

HOARE, B. G. (1968). Scientific note on *Zonocerus variegatus* damaging cassava. *Nigerian entomol. Mag.* **1**, 54.

HORBER, E. (1971). Host plant resistance to insects. *Ford Foundation/IRAT/IITA seminar on plant protection of tropical food crops, Ibadan.*

IARI (1971). *Investigations on insect pests of sorghum and millets.* (1965-1970) Final Tech. Report, Division of Entomology, Indian Agricultural Research Institute, New Delhi.

IITA (1971). *Annual Report, 1971.* International Institute of Tropical Agriculture, Ibadan, Nigeria.

IITA (1972a). Cereals improvement Program—Entomology. *Annual Report, 1972,* pp. 32-4. International Institute of Tropical Agriculture, Ibadan, Nigeria.

IITA (1972b). Grain legume improvement Program—

Entomology. *Annual Report, 1972,* pp. 28-40. International Institute of Tropical Agriculture, Ibadan, Nigeria.

IITA (1972c). Root, tuber and vegetable improvement Program. *Annual Report, 1972,* pp. 10-12, 18-22. International Institute of Tropical Agriculture, Ibadan, Nigeria.

INERA (1971). *Rapport Annuel Exercise 1971.* Institut National pour l'Étude et la Recherche Agronomique, Republique du Zaire.

INGRAM, W. R. (1958). The lepidopterous stalk borers associated with Graminae in Uganda. *Bull. ent. Res.* **49**, 367-83.

IRRI (1967). Chemicals against insects. *The IRRI Reporter* (International Rice Research Institute, Los Baños, Philippines) **3**, 1-3.

ISA, A. L. and BISHARA, M. A. (1966). *Chemical control of corn borers.* Plant Protection Department, Ministry of Agriculture, United Arab Republic.

JAMESON, J. D. (ed.) (1970). *Agriculture in Uganda.* 2nd edn. Oxford University Press, London.

JANSEN, W. P. and STAPLES, R. (1970). Transmission of cowpea mosaic virus by the Mexican bean beetle. *J. Econ. Entomol.* **63**, 1719-20.

—— —— (1971). Specificity of transmission of cowpea mosaic virus by species within the subfamily Galerucinae, family Chrysomelidae. *J. Econ. Entomol.* **64**, 365-7.

JENNINGS, P. R. and PINEDA, A. (1970a). Screening for resistance to the planthopper, *Sogatodes oryzicola* (Muir). *Crop Sci.* **10**, 687-9.

—— (1970b). Effect of resistant rice plants on multiplication of the planthopper, *Sogatodes oryzicola* (Muir). *Crop Sci.* **10**, 689-90.

JENSEN, R. L. (1972). Evaluation of soybean germ plasm for resistance to the bean leaf beetle *Cerotoma trifurcata* (Forster), the southern green stink bug *Nezata viridula* (L.) and the soybean looper, *Pseudoplasia includens* Walker. Ph.D. Dissertation, Louisiana State University, Baton Rouge, Louisiana.

—— and NEWSOM, L. D. (1972). Effect of stink bug-damaged soybean seeds on germination, emergence and yield. *J. Econ. Entomol.* **65**, 261-4.

JEPSON, W. F. (1954) *A critical review of the world literature on the lepidopterous stalk borers of tropical graminaceous crops.* Commonwealth Institute of Entomology, London.

JERATH, M. L. (1965a). Note on the biology of *Zonocerus variegatus* (L.) from Eastern Nigeria. *Rev. Zool. Bot. Afr.* **72**, 243-51.

—— (1965b). *Rice pests and their known parasites and predators from Nigeria.* Nigeria Dept. Agric. Res. Mem. No. 86.

—— (1965c). *Yam pests and their known parasites and predators in Nigeria.* Nigeria Dept. Agric. Res. Mem. No. 83.

JERATH, M. L. (1967). In *Fifty years applied entomology in Nigeria*, pp. 5-12. Proc. Entomol. Soc. Nigeria. University of Ife, Ibadan, 57 pp.

—— (1968). Insecticidal control of *Maruca testulalis* on cowpeas in Nigeria. *J. Econ. Entomol.* **61**, 413-16.

JOHNSTON, H. B. (1956). *Annotated catalogue of African grasshoppers.* Cambridge University Press and the Anti-Locust Research Centre, London.

—— (1968). *Supplement to annotated catalogue on African grasshoppers.* Cambridge University Press and the Anti-Locust Research Centre, London.

JOTWANI, M. G. and YOUNG, W. R. (1972). Recent developments on chemical control of insect pests of sorghum. In *Sorghum in seventies*, pp. 377-98. Oxford. University Press and IBN Publishing Co., New Delhi.

KAUFMANN, T. (1965). Observations on aggregation, migration, and feeding habits of *Zonocerus variegatus* in Ghana (Orthoptera: Acrididae) *Ann. entomol. Soc. Am.* **58**, 426-36.

KHAN, M. Q. (1967). Control of paddy stem borers by cultural practices. In *The major insect pests of the rice plant*, pp. 369-89. Johns Hopkins Press, Baltimore.

KINCADE, R. T., LASTER, M. L., and HARTWIG, E. E. (1971). Simulated pod injury to soybeans. *J. Econ. Entomol.* **64**, 984-5.

KOEHLER, C. S. (1971). Principles of natural and integrated control. *Ford Foundation/IRAT/IITA seminar on plant protection of tropical food crops, Ibadan.*

—— MEHTA, P. N. (1970). Effects of attack by *Melanagromyza chalcosoma* Spencer on germination and development of cowpea. *E. Afr. agric. for. J.* **36**, 83-7.

—— —— (1972). Relationships of insect control attempts by chemicals to components of yield of cowpeas in Uganda. *J. Econ. Entomol.* **65**, 1421-7.

—— RACHIE, K. O. (1971). Notes on the control and biology of *Heliothis armigera* (Hubn.) on pigeon pea in Uganda. *E. Afr. agric. for. J.* **36**, 296-7.

KOGAN, M. (1972). Feeding and nutrition of insects associated with soybeans. 2. Soybean resistance and host preferences of the Mexican bean beetle, *Epilachna varivestis. Ann. entomol. Soc. Am.* **65**, 675-83.

KOVITVADHI, K. and LEAUMSANG, P. (1971). Rice gall midge. Paper presented at the *International Rice Research Conference, IRRI, Philippines.*

KRISHNA, K. and WEESNER, F. M. (eds.) (1969 and 1970). *Biology of termites*, Vols. I and II, 598 and 643 pp. Academic Press, New York.

LANGHAM, R. M. (1968). Inheritance and nature of shoot fly resistance. Thesis. Ahmadu Bello University, Zaria, Nigeria.

LEE, S. Y. (1962). The mode of action of endrin on the bean stem miner. *Melanagromyza sojae*, with special reference to its translocation in soybean plants. *J. Econ. Entomol.* **55**, 956-64.

LE PELLEY, R. H. (1959). *Agricultural insects of East Africa.* East African High Commission, Nairobi.

LEVER, R. J. A. W. (1969). Do armyworms follow the rain? *Wld Crops* **21**, 351-2.

—— (1971). Losses in rice and coconuts due to insect pests. *Wld Crops* **23**, 66-7.

LIBBY, J. L. (1968). *Insect pests of Nigerian crops.* University of Wisconsin Res. Bull. 269.

MACCUAIG, R. D. (1963). Recent development in locust control. *World Rev. Pest Control* **2**, 7-17.

MCEWEN, F. L. and DAVIS, A. C. (1965). Tests with insecticides for seed-corn maggot control in lima beans. *J. Econ. Entomol.* **58**, 369-70.

MARAMOROSCH, K. (1971). The relationship between viruses and vectors. *Ford Foundation/IRAT/IITA seminar on plant protection of tropical food crops, Ibadan.*

MATERU, M. E. A. (1970). Damage caused by *Acanthomia tomentosicollis* Stal and *A. horrida* Germ. (Hemiptera, Coreidae). *E. Afr. agric. for. J.* **35**, 429-35.

MEHTA, P. N. (1971). The effect of defoliation on seed yield of cowpeas (*Vigna unguiculata*) (L.) Walp.) and analysis of the leaf harvest for dry matter and nitrogen content. *Acta Hort. (International Society for Horticultural Science Tech. Comm.)* **21**, 167-71.

MINER, F. D. (1966). *Biology and control of stink bugs on soybeans.* Arkansas Agric. Exp. Stn Bull. 708.

NICKEL, J. L. (1964). *Biological control of rice stem borers: A feasibility study.* International Rice Research Institute Tech. Bull. 2. Las Baños, Philippines.

NORTH CAROLINA STATE UNIVERSITY. (1968). *Pesticide Manual.* Division of Continuing Education, North Carolina State University, Raleigh.

NYASALAND DEPARTMENT OF AGRICULTURE (1961). Control of maize stem borer *Busseola fusca. 1959-1960 Report of Department of Agriculture*, pt. 2, p. 26.

NYIIRA, Z. M. (1970). Infestation of cereals and pulses in the field by stored products insects and two new records of stored products Coleoptera in Uganda. *E. Afr. agric. for. J.* **35**, 411-13.

OBOITE, A. U. (1968). *An index of parasites of agricultural insects of the Ibadan area.* Occ. Pub. entomol. Soc. Nigeria, Ibadan. No. 1.

OKWAKPAM, B. A. (1967). Three species of thrips (Thysanoptera) in cowpea flowers in the dry season at Badeggi, Nigeria. *Nigerian entomol. Mag.* **1**, 45-6.

—— (1971). Outbreak of blister beetles in rice fields at Badeggi. *Nigerian entomol. Mag.* **2**, 100-2.

ONAZI, O. C. (1968). Locust control in Nigeria. In *Fifty years applied entomology in Nigeria*, pp. 88-93. Proc. Entomol. Soc. Nigeria. University of Ife, Ibadan, 57 pp.

PAINTER, R. H. (1951). *Insect resistance in crop plants.* Macmillan, New York.

—— (1958). Resistance of plants to insects. *A. Rev. Entomol.* **3**, 267-90.

PALIWAL, Y. C. (1971). Increased susceptibility to maize borer in maize mosaic virus-infected plants in India. *J. Econ. Entomol.* **64**, 760-1.

PATANAKAMJORN, S. and PATHAK, M. D. (1967). Varietal

resistance of rice to the Asiatic rice borer, *Chilo suppressalis* (Lepidoptera: Crambidae), and its association with various plant characters. *Ann. entomol. Soc. Am.* **60**, 287-92.

PATHAK, M. D. (1967). Significant developments in rice stem borer and leafhopper control. *Pest. Artic. News Summ. (PANS)* A **13**, 45-60.

—— (1968). Ecology of common insect pests of rice. *A. Rev. Entomol.* **13**, 257-94.

—— (1969). Stem borer and leafhopper-planthopper resistance in rice varieties. *Entomol. exp. appl.* **12**, 789-800.

—— (1970). Genetics of plants in pest management. In *Concepts of pest management* (eds R. L. Rabb and F. E. Guthrie), pp. 138-57. North Carolina State University Press, Raleigh.

—— ANDRES, F., GALACGAC, N., and RAROS, R. (1971). *Resistance of rice varieties to striped rice borers*. International Rice Research Institute Tech. Bull. No. 11. Las Baños, Philippines.

PHIPPS, J. (1970). Notes on the biology of grasshoppers (Orthoptera: Acridoidea) in Sierra Leone. *J. Zool. (Lond.)* **161**, 317-49.

PILLAI, K. S. and MAGOON, M. L. (1969). Studies on chemical control measures for sweet potato weevil *Cylas formicarius* F. *Ind. J. Hort.* **26**, 202-8.

PREVETT, P. F. (1961). Field infestation of cowpea (*Vigna unguiculata*) pods by beetles in the families Bruchidae and Curculionidae in Northern Nigeria. *Bull. ent. Res.* **52**, 635-45.

RACHIE, K. O. and WURSTER, R. T. (1971). The potential of pigeon pea (*Cajanus cajan* Millsp.) as a horticultural crop in East Africa. *Acta Hort.* (*International Society for Horticultural Science Tech. Comm.*) **21**, 172-8.

RATCLIFFE, R. H., DITMAN, L. P. and YOUNG, J. R. (1960). Field experiments on the insecticidal control of insects attacking peas, snap and lima beans. *J. Econ. Entomol.* **53**, 818-20.

REDDY, D. B. (1967). The rice gall midge *Pachydiplosis oryzae* (Wood-Mason). In *The major insect pests of the rice plant*, pp. 457-91. Johns Hopkins Press, Baltimore.

REHR, S. S., JANZEN, D. H. and FEENY, P. O. (1973). *L*-dopa in legume seeds: a chemical barrier to insect attack. *Science, N.Y.* **181**, 81-2.

RIHERD, P. T. (1949). Chlorinated insecticides for control of cowpea insects. *J. Econ. Entomol.* **42**, 991-2.

ROBERTSON, I. A. D. (1969). Insecticide control of insect pests of soya bean (*Glycine max* (L.)) in eastern Tanzania. *E. Afr. agric. for. J.* **35**, 181-4.

SANDS, W. A. (1962). The evaluation of insecticides as soil and mound poisons against termites in agriculture and forestry. *Bull. ent. Res.* **53**, 179-92.

—— (1971). Termites as pests of tropical food crops. *Ford Foundation/IRAT/IITA seminar on plant protection of tropical food crops, Ibadan*.

SAXENA, H. P., KUMAR, S., and PRASAD, S. K. (1969). Screening of germplasm of pulse crops for insect resistance. *3rd Annual Workshop Conference on pulse crops, 1969*, pp. 138-42. Indian Agricultural Research Institute, New Delhi.

SCHEIBELREITER, G. (1971). Biological control projects in Ghana and West Africa. *Ford Foundation/IRAT/IITA seminar on plant protection of tropical food crops, Ibadan*.

SCHMUTTERER, H. (1969). *Pests of crops in Northeast and Central Africa with particular reference to the Sudan*. Gustav Fischer Verlag, Stuttgart.

SCHWARTZ, P. H., OSGOOD, C. E., and DITMAN, L. P. (1961). Experiments with granulated systemic insecticides for control of insects on potatoes, lima beans, and sweet corn. *J. Econ. Entomol.* **54**, 663-5.

SELLSCHOP, J. P. F. (1962). Cowpeas, *Vigna unguiculata* (L.) Walp. *Fld Crop Abstr.* **15**, 1-8.

SEPSAWADI, P., MEKSONGSEE, B., and KNAPP, F. W. (1971). Effectiveness of various insecticides against a sorghum shoot fly. *J. Econ. Entomol.* **64**, 1509-11.

SERGHIOU, C. S. (1971). Laboratory and field evaluation of insecticides against *Spodoptera littoralis* larvae. *J. Econ. Entomol.* **64**, 115-16.

SHASTRY, S. V. S. and SESHU, D. V. (1971). Current status of research in India on rice gall midge resistance. AICRIP Pub. No. 25 presented at the *International Rice Research Conference, 1971, IRRI, Philippines*.

SHOREY, H. H., DEAL, A. S., and SNYDER, M. J. (1965). Insecticidal control of lygus bugs and effect on yield and grade of lima beans. *J. Econ. Entomol.* **58**, 124-6.

SIMMONDS, F. J. (1971). Commonwealth Institute of Biological Control report on work carried out during 1970. Trinidad.

SINGH, S. R. et al. (1968). Resistance to stem borer, *Chilo zonellus* (Swinhoe), and stem fly, *Atherigona varia soccata* Rondani, in the world sorghum collection in India. *Mem. entomol. Soc. India* No. 7.

SMITH, C. E. (1924). Transmission of cowpea mosaic by the bean leaf beetle. *Science, N.Y.* **50**, 268.

SMITH, R. H. and BASS, M. H. (1972a). Soybean response to various levels of podworm damage. *J. Econ. Entomol.* **65**, 193-5.

—— —— (1972b). Relationship of artificial pod removal to soybean yields. *J. Econ. Entomol.* **65**, 606-8.

STARKS, K. J. (1969). *Some cereal crop insects in East Africa*. E. Afr. Agr. For. Res. Org., Serere Res. Sta., Uganda (Mimeo).

—— DOGGETT, H. (1970). Resistance to a spotted stem borer in sorghum and maize. *J. Econ. Entomol.* **63**, 1790-5.

—— SCHUMAKER, G., and EBERHART, S. A. (1971). Soil fertility and damage by *Chilo zonellus* to grain sorghum. *J. Econ. Entomol.* **64**, 740-3.

SU, H. C. F., SPEIRS, R. D., and MAHANY, P. G. (1972). Toxic effects of soybean saponin and its calcium salt on the rice weevil. *J. Econ. Entomol.* **65**, 844-7.

TAMS, W. H. T. and BOWDEN, J. (1953). A revision of the African species of *Sesamia* Guenee and related genera (Agrotidae-Lepidoptera). *Bull ent. Res.* **43**, 645–78.

TARDIEU, M. (1961). Imperatifs culturaux et dolique de Chine. *Agron. Trop.*, *Nogent* **16**, 387–92.

TAYLOR, T. A. (1964*a*). *Blaesoxipha filipjevi* Rohd. (Diptera, Sarcophagidae) parasitizing *Zonocerus variegatus* (L.) (Orthoptera, Acridoidea) in Nigeria. *Bull. ent. Res.* **55**, 83–6.

—— (1964*b*). The field pest problems on cowpeas (*Vigna sinensis* L.) in southern Nigeria. *Nigerian Grower and Producer* **3**, 1–4.

—— (1964*c*). Studies on the Nigerian yam beetles: I. Systematic notes on the common Nigerian yam beetle species (Coleoptera: Dynastidae). *J. W. Afr. Sci. Ass.* **8**, 180–9.

—— (1964*d*). Studies on the Nigerian yam beetles: II. Bionomics and control. *J. W. Afr. Sci. Ass.* **9**, 13–31.

—— (1965). Observations on the bionomics of *Laspeyresia ptychora* Meyr. (Lepidoptera, Encosmidae) infesting cowpea in Nigeria. *Bull. ent. Res.* **55**, 761–74.

—— (1967*a*). *Locusts and grasshoppers in Africa*. The text of a lecture given to the Philosophical Society, University of Ibadan, Nigeria (Mimeo).

—— (1967*b*). In *Fifty years applied entomology in Nigeria*, pp. 20–4. Proc. Entomol. Soc. Nigeria, University of Ife, Ibadan, 57 pp.

—— (1967*c*). The bionomics of *Maruca testulalis* Gey. (Lepidoptera: Pyralidae), a major pest of cowpea in Nigeria. *J. W. Afr. Sci. Ass.* **12**, 111–29.

—— (1968*a*). The control of armyworm (*Spodoptera exempta* (Wlk.) Lepidoptera: Noctuidae) using carbaryl and *Bacillus thuringiensis*. *Nigerian entomol. Mag.* **1**, 60–1.

—— (1968*b*). The effects of insecticide applications on insect damage and the performance of cowpea in southern Nigeria. *Nigerian agric. J.* **5**, 29–37.

—— (1968*c*). The pathogenicity of *Bacillus thuringiensis* var. *thuringiensis* Berliner for larvae of *Maruca testulalis* Geyer. *J. Invert. Path.* **11**, 386–9.

—— (1969*a*). Preliminary studies on the integrated control of the pest complex on cowpea, *Vigna unguiculata* Walp., in Nigeria. *J. Econ. Entomol.* **62**, 900–2.

—— (1969*b*). On the population dynamics and flight activity of *Taeniothrips sjostedti* (Trybom) (Thysanoptera: Thripidae) on cowpea. *Bull. entomol. Soc. Nigeria* **2**, 60–71.

—— (1971*a*). Studies on the control of yam beetles. *Ford Foundation/IRAT/IITA seminar on root and tuber crops in West Africa, Ibadan.*

—— (1971*b*). Insect pests of tropical grain legumes. *Ford Foundation/IRAT/IITA seminar on plant protection of tropical food crops, Ibadan.*

TODD, J. W. and CANERDAY, T. D. (1968). Resistance of southern peas to cowpea curculio. *J. Econ. Entomol.* **61**, 1327–9.

—— (1972). Control of soybean insect pests with certain systemic insecticides. *J. Econ. Entomol.* **65**, 501–4.

—— MORGAN, L. W. (1972). Effect of hand defoliation on yield and seed weight of soybeans. *J. Econ. Entomol.* **65**, 567–70.

TURNIPSEED, S. G. (1967). Systemic insecticides for control of soybean insects in South Carolina. *J. Econ. Entomol.* **60**, 1054–6.

—— (1972). Response of soybeans to foliage losses in South Carolina. *J. Econ. Entomol.* **65**, 224–9.

US DEPARTMENT OF AGRICULTURE. (1968). Suggested guide for the use of insecticides to control insects affecting crops, livestock, households, stored products, forests and forest products—1968. *Agriculture Handbook* No. 331.

USUA, E. J. (1967). Observations on diapausing larvae of *Busseola fusca* (Fuller). *J. Econ. Entomol.* **60**, 1466–7.

—— (1968). Effect of varying populations of *Busseola fusca* larvae on the growth and yield of maize. *J. Econ. Entomol.* **61**, 375–6.

VAISHAMPAYAN, S. M. and SINGH, Z. (1969). Comparative effectiveness of some modern insecticides against the blossom thrips of red-gram (*Cajanus cajan* (L.) Millsp.) *Ind. J. agric. Sci.* **39**, 52–6.

VAN DUYN, J. W. (1972). Investigations concerning host parasite resistance to the Mexican bean beetle. *Epilachna varivestis* Mulsart in soybeans, *Glycine max* (L.) Merrill. Ph.D. Thesis, Clemson University, South Carolina.

—— TURNIPSEED, S. G., and MAXWELL, J. D. (1971). Resistance in soybeans to the Mexican bean beetle. I. Sources of resistance. *Crop Sci.* **11**, 572–3.

VAN EIJNATTEN, C. L. M. (1965). Towards the improvement of maize in Nigeria. Insect Pests. *Meded Landb. Hogesch. Wageningen* **65**, 59–66.

VAN EMDEN, H. F. (1971). Plant resistance to insects. Developing 'risk-rating' methods. *SPAN* **15**, 71–4.

VAN HALTEREN, P. (1971). Insect pests of cowpea, *Vigna unguiculata* (L.) Walp., in the Accra plains. *Ghana J. agric. Sci.* **4**, 121–3.

VAN HOOF, H. A. (1962). List of virus diseases observed and transmitted in Surinam. *De Surinaamse Landbouw* **10**, 36.

WALKER, P. T. (1959). The progress of stalk borer control in East Africa. *Proceedings of the 4th International Congress on Crop Protection, Hamburg, 1957*, Vol. I, pp. 761–3.

—— (1960*a*). The relation between infestation by stalk borer, *Busseola fusca*, and yield of maize in East Africa. *Ann. appl. Biol.* **48**, 780–6.

—— (1960*b*). Insecticide studies on the maize stalk borer, *Busseola fusca* (Fuller) in East Africa. *Bull. ent. Res.* **51**, 321–51.

—— (1960*c*). Insecticide studies on East African agricultural pests. III. Seed dressings for the control of the bean fly, *Melanagromyza phaseoli* (Coq.) in Tanganyika. *Bull. ent. Res.* **50**, 781–93.

WALKER, P. T. (1963). *The relation between height of maize and attack by maize stem borer*, Busseola fusca, *in Tanganyika*. Report of Tropical Pesticides Research Unit, Porton, No. 257.

—— (1967). A survey of losses of cereals to pests in Kenya and Tanzania. Paper presented in *FAO Symposium on Crop Losses, Rome*, pp. 79-88.

WALTERS, H. J. and HENRY, D. G. (1970). Bean leaf beetle as a vector of the cowpea strain of southern bean mosaic virus. *Phytopathology* **60**, 177-8.

WENE, G. P. and GRIFFITHS, F. P. (1954). Effect of benzene hexachloride and lindane on the flavor of purple hull peas. *Proc. Am. Soc. hort. Sci.* **64**, 390-2.

WHEATLEY, P. E. (1961). The insect pests of agriculture in coast provinces of Kenya. 5—Maize and sorghum. *E. Afr. agric. for. J.* **27**, 105-7.

WHITNEY, W. K. (1970). Observations on maize insects at the International Institute of Tropical Agriculture (IITA) Ibadan, Nigeria. *Bull. entomol. Soc. Nigeria* **2**, 146-55.

—— (1972). Observations on cowpea (*Vigna unguiculata* (L.) Walp. (Leguminosae: Papilionaceae) insects at the International Institute of Tropical Agriculture. Paper presented at the *14th International Congress of Entomologists, 1972, Canberra*. (Mimeo)

—— GILMER, R. M. (1974). Insect vectors of cowpea mosaic virus in Nigeria. *Ann. appl. Biol.* **77**, 17-21.

WICKRAMASINGHE, N. (1971). Studies in Ceylon on the resistance of the rice plant to the gall midge. Paper presented at the *International Rice Research Conference, 1971, IRRI, Philippines*. (Mimeo)

WILL, H. and JONES, R. A. D. (1970). Three and a half decades of rice research at the Rice Research Station, Njala University College, University of Sierra Leone, Rokupr, Sierra Leone (Mimeo).

WISEMAN, B. R. and McMILLIAN, W. W. (1970). *Preference of sorghum midge among selected sorghum lines, with notes on over-wintering midges and parasite emergence.*

Production Res. Rept. No. 122, Agric. Res. Serv., US Department of Agriculture.

WOLFENBARGER, D. O. (1963). Control measures for the leafhopper *Empoasca kraemeri* on beans. *J. Econ. Entomol.* **56**, 417-19.

—— (1964). Effects of insecticides, rates, intervals between, and number of applications and insecticide-oil and surfactant combinations for insect control on southern peas. *J. Econ. Entomol.* **57**, 966-9.

—— CORREA, R. T. (1963). *Variations in southern pea varieties to cowpea curculio infestations*. Texas Agri. Exp. Stn Prog. Rept. 2286.

—— SCHUSTER, M. F. (1963). Insecticides for control of the cowpea curculio, *Chalcodermus aeneus*, on southern peas. *J. Econ. Entomol.* **56**, 733-6.

—— SLEESMAN, J. P. (1961a). Resistance to the Mexican bean beetle in several bean genera and species. *J. Econ. Entomol.* **54**, 1018-22.

—— —— (1961b). Resistance to the potato leafhopper in lima bean lines, interspecific *Phaseolus* crosses, *Phaseolus* spp. the cowpea, and the Bonavist bean. *J. Econ. Entomol.* **54**, 1077-9.

—— —— (1963). Variation in susceptibility of soybean pubescent types, broad bean and runner bean varieties and plant introductions to the potato leafhopper. *J. Econ. Entomol.* **56**, 895-7.

WYNIGER, R. (1962). *Pests of crops in warm climates and their control*. Acta Tropica Suppl. 7. Verlag für Recht und Gesellschaft AG, Basel.

—— (1968). *Pests of crops in warm climates and their control, V. Control measures*. Appendix to suppl. 7. Acta tropica (2nd edn.) Verlag für Recht und Gesellschaft AG, Basel.

ZEID, M., SAAD, A. A., AYAD, A. M., TANTAWI, G., and ELDEFRAWI, M. E. (1968). Laboratory and field evaluation of insecticides against the Egyptian cotton leafworm. *J. Econ. Entomol.* **61**, 1183-6.

11. Disorders associated with fungi, bacteria, viruses, and nematodes and their control

M. DELASSUS

11.1. Introduction

11.1.1. General

Research into plant protection should provide information that can be used to reduce economically the damage caused by pathogens to cultivated plants. As a guide to what research is most important it is necessary to obtain quantitative estimates of the losses caused by each pathogen, so that efforts can be concentrated on those of major economic importance. Strickland (1971) has indicated the requirements for obtaining a correct assessment of crop losses.

Correct identification of the pathogen is the first logical step in controlling it. Measurement of the incidence of the disease comes second, while the third step is the assessment of the reactions of individual plants, and the crop as a whole, to attack.

11.1.2. Correct identification of the cause

Although the causal organisms responsible for much of the crop damage in West Africa are known, there are still some symptoms associated with reduction in yields, such as rice yellowing, for which the cause is uncertain. Complex diseases may involve insects creating infection courts for pathogenic fungi or bacteria. In such instances, the role of the insects has probably often been underestimated. This is exemplified by the experiments of Landis (1971), who showed the extent to which carbofuran, an insecticide, reduced fungal stalk rot of maize.

11.1.3. Measurement of incidence

Estimates of incidence must be sufficiently precise (Chiarappa and Chiang 1971). Methods of assessment necessarily vary for different kinds of disease; and assessments will also tend to differ between experimenters as well as between years and locations.

Note must be taken of the plant cultivars concerned, of the conditions of the experiment, and of the area of the plots, particularly for air-borne disease.

11.1.4. Yield responses to diseases at recorded levels of incidence

For some of the most important diseases of tropical food crops, experiments have been conducted with sufficient statistical strictness to enable the intensity of the attack and the crop loss to be correlated. After obtaining a large number of data and finding a significant correlation, the regression equation and its graphical presentation reveal data of economic importance more clearly than does the correlation coefficient (Chester 1950). Numerous rules and techniques have been prescribed for the precise and valid determination of disease incidence and associated losses. The FAO publication *Crop loss assessment methods* (Food and Agricultural Organization 1971) explains both the general principles for estimation and more than 80 special methods, each applying to particular pathogens.

Allowing for the number of pathogens, the number of methods so far considered to be valid is very low, which implies that the real effects of diseases on crop performance is often not known.

11.1.5. Economic control of diseases

In addition to estimating losses by quantity and/or quality, consideration must then be given to the ratio between the cost of the treatments and the expected value of the gain in yield; a ratio which should be as low as possible. Since there will be concealed economic costs in addition to the direct price of the control method, the ratio may have to be substantially less than one before control methods can be justified. Hartley and Rathbun-Gravatt (1937) draw a distinction on economic grounds between diseases that attack weakened plants and those that attack otherwise healthy plants. The former tend to increase variations in yield between crops, while the latter reduce them. These relations can be expressed in the form of a correlation coefficient *r* established over a number of seasons between the loss due to the disease and the potential yield, under the same environmental conditions, obtained in the absence of disease. If *r* is negative, the disease increases the annual variability of yield; if *r* is positive, the variability is reduced. The economic importance of the former is greater than of the latter because of consequent instability in the marketing sector.

11.2. Main hazards to crops

11.2.1. Diseases not caused by pathogenic organisms

Many pathological conditions of plants are not caused by pathogenic organisms but are due to a poor environment; they are often called physiological disorders.

Physiological disorders of rice. Because of its method of cultivation, physiological disorders are particularly common on swamp rice. Akiochi (Ou 1972) is common in Japan and West Africa. Brown spots develop on the leaves (and may be colonized by *Helminthosporium oryzae* Van Breda der Haan), on the culms, drying the lower leaves, and the roots rot. In Africa, akiochi is found in partially anaerobic swamp soils rich in organic matter and producing hydrogen sulphide. It can be effectively controlled by applying blast-furnace slag with a high content of silica, calcium, and trace elements; splitting the nitrogen and potash applications; using fertilizers without sulphate; and planting resistant cultivars such as IR 442 (Delassus 1972).

Bronzing, studied in Ceylon (Ponnamperuma 1958), Senegal (Beye 1972), and Liberia (Carpenter 1973) is characterized by yellowish, bronzed, or purple leaf discoloration with numerous very small brown spots. Bronzing is due to iron toxicity associated with low pH. Dressing with lime, drainage of the soil, nitrogen application, and the growing of resistant cultivars are possible control methods.

Other physiological disorders of irrigated or upland rice include Akagare, erect panicles; Aogare; Khaira disease (Nene 1972) (chlorosis of the base of the leaves, which then turn reddish-brown), which is due to zinc deficiency in soils with a high pH; and copper deficiency (Cunha and Baptista 1958).

Physiological disorders of other crops. Low pH causes disorders such as yellow stunt disease of groundnuts in Senegal (Blondel 1970) and induces manganese toxicity on groundnuts and aluminium toxicity on maize (Beye 1972).

The most common symptoms of nutrient deficiencies reported in Africa include the following:

1. Boron deficiency (Gillier and Silvestre 1969) on groundnuts takes the form of canker on the stems, hollow heart, and internal blackening of the nuts. This can be overcome by applying borax at 5 kg/ha.
2. Molybdenum deficiency on groundnuts restricts vegetative growth and inhibits nodulation. It can be corrected by applying 28 g of ammonium molybdate per ha by seed-treatment.
3. Zinc deficiency occurs on maize on light soils in Western Nigeria. This has been corrected by the application of zinc chelate (1 kg/ha) (Kang and Osiname 1972).
4. Sulphur and calcium are sometimes deficient, especially on groundnuts. Adequate sulphur is conveniently applied by the use of thiocarbamate fungicides intended primarily to control *Cercospora* leafspot.

The effects of trace elements and their corresponding deficiency symptoms are of great variety, such as, for example, that of boron at 3 p.p.m. and also molybdenum, which foster the formation of female flowers on various cucurbits (Choudhury 1971).

Magnesium deficiency on tannia (*Xanthosoma saggitifolia*) produces such pronounced symptoms that this crop has been suggested as the most useful indicator plant for soil deficiency of magnesium (Bull 1960).

Water plays a vital role in plant growth. In addition to ordinary wilting, a shortage of water can show itself in special symptoms such as the rolling-up of individual sorghum leaves in certain vertisols of the Niger (Jouan and Delassus 1971), in blossom-end rot of tomatoes, and in marginal necrosis of lettuce leaves (these two disorders are also associated with a shortage of calcium and, in the case of lettuce, with alternating dull and sunny days (Messiaen, Fournet, Beyries, and Quiot 1971). An excess of water, with or without prolonged submersion, leads to generally poor growth of maize, groundnuts, and many market-garden crops.

Too high or too low a temperature also adversely affects yields. At the IITA in Ibadan, reduction of the soil temperature by means of a surface mulch-straw was associated with appreciable increases in yield with maize and yams (IITA 1973).

In certain soils in Senegal, it has been suggested that sorghum produces a toxic compound that is very harmful to the growth of an immediately-following second crop of the same cereal (Chopart and Nicou 1971).

11.2.2. Diseases caused by pathogens

Diseases caused by pathogenic fungi. Most cultivated plants have, in one phase or other, been reported to be attacked by several dozen pathogenic fungi, and it is inappropriate to list all of these in the present review. Some are of almost trivial importance, and their listing, for the mycological record, can give a quite false impression of the number of real pathological problems involved. Here we shall note only the principal pathogens according to the category of fungus rather than by crops, because often similar control measures will be appropriate for diseases caused by similar fungi.

Fungi causing damping-off diseases (seedling necroses). These diseases are common on food legumes such as groundnuts and beans and on cucurbitaceous market-garden crops, but also occur on cereals, especially when germinating in over-moist soils. Common pathogens include *Pythium* spp., *Sclerotium rolfsii* Sacc., *Aspergillus* spp., especially *A. niger* van Tiegh. (Gillier and Silvestre 1969), *Rhizoctonia solani* Kühn, *Rhizopus* spp., and *Fusarium* spp. including *F. roseum* f.s. *cerealis* (Cooke) Snyder and Hansen of *F. moniliforme* Sheldon on cereals (Futtrell and Webster 1967). The primary infection in tropical soils may come either from the soil itself or from the seeds.

Downy mildews. This group of diseases include several important mildews of cereals caused by *Sclerospora* spp. Large chlorotic leafspots or streaks occur with a whitish down appearing at night in cool weather, but soon disappearing after dawn. Deformation of the plants, including dwarfism and serious structural modification of the ear, are common symptoms. *Sclerospora graminicola* (Sacc.) Schroet. infects pearl millet (King 1970). Tift 23, a parent commonly used to produce dwarf hybrid millets, is particularly sensitive to it. Very moist ground encourages infection of seedlings by the oospores which have remained dormant in the soil since a previous crop. Secondary infection by sporangia

occurs on young leaf tissue under wet conditions. Seeds can also harbour the pathogens. On sorghum, *Sclerospora sorghi* (Kulk.) Weston and Uppal, although reported in Asia, America and Europe (Tarr 1962) is still found only to a limited extent in West Africa in Nigeria and Cameroun (King 1970).

No downy mildew has yet been reported on maize in West Africa, although Ullstrup (1970) reports that eight species of *Sclerospora* and *Sclerophthora* can cause downy mildews on this crop in other places, and at least two of the pathogens concerned can also infect *Sorghum*. Measures to avoid the accidental introduction of these pathogens are clearly of very great importance. For rice, likewise, West Africa is in a fortunate position. *Sclerospora macrospora* Sacc., which is also capable of attacking numerous other grasses including maize, has been reported so far only in Italy, Australia, Asia, and the USA. Of the other pathogenic Phycomycetes reported in various parts of the world, mention may be made of *Phytophthora cryptogea* Pethybridge and Lafferty, which causes cassava rot in Zaire (Buyckx 1962), *P. infestans* (Mont.) De Bary which is endemic on the potato, and *Pseudoperonospora cubensis* (Berk. & Curt.) Rostov. on cucumbers and other cucurbits.

Smuts. Smuts are frequent on *Sorghum*, and four species of smut pathogen occur on this crop in West Africa; *Sphacelotheca sorghi* (Link) Clinton, which causes covered smut is the most frequent (Tarr 1962). Each grain is replaced by a small conical-shaped smut sorus. *S. cruenta* (Kühn) Potter causes loose smut, *Tolyposporium ehrenbergii* (Kühn) Pat. causes long smut with a whitish smut sorus, much elongated in comparison with the sori in covered and loose smuts. *S. reiliana* (Kühn) Clinton causes head smut, in which the whole panicle is transformed into a voluminous sorus. This fungus has one strain differing from that on *Sorghum*, which is confined to maize, and this in tropical Africa has been reported only in Cameroun (Delassus 1968). In loose and covered smut diseases infection occurs mainly at the time of germination of the seed. Long-smut infection occurs during flowering. *S. reiliana*, causing head smut, infects the plant mainly systemically during the first part of its growth and remains latent in the terminal meristem. Young seedlings can easily be infected experimentally. This is in contrast with the situation in the other smut diseases.

On pearl millet, flowers infected by *Tolyposporium penicillariae* Bref. develop small sori in the place of grains. The sori are at first bright green, later becoming brownish. Some other smuts of food crops are uncommon in West Africa: e.g. *Ustilago maydis* (DC) Corda which causes a gall on maize and *Tilletia barclayana* (Bref.) Sacc. & Syd. which may completely or partially replace some grains in rice with a mass of blackish spores.

Rusts. Rust diseases are numerous. These fungi form pustules, usually easily recognizable as rust lesions, or induce hypersensitive reactions which may be difficult to diagnose as being due to a rust pathogen. Where two species occur on the same host plant it may, in practice, be difficult to distinguish them by the macroscopic symptoms. This is so even for *Puccinia sorghi* Schw., which is said to produce an elongated uredosorus tending towards dark red, and dehiscent at maturity, and *P. polysora* Underw., which should have sori of a yellowish colour and rounded form (Cummings 1941). On maize, *P. polysora* was first reported in Africa in 1949. It spread rapidly and has caused extensive losses throughout the low-altitude humid regions of West Africa (Cammack 1958, 1959). The cultivars now grown have been bred for resistance to it. Rice rust is so far an economically unimportant disease (Ou 1972).

On beans and cowpeas, *Uromyces appendiculatus* (Pers.) Unger (= *U. Vignae* Barcl.) (Laundon and Waterston 1965) is the cause of rust disease in which reddish-brown sori are surrounded by a yellow halo in some cultivars. Many cultivars of both crop species are resistant.

On groundnuts, *Puccinia arachidis* Speg. (Chahal and Chohan 1971), and on soya beans, *Phakopsora pachyrhyzae* Syd. (some pathotypes), which are both serious pathogens in some tropical areas causing severe leaf drop, have not yet been reported in Africa; since its introduction *Puccinia arachidis* has very rapidly spread in Asia (Commonwealth Mycological Institute Map 160 Ed. 3). These two pathogens are major disease threats depending on phytosanitary measures for continued 'control' by exclusion from West Africa.

Powdery mildew diseases. Powdery mildew diseases caused by members of the Erysiphaceae are often more prominent in hot dry weather conditions than in cool and cloudy conditions. Species with conidial *Oidium* forms occur on some leguminous crops and notably on bambara groundnuts and on cucurbitaceous fruits. In oidial diseases the mycelium is largely superficial. *Leveillula* conidial forms are also quite common, and in these the conidiophores arise through the stomata of the host plant.

Fungi Imperfecti. The Fungi Imperfecti, for which the perfect or sexual forms are, with few exceptions, rarely found in tropical regions, comprise a very large group of pathogenic fungi among which are some of the most destructive. *Pyricularia oryzae* Cav. on rice causes rice blast disease. Leafspots occur which are typically elongated; on stems the necrotic lesions develop on the nodes and the internode next below the panicle, causing breakage of the neck and blind ears.

Helminthosporium oryzae is probably primarily responsible for rice brown spot disease in West Africa (Aluko 1970). *H. turcicum* Pass. causes large spreading grey-brown leafspots and *H. maydis* Nisik. & Miyake, much smaller leafspots on maize. Most cultivars grown in Nigeria are susceptible to this pathogen (van Eijnatten 1961) *Rhynchosporium oryzae* Hashioka & Yokogi, which has only recently been introduced into Africa (Lamey and Williams 1972), causes zonate leafspots on the leaves of rice; the disease may also be known as 'scald' by analogy with a similar disease of barley.

Cercospora oryzae Miyake causes elongated spots on rice, and on groundnuts, *C. personata* (Berk. & Curt.) Ell. & Ev. and *C. arachidicola* Hori both cause brown necrotic leafspots, and pod yields may be reduced by half (Gillier and Silvestre 1969). *C. cruenta* Sacc. and *C. canescens* Ell. & Mart. both cause damage on cowpeas (IITA 1972a) and *C. henningsii* Allesch. (brown spot) and *C. caribaea* Cif. (light spot) on cassava are common but probably not serious pathogens in humid regions.

Pathogenic *Alternaria* spp. include *A. solani* Sorauer, causing leafspots and stem lesions on potato and tomato, and *A. porri* (Ell.) Cif., which can be a very serious pathogen, causing purple leafspot on onions.

The main economically important representatives of *Fusarium* are *F. roseum* on maize and sorghum, and *F. moniliforme* on rice, maize, sorghum, and pearl millet. Attacks may either be general or limited to the grain. On tomato, *F. oxysporum* f. *lycopersici* (Sacc.) Snyder & Hansen causes wilting. *Cephalosporium maydis* Samra, Sabet & Hingorani and *C. acremonium* Corda attack maize causing wilting and rot of the ears in India (Payak, Sangam, Janki, and Renfro 1970).

Colletotrichum spp. (Roger 1953) cause leafspot,

canker on the stems, and rots and necrosis on many fruits. The main economically important species in the lowland tropics in West Africa are *C. linde-muthianum* (Sacc. & Magn.) Briosi & Cav. causing anthracnose of beans and cowpeas, *C. gloeosporioides* (Penz.) Sacc. on yams, and *C. graminicola* (Ces.) Wils. on cereals.

Phaeoisariopsis griseola (Sacc.) Ferraris is encountered on beans.

Fungal mycotoxicoses. In addition to the damage they cause to plants, some fungi are particularly harmful because their growth on plant material is accompanied by the production of toxins dangerous to man or domestic animals. Many species secrete toxins, but the most important and notorious is *Aspergillus flavus* Link ex Fr. (Moreau 1968) which often produces aflatoxins on groundnuts (see also Chapter 4, p. 48) and to a lesser extent also on maize and sorghum when these are stored in moist conditions, and also on other foodstuffs. A content in food or feed, as low as 20 p.p.b. is considered the maximum limit that must not be exceeded. *A. ochraceus* Wilh. secretes ochratoxins, especially on maize (Moreau 1968). Several *Fusarium* spp. (including various strains of *F. roseum*) produce an oestrogenic compound, zearalenone, known as F2, especially during storage in excessively humid conditions (Caldwell and Tuite 1970). Several toxic alkaloids have been found in the ergots of cereals (Kannaiyan, Vidhyasekaren, and Kandaswany 1971).

Diseases caused by bacteria. The number of bacteria attacking food crops in tropical regions is relatively small in comparison to the number of fungi, and Harris (1971) considered only about 30 species to be involved. But some bacterial infections are very important. They can be divided into those causing systemic infections and those causing localized infection within their host plant.

Of the former, special mention must be made of *Pseudomonas solanacearum* (E. F. Smith) E. F. Smith which is sometimes considered the most serious of all tropical plant pathogenic organisms. The existence of different strains of this bacterium has been recognized according to biochemical and pathogenic criteria (Hayward 1964; Buddenhagen and Kelman 1964). Strains can be divided into two groups: the first group attacks members of the Solanaceae and other plants, but not triploid *Musa* cvs.; the second group attacks only the triploid *Musa* cultivars and *Heliconia* spp. and is found only in part of the New World. Throughout West Africa at present tomatoes and potatoes are very seriously affected, and aubergine, to a smaller extent. Losses on groundnuts in West Africa from 'slime diseases' caused by this pathogen are very small. The strain attacking *Musa* triploids is transmitted by insects that infect the floral scar of attachment and is spreading rapidly in South America, where it is causing extensive losses.

Bacterial wilt of rice (Ou 1972; Mizukami and Wakimoto 1969) caused by *Xanthomonas oryzae* (Uyeda & Ishiyama) Dowson is one of the most damaging rice diseases in Asia and Indonesia. The bacteria gain entry mainly through the hydathodes on the leaf edges and to a lesser extent through the roots. At first confined to the edges of the leaves, the lesions later spread and merge to cover the whole leaf, which turns white. A special facies of the disease known as 'kresek' is reminiscent of attack by borers. The pathogen which is internally seed-borne (Srivastana and Rao 1966) is still believed to be absent from Africa and is one of the most important pathogens justifying the maintenance of strict phytosanitary, including quarantine, measures. *X. manihotis* (Arthand-Berthot) Starr, which is responsible for bacterial wilt of cassava long known in South America, was only recently observed in Nigeria (IITA 1972b). It has also now been found in Zaire (Williams, personal communication). The first symptom is the wilting of one of the leaves at the top of the plant; then other leaves dry up, fall, and the end of the stem dries up.

X. phaseoli (E.F.S.) Dowson is a serious pathogen on beans and cowpeas, and *Erwinia tracheiphila* (Smith) Holland on cucurbits. Of the bacteria generally confined to the parenchyma, i.e. not systemic in their hosts, the most common are *X. oryzicola* Fang *et al.*, the cause of bacterial streak disease of indica types of rice, *X. phaseoli* var. *sojense* (Hedges) Starr & Burkh. on soya beans, *X. vignicola* Burkh. on cowpeas, *Pseudomonas phaseolicola* (Burkh.) Dowson and *X. phaseoli* var. *fuscans* (Burkh.) Starr & Burkh. on beans, *P. lachrymans* (E.F.S. & Bryon) Carsner on cucurbits, and *X. vesicatoria* (Doidge) Dowson on tomatoes and peppers.

Diseases caused by viruses and mycoplasmata. The Mycoplasmatales differ from viruses by being able to be cultivated on inert media and by their nucleic acids, and from bacteria by the absence of a rigid cell wall. They can cause certain plant diseases

(Darpoux 1971; Davis and Whitcomb 1971). Recently, corn stunt was attributed to a spiroplasm, a similar micro-organism of spiral form (Davis 1973).

Virus diseases in West Africa have been subjected to little systematic study, but are fairly numerous, being economically significant mainly on plants that multiply vegetatively and on perennials. It appears that damage is very limited on the cereals. For example, for rice, although some dozen virus and mycoplasma diseases are reported in Asia (Ou 1972), only one virus disease (which appears to differ from any of these)—rice yellow mottle disease—has been reported in Africa, and that only in Kenya (Bakker 1970). Although the presence of virus diseases resembling tungro has from time to time been suspected in West Africa, no confirmation has been obtained and losses, if indeed they exist at all, appear to be small.

On rice, two mycoplasmal diseases (Nasu, Sigiura, Wakimoto, and Iida 1967), dwarf and yellow dwarf, and two viral diseases, stripe and black streaked dwarf, occur in temperate parts of Asia. Eight other viral or mycoplasmal diseases have been described in tropical Asia. These are orange leaf, leaf yellowing, 'Penyakit merak', transitory yellowing, 'tungro', yellow-orange leaf, grassy stunt (which is certainly due to a mycoplasma), and mosaic. White leaf (hoja blanca) disease occurs in the USA, Central America, and some parts of South America. Several of the diseases mentioned above, i.e. tungro, 'Penyakit merak', are very similar if not identical, and are all transmitted by the leafhopper *Nephotettix impicticeps*. Recently in Indonesia 'mentek', wrongly attributed to nematodes, was found to be a virus disease probably closely similar to tungro (Ou 1972).

Apart from mosaic, which is transmitted mechanically, all the rice virus diseases of Asia and America are transmitted by Cicadellidae or Jassids, which can be divided into two groups: the Deltocephalidae, including the genera *Nephotettix* and *Inazuma* and the Delphacidae, including *Delphacodes* and *Sogata*. Some of the pathogens, e.g. dwarf, white leaf, and stripe, are transmittible through eggs. Some are persistent in their vectors, e.g. black streaked dwarf, orange leaf, and transitory yellowing, while others, such as the viruses of tungro and yellow-orange leaf, are not persistent.

Maize virus diseases. Numerous virus diseases have been reported on maize in different parts of the world (Granados 1969). Mosaic virus of sugar cane can also infect maize; it can be transmitted mechanically

and by aphids, including *Rhopalosiphum maidis*. Dwarf mosaic virus of maize, known in the USA, is transmitted to a slight extent through seed and also by aphids. Maize leaf-fleck disease is transmitted solely by aphids. Other diseases are transmitted by Cicadellidae: species of *Dalbulus* transmit corn stunt; *Peregrinus maydis* transmits mosaic; and *Cicadulina* spp. transmit streak. Various cicadellids transmit maize rough dwarf disease. Wheat streak mosaic disease which is transmitted by the mite *Aceria tulipae* and is also encountered on maize. A disease occurring in Madagascar and resembling maize wallaby ear (Schindler 1942), which so far is confined to Australia, is probably not due to a pathogen but to the action of toxins secreted by insects (Maramorosch 1959).

In Africa, probably the most important virus disease of maize is known as streak disease, and is transmitted only by the Jassid leafhopper *Cicadulina mbila* (Storey and McLean 1930; Storey 1936; Storey and Howland 1967). In India, a possibly related virus causes leaf streaks and poor ears on pearl millet (Seth, Raychaudhury, and Singh 1972). Other symptoms suggesting possible virus causation have been observed on maize in tropical West Africa, but no precise causal determinations have been made. In East Africa, Kulkarni (1973) has characterized maize stripe, and maize line viruses transmitted by the delphacid leafhopper *Peregrinus maydis*.

Groundnut virus diseases. On the groundnut, the rosette complex comprises the main virus problem in Africa. The complex in West Africa (Okusana and Watson 1966) probably differs from that in East Africa (Storey and Ryland 1957). There are two related strains of the main virus causing chlorotic rosette and green rosette (Feakin 1967; Gillier and Silvestre 1969). The viruses are transmitted mainly by *Aphis craccivora*, and to a lesser extent by *A. gossypii*. Other unrelated groundnut viruses causing mottle (Kühn 1964) and stunt (Miller and Troutman 1966) diseases occur in the USA. Mottle can be transmitted by seed and could therefore easily be accidentally introduced to new areas. Other virus or virus-like diseases have been reported on groundnuts, including a stunt, which is common in Senegal and Upper Volta, and a mosaic disease and other conditions in India (Sharma 1966). Tomato spotted wilt virus produces annular spots on groundnuts (Gillier and Silvestre 1969).

Cowpea virus diseases. In West Africa, many disease symptoms similar to those associated with viruses have been observed on cowpeas; the most commonly involved virus-induced mosaic is the beetle-transmitted cowpea yellow mosaic virus (Chant 1959; Bock 1971), which occurs with or without associated leaf deformation (IITA 1972). Hardcastle (1963) reported three additional distinct viruses unrelated to cowpea yellow mosaic. One of these has now been fully characterized as cowpea mild mottle virus (Brunt and Kenten 1973). The presence of the aphid-borne cowpea mosaic virus, which is a strain of potato virus Y (Bock 1973) has not yet been confirmed in West Africa, though it is common in East Africa and is seed-borne.

Cassava virus diseases. On cassava, the symptoms of mosaic are found in nearly all crops in the tropical regions of Africa. Under certain environmental conditions, especially in the arid regions, these symptoms are transitory. It has been shown that cassava mosaic in Africa and common mosaic in America are caused by different viruses (Costa and Kitajima 1972). The former is transmitted by *Bemisia* spp. but not mechanically (Storey and Nichols 1938), while the latter is transmitted mechanically. Witches' broom disease, prevalent in South America, is a yellowing disease associated with a mycoplasma. Brown streak disease is a virus disease described in East Africa (Nichols 1950; Jennings 1960), but it is not known elsewhere.

Sweet-potato virus diseases. Sheffield (1953, 1957) has reported two virus diseases on the sweet potato in East Africa. Similar symptoms to those she described have been observed in West Africa, with either annular spots or vein clearing, but no proper causal diagnosis has been made. The possible relationships between the virus-like condition in West Africa and those in East Africa and in the New World (Martin 1967; Alconeiro 1971, 1972) need to be determined.

Yam virus diseases. On yams, a green-banding virus (Ruppel, Delphin, and Martin 1966) and a mosaic disease (Adsuar 1955) have been described from Puerto Rico. Similar symptoms to those described for mosaic are frequently observed in Africa, particularly on *Dioscorea alata*, but from Africa no diagnostic work has yet been reported.

Virus diseases of aroids. On taro and tannia, symptoms similar to those of virus diseases observed in the Caribbean (Alconeiro and Zettler 1971) have been seen in West Africa. However, no virus could be detected as a cause of lethal root-rot of coco-yams (Posnette 1945; Kenten and Woods 1973).

Virus diseases of Solanaceae. The Solanaceae can be attacked by numerous virus and mycoplasma diseases such as little leaf disease of the egg plant in India (Thomas and Krishnaswami 1939) induced by tomato big-bud virus (Martyn 1968).

Nematode-induced diseases

1. *Polyphagous nematodes.* Numerous problems associated with nematode infestation have been encountered on food crops in West Africa. Too little is known about these pathogens and the associated losses on crops in tropical Africa. One of the most common and most important endoparasitic nematode genera is *Meloidogyne*, which causes small galls known as root knots on roots and tubers of the infested plants.

According to Netscher (1971) *M. incognita* (Kofoid & White) Chitwood, *M. javanica* (Traub) Chitwood, and *M. arenaria* (Neal) Chitwood are the most frequent species. Except for resistant cultivars, tomatoes, egg plants, *Phaseolus* beans, carrots, lettuces, potatoes, and yams are all very sensitive; peppers, however, are not very sensitive, and groundnuts seem rather tolerant. Netscher also reported that *Trichodorus minor* Colbran (stubby root nematode) occurs on market-garden crops in Senegal associated with truncated roots. *Rotylenchus reniformis* Linford & Oliveira has been observed by Peacock (1956) on numerous plants in Ghana, including tomato, egg plant, sweet potato, soya-bean, carrot, and cassava.

2. *Nematode pests on specific crops.* On rice (Luc 1971) *Aphelenchoides besseyi* Christie causes white leaf-tip disease. This has only recently been reported in numerous regions of West Africa but may well have been previously overlooked rather than been recently introduced. Losses so far are small but according to Atkins and Todd (1959), this pathogen has been associated with field losses of the order of 17 per cent in susceptible cultivars. *Ditylenchus angustus* (Butler) Filipjev is found in Asia, Madagascar, and Egypt, and mainly affects the stems and panicles. Two species of *Hirschmanniella*, *H. oryzae* (van Breda de Haan) Luc & Goodey, and *H. spinicaudata* (Sch. & Stek.) Luc & Goodey have been found on rice in West Africa, but little is yet known of their importance (Feakin 1970). *Heterodera oryzae* Luc & Burdon, the rice cyst nematode, is found in the Ivory Coast and causes crop losses of about 25–30 per cent. The ectoparasitic nematode *Tylenchorhynchus martini* Fielding is reported as a

rice pathogen in Sierra Leone and Senegal (Feakin 1970).

In addition to that caused by *Meloidogyne* (Merny 1971) mentioned earlier, damage caused by nematodes is extensive on yams: *Pratylenchus brachyurus* (Godfrey) Filipjev & Stekhoven, a lesion nematode, and *Scutellonema bradys* (Steiner & Lettew) Andrassy, the latter known as yam nematode but capable also of attacking cowpeas, are migrating endoparasites, causing fairly similar symptoms: cracks in the skin and yellowish-brown spots turning to brown in the outer parts of the flesh.

Nematodes infesting cassava are practically the same as for yams.

On maize (Caveness 1967) the most dangerous nematodes belong to the genus *Pratylenchus*, in particular *P. brachyurus* (Godfrey) Filipjev & Stekhoven. In control trials Caveness obtained very large increases in yields in Nigeria by nematicide treatments (DGCP at about 15 l/ha).

Several nematodes have been reported on groundnuts in the world. In Upper Volta, a groundnut chlorosis has been related to the presence of *Aphasmatylenchus straturatus* Germani (Germani 1971).

Little critical work has been done on pathogenic nematodes of sorghum. In India, the genera *Hoplolaimus*, *Helicotylenchus*, *Pratylenchus*, *Rotylenchus*, *Heterodera*, *Tylenchorhynchus*, and *Longidorus* have all been considered to be potential pathogens of this crop (House 1970).

Parasitic spermatophyta.

Numerous crops in dry tropical regions are attacked by *Striga* spp. belonging to the family Scrophulariaceae (Ogborn 1971). Fifteen species have been identified, of which *S. hermonthica* (Del.) Benth. and *S. asiatica* (L.) O. Ktze. are the most important. Cereals such as pearl millet, sorghum, maize, and upland rice and many Gramineae support the full life-cycle of the parasite, i.e. the seed germinates, parasitizes the root, and puts out leafy aerial shoots which then bear flowers and distribute seeds. On other semi-host plants, such as groundnuts, cowpeas, Egyptian beans, soya beans, and cotton, *Striga* seeds germinate in contact with the roots and become parasitic, but most of the shoots remain underground and almost no reproduction occurs. Even in this situation, however, considerable crop losses can occur. On irrigated rice, *Rhamphicarpa* spp. are sometimes encountered, e.g. in Mali.

11.3. Control methods

A distinction must be made between diseases of physiological and pathogenic origins. In the former, two approaches to control can be envisaged: improvement of environmental conditions or choice of cultivars tolerating the adverse environmental conditions. The use of tolerant crops should be envisaged when it is too costly (e.g. swamp rice-fields) or materially impossible (e.g. temperature too high or too low, or prevalence of drought) to influence the cause of the trouble more directly. Although a great amount of work has been carried out on the inheritance of resistance to pathogens, work on the genetics of susceptibility to adverse environment appears to be rare. It should be possible to obtain stable resistance, since the adverse factor, although it may vary in intensity, remains the same qualitatively.

11.3.1. *Control of diseases caused by pathogenic organisms*

For diseases of pathogenic origin there are many approaches to control:

(1) reducing or eliminating the propagules of the pathogen, or taking precautions to keep the region free from alien pathogens, i.e. phytosanitary restrictions and quarantine measures;

(2) application of cultivation techniques reducing any predisposition to disease;

(3) field control by chemical or biological methods;

(4) control by breeding and selection, i.e. using cultivars that are immune, resistant, or tolerant to specific pathogens.

Reduction of sources of inoculum. Treatment of the seed or vegetative propagating material, either chemically with biocides or physically (especially by heat), can reduce or eliminate an initial source of infection whether it is by fungi, bacteria, viruses, or nematodes. For example, *Aphelenchoides besseyi* (Luc 1971) can be controlled by easy and safe treatment of rice seed by heat (dry seed treated with hot water at 56–57 °C for 15 minutes) or by seed-treatment with the fungicide and nematicide benomyl at 4 g per kg of seed. Other more toxic, and hence hazardous, methods have also been reported (Feakin 1970). In Nigeria (Ekanden 1970), cassava mosaic virus is rendered inactive by heat-treatment of the young plants. Temperatures of 37–39 °C are sufficient if the treatment is continued for 4–6 weeks. Particularly careful treatment is necessary for seed brought in from foreign countries. It must be ensured that the seed comes as far as possible only from healthy plants, but strict disinfection after arrival is an additional precaution. For the pathogen of bacterial wilt of rice (*Xanthomonas oryzae*), hot-water treatment (53 °C for 30 minutes with pre-liminary soaking for 12 hours) will completely eradicate seed-borne infection, according to Sinha and Nene (1967). Heat-treatments require precise equipment and are very difficult to carry out satis-factorily on tubers such as yams, where nematode control is recommended by soaking in water at 50 °C for 30–40 minutes (Merny 1971). Culture of sufficiently small meristems, whether or not accom-panied by heat-treatment, usually gives rise to healthy scions or plantlets. Healthy seed can be obtained by growing parent plants in regions un-favourable to any particular seed-borne pathogen or by rigorous chemical disease control during growth of the seed-producing crop. In this way, for example, bean seeds can be obtained with only a low inoculum level of bacterial disease and anthracnose pathogens (Zaumeyer and Thomas 1957).

Soil-borne inoculum can be reduced by other techniques. Incorporating harvest residues to a depth of at least 20 cm greatly reduces attacks by *Sclerotium rolfsii* the cause of southern blight of groundnuts (Feakin 1967). Prolonged flooding, either natural or artificial, has been reported to reduce soil infesta-tion by *Pseudomonas solanacearum* and by nematodes (Netscher 1971). Destruction of volunteer ground-nuts is recommended both for rosette control (Feakin 1967) and for the *Cercospora* leafspot diseases (Fowler 1970). For other virus diseases, eradication of the wild alternative hosts may reduce attack on crops, e.g. the proscription of many Malvaceae in cotton-growing areas of Sudan (Tarr 1951, 1955) for the control of leaf curl. Throughout cultivation, removal and destruction of diseased plants is often recom-mended, but is often impracticable except in special crops being grown for seed. For this technique to succeed, the pathogens concerned must have a limited dissemination potential in quantity or area. The operation must be carried out with great care as soon as possible after diseased plants are noticed.

In the soil, the pathogenic inoculum may be reduced by using biocides. This technique is costly, and for the time being can only be envisaged in certain nurseries where heat-treatment can also be used.

Crop rotation is the most important method of limiting the inoculum potential of pathogens of restricted host range. In some circumstances the inclusion of a specific crop antagonistic to a particular pathogen may even be desirable. Several studies have indicated that reduction of *Meloidogyne* in the soil can be significantly reduced with a crop rotation employing species such as *Cynodon dactylon* (Chheda 1971), *Amaranthus* spp. (IITA 1972), *Eragrostis curvula* c.v. *Ermelo*, certain *Tagetes* spp., and *Crotalaria* spp. To reduce numbers of *Pratylenchus* spp. the growing of *Stylosanthes gracilis*, *Cajanus cajan* and *Crotalaria juncea* has been advocated (Luc 1971).

To control *Striga*, the practice of planting and then ploughing-in a sensitive cereal when the leafy shoots of the parasite first appear has the practical and psychological disadvantage of wastefully destroy-ing a crop and cannot be seriously considered in existing cultivation systems in Africa (Ogborn 1971).

The use of cultivation techniques. A thorough under-standing of the biology of both the pathogens and their hosts is necessary if 'good agricultural practices' are to be employed to reduce the incidence of disease. The factors of free water and aerial humidity both play a vital role. Crop conditions chosen to dis-courage one pathogen may well predispose to another, e.g. in the soil, *Phytophthora cinnamoni* Rands, *Rhizoctonia solani*, and *Thielaviopsis basicola* (Berk. & Br.) Ferraris show optimum growth at a water potential of 5 bars, whereas for *Fusarium solani* and *Fusarium roseum* 30 bars is optimal (Cook 1973). If drought conditions occur during the germination of groundnut or sorghum, susceptibility to disease attacks may be severe and consequently sowing should only be done in sufficiently moist soil.

Concerning fungi which are not soil-borne, many species among the Phycomycetes, Uredineae, and certain Fungi Imperfecti benefit for dispersal and/or infection from copious rainfall and high humidity. On the other hand, the powdery mildews caused by fungi of the Erysiphaceae frequently develop severely under relatively dry conditions.

It is not possible directly to influence atmospheric or soil moisture without irrigation. Indirect action to escape unfavourable conditions can be taken by choosing the date of sowing, the length of the growth season of a selected cultivar, and the method of cultivation (e.g. on the flat or on ridges), and by choosing a cultivar suited to local water conditions.

Temperature also affects the incidence of most diseases; e.g., *Pyricularia oryzae*, the pathogen causing rice blast disease, is favoured by temperature regimes with relatively low night temperatures (below 20 °C).

Chemical nutrients also affect the incidence of disease. High soil nitrogen levels predispose to rice blast disease and scald caused by *Rhynchosporum oryzae*, while soil-dressing with ammonium nitrate has been claimed greatly to reduce maize panicle smut (Radulescu, Persica, and Popescu 1959).

Sufficiency of soil potassium often improves resistance to attacks of pathogens which flourish under dry conditions (Arnoux and Franquin 1960). Silicon reduces attacks of rice blast (*Pyricularia*) disease in silicon poor soils (Ou 1972).

In the West Indies planting in soils with a high pH, overlying limestone, appears to prevent the development of bacterial wilt of the Solanaceae caused by *Pseudomonas solanacearum* (Messiaen, Fournet, Beyries, and Quiot 1971). In India, soil application of zinc salts has been claimed to reduce susceptibility to downy mildew caused by *Sclerospora sacchari* on maize (Singh, Chaube, Singh, Asnani, and Singh 1970).

Damage by nematodes, and also by bacterial wilts of solanaceous and other hosts, is more extensive in light than in heavy soils. The addition of various organic compounds such as organic manures and oilseed residues may greatly reduce the severity of attacks by parasitic nematodes. Covering the soil with aluminium sheets has proved effective against virus diseases transmitted by aphids through the discouragement of alighting vectors (Smith and Webb 1969).

Some diseases can be substantially reduced by very simple cultivation techniques, e.g. groundnut rosette is of little importance where early planting and dense sowing are practised.

Chemical and biological field control. In tropical agriculture, as in temperate agriculture, it is possible to reduce parasitic attacks by using fungicides, antibiotics, nematicides, and insecticides (the latter by control of the vectors of viruses) (Broadbent 1969).

Of the main groups of fungicides, according to Sinclair (1971), copper compounds are effective against mildews, numerous Fungi Imperfecti (except *Oidium* spp.), and several bacteria, but in normal rates of application are toxic to several crops including rice and potatoes; inorganic sulphur is effective against oidiums and the two *Cercospora* leafspot

diseases of groundnuts; the dithiocarbamates— thiram, zineb, maneb, and mancozeb—can be used for seed-treatments as well as to control mildews and numerous Fungi Imperfecti by spraying crops. Dinocap is generally effective as a spray against powdery mildews. Antibiotics (blasticidine, kasugamycin) are used against rice blast disease, as is ediphensoph (organophosphorus) and pentachloronitrobenzene (PCNB) is used against *Rhizoctonia*.

Organomercury compounds, which used to be much used, especially for seed-treatment, are likely to be prohibited in the fairly near future on account of their danger, especially for treating seed of food crops which might be accidentally eaten.

Systemic fungicides, which have been developed in the last decade, are often very effective, providing control for 10–25 days or more. They have extremely low toxicity for humans or other animals, and thiabendazole indeed is also used as a vermifuge for cattle. The two main compounds used, thiabendazole (TBZ) and benomyl, are both active against a broad spectrum of fungi. However, they are ineffective against downy mildews and rust fungi. The properties of the thiophanates are similar to TBZ and benomyl. Oxathiins are effective against rusts; dimethirimol against certain oidiums; triforine against oidiums and rusts; chloroneb against mildews; and kitazine against *Pyricularia oryzae*. Many of these systemic fungicides are effective in disinfecting seed (Cremlyn 1973).

In West Africa, fungicides are at present used mainly for market-garden crops, to disinfect seed, and, to a minor extent, against rice blast disease and groundnut leafspots caused by *Cercospora* spp.

For use as seed-treatments, economic and health-hazard considerations have led to the introduction of formulae containing too low a content of the active ingredients, especially of insecticides, to retain the necessary effectiveness. Psychologically, since in certain countries peasants judge the action of these compounds on their effectiveness in relation to birds, the use of products of low toxicity is at present a serious handicap in convincing them of the value of the products used.

Whereas nematicides are widely used on bananas grown as a plantation crop for export from West Africa, their use is very limited on this or other crops grown for local food use. Nematicides comprise organohalogen compounds: organochloride compounds such as chloropicrin and the isomers of dichloropropene (D.D.); organobromide compounds such as ethylene dibromide, dibromochloropropane, and methyl bromide; thiocyanates such as dazomet and metam-sodium; and the organophosphorus compounds, parathion and thionazine. New nematicidal products have recently appeared: phenamiphos, aldicarb, fensulfothion, prophos, and methomyl, but have not yet been evaluated.

Two herbicides, salts of 2,4-D and ametryne, can destroy *Striga* after it has sprouted on sorghum and pearl millet without injuring the crop host plant (Ogborn 1971).

Biological control by the use of a virulent strain, particularly of viruses, to induce resistance, is the subject of active research, and indications are that this method may become practicable in the future.

Control by breeding. The selection of plants resistant to pests and diseases should be a normal part of any general plant-breeding programme aimed at improving quality and quantity (Smith 1968). When a new pathogen is suddenly introduced (as occurred when *Puccinia polysora*, the causal organism of American maize rust, first reached Africa in 1950), resistance breeding immediately becomes the main objective of a breeding programme. In all circumstances a fair balance must be drawn between improvement of yield potential and resistance. In the past, too much importance has sometimes been attached to resistance, at the expense of yield improvement.

11.3.2. Breeding for resistance to diseases

Van der Plank (1968) called two types of genetic resistance 'vertical' and 'horizontal'. Vertical resistance is present when there is a differential interaction between the cultivars of the host plant and the races of the pathogen. Horizontal resistance is present when a given level of genetic resistance in the host plant operates uniformly against all races of a pathogen.

Vertical resistance, which may appear stable in the absence of pathogen races able to overcome it, is lost suddenly when a new race appears with matching genes to overcome it.

Robinson (1971) presented a series of 'rules' defining general principles for the proper use of vertical resistance. They include:

1. Vertical resistance is not appropriate for use with perennial crops.
2. Vertical resistance is not desirable to use for plants which are difficult to hybridize.
3. Vertical resistance should preferably be used for diseases of 'simple interest' characteristics of which the pathogens lack strong 'vertical mutability'. This will exclude *inter alia* therefore *Phytophthora infestans*, *Puccinia polysora*, and *Pseudomonas solanacearum*, which mutate frequently.
4. Vertical resistance is also not recommended when the host population is genetically uniform, i.e. when a single clone, or pure-line cultivar of an inbreeding crop, occupies an extensive continuous area.
5. Vertical resistance is more likely to be effective if several 'vertical pathodemes' are cultivated either simultaneously (i.e. cultivars of the same crop possessing different vertical resistance) or successively. Under such circumstances 'stabilizing selection' may act to prevent an overwhelming build-up of any one virulent pathotype.
6. Vertical resistance is likely to be relatively ineffective against a disease transmitted in the seed or vegetative propagation material of the host plant since this provides for the very effective multiplication of any corresponding vertical pathotype.
7. Vertical resistance is recommended when there are clear closed seasons for crops which can correspondingly reduce the effective population of the pathogen and will be particularly effective if it is possible legally to ban the sowing or planting of the old sensitive cultivars once resistant cultivars begin to be cultivated.
8. It is important to reinforce vertical resistance as much as possible by horizontal resistance.

While 'vertical' and 'horizontal' resistance are terms used to describe the behaviour of different kinds of resistance in the epidemiological context, it is probable that the two forms of resistance also have different underlying mechanism of action. Vertical resistance is usually associated with immunity or hypersensitivity, and these phenomena may depend on a variety of physical and chemical mechanisms (Littlefield 1973).

Horizontal resistance, however, takes the outward form of a reduced severity either, for example, in the number of spots in the case of leaf diseases or the longer time taken by a new lesion to first form spores and slower production of spores thereafter. A cultivar having a horizontal resistance has a lower infection rate r, being a measure of the speed at which the epidemic spreads in relation to time under given environmental conditions, than that of a cultivar having lower horizontal resistance. r is measured by use of the formula after van der Plank (1968),

$$r = \frac{1}{t_2 - t_1}\left(\ln\frac{x_2}{1-x_2} - \ln\frac{x_1}{1-x_1}\right),$$

in which x_1 and x_2 are the proportions of diseased tissues at times t_1 and t_2. In general, r is the regression coefficient of $\ln x/(1-x)$ referred to time, where $1-x$ is the proportion of healthy tissue.

As a result of breeding work, many authors consider that vertical resistance corresponds to resistance governed by one or only a small number of genes, i.e. showing simple Mendelian inheritance. Horizontal resistance, however, often appears to be governed by a complex genetic system, which may or may not be due to polygenes (in the sense of the mathematical models of qualitative genetics, i.e. numerous genes of small and similar effect) but which in any case makes genetic manipulation difficult. Killick and Malcolmson (1973) pointed out that the use of specific combining ability is valuable in assessing clone-by-clone crosses in breeding for field resistance in potatoes. Field resistance as used by these authors being the same as horizontal resistance in the sense of van der Plank. Horizontal resistance by a given genotype depends for its expression to a much greater degree than vertical resistance on environmental conditions, including the rates of mineral fertilizer use, temperature regime, and water balance of the plant. In other words, it is often related to the general physiological and metabolic characteristics of the plant.

Although these principles can be considered generally relevant, each pathogen must be examined individually. Indeed, even for fungi of the same genus, e.g. *Fusarium*, the situation can be quite different, as in the vascular wilt diseases of flax, tomato, and cabbage. For flax, there are vertical resistance genes, but none of them is strong. Consequently the resistance tends to break down, and new cultivars must constantly be bred. For the tomato, there is a 'strong' resistance gene, and the gene, introduced from *Lycopersicon pimpinellifolium*, has been used successfully throughout the world for many years, although a new race has appeared recently able to overcome it. For cabbage, monogenic resistance has remained stable up to now (Van der Plank 1968).

11.3.3. *Examples of success in breeding for resistance to fungi and bacteria in food crops*

For the food crops with which we are concerned, the use of varietal resistance has given useful results. Some rice cultivars (Ou and Jennings 1969) have horizontal resistance to blast disease expressed in the small number of leaf lesions of type 3–4 (lesions of types of average sensitivity to average resistance), whereas the determination of vertical resistance genes has so far demonstrated 12 resistance genes (Toriyama 1971).

For maize, recurrent selection in composites has given good rust resistance, and genes for resistance to *Helminthosporium turcicum* can be introduced (Jenkins and Robert 1961). The use of the cytoplasmic male-sterility factor T as a tool for breeding hybrid maize has had, however, to be abandoned because it simultaneously introduced sensitivity to *H. maydis* (Ullstrup 1972).

In the groundnut, differences between cultivars in sensitivity to *Cercospora* disease have frequently been observed. In general, it appears that runner or semi-runner cultivars are more resistant, but with notable exceptions such as Tifton 1108 in India (Chahal and Sandhu 1972).

Resistance to groundnut rosette virus is controlled by two recessive genes. By back-crossing it has been possible to obtain a new cultivar both resistant to rosette and tolerant to the *Cercospora* leafspot diseases (Mauboussin 1970). Two other lines, PI 337394f and PI 337409, have been selected that are only slightly affected by *Aspergillus flavus* (Mixon and Roger 1973).

On market-garden crops, lines have been bred that are tolerant to *Pseudomonas solanacearum*, e.g. the Saturn and Venus cultivars of tomato have shown good resistance in several regions (Daly 1973a). Egg plant lines resistant to *P. solanacearum* have also been developed (Daly 1973b). The 'are' gene is efficient against all races of anthracnose caused by *Colletotrichum lindemuthianum* in the bean.†

Methods used to control viruses include the use of genetic immunity, which, against viruses, is often controlled by a simple recessive gene or hypersensitivity which is often dependent on a major dominant gene (Pochard 1972).

Messiaen and Lafon (1970) have selected a tomato cultivar combining resistance to root-knot nematode

† This is sadly no longer true. Ed. (1975).

(Mi gene) and the vascular wilt caused by *Fusarium oxysporum* (I gene).

Crossing cassava (*Manihot esculenta*) with ceara rubber (*M. glaziovii*) has given resistance to cassava mosaic disease, and this resistance was successfully incorporated into high-yield lines (Nichols 1947; Jennings 1957). However, good resistance was also obtained by crosses within *M. esculenta* (Jameson 1964; and see also Chapter 5, pp. 82–3, this book).

Tolerance, in the strict meaning of the word, exists when, despite severe parasitic infection, with or without accompanying disease symptoms, a clone or other cultivar nevertheless gives a higher yield than others which are apparently affected less severely. In Nigeria some cassava clones show severe symptoms of the mosaic disease, but nevertheless produce 2·4–4 times more than the average of the other clones (Beck 1971).

11.4. Desirable trends

Plant pathology has many complex problems which interact with those of breeders, physiologists, entomologists, and agronomists. Therefore plant pathological studies should be planned when possible on an interdisciplinary basis, i.e. involving teams of people of different expertise brought together to solve production problems. Team-work should not consist merely of bringing together the various specialists required, but should also involve a constant interchanging of ideas and joint examination of the planning and progress of the work. This co-operation should be particularly close between the breeder and the plant pathologist, and a common background of general agricultural understanding is beneficial to such co-operation.

To obtain high and stable yields of good-quality produce is the usual objective. Special attention must be paid to losses accounting for the difference between known potential and actual field performance. Although losses are well enough known to justify control programmes for the most obvious problems, such as rice blast disease and groundnut leafspot disease, in many other instances losses need to be estimated much more thoroughly and regression equations established to link the intensity of the infection at critical periods of the crop ontogeny with losses in yield for a range of cultivars.

In the current state of agriculture in West Africa, there can be no realistic prospect, at least in the near future, of using fungicide treatments (apart from seed-treatments) for cereals such as pearl millet, sorghum, upland rice, and root and tuber crops or for grain legumes such as cowpeas and soya beans. It also appears difficult to envisage justification for chemical treatments for irrigated rice and groundnuts, except in very special circumstances; although for certain market-garden crops, chemical control may be developed. Consequently, in general, efforts must be directed to the selection of resistant crop cultivars.

For almost all crops, studies must be centred on the search for resistant or tolerant genotypes, and this property must then be incorporated by breeding into otherwise advantageous cultivars. This work will clearly have to be done in close co-operation between a pathologist and a breeder. The plant pathologist will be required to advise on the kind of resistance that is desirable, the location of sources of resistance, and on suitable methods for the testing that is routinely necessary in carrying out a practical breeding programme. There must also be the most widely based co-operation between countries to avoid costly and unnecessary duplication of work. For this purpose it must be possible for breeding lines and improved cultivars to move from one continent to another for research and development purposes; but the transfers must be carried out with all due care, so as to avoid introducing alien pathogens or new races of those already present. The conflicts that may arise between the interests of germplasm movement and plant protection through phytosanitary regulations usually need very careful consideration to arrive at a reasoned judgement of the wisest course to adopt in any particular situation.

The plant pathologist must also be in contact with physiologists and biochemists. Host-specific toxins have been found for several parasites (Scheffer and Yoder 1972) (*Helminthosporium victoriae* Meehan & Murphy on oats, *H. carbonum* Ullstrup on maize, *Periconia circinata* (Mangin) Sacc. for sorghum, etc.). Some of these toxins can be conveniently used to test cultivars.

Work on the toxins secreted by the pathogens and on the biochemistry of resistance is also of course important (Blaha 1973), but may be more appropriately carried out in countries having better laboratory facilities and larger numbers of trained professional and technical staff than are yet currently found in most countries in the lowland tropics. As more immediate problems, such as the need to assess the importance of diseases and find economic solutions, begin to become fewer, we may hope for the development of a full range of plant pathological research, provided with all the necessary facilities and equipment, in lowland tropical countries.

References

ADSUAR, J. (1955). A mosaic disease of the yam, *Dioscorea rotundata* in Puerto Rico. *J. Agric. Univ. P. Rico* **39**, 111-13.

ALCONEIRO, R. (1971). Sweet potato virus infection in Puerto Rico. *Plant Dis. Reptr* **55**, 902-6.

—— (1972). Effects of plant age, light intensity and leaf pigments on symptomology of virus-infested sweet potatoes. *Plant Dis. Reptr* **56**, 501-8.

—— ZETTLER, F. W. (1971). Virus infections of *Colocasia* and *Xanthosoma* in Puerto Rico. *Plant Dis. Reptr* **55**, 506-8.

ALUKO, M. O. (1970). *Helminthosporium* leaf spot of rice in Nigeria. *Ford Foundation/IITA/IRAT seminar on research on rice, Ibadan.*

ARNOUX, M. and FRANQUIN, P. (1960). Bilan de dix années de recherches sur les fibres jutières à la station de l'IRCT de Madingou, 1949-59. *Coton Fibres trop.* **15**, 69-80.

ATKINS, J. G. and TODD, E. H. (1959). White tip disease of Rice 111—yield tests and varietal resistance. *Phytopathology* **49**, 189-91.

BAKKER, W. (1970). Rice yellow mottle, a mechanically transmissable virus disease of rice in Kenya. *Neth. J. Plant Path.* **76**, 53-63.

BECK, B. (1971). Cassava production in West Africa. *Ford Foundation/IITA/IRAT seminar on root and tuber crops in West Africa, Ibadan.*

BEYE, G. (1972). Studies on tropical soils: acidification, toxicity and amendments. *Ford Foundation/IITA/IRAT seminar on tropical soils in West Africa, Ibadan.*

BLAHA, G. (1973). Les supports biochimiques de la sensibilité et de la résistance du cacaoyer à *Phytophthora palmivora*. Orientation des recherches au Cameroun. Deuxième réunion du sous-groupe Afrique sur *P. palmivora*, Brazzaville.

BLONDEL, D. (1970). Relation entre le nanisme jaune de l'arachide en sol sableux (dior) et le pH; définition d'un seuil pour l'activité du *Rhizobium*. *Agron. Trop.*, Nogent **25**, 589-95.

BOCK, K. R. (1971). Notes on East African plant virus diseases. 1. Cowpea mosaic virus. *E. Afr. agric. for. J.* **37**, 60-2.

—— (1973). East African strains of cowpea aphid-borne mosaic virus. *Ann. appl. Biol.* **74**, 75-83.

BROADBENT, L. R. (1969). Disease control through vector control. In *Viruses, vectors and vegetation*, pp. 593-630. Interscience Publishers, New York.

BRUNT, A. A. and KENTEN, R. H. (1973). Cowpea mild

mottle, a newly recognized virus infecting cowpeas (*Vigna unguiculata*) in Ghana. *Ann. appl. Biol.* **74**, 67-74.

BUDDENHAGEN, I. and KELMAN, A. (1964). Biological and physiological aspects of bacterial wilt caused by *Pseudomonas solanacearum*. *A. Rev. Phytopath.* **2**, 203-30.

BULL, R. A. (1960). Micronutrient deficiency symptoms in the cocoyam (*Xanthosoma* sp.). *Jl W. Afr. Inst. Oil Palm Res.* **3**, 181-6.

BUYCKX, E. J. E. (1962). *Précis des maladies et des insectes nuisibles rencontrés sur les plantes cultivées au Congo, au Rwanda et au Burundi*, p. 708. Publications de l'Institut National pour l'Étude Agronomique de Congo—hors série.

CALDWELL, R. W. and TUITE, J. (1970). Zearalenone production in field corn in Indiana. *Phytopathology* **60**, 1696-7.

CAMMACK, R. H. (1958). Studies on *Puccinia polysora* Underw. I. The world distribution of forms of *Puccinia polysora*. *Trans. Br. mycol. Soc.* **41**, 89-94.

—— (1959). Studies on *Puccinia polysora* Underw. II. A consideration of the method of introduction of *Puccinia polysora* into Africa. *Trans. Br. mycol. Soc.* **42**, 27-32.

CARPENTER, A. J. (1973). *Crop losses affecting rice in Liberia*. West Africa Rice Development Association S/P/73/20, Monrovia.

CAVENESS, F. F. (1967). *End of tour progress report on nematology project*. Ministry of Agriculture, Nigeria and USAID.

CHAHAL, D. S. and CHOHAN, J. S. (1971). Rouille de l'arachide. *Bull. phytos.*, *FAO* **19**, 90.

—— SANDHU, R. S. (1972). Reaction of groundnut varieties against *Cercospora personata* and *C. arachidicola*. *Plant Dis. Reptr.* **56**, 601-3.

CHANT, S. R. (1959). Virus diseases of cowpea, *Vigna unguiculata*, in Nigeria. *Ann. appl. Biol.* **47**, 565-72.

—— (1960). The effect of infection with tobacco mosaic virus and cowpea yellow mosaic virus on the growth rate and yield of cowpea in Nigeria. *Emp. J. exp. Agric.* **28**, 114-20.

CHHEDA, H. R. (1971). Cynodon improvement in Nigeria. *Ford Foundation/IITA/IRAT seminar on forage crop research, Ibadan.*

CHESTER, K. S. (1950). Plant diseases losses: their appraisal and interpretation. *Plant Dis. Reptr* Suppl. 193.

CHIARAPPA, L. and CHIANG, H. C. (1971). The need, value and procurement of reliable crop loss information. *Ford Foundation/IITA/IRAT seminar on plant protection of tropical food crops, Ibadan.*

CHOPART, J. L. and NICOU, R. (1971). Effet dépressif de cultures répétées du sorgho dans les sols sableux du Sénégal. Premiers essais d'explication. *Seminaire CSTR/OUA sur les facteurs du milieu qui influencent les rendements des cultures tropicales, Dakar.*

CHOUDHURY, B. (1971). Research on improvement of

cucurbits in India. *Ford Foundation/IITA/IRAT seminar on vegetable crop research, Ibadan.*

COOK, R. J. (1973). Influence of low plant and soil water potential on diseases caused by soilborne fungi. *Phytopathology* **63**, 451-8.

COSTA, A. S. and KITAJIMA, E. W. (1972). *Cassava common mosaic virus.* CMI/AAB Descriptions of plant viruses No. 90, pp. 1-4.

CREMLYN, R. J. W. (1973). Systemic fungicide. A widening choice for agriculture. *Int. Pest Control* **15**, 8-16.

CUMMINGS, G. B. (1941). Identity and distribution of three rusts of corn. *Phytopathology* **31**, 856-7.

CUNHA, J. M. DE A. and BAPTISTA, J. E. (1938). Estudo da branca do Arroz I. Combate da doenca. *Agron. luset.* **20**, 17-64.

DALY, P. (1973*a*). Obtention d'une variété d'aubergine tolérante au *Pseudomonas solanacearum. Agron. Trop., Nogent* **28**, 23-7.

—— (1973*b*). Étude de trois variétés de tomates tolérantes au *Pseudomonas solanacearum. Agron. Trop., Nogent* **28**, 28-33.

DARPOUX, H. (1971). Les maladies végétales occasionnées par des microorganismes de type 'mycoplasmes'. *Phytoma* **23**, 16-23.

DAVIS, R. E. (1973). Occurrence of a spiroplasm in corn stunt infected plants in Mexico. *Plant Dis. Reptr* **57**, 333-7.

—— WHITCOMB, R. F. (1971). Mycoplasmas, Rickettsiae and Chlamydiae: possible relation to yellow diseases and other disorders of plants and insects. *A. Rev. Phytopath.* **9**, 119-54.

DELASSUS, M. (1968). Principales maladies du maïs dans l'Ouest Cameroun. *Agron. Trop., Nogent* **23**, 429-34.

—— (1972). *Rapport de synthèse Phytopathologie— Aménagement hydroagricole pilote dans la vallée de l'Ouémé.* SEDAGRI, Paris.

EKANDEN, M. J. (1970). *Cassava research in Nigeria before 1967.* Memorandum 103—Federal Department of Agricultural Research, Ibadan.

FEAKIN, S. D. (1967). *Pest control in groundnuts.* PANS Manual No. 2. Ministry of Overseas Development, London.

—— (1970). *Pest control in rice.* PANS Manual No. 3. Ministry of Overseas Development, London.

FOOD AND AGRICULTURE ORGANIZATION (1971). *F.A.O. Manual on crop loss assessment methods.* Commonwealth Agricultural Bureaux, Slough, England.

FOWLER, A. M. (1970). The epidemiology of *Cercospora* leaf spot disease of groundnuts. *Ford Foundation/ IITA/IRAT seminar on research on grain legumes, Ibadan.*

FUTTRELL, M. C. and WEBSTER, O. J. (1967). Fusarium scab of sorghum in Nigeria. *Plant Dis. Reptr* **51**, 174-8.

GERMANI, G. (1971). Une chlorose des légumineuses de Haute Volta liée à la présence d'un nématode. *C. r. hebd. Séanc. Acad. Agric. Fr.* **58**, 202-5.

GILLIER, P. and SILVESTRE, P. (1969). *L'arachide.* G.P. Maisonneuve et Larose, Paris.

GRANADOS, R. R. (1969). Maize viruses and vectors. In *Viruses, vectors and vegetation,* pp. 327-59. Interscience Publishers, New York.

HARDCASTLE, J. Y. (1963). Outbreaks and new records— Nigeria. *FAO Plant Prot. Bull.* **11**, 115-16.

HARRIS, D. C. (1971). A survey of bacterial diseases of tropical food-crop plants. *Ford Foundation/IITA/IRAT seminar on plant protection of tropical food crops, Ibadan.*

HARTLEY, C. and RATHBUN-GRAVATT, A. (1937). Some effects of plant diseases on variability of yields. *Phytopathology* **27**, 159-71.

HAYWARD, A. C. (1964). Characteristics of *Pseudomonas solanacearum. J. appl. Bact.* **27**, 265-77.

HOUSE, L. R. (1970). A world view of sorghum research including the economics of production. *Ford Foundation/ IITA/IRAT seminar on sorghum and millet research in West Africa, Bambey.*

IITA (1971). Report of the drafting committee of the *Ford Foundation/IITA/IRAT seminar on vegetable crop research, Ibadan.*

IITA (1972*a*). Grain legume improvement programme. Report, p. 83.

IITA (1972*b*). Root, tuber and vegetable improvement programme. Report, p. 48.

IITA (1973). Letter 2. Ibadan, May 1973.

JAMESON, J. D. (1964). Cassava mosaic disease in Uganda. *E. Afr. agric. J.* **29**, 208-13.

JENKINS, M. T. and ROBERT, A. L. (1961). Further genetic studies of resistance to *Helminthosporium turcicum* Pass. in maize by means of chromosomal translocations. *Crop Sci.* **1**, 450-5.

JENNINGS, D. L. (1957). Further studies in breeding cassava for virus resistance. *E. Afr. agric. J.* **22**, 213-19.

—— (1960). Observations on virus diseases in resistant and susceptible varieties. II—Brown streak disease. *Emp. J. exp. Agric.* **28**, 261-70.

JOUAN, B. and DELASSUS, M. (1971). Principales maladies des mils et sorgho observées au Niger. *Agron. Trop., Nogent* **26**, 830-60.

KANG, B. T. and OSINAME, O. A. (1972). Micronutrient investigations in West Africa. *Ford Foundation/IITA/ IRAT seminar on tropical soil research, Ibadan.*

KANNAIYAN, J., VIDHYASEKAREN, P., and KANDASWANY, T. K. (1971). Mammalian toxicity of ergot of bajra. *Curr. Sci.* **40**, 557-8.

KENTEN, R. J. and WOODS, R. D. (1973). Viruses of *Colocasia esculenta* and *Xanthosoma saggitifolium. Pest Artic. News Summ. (PANS)* **19**, 38-41.

KILLICK, R. U. and MALCOLMSON, J. F. (1973). Inheritance in potatoes of field resistance to late blight (*Phytophthora infestans* (Mont.) de Bary). *Physiol. Plant Path.* **3**, 121-31.

KING, S. B. (1970). Millet diseases. *Ford Foundation IITA/IRAT seminar on sorghum and millet research in West Africa, Bambey.*

M

KÜHN, C. W. (1964). Mechanical transmission of a virus causing leaf mottle of peanuts. *Phytopathology Abst.* 54, 624.

KULKARNI, H. Y. (1973). Comparison and characterisation of maize stripe and maize line viruses. *Ann. appl. Biol.* 75, 205-16.

LAMEY, H. A. and WILLIAMS, R. J. (1972). Leaf scald of rice in West Africa. *Plant Dis. Reptr* 56, 106-7.

LANDIS, W. R. (1971). The effect of carbofuran on stalk rot of corn. *Plant Dis. Reptr* 55, 634-8.

LAUNDON, G. F. and WATERSTON, J. M. (1965). *Uromyces appendiculatus.* C.M.I. Descriptions of Pathogenic Fungi and Bacteria No. 57.

LITTLEFIELD, L. J. (1973). Histological evidence for diverse mechanisms of resistance to flax rust, *Melampsora lini* (Ehrenb.) Lev. *Physiol. Plant Path.* 3, 241-7.

LUC, M. (1971). Nematode parasite of food crops in West Africa. *Ford Foundation/IITA/IRAT seminar on plant protection of tropical food crops, Ibadan.*

MARAMOROSCH, K. (1959). An ephemeral disease of maize transmitted by *Dalbulus elimatus. Entomol. exp. appl.* 2, 169-70.

MARTIN, N. J. (1967). Sweet potato diseases and their control. *International Symposium on tropical root crops, Trinidad,* Pt. IV, pp. 1-8.

MARTYN, E. B. (1968). *Plant virus names—an annotated list of names and synonyms of plant viruses and diseases.* Commonwealth Mycological Institute, Kew.

MAUBOUSSIN, J. C. (1970). Groundnut breeding by IRAT. *Ford Foundation/IITA/IRAT seminar on research on grain legumes, Ibadan.*

MERNY, G. (1971). Parasitic nematodes associated with root crops in West Africa. *Ford Foundation/IITA/IRAT seminar on research on root and tuber crops in West Africa, Ibadan.*

MESSIAEN, C. M. and LAFON, R. (1970). *Les maladies des plantes maraîchères* (2nd edn.). I.N.R.A., Paris.

—— FOURNET, J., BEYRIES, A., and QUIOT, J. B. (1971). Some major plant disease problems in tropical market gardening. *Ford Foundation/IITA/IRAT seminar on vegetable crop research, Ibadan.*

MILLER, L. I. and TROUTMAN, J. L. (1966). Stunt disease of peanuts in Virginia. *Plant Dis. Reptr* 50, 139-43.

MIXON, A. C. and ROGER, K. M. (1973). Peanuts resistant to seed invasion by *Aspergillus flavus. Oléagineux* 28, 23-7.

MIZUKAMI, T. and WAKIMOTO, S. (1969). Epidemiology and control of bacterial leaf blight of rice. *A. Rev. Phytopath.* 7, 51-72.

MOREAU, C. (1968). *Moisissures toxiques dans l'alimentation.* Lechevalier, Paris.

NASU, S., SIGIURA, M., WAKIMOTO, S., and IIDA, T. T. (1967). Pathogen of rice yellow dwarf disease. *Ann. phytopath. Soc. Japan* 33, 343.

NENE, Y. L. (1972). Khaira disease of rice (*Oryza sativa* L.). *Ind. J. agric. Sci.* 42, 87-95.

NETSCHER, C. (1971). Problems caused by the nematodes in vegetable crop production in the inter-tropical zone. *Ford Foundation/IITA/IRAT seminar on vegetable crop research, Ibadan.*

NICHOLS, R. W. F. (1947). Breeding cassava for virus resistance. *E. Afric. agric. J.* 12, 184-94.

—— (1950). The brown streak disease of cassava. *E. Afr. agric. J.* 15, 154-60.

OGBORN, J. E. A. (1971). Methods of controlling *Striga hermonthica* for West African farmers. *Ford Foundation/IITA/IRAT seminar on sorghum and millet research in West Africa, Bambey.*

OKUSANA, B. A. M. and WATSON, M. A. (1966). Host range and some properties of groundnut rosette virus. *Ann. appl. Biol.* 58, 377-87.

OU, S. H. (1972). *Rice diseases,* p. 360. Commonwealth Agricultural Bureaux, London.

—— JENNINGS, P. R. (1969). Progress in the development of disease resistant rice. *A. Rev. Phytopath.* 7, 383-410.

PAYAK, M. H., SANGAM, L., JANKI, L., and RENFRO, B. O. (1970). *Cephalosporium maydis.* A new threat to maize in India. *Ind. Phytopath.* 23, 562-9.

PEACOCK, F. C. (1956). The reniform nematode in the Gold Coast. *Nematologica* 1, 307-10.

POCHARD, E. (1972). Problèmes posés aux sélectionneurs par les maladies à virus. *Le sélectionneur français* 14, 27-33.

PONNAMPERUMA, F. N. (1958). Lime as a remedy for physiological disease of rice associated with excess iron. *Int. Rice Comm. Newslett.* 7.

POSNETTE, A. F. (1945). Root rot of cocoyams (*Xanthosoma saggitifolium* Schott). *Trop. Agric., Trin.* 22, 164-70.

RADULESCU, E., PERSICA, E., and POPESCU, I. (1959). Influenta applicarii ingrasamintelor la Porumb asupra atacului cuipercii. *Sorosporium holci-sorghi* (Riv.) Moesz f. zeae. *Stud. Cercet. Agron. Aced. Repub. cbv.* 10, 161-8. (Seen in *Rev. appl. Mycol.* (1962), p. 301.)

ROBINSON, R. A. (1971). Vertical resistance. *Rev. Plant Path.* 50, 233-9.

ROGER, L. (1953). *Phytopathologie des pays chauds,* Vol. 2, pp. 1129-2256. Lechevalier, Paris.

RUPPEL, E. G., DELPHIN, H., and MARTIN, F. W. (1966). Preliminary studies on a virus disease of a sapogenin-producing *Dioscorea* in Puerto Rico. *J. Agric. Univ. P. Rico* 50, 151-7.

SCHEFFER, R. P. and YODER, O. C. (1972). Host specific toxins and selective toxicity. In *Phytotoxins in plant diseases,* pp. 251-72. Academic Press, London.

SCHINDLER, A. J. (1942). Insect transmission of wallaby ear disease of maize. *J. Aust. Inst. agric. Sci.* 8, 35-7.

SETH, M. L., RAYCHAUDHURY, S. P., and SINGH, D. V. (1972). Bajra (pearl millet) streak: a leafhopper-borne cereal disease in India. *Plant Dis. Reptr* 56, 424-8.

SHARMA, D. C. (1966). Studies on 'bunchy top', 'chlorosis' and 'ring mottle' virus diseases of groundnuts (*Arachis hypogaea* L.). *Phytopath. Z.* 57, 127-37.

SHEFFIELD, F. M. L. (1953). Virus diseases of the sweet potato in parts of Africa. *Emp. J. exp. Agric.* **21**, 184-9.

—— (1957). Virus diseases of sweet potato in East Africa. *Phytopathology* **57**, 582-90.

SINCLAIR, J. B. (1971). Fungicides for use on tropical food crops. *Ford Foundation/IITA/IRAT seminar on plant protection of tropical food crops, Ibadan.*

SINGH, R. S., CHAUBE, H. S., SINGH, N., ASNANI, V. L., and SINGH, R. (1970). Observations on the effect of host nutrition and seed, soil and foliar treatments on the incidence of downy mildews. *Ind. Phytopath.* **23**, 209-15.

SINHA, S. K. and NENE, Y. L. (1967). Eradication of the seedborne inoculum of *Xanthomonas oryzae* by hot water treatment of paddy seeds. *Plant Dis. Reptr* **51**, 880-3.

SMITH, F. F. and WEBB, R. E. (1969). Repelling aphids by reflecting surfaces, a new approach to the control of insect-transmitted viruses. In *Viruses, vectors and vegetation*, pp. 631-9. Interscience Publishers, New York.

SMITH, H. C. (1968). Breeding crops for resistance to diseases. *Pest Artic. News Summ. (PANS)* **11**, 89-91.

SRIVASTANA, D. N. and RAO, Y. P. (1966). Symptoms and diagnosis of the bacterial blight disease of rice. *Curr. Sci.* **36**, 60-1.

STOREY, H. H. (1936). Streak disease of maize. *E. Afr. agric. for. J.* **1**, 471.

—— HOWLAND, A. K. (1967). Inheritance of resistance in maize to the virus of streak disease in East Africa. *Ann. appl. Biol.* **59**, 429-36.

—— McLEAN, P. D. (1930). The transmission of streak disease between maize, sugar cane and wild grasses. *Ann. appl. Biol.* **17**, 601-719.

—— NICHOLS, R. F. W. (1938). Studies of the mosaic disease of cassava. *Ann. appl. Biol.* **25**, 790-806.

—— RYLAND, A. K. (1957). Viruses causing rosette and other diseases in groundnuts. *Ann. appl. Biol.* **45**, 318-26.

STRICKLAND, A. H. (1971). The actual status of crop loss assessment. *EPPO Bull.* **1**, 39-51.

TARR, S. A. J. (1951). *Leaf curl disease of cotton.* Commonwealth Mycological Institute, Kew.

—— (1955). *Fungi and plant diseases of the Sudan.* Commonwealth Mycological Institute, Kew.

—— (1962). *Diseases of sorghum, sudangrass and broom corn.* Commonwealth Mycological Institute, Kew.

THOMAS, K. M. and KRISHNASWAMI, C. S. (1939). Little leaf—a transmissable disease of Brinjal. *Proc. Ind. Acad. Sci.* B **10**, 201-12.

TORIYAMA, K. (1971). Recent progress of studies on horizontal resistance in rice breeding for blast resistance in Japan. *Seminar on horizontal resistance to the blast disease of rice.* C.I.A.T., Colombia.

ULLSTRUP, A. J. (1970). Opportunities for international cooperative research on downy mildews of maize and sorghum. *Ind. Phytopath.* **23**, 386-8.

—— (1972). The impacts of the southern corn leaf blight epidemics of 1970-71. *A. Rev. Phytopath.* **10**, 37-50.

VAN DER PLANK, J. E. (1968). *Disease resistance in plants.* Academic Press, New York.

VAN EIJNATTEN, C. L. M. (1961). Susceptibility to leaf blight caused by *Cochliobolus heterostrophus* in Nigerian varieties of maize. *Nature, Lond.* **191**, 515-16.

ZAUMEYER, W. J. and THOMAS, H. R. (1957). *A monographic study of bean diseases and methods for their control.* Tech. Bull. U.S. Dep. Agric. 868.

12. Grain Storage

D. W. HALL

12.1. Introduction

In areas where dense populations live at subsistence levels, local failure of a basic food crop can bring famine and suffering; while in other parts of the same country there may be food surpluses in store. The loss of food, both in quality and quantity, from lack of use of existing knowledge on the control of pests, inadequate transport, and poor dissemination of information on current stocks and future requirements all contribute to the risk of famine. Much handling of stored products is carried out with unreliable equipment.

Increases in production of food grains provide opportunities for communities to utilize the increased supplies as food and animal feed (to produce yet more food) and for the development of agro-industries. Such is not automatically the case, however, as there are difficulties in integrating increased production into an economy that is geared to utilize only subsistence or marginal production.

The following objectives must be reached if increased food production is to be used to maximum benefit:

(1) the reduction of human drudgery through improvement of the economic situation of farmers and a raising of living standards;

(2) the reduction of food losses to a minimum and the maintenance of high quality during storage, to foster the economic development and reduce human suffering from famine or malnutrition;

(3) an increase in the efficiency in the handling and distribution of food grains from field to consumer.

The main harvested crops are cereals, oilseeds, pulses, root crops, vegetables, and fruits, all of which deteriorate in quality after harvest. General or specific treatises of these problems have been given by Hall (1970), the Asian Productivity Organization (1970), Majumder and Venugopal (1969), and Munro (1966). Cereals, oilseeds, and pulses comprise the grains which are stored.

There is an urgent need at the farmer, trader, and government levels to protect harvested crops from deterioration and to promote their full utilization to the benefit of all in rural and urban areas. To maximize utilization of environmental resources requires the integration of objectives in agriculture, sociology, and economics. The major constraints on this integration within an agricultural and marketing complex at present are:

(1) lack of implementation of research results through extension;

(2) lack of appreciation of the role of the primary buyer;

(3) inadequate linkage of the appropriate activities of agricultural extension staff and of produce inspectorate staff at primary and secondary buyer levels;

(4) inadequate communication between agriculturalists, produce inspectors, pest-control staff, and community-development officers.

The range of post-harvest problems encountered varies with the local customs and prevailing agricultural systems. In all regions of the world four important criteria determine the problems of grain storage:

(1) climate;

(2) period of harvesting;

(3) temperature regime;

(4) relative humidity.

For each country, region, or area, a storage climatic pattern (SCP) can be prepared. For the humid lowland tropics there are two main forms of SCP:

1. Areas with single rainy season and the crops harvested during the rains: during the storage period which follows the relative humidity is less than 75 per cent except during a period of 4 months when it is between 75 per cent and 80 per cent and the temperature range is generally 23–30 °C

but may drop below 23 °C and rise to 32 °C. In such areas there are drying and insect problems to be overcome.

2. Areas with two rainy seasons each year and the crops harvested following the two main peaks. During the storage periods the relative humidity is generally 70–75 per cent for about 10 months with the stored crop exposed to an increase in relative humidity during the second rains, and the temperature is generally 23–27 °C with a minimum of 16 °C. In such areas there are drying, moisture re-absorption, and insect problems.

Using the SCP concept it seems possible to predict the major problems to be encountered for each area and, therefore, from the technology available, to suggest technically possible solutions appropriate to different socio-economic conditions.

12.2. Storage considerations

12.2.1. Economics

Losses. In many countries the extent of losses after harvest have not been fully assessed. Those assessments that have been made indicate a serious wastage. Assessment of loss of quantity and quality of produce should be undertaken on the farm, in traders' stores, and at central storage depots, as well as in processing plants and in the export chain.

An FAO expert committee (Food and Agriculture Organization 1968) estimated in 1946 that world storage losses for cereals, pulses, and oilseeds, resulting from attack by insects, mites, rodents, and moulds, were of the order of 10 per cent. For cereals alone this is equivalent to an annual storage loss of more than 100 million t. For the tropics and sub-tropics post-harvest losses may average some 20 per cent, which for harvested grains in Africa, Asia, and Latin America could represent about 77 million t (Hall 1969).

For post-harvest losses caused by insects, reliable estimates on a country basis are difficult to obtain because of the variety of storage methods used, the continually changing quantities in store, and the trading practices adopted. Data which can be collected from the records of central depots are seldom sufficiently accurate or reliable to give a real value for the loss of food material. Subsistence farmers do not keep relevant records, but data collected from actual sample farms show that in some specific situations losses exceed 30 per cent (Hall 1969).

In Ghana (Food and Agriculture Organization 1969) cob maize stored in the sheath by farmers suffered losses from 8 per cent to 16 per cent over storage periods of 8–23 weeks, whilst cob maize stored without the sheath suffered a loss of 26 per cent after 23 weeks. Shelled maize stored for a similar period showed a loss in weight of 34 per cent. The author estimated annual losses for maize based on 1966 Ministry of Agriculture production figures. Using an estimated average weight loss of 15 per cent and annual production of 350 000 t of maize the financial loss was calculated as N₵ 4 200 000.† A similar exercise for cowpea led to an estimated loss of N₵ 280 000.

Considerable work has been done on damage and loss to stored cowpeas in the northern states of Nigeria. Caswell (1971) has estimated 4·5 per cent of annual production of 1 200 000 t is lost, i.e. over 55 000 t, equivalent to about 14 000 t of protein.

In Asia, average estimates of grain losses at different steps in the marketing process have been given by Stevenson (1971) as: field transportation 2·0 per cent; drying 2·0 per cent; storage transportation 1·5 per cent; storage 4·0 per cent; processing transportation 1·0 per cent; processing 5·0 per cent; packaging and transportation 1·0 per cent.

Although insects can be regarded as the main agents of loss, rodents can also cause significant losses of stored foodstuffs and contamination by excreta. Majumder (1968) cites 130 diseases as being transmitted by rats, and the Indian rat population is at least 2400 million. Each rat consumes about 26 g of food per day (Deoras 1967) and contaminates a further 200 g; the daily human intake of cereals is about the same as the quantity which a single rat in a day renders unfit for consumption.

In general, loss of produce due to rodents after harvest is less than that before harvest. Meaningful estimates of post-harvest loss can be obtained only in relation to the population densities of rodents in specific townships or warehouses. Extrapolation to other areas, and particularly to a whole country, is almost meaningless.

Losses due to moulds are very difficult to quantify. Moulds lead to both qualitative change and quantitative loss. Of particular importance is the production, by certain species of fungi, of toxic metabolites or mycotoxins (e.g. aflatoxin which is produced by *Aspergillus flavus*), the presence of which may render

† The currency used is the Ghanaian 'new cedi'.

the crop unsaleable in some markets and for some uses.

Basic foodgrain production (million tonnes) in Africa is about 51 for cereals, 8 for oilseeds, and 4 for pulses, all of which is handled on the farm and stored for varying lengths of time; some 85 per cent, or 53 million t, of foodgrains remain in farm storage, which emphasizes the necessity for rural-level attention to storage. Lack of improved methods for handling and storing harvests is one of a number of constraints on the expansion of production.

Hall (1971*a*) enumerated the following as causing losses to crops immediately prior to and after harvest.

1. Soil pests eating the grains, contributing to contamination by moulds and chemical poisons.
2. Plant pests living on the grains, causing distortion and discoloration.
3. Birds, rodents, and monkeys consuming the crop prior to harvest.
4. Mould development on crops if harvesting occurs during wet weather.
5. Premature drying of the crop before maturity, resulting in shattering, cracking, or scorching of the grain.
6. Delay in harvesting after reaching peak maturity, resulting in heavier insect infestation.
7. Inadequate drying after harvesting, resulting in mould and toxin development.
8. Rats, and chickens or other birds consuming produce spread on the ground to dry or stored in open-sided cribs.
9. Theft by man.
10. Inefficient threshing and shelling methods resulting in breakage, cracking, and lack of cleaning. This may in turn accelerate pest attack in storage.
11. Inadequately constructed containers, permitting rain to penetrate and wet the stored produce, thus allowing mould to develop. This occurs particularly with store designs where a thatch roof has to be moved whenever produce is loaded or unloaded.
12. Re-absorption of moisture from humid air during storage through use of inappropriately designed container for climatic conditions, e.g. continuing storage in an open-type crib when produce is already dry and a subsequent rainy season is beginning.
13. Unloading, which requires a person to enter the crib, may result in shattering of grains from panicles with consequent loss (e.g. from traditional cribs constructed of plant material and with top opening for entry).
14. Beetle and moth pests eating and multiplying in the stored grain, resulting in considerable loss of foodstuff (and lowering of grade obtained when produce is sold in a discriminating market).
15. Design of traditional storage containers which necessitates the first-loaded grain being the last unloaded.
16. Unawareness of the value of solid-walled containers and the admixture with foodgrains of sand, earth, ash, or the less toxic chemicals such as 'blue-cross' formulations of lindane or malathion.
17. Lack of attention by plant breeders to introducing characteristics which will (a) accelerate speed of drying of mature foodgrains, and (b) provide resistance to insect attack and breeding.
18. Inadequate inspection of produce with rejection of poor-quality samples to ensure storage of only clean, good-quality grains.
19. Inefficient use of modern insecticides, such as lindane, malathion, and pyrethrins, even when details are known locally.
20. Non-availability of modern insecticides in appropriately sized packs.
21. Lack of encouragement of use of low storage temperatures (this method would be especially useful in upland areas) or periodic aeration of stored foodgrains.
22. Lack of adequate bulk storage facilities and warehouse practices in the major collecting centres.

With extension of known simple technology, farmers and traders would be able to implement the new knowledge and see and understand that good storage is a major benefit in the rural sector. At centralized storage centres technically trained staff should be employed on the effective management of warehouses and silo complexes.

A detailed consideration of the assessment of the different types of losses is given by Hall (1970) and present needs have also been highlighted in a recent seminar (GASGA 1973).

Economic aspects of alternative storage methods. The cost of storage varies with the quantity and type of grain, the period over which it is kept, the type of handling facilities required, and, in towns, the value of the space. For the small farmer, where satisfactory structures can be constructed using locally available material, the only extra inputs required will be concerned with improved drying and insect control methods.

Caswell (1961) showed that if Nigerian farmers were able to store cowpeas in good condition for 6 months they could expect a gross profit of US$12–24 per ton with a capital outlay of about US $360 for a bin holding 25 tons and fitted with an auger. Such an installation could be expected to last for 10 years, giving an annual cost of about US $1·50 per ton.

Anthonio (1971) discussed the economics of storage with particular reference to market prices. He concluded that rapid improvement in storage methods in Nigeria, and other developing countries in Africa, was difficult partly because individual production was too low for economic use of modern storage facilities on the farm. For efficient larger-scale storage systems the following points require consideration in a thorough pre-investment feasibility study for individual commodities, including consideration of the following factors:

(1) nature of the supply of the commodity with regard to quantity, quality and sources, with due emphasis on the marketable surplus;
(2) total effective demand with respect to quantity, quality, and time (or seasonal) requirements;
(3) the nature and capacity of existing storage;
(4) the location and capacity of desired storage;
(5) the type of storage appropriate to a given situation;
(6) the decision-making and management of the storage system with respect to movement or inventory of the stored commodities.

Any effective storage policy must reflect and be integrated with the over-all agricultural programme. In areas where there is a tradition of communal or co-operative activity there is the possibility of introducing larger-scale storage and improved pest control at the village or rural community level.

During 1959–60 a detailed study of the economics of storage and drying of maize, sorghum, and cowpea, using equipment appropriate for farmer–trader level, was carried out at Ibadan, Nigeria (Upton 1962a, 1962b; Anthonio 1962). The cost of drying maize from an initial moisture content of 19 per cent to 13 per cent, using an oil-fired batch drier, was calculated as US $7·0 per ton. However, this was based on the use of the drier to dry only 40 tons of maize at the rate of a ton per day; clearly, had the drier

been used at other times, the depreciation cost per ton would significantly reduce. Total storage costs for 40 tons over an 8-month storage period, including cost of fumigation, were US $2·20 per ton. The grain was stored in two concrete block bins and one steel bin, and this figure is based upon an estimated life of 10 years for the former and 25 years for the latter. At the end of the storage period the maize was sold, realizing a profit of US $21·90 per ton over the purchase price. Similarly, for sorghum a profit of US $10·20 per ton was realized after 9 months storage (of 6·2 tons). The sale value exceeds the purchase price by virtue of the seasonal variation in price. Under normal conditions of bag storage, without the use of insect-control measures, an over-all loss could be anticipated. The sale values referred to take into account any loss that occurred due to moisture loss or infestation.

During 1961–2 a comparative study of the cost of storage of millet, sorghum, paddy, and cowpeas in bags, individual silos, and co-operative silos was carried out in Senegal (Bonlieu, Nicou, and Tourte 1964); this showed a comparison of 66, 40, and 110 CFA francs per quintal per year respectively (a quintal = 100 kg).

Primary traders and co-operatives usually hold produce in bags and in bulk. Grains in bags are often kept in rooms of similar construction to the trader's dwelling house. The economics of co-operative storage can vary with the size of the installation and the amount of drying and bulk-handling facilities as has been indicated by Corbett (1971) and WASPRU (1962).

For central storage, two types of buildings or building complexes are considered: warehouses for storage in bags and silos for storage in bulk. Storage units with minimum drying and cleaning machinery, and designed for about an annual turnover of stocks, can cost up to about £10–£15 per tonne stored. The lowest cost possible is about £2 per tonne for temporary inflatable bag-storage warehouses. Storage and handling complexes with cleaning, drying, elevating facilities, and fumigatable units to handle three or four complete turnovers of stock per year can require a capital outlay of up to some £30 per tonne. Both Hall (1970) and Takasu (APO 1971) have provided costs for different types of warehouse construction in Asia. Also, commercial warehouses may be rented, the storage fee in some countries consisting of the rate calculated from the value and weight of the grain (APO 1971).

The prevention of pest damage can be achieved through the design and construction of sealed storage containers and by the use of drying techniques. In non-airtight containers, however, insects must be controlled by using chemicals. Under tropical conditions there are few reliable guides to the economics of pest control other than those originating from experimental stations.

In francophone Africa, Pointel (1967) has reported on grain storage in Niger and Togo. Expenditure of 2·50 CFA francs per kg of grain stored to buy a plastic sack and capsules of carbon tetrachloride which prevented 20–40 per cent loss of grain, increased the value over a 4-month storage period from 17 francs per kg to 40 francs per kg. In Ghana, Rawnsley (Food and Agriculture Organization 1969) reported on the economics of polyethylene plastic sacks and ampules of ethylene dibromide to maintain the quality of beans. Farmers received 6 pesewas* per lb wholesale, and the retail price for unprocessed beans was 18 pesewas. By spending 2 pesewas per lb on the packaging–fumigation procedure (including labour) farmers received a wholesale price of about 15 pesewas per lb. In Nigeria, a net gain of 14·4 per cent followed fumigation of cowpeas and their storage in a concrete bin; the cost of the chemical treatment was less than 20 kobo† per tonne (Caswell 1961; WASPRU 1962).

Cheap control of maize-storage pests in farmers' maize cribs in Kenya was described by Coyne (1971). The empty cribs were sprayed using 50 per cent lindane in a wettable-powder formulation. Maize in cob was dusted with 1·0 per cent lindane or 2·0 per cent malathion dust at a rate of 50 g per bag of cobs (9 ft³) before filling the cribs. This treatment, which cost 8 Kenya shillings per 20 bags of shelled maize, avoided the loss of 3 bags of maize, valued 6–9 months later at about 40 Kenya shillings per bag.

12.2.2. *Inspection and quality assessments*

In many countries inspection of produce for quality is practised only for export crops. There is a welcome trend, in Nigeria for example, to broaden the terms of reference of produce inspection services to include attention to local food crops (Okwelogu 1969).

Pattinson (1971) has outlined existing quality standards for produce in tropical Africa. International commodity contracts have now been established for a wide range of African crops, and the

effects of the externally imposed requirements on the sellers have been discussed. Developed countries are demanding increasingly high standards of quality for imported raw food products. Requirements for certified freedom from aflatoxin, for example, impose a particular burden on producers and quality-inspection services.

Pattinson (1971) and Ashman (1966, 1971) have discussed the various aspects of quality. These may be broadly enumerated as:

(1) freedom from adulteration by other seeds and foreign matter;

(2) correct maturity, which will influence the chemical constituents of the grain;

(3) moisture content, which influences the susceptibility of produce to attack by pests, and materially affects the apparent weight of produce bought;

(4) cultivar characteristics, e.g. in East Africa white maize attracts a higher price than yellow maize; cultivar characteristics will also influence processing quality and susceptibility to insect attacks;

(5) freedom from insect contamination and resulting losses and effect upon appearance;

(6) freedom from damage (for which tolerance limits exist in the marketing of cocoa, coffee, and tea) and the resulting problem of contamination by mycotoxins.

Ashman (1971) suggested that anticipation and detection of quality changes are the primary roles of an alert produce inspection service. Early detection should be followed by sound advice on prevention of further quality deterioration and recommendations for treatment and safe storage. He also discusses in detail methods to be used in the inspection of produce.

In addition to direct inspection of any commodity in store, when inspection is carried out it is important to record the fullest details possible of the storage environment, e.g. the nature and condition of store construction, level of store organization and hygiene, insect infestation, information on pest-control treatments being carried out, and temperature and relative-humidity data.

In considering the commodity itself, mechanical and insect damage will be indicated by the quantity of dust falling from packages during handling. Temperature-measuring devices placed in stacks or bulks of grain will show whether there are differences in temperature not attributable to ambient

---

* 6 Ghanaian pesewas are approximately equal to 5¢ US (cents).

† 20 Nigerian kobo are approximately equal to 40¢ US (cents).

effects. In order to assess insect or mould infestation, moisture content, and chemical changes, it is necessary to withdraw a sample for laboratory examination. The difficulty in obtaining a representative sample from a batch of grain is the biggest single problem facing the produce inspector.

Many of the important chemical, physical, and biological defects will be irregularly distributed throughout a batch of grain, and Ashman (1971) discusses in some detail the problem of sampling. He expresses the view that 'the sampling spear is the demon symbol of the so-called expert' and says that the sack should, ideally, be taken as the sample unit (or at least 100 kg if the grain is in bulk) and that the sample should comprise the square root of the total number of units in the batch (i.e. 10 sacks from a batch of 100). Each of these units should be examined in the following manner. The whole content of the sack should be emptied onto a sloping sieve mesh (a 'sack sieve'—see Ashman (1966)) and moved slowly forward over the mesh. It is thereby possible to gain a visual assessment of the entire quantity of grain and to sample the grain as it flows off the sieve into a sack. The dust, grain particles, and insects falling through the sieve are collected in a tray and examined in detail. Thus, the sieve will give, in one operation, the following information for each sack-sample unit in a few minutes:

(1) a visual assessment of quality;
(2) a sample from all parts of the sack;
(3) the quantity of dust and broken pieces;
(4) an efficient separation of insects from the contents of the sack.

Such a device can be easily made and will fit into the flow pattern for most bagged produce at trader- and central-storage level. The method is, in fact, already established in some African countries as a basic feature of the inspection system.

Having obtained a series of samples representing the batch it is common practice to bulk these and, after mixing, extract a sub-sample for detailed examination. If this is done, however, the extent of any variability within the batch has not been sampled. If the number of samples taken is too great for individual attention, adjacent samples may be bulked, but it is suggested that there should be at least four replicates of bulked samples for each zone. Ashman also discussed methods which can be used to break down samples into a series of smaller samples.

The need for improvement of quality standards

at the producer level was emphasized by Pattinson (1971). Such standards are unfamiliar to African producers and are not applied in village markets. Without cash incentives, rural education programmes are vital in order to encourage farmers to take extra care with the growing, harvesting, and storage of crops, whether they are intended for consumption on the farm or are introduced into a marketing pattern.

12.2.3. *Pre-storage and storage*

Systems. In the past there has been little attempt made to integrate cropping practices with storage requirements. For successful productivity, the complete operation has to be considered, its objectives decided, and then procedures which result in the minimum over-all cost must be selected for each part of the operation. The consumer may be the farmer himself, an urban consumer, a processor for human or animal feed, or an exporter. The operation involves some or all of the following: harvesting; threshing; drying; cleaning; grading; weighing; storage at farm, intermediate or central level; and transport between farm, storage, and distribution points. Corbett (1971) and Hall (1971*a*) discussed operations at farm and marketing levels separately.

A typical farmer in a bimodal rainfall SCP will have 1–2 ha under manual cultivation, predominantly for maize, yielding up to 1000 kg/ha or up to 2000 kg/ha with local and improved cultivars respectively. Maize grown in the first rains matured in July and, after being left in the field to dry for 4–6 weeks, is harvested in August when rainfall is reduced, but there is little sun and relative humidity is still high. It is then stored through the second rains and the succeeding dry season. Traditionally the farmer places his cobs, with or without sheaths, in a crib on the farm or in the village, from which small amounts are removed periodically and shelled by hand for consumption or sale.

The problems in this system are the familiar ones of moisture and insects. Field drying rate is slow, particularly in seasons with higher than average rainfall, when there is a danger of mouldy cobs. Insects infest the drying crop in the field resulting in heavy losses in the farm store unless control measures are taken.

Corbett (1971) indicates the technical solutions to this problem to be:

(1) harvesting of cobs as soon as they are mature, when the moisture content is still 30–35 per cent;

(2) immediate drying, either before or after shelling;

(3) storage of clean shelled grain at safe moisture content in a closed container with effective insect control.

The problem is to provide economic and suitable equipment to dry, shell, clean, and store maize for the 1–2 ha farmer. An obvious alternative is the joint use by a number of farmers of relatively large-scale equipment. In Ghana (Forsyth 1962) and Nigeria (Cornes and Adeyemi 1964; Williams 1971) village co-operatives, under Government supervision, mechanically shell and dry maize, using simple sack or tray driers, and store it in bulk in wooden or metal bins. Few of these schemes are operating, for organizational rather than technical reasons.

Corbett (1971) points out that if we could assume that farmers, individually or jointly, were able to store all their grain in good condition, then the marketing operation would simply be one of transporting this grain to consumers as required. However, many farmers prefer to sell for cash at harvest time, thus avoiding the drying and storage operations. Marketing must therefore cater both for substantial amounts of wet grain at harvest time and for regular amounts of dry, but often infested, grain thereafter. As a marketing system develops it needs to establish standards of weight (instead of volume), quality, moisture, and freedom from insects and moulds. Corbett discussed the equipment needed to maintain such standards at every point in the marketing chain.

At the urban, central, and reserve storage levels, storage must be appropriate, and this is discussed by Corbett (1971) with respect to engineering requirements.

Pre-drying. The importance of harvesting and threshing in determining the susceptibility of dried grain to deterioration during storage cannot be over-emphasized.

When mechanical harvesters are used, the plant is cut and the grain and straw are usually separated immediately. The use of combine harvesters can lead to a higher proportion of green or immature grains and more cracks in the glumes of paddy grains than does hand-harvesting. This provides easy access for fungal spores and insects (Breese 1964). Threshing grain that is still wet is more difficult than threshing grain that has been allowed to dry in the ear after harvesting.

Harvesting the crop before maturity usually results in a lower yield and a higher proportion of immature seeds. Conversely, the longer produce is left in the field, exposed to alternate periods of wetting and drying (e.g. by dew at night and hot sun by day), the greater is the damage from cracking and from insect pests, birds, and premature shattering. Craufurd (1961) has discussed this.

Moisture- and temperature-reduction. The drying of grain before storage is of paramount importance. Reviews of drying methods have been presented by Hall (1970) and Muckle and Stirling (1971); Greig (1971) discusses the conditions necessary for the successful application of various drying systems.

There is little published information on the efficiency of traditional maize cribs in reducing moisture in the stored crop, but such natural drying presents severe difficulties especially in humid climates.

In 1961 a study was undertaken in western Nigeria, at three locations (Cornes and Riley 1961), to compare rates of drying of cob maize stored in various traditional cribs of $\frac{1}{3}$ ton and $\frac{2}{3}$ ton capacities. The maize was stored without sheaths. Mould damage was observed during the first month of storage, suggesting that the types of cribs studied did not allow sufficiently rapid initial drying of maize under the prevailing environmental conditions. No significant differences were observed in the rates of drying between outer and inner samples. At Ildro the cribs were less well ventilated than at the other sites, and drying-out from an initial 19·5 per cent moisture content to 15 per cent took 3 months, compared with 1 month at Ilora, which had the lowest rainfall. At Akure, drying from 24·5 per cent to 15 per cent took 3 months.

In a further experiment (Cornes and Riley 1962) a comparison was made between a traditional crib at Ilora and 8 experimental cribs, all of 1 ton capacity; the latter each comprised two longitudinal compartments 9 ft long, 2 ft wide, and 3 ft deep, separated by a 1 ft wide air-space, for improved ventilation, and covered by a galvanized iron roof. The initial moisture content of the maize was 21·5–22·0 per cent. A difference in drying rate between the local crib (1·5 per cent in 10 days) and the experimental cribs (2·0 per cent in 10 days) was apparent within the first 20 days.

A simple and effective artificial drier for drying groundnuts, based on the Samoan cocoa drier, was developed at Mokwa, Nigeria, during the period

1961–3 (A'Brook 1964*a*, *b*). Subsequently, the extension service of the Western State Ministry of Agriculture and Natural Resources have promoted the use of this drier to dry cob maize (Webb 1969; Rambo 1971).

The drier consists of a rectilinear pit. The laterite spoil from excavating the pit is used to build walls which support the drying floor. The enclosed space forms the plenum chamber into which the gastight flue made of oil-drums is placed. The firebox is located outside the plenum chamber. The drier burns wood fuel and has a capacity of 1 ton of maize (shelled equivalent).

In Ghana (Food and Agriculture Organization 1969), Rawnsley dried maize effectively in a locally constructed drier based on a traditional fish smoking–drying oven. This is a clay cylinder, thatched, with a clay heat exchanger separating the lower firing chamber from the upper chamber. Advice is given in this paper on construction and use of visual aids on storage to assist village education programmes.

The use of 'in-bin' drying, where the storage container (normally a metal silo) is the drier, particularly lends itself to community- or co-operative-level storage systems. Bins equipped with a false floor of perforated steel have been used in this way in Nigeria. They can be used for aeration (or ventilation) of grain with ambient air during subsequent storage, as discussed in the next section. However, the successful operation of an in-bin drying system calls for a high degree of technical management, and the operation of such a system is discussed by Greig (1971) and Smith (1969).

An artificial drier using solar heat was developed a few years ago in Malaysia (Williams, Beeney, and Webb 1969). Two layers of corrugated aluminium or galvanized iron, which cover the material to be dried, are laminated together so as to create a series of parallel tubes formed by the corrugation of the sheets. Air was drawn between the laminations by means of a fan and the heated air passed to a drying chamber beneath.

As low a temperature as possible should be maintained in storage containers, buildings, or silos, as spoilage organisms multiply less rapidly at lower temperatures and thus less deterioration in quality will occur. Fluctuations in temperature of stored grain can be caused by insects and moulds or by ambient temperatures that may heat or cool the periphery of the container and the outer layers of grain. This may cause moisture movement by evaporation and condensation and favour mould growth.

Useful information on temperature in 20-ton aluminium silos has been recorded in the north of Nigeria (Prevett 1961*a*), but condensation problems did not occur due to low grain moisture content and low relative humidity.

The technology of grain aeration to prevent insect attack has been surveyed by Calderon (1971) who defined aeration as 'the movement of selected ambient air through a bulk of grain, for the improvement of its storability'. He discussed the benefits and limitations of grain aeration in warm climates and gave data from experimental work in sub-tropical climates.

The use of refrigerated air ('grain chilling') for grain aeration has been practised in Europe since the 1930s (Burrell and Burges 1964). Considerable research has been done in Britain in recent years on the use of this technique for the safe storage of damp grain and the use of chilling in a warm climate is reported by Sutherland, Pescod, Airah, and Griffiths (1970) in Queensland, Australia. Grain of 11 per cent moisture content was cooled from 32 °C to 13 °C in 17–18 days, and this temperature was maintained for 10 months. Their methods of cooling grain, while quite feasible, has high energy consumption and costs. Similar studies have been undertaken in Israel by Calderon (1971), who expressed the view that a sensible use of ambient or chilled air for aeration of grain offers new possibilities in many parts of the world for the preservation of grain reserves, without (or with very little) use of chemicals.

It is equally important that buildings for bag storage should be designed and constructed in such a way as to create the most favourable physical environment (i.e. low temperature and relative humidity) to maintain produce quality. Provision should be made for controlling the amount of ventilation according to the weather and for controlling insects (see later section).

Storage structures (rural and central). The basic requirements of structures for storage were discussed by Hall (1970). The choice between storage in bags or in bulk depends upon the following factors:

(1) type of produce;
(2) duration of storage;
(3) value of produce;
(4) climate;
(5) transport system;
(6) cost and availability of labour;

(7) cost and availability of sacks;
(8) incidence of rodents and certain types of insect infestation.

Farmers normally store grain unthreshed in some form of bulk container. When produce is moved from the farm it is purchased and stored by traders or co-operatives. Bag storage is most common at this level, although attempts have been made to introduce bulk storage in a number of countries at the co-operative level (Forsyth 1962; Williams 1971). As suggested by Corbett (1971) bag storage should be retained in areas where road and transport facilities are poorly developed, where small lots of grain and a wide variety of products is handled. Bulk storage may be considered where the grain to be handled is of one type, where roads are constructed to handle trucks throughout the year, and especially where storage is likely to extend throughout the year.

Much attention has been focused in recent years on the need for reduction of storage losses in rural areas and the possibility of achieving this through the introduction of improved storage.

In humid areas, if grain has been adequately dried, the storage container should be of such construction as to avoid re-absorption of moisture, i.e. there should be minimum air movement throughout the container. The simplest way of achieving this is for the container to have solid walls; thus wicker-type cribs should have the inner and outer surfaces mud-plastered, and mud-brick containers should have any cracks patched. Mud-plastering of granaries constructed from woven plant materials was satisfactory in Malawi (Schulten and Westwood 1972). Hermetic sealing of the crib or silo, though difficult, offers considerable advantages such as avoiding the need for pesticide chemicals for protecting the grain from re-infestation by insect pests.

During the period 1958–60 a trial was undertaken at Mokwa, Nigeria (Ozburn, Ward, George, Riley and Corres 1960) on the hermetic storage of sorghum, at 10 per cent moisture content, in 44-gallon drums. The trial, which extended over a storage period of $2\frac{1}{2}$ years, gave satisfactory results both from the point of view of insect control and maintenance of viability. Pattinson (1971) reports the successful introduction of this method to Senegal. In the north of Nigeria, O'Dowd (1971b) concluded from trial data that the most satisfactory improvement for storage of cowpeas would be to line the traditional mud granary (the 'rhumbu') with polythene. In order to prevent

emerging Brucidae from penetrating the polythene liner it was recommended that the polythene should be provided with an inner liner of cotton cloth.

Trials have been carried out in a number of countries of small weld-mesh silos lined with butyl rubber for hermetic storage. O'Dowd (1971a, 1971b) tested silos of $\frac{1}{2}$-ton and 10-ton capacity in Nigeria for storage of cowpeas. In the $\frac{1}{2}$-ton silos insect control was achieved and cowpeas maintained their quality. The liners did not deteriorate after two season's use. However, it was thought that the cost was likely to make this type of silo too expensive for farmers. Synthetic-rubber-lined silos of 10-ton capacity were tested at Samaru for storage of cow-pea (5 months) and sorghum (8 months). Some degradation of the butyl rubber (later established to be below specification) was experienced, but the grains remained in good condition. The silos were easy to transport, construct, and use, but careful supervision was essential, especially in manufacture, to obtain good results. Newer forms of liners will probably give good results.

In Malaysia commercial paddy at a moisture content of 13 per cent was stored for 12 months in two tightly closed butyl rubber silos of 1-ton capacity (Beeny, Jensen, Varghese, and Mohd Sanusi Bin Jangi 1972). The silos were exposed over a period of 12 months to the open tropical sky, under conditions of intense sun and rainfall. The stored paddy was monitored for temperature fluctuations, and samples were assessed each month for moisture content, germination, and fungal and insect activity. Milling and nutritional quality were found to be satisfactory at the end of the storage period, and the grain was readily marketable. The method is recommended as a low-cost and versatile system suitable for tropical cereal-growing areas.

A different approach to the introduction of small-scale hermetic storage structures has been the development, in Thailand, of ferro-cement bins designed to hold 4 tons of grain (Smith and Boonlong 1970). The materials needed are cheap and readily available in many developing countries. In view of the need to have properly dry grain for storage in such a 'closed' structure, consideration can be given to the possibility of in-bin drying as an alternative to drying prior to storage intake in these hermetic concrete 'Thailo' structures (Anon. 1973).

Recently, interest has been shown in some tropical countries in the possible use of butyl-rubber silos for the hermetic storage of high-moisture grain.

Hermetic storage is now a widely established procedure in Europe and North Africa for the storage of high-moisture grain for subsequent use for animal feed (Hyde 1962). However, this technique may be very hazardous for grain for human consumption when used in tropical countries for the following reasons (Hyde 1969).

1. Most of the grain produced in the tropics is for human consumption. The changes which take place, including the taint, may make the grain unacceptable.
2. At the higher temperatures the fermentation may be such that the grain is not even acceptable to animals. Also, it is likely that some of the grain will go mouldy leading to mycotoxic hazards.
3. The degree of airtightness needed to prevent mould growth is much more critical than for insect control in dry grain—where a very slight leak may sometimes be tolerated. The oxygen level must be maintained at less than 0.5 per cent—even under the carefully controlled use of this technique in temperate countries this is often difficult. In the tropics the problem will be accentuated by the higher temperatures, by the possibility of undetected damage by rodents, termites, etc., and by a lack of informed personnel.
4. The 'shelf life' of damp grain is very short, whether before or after storage. The silo must be filled quickly and, equally, the grain must be used immediately after removal from the silo, in order to prevent mould growth and possible toxin production. After storage this will proceed more rapidly since the grain will now be dead.

Clearly, there is a need for more study of this technique under tropical conditions before it can be recommended in any form. A preliminary experiment has been undertaken in Nigeria in which maize at 20 per cent moisture content was stored for 8 weeks in polythene bags in sealed 44-gallon drums (Broadbent and Oyeniran 1968). After storage the maize had a sour smell, suggesting fermentation, but had retained its original bright and clean appearance. It was free of insects; in contrast, maize stored unsealed as a control was superficially very mouldy and contained living insects.

12.2.4. Biology of organisms causing losses in storage
Factors influencing populations. Bacteria, fungi, insects, and mites are a problem only in the presence of moisture. Generally speaking, 70 per cent relative humidity is taken as the level above which development by micro-organisms will take place, the moisture content of any product which is in equilibrium with 70 per cent relative humidity being referred to as its 'safe' level of moisture content for storage. Development by the majority of insect pests of stored produce is not possible below 40 per cent relative humidity. However, it is neither economic nor 'good for the product' to dry down to this level, and therefore it is often necessary to resort to other methods for insect control.

Micro-organisms and arthropods become more active with increase in temperature over the range likely to be encountered in storage. The relative humidity for a given absolute moisture content decreases with rising temperature. Roberts (1962) has presented data on the physical limits for multiplication of fungi bacteria, mites and insects, whilst Howe (1965) has summarized estimates of optimal and minimal conditions for population increase for some storage insect pests.

The characteristics of the produce itself play an important part in determining susceptibility to attack. Undamaged shells of groundnuts, sheaths of maize cobs (Giles and Ashman 1971), and husks of paddy grains reduce penetration by most species of insects. Storage of unthreshed grain is often more successful than storage of threshed grain because of physical damage during threshing. Reference has already been made to the observations of Breese (1964) in relation to paddy.

Genetically controlled cultivar characteristics affect susceptibility to insect attack. Inherent susceptibility is due to both physical and chemical properties; there is the degree of protection given to the grain by the form of outer or sheathing coverings, and the chemical composition of the grain itself either encourages or discourages insect attack and penetration. This has been demonstrated for paddy in Sierra Leone (Prevett 1959), sorghum in East Africa (Davey 1965), and maize generally. More recently, further attention has been given to the difference in susceptibility between cultivars of maize to attack by *Sitophilus zeamais* (Wheatley 1971). The practical significance of variation in susceptibility is that improved cultivars of maize which are being progressively introduced to farmers in tropical countries exhibit a greater susceptibility to *Sitophilus* attack than those being replaced. Insect preferences tend to be similar to those of the intended consumer.

Prior infestation by field pests can also be of significance in influencing susceptibility to attack by storage pests. For example, it has been demonstrated by Ashman (1968) in Cyprus that pre-harvest damage to carobs by the carob moth (*Ectomyelois ceratoniae*) accentuates infestation in storage by *Ephestia* spp.

Micro-organisms. Christensen and Kaufmann (1965) have reviewed stored products fungi, with particular reference to the deterioration caused. The same authors (1969) also discuss methods of preventing deterioration. Clarke (1968) gives a useful account of fungi in stored products, which included a discussion of methods for isolating fungi.

Most storage fungi grow most quickly at some point in the temperature range 20–40 °C and above 90 per cent relative humidity, and some are practically confined to these conditions. The fungi which grow on products stored at higher and lower temperatures or at lower humidities fall into fairly clearly defined groups. This makes it possible to predict the degree of drying necessary to prevent growth of specific fungi at existing temperature conditions.

A detailed study of microbiological deterioration of cocoa, groundnuts, palm products, maize, gari, and yams has been carried out at the Nigerian Stored Products Research Institute, Ibadan (Broadbent 1971).

Of the cereals, maize is among the most liable to post-harvest mould deterioration, since the main crop is harvested during the rainy season and is an excellent substrate (cf. corn meal agar). However, Broadbent pointed out that farmers who store maize for human consumption for long periods in sheath at above the 'safe' level of moisture observed little deterioration. On the other hand, livestock-feed maize stored shelled in bulk was considerably worse affected. At three livestock farms near Ibadan, between June and September 1966, Broadbent (1966) found most of the feed maize visibly mouldy; the moisture content ranged from 16·2 per cent to 24·4 per cent. A wide variety of fungi was isolated, including six (*Aspergillus flavus*, *A. fumigatus*, *A. niger*, *Fusarium moniliforme*, *Paecilomyces varioti*, and *Penicillium variabile*) known to produce mycotoxins affecting farm animals.

Insects. Surveys of the insects attacking stored food-grains have been made in a number of African countries, e.g. Sierra Leone (Prevett 1959), Uganda (Davies 1960), Ghana (Forsyth 1966), and Nigeria (Cornes 1964). Such lists mention a large number of different species, but fortunately not all of these are serious pests.

Information on the basic biology of the more important pests is summarized by Hall (1970) and Prevett (1971a). The information on the ecology and behaviour of the pests in their natural environment is of particular significance. Reference has already been made to the influence of damage by field pests upon subsequent attack by storage species of carobs in Cyprus (Ashman 1968). In Togo, a positive correlation was found between infestation during storage by *Sitophilus zeamais* (Motsch.) and cobs which were damaged by caterpillars during field-drying (Pointel 1969).

For cereal grains, it is well known that such important primary storage pests as *Sitophilus zeamais* and *Sitotroga cerealella* (Oliv.) begin their attack in the field. Schulten (1971) considered in some detail the initiation and progress of infestation of maize, with particular reference to *Sitophilus zeamais*. This species flies readily (Giles 1969), and the greatest field infestation is found near maize stores; there is a gradient of infestation from the edge of the field to the centre (Giles and Ashman 1971). Giles and Ashman demonstrated that eggs which were deposited in grains at about 60 per cent moisture content developed into adults, the shortest development time observed being 46 days. Eggs deposited at higher moisture contents did not produce adults, but weevils were able to survive on the grains until they could produce progeny. Similarly in Dahomey, Le Conte and Bossou (1963) showed that field infestation took place about 30 days before harvest.

Much less information is available on the moth *Sitotroga cerealella*. Caswell (1971) points out the need for further study of this species, which appears to be more important than *Sitophilus* in the drier northern zone of West Africa. A recent laboratory study of strains of the moth *Plodia interpunctella* (Hubn) from Nigeria and the Republic of South Africa (Prevett 1971b) demonstrated a difference between strains in the incidence of diapause, and this was discussed in relation to climatic differences. The moth *Ephestia cautella* (Wlk.) has peak activity at dusk, as first noted by Rawnsley (1968) in Ghana and more recently confirmed in Kenya (Graham 1970).

Whereas *Bruchidius atrolineatus* (Pic), which attacks cowpea pods in the field, dies out in store, *Callosobruchus* spp., which do not attack pods in the field until they are almost dry, become major storage pests (Caswell 1961; Prevett 1961*b*). Prevett has demonstrated that prompt harvesting greatly assists post-harvest control of Bruchidae.

Rodents. A detailed study of rodent biology and control in India was undertaken by Krishnamurthy and co-workers (Krishnamurthy 1971; Krishnamurthy, Ramasivan, and Uniyal 1971). The population of *Rattus rattus* in villages around Hapur was, on average, 1057 rats per village, 9·8 per house, or 1·3 per person. It was estimated that they caused, on average, a loss of 2·34 tons of grain per village annually, equal to 1·69 per cent of the total food grains stored for consumption and seed purposes. The rat population varied with changes of weather and food; in one village the estimated population varied from 8·5 rats to 18·5 rats per house during 1 year. Breeding occurred throughout the year, with a peak during the period February–April.

12.3. Pest-control methods

12.3.1. Physical

A comprehensive review of non-chemical methods for the control of storage insects has been presented by Whitney (1971). More recently physical methods have been reviewed by Watters (1972). Control in any form depends primarily upon making the environment unfavourable for the survival of pest species. In addition to the design of storage buildings to create a favourable environment for the safe keeping of food products it is important to pay strict attention to storage hygiene and management.

Perhaps the simplest form of physical control is the exclusion of insects from the stored food in ways that have already been discussed. Where a moth pest is a particular problem on grain it is advantageous to thresh the grain before storage, since this limits infestation to the surface layer of produce in the store, by preventing access to deeper layers (Prevett 1961*a*). Whitney (1971) points out that the traditional farmer's practice of mixing fine-seeded millet or fine sand with larger-seeded sorghum or maize is effective against moth pests because it similarly prevents access by filling up the inter-granular spaces.

In an airtight container insect infestation will be controlled when the oxygen level declines to 2 per cent by volume. The rate at which this occurs will clearly depend upon the initial level of infestation present. However, even if complete control is not achieved, there will be a significant reduction in the

rate of increase of insects, and, therefore, of damage. Reference has been made in an earlier section of this chapter to different types of airtight container.

Although various forms of radiant energy are known to be effective against storage insects, they have not yet been widely used commercially. Watters (1972) has considered in some detail the use of electromagnetic energy, radio-frequency heating, ionizing radiation, infra-red, visible, and ultra-violet radiation, and sound. He concluded that the use of physical means of controlling storage insects will probably expand as the limitations of other methods of control become more apparent. Watters emphasized the desirability for less reliance on persistent insecticides and a greater appreciation of the value of hygiene and sanitation.

12.3.2. Chemical methods

There is still no safe and effective fungicide available for the control of fungi in stored grain (Delassus 1971). For insect pests, however, the use of residual chemicals to control and prevent infestation of stored grain is now a well-established practice throughout the world. A great deal of work has been done in recent years to evaluate the effectiveness of insecticides applied as dilute dusts to maize and other cereals stored in 'traditional' containers. In Malawi (Schulten 1971; Schulten and Westwood 1972) and a number of countries in East and West Africa (Kockum 1965; Riley and Matheson 1970) a 0·5 per cent or 1·0 per cent lindane dust has been used to protect maize cobs stored in local cribs. Malathion, which breaks down if mixed with grain of high moisture content (Watters 1971), may be mixed with maize after drying and threshing.

An alternative approach, which has been evaluated in a number of countries, is to use plastic sacks for storage and a fumigant chemical for insect control (Pointel 1971). In Ghana, Rawnsley (Food and Agriculture Organization 1969) has used ethylene dibromide for the control of pests of stored maize, and in francophone Africa, Pointel (1967) has used carbon tetrachloride for protection of stored cowpea. In the north of Nigeria, however, Caswell (private communication) has shown that adequate control of Bruchidae in cowpeas can be achieved through the use of plastic sacks without the addition of any chemical. Proctor and Ashman (1972) successfully used phosphine for storing confectionery-grade Zambian groundnuts in polythene-lined jute or woven-polypropylene sacks. There is a need for further work to determine the advantages and disadvantages of various alternative methods. A comparison of phosphine and ethylene dibromide for the control of *Callosobruchus maculatus* (F.) in cowpeas and *Sitophilus zeamais* and *Tribolium castaneum* (Herbst) in maize stored in polythene-lined sacks has been undertaken in Nigeria (Cornes, Adeyemi, and Qureshi 1967; Cornes and Adeyemi 1969). In both tests phosphine gave good control, but only partial control was obtained with ethylene dibromide.

Ethylene dibromide has also been used in India, in combination with methyl bromide, for large-scale fumigation under gas-proof sheets and as a single fumigant for small-scale fumigation in household storage containers (Pillai, Muthu, Majumder, Sharangapani, and Amla 1970). For the latter a fumigant 'tablet' has been prepared by soaking 4 cm squares of strawboard with ethylene dibromide and packing them in polycellular formaldehyde-treated gum paper. More recent work in India (Girish, Tripathi, Srivastava, and Krishnamurthy 1971) has demonstrated improved penetration of a mixture of ethylene dibromide and carbon tetrachloride. In a comparison of this mixture with ethylene dibromide and phosphine (Girish *et al.* 1972) all gave effective control, but the mixture was most effective for controlling hidden infestation; phosphine was the cheapest of the three treatments.

For large-scale fumigation of bagged produce under gas-proof sheets methyl bromide is generally used, although in situations where 5 days can be allowed, phosphine is increasingly replacing methyl bromide. Detailed information on the use of these fumigants was given by Monro (1971).

In Nigeria, investigations on the fumigation of bagged cocoa beans and groundnuts have indicated the importance of the arrangement of piping for introduction of gas (Riley 1963) and of the shape of the stack (Halliday and Prevett 1963). Fumigation of bagged produce is possible in warehouses which are gastight. This has the advantage that infestation present in the fabric of the warehouse is also killed, thereby minimizing the possibility of re-infestation. Such warehouses have been built in a number of African countries. Studies have been undertaken in Nigeria on the fumigation of cocoa in specially designed warehouses, using phosphine (Riley and Simmons 1967*a*) and methyl bromide (Riley and Simmons 1967*b*).

Watters (1971) has reviewed new insecticides for the insect pests of stored products. Dichlorvos shows considerable promise, particularly for control of such species as *Ephestia cautella*. Because of the dusk flight activity of this species a method has been developed (Anon. 1972) for the dispersal of a mist of dichlorvos into the free space of a warehouse using an electrically operated sprayer controlled by time-clocks to disperse the insecticide at the time of maximum flight activity. This is an excellent example of an effective space treatment which can be used in a warehouse with completely controllable ventilation.

Much attention has been focused in recent years on the problems of insecticide resistance in the insect pests of stored products. Dyte (1970) has pointed out that resistant strains of at least 13 species were known in 1970. Resistance in *T. castaneum* was widespread, with malathion resistance occurring in at least 12 countries, and lindane resistance in at least 10 countries. The rapid spread of resistant *T. castaneum* through international trade is indicative of a problem likely to occur with other pests of stored products, and a joint FAO/CSIRO, Australia/UK Ministry of Agriculture, Fisheries and Food, Pest Infestation Control Laboratory project is currently in progress, with the objective of evaluating the situation.

Studies have been carried out in India (Krishnamurthy 1971) to evaluate different poisoning methods for use in large-scale rat-eradication programmes. Continuous baiting with anticoagulants was found to give more satisfactory control than zinc phosphide; at the same time the importance of improved sanitation and rat-proofing of houses and storage structures was emphasized. For control of field rats excellent results were obtained by fumigating burrows with phosphine.

12.4. The impact of research

Implementation of research findings is achieved only by the effective management of research, development, extension, and information dissemination. Modern techniques of storage require high inputs of skill. Training staff to use machinery matched to the local facilities available for its care and maintenance is most important. There is only limited value in transferring knowledge without a local matching understanding of the processes involved on the part of staff with the ability to take any necessary day-to-day decisions.

Basic research has had limited impact on grain storage over the past 50 years, but applied research towards practical objectives, carried out in universities, government departments, and commercial organizations, has given rapid advances in grain-storage technology during the past 25 years in technically advanced countries. Technologies for the socio-economic conditions of developing countries have had, in comparison, little attention. Decision-makers in developing countries are frequently unaware of the technological choices available, and the possibilities for 'intermediate' technology are often overlooked.

There is already a wealth of research data on grain storage, and further extensive programmes of research may be unnecessary. There is now a need to re-allocate the limited resources available in favour of dissemination of technical information and development of its use through extension and training.

It is not unknown (or even uncommon) for people in different regions or areas within one country to be unaware of the methods being used successfully by those with similar problems in another; and this is especially true between countries. In an attempt to promote greater awareness, the FAO has established an African Rural Storage Centre in West Africa to assist interested governments in learning the range of techniques already available and to promote the introduction of relevant techniques to overcome specific problems (Hall 1971*b*).

There is a *Journal of Stored Products Research*. In addition, technical information is made available to specialists working in departments in developing countries through the FAO and the major donors, especially from France (IRAT publications) and from the UK (Tropical Products Institute: *Tropical Stored Products Information* and *Tropical Storage Abstracts*). However, there are few local information sheets, pamphlets, and booklets for the local trader, farmer, and administrator. Exceptions are the literature of the extension programmes in Senegal and, in Nigeria, literature from the Economy Research Unit at Zaria. There is little organized in the way of a retrieval service for the extension worker, the trader, or the farmer. At these levels practical demonstrations are required to provide proof of the detail and value of the range of methods, some of which could be introduced as innovations.

Sociological, as well as technical, conditions in a community have to be taken into account for any innovation to succeed. Emphasis has to be on socially acceptable incentives communicated in the appropriate fashion to the most absorptive section of rural and urban communities, which in many countries is mainly the women. The role of the village elders and teachers in disseminating information cannot be overstressed.

Although much is being achieved, failures to improve storage in some areas are due to political problems, over-emphasis on administration and not enough emphasis on management, to lack of finance, and lack of effective extension services.

Grain-storage problems (except for seed storage and distribution) on the farm and in traders' stores have been peripheral to the interests of the agricultural extension service, which is normally over-stretched in dealing with the pre-harvest activities of the farmer. Such problems are also somewhat outside the scope of the community development or welfare officer or of the Health Inspector, who normally has insufficient training in the subject to recognize opportunities for introducing relevant improved methods of handling produce. The existence of produce inspectors in many countries has effected improvement in the quality of grain entering the major trading channels, but there is little, if any, co-ordination of activity to protect grain from deterioration from the time of harvest to the time of consumption of the grain by man or domestic animals.

Personnel in universities, government departments, and commercial organizations in developing countries must be trained in grain-storage techniques. Such informed personnel can then continue the training of local administrators, traders, and farmers, so that at all levels in a community there is an awareness of why grain deteriorates and what can be done to prevent it.

Acknowledgements

I would like to express my grateful thanks to Dr. P. F. Prevett for his assistance in the preparation of the subject-matter, and I gratefully acknowledge the permission of the Ministry of Overseas Development to prepare this chapter.

References

A'BROOK, J. (1964*a*). A cheap crop drier for the farmer. *Trop. stored Prod. Inf.* 7, 257-68.
—— (1964*b*). A cheap crop drier for the farmer: results and recommended design. *Trop. stored Prod. Inf.* 8, 301-7.

ANON. (1972). Dichlorvos sprayers. *Trop. stored Prod. Inf.* **23**, 6.

—— (1973). Ferrocement food-storage silos in Thailand. Appendix B in *Ferrocement: applications in developing countries*, pp. 55-9. Report of the National Acedemy of Sciences, Washington DC.

ANTHONIO, Q. B. O. (1962). WASPRU/UCI Grain Storage Project. The economics of cowpea storage in a concrete bin. A. Rep. W. Afr. stored Prod. Res. Unit, Tech. Rep. No. 19, pp. 94-5.

—— (1971). Economics of storage and effects on prices. *Ford Foundation/IRAT/IITA seminar on grain storage in the humid tropics, Ibadan.*

ASHMAN, F. (1966). Inspection methods for detecting insects in stored produce. *Trop. stored Prod. Inf.* **12**, 481-94.

—— (1968). The control of infestation in carobs (*Ceratonia siliqua* L.) with special reference to Cyprus. Rep. Int. Conf. on the Protection of Stored Products, Lisbon. EPPO Publ. Series A, No. 46—E, pp. 117-20.

—— (1971). Methods and techniques of assessing quality in stored foods. *Ford Foundation/IRAT/IITA seminar on grain storage in the humid tropics, Ibadan.*

APO (1971). Training in storage and preservation of food grains. Asian Productivity Organization Project TRC/IV/68.

BEENEY, J. M., JENSEN, L. A., VARGHESE, G., and MOHD SANUSI BIN JANGI (1972). Use of butyl rubber silos for paddy storage in the tropics. *Trop. Agric., Trin.* **49**, 151-60.

BONLIEU, A., NICOU, R., and TOURTE, R. (1964). La conservation des récoltes au Sénégal. Essais sur le mil, le sorgho, le paddy, le niébé. *Agron. Trop., Nogent* **19**, 7-44.

BREESE, M. H. (1964). Factors governing the infestibility of paddy and rice. *Trop. stored Prod. Inf.* **8**, 289-99.

BROADBENT, J. A. (1966). Microbiological deterioration of maize used as poultry and livestock feed at farms near Ibadan during the wet season. A. Rep. Nigerian stored Prod. Res. Inst., Tech. Rep. No. 16, pp. 115-18.

—— (1971). Microbiological deterioration of foodstuffs during storage in Nigeria. Paper presented at the *Ford Foundation/IRAT/IITA seminar on grain storage in the humid tropics, Ibadan.*

—— OYENIRAN, J. O. (1968). A preliminary experiment on the airtight storage of damp maize. *A. Rep. Nigerian stored Prod. Res. Inst.*, Tech. Rep. No. 9, pp. 71-5.

BURRELL, N. J. and BURGES, H. D. (1964). Cooling bulk grain by aeration. *Wld Crops* **16**, 33-8.

CALDERON, M. (1971). Aeration of grain: benefits and limitations. *Ford Foundation/IRAT/IITA seminar on grain storage in the humid tropics, Ibadan* (and published in EPPO Bull. No. 6 (1972), pp. 83-94).

CASWELL, G. H. (1961). The infestation of cowpeas in the Western Region of Nigeria. *Trop. Sci.* **3**, 154-8.

—— (1971). The impact of infestation on commodities. *Ford Foundation/IRAT/IITA seminar on grain storage in the humid tropics, Ibadan.*

CHRISTENSEN, C. M. and KAUFMANN, H. H. (1965). Deterioration of stored grain by fungi. *A. Rev. Phytopath.* **3**, 69-84.

—— —— (1969). *Grain storage. The role of fungi in quality loss.* University of Minnesota Press, Minneapolis.

CLARKE, J. (1968). Fungi in stored products. *Trop. stored Prod. Inf.* **15**, 3-14.

CORBETT, G. G. (1971). Grain handling systems for the lowland humid tropics. *Ford Foundation/IRAT/IITA seminar on grain storage in the humid tropics, Ibadan.*

CORNES, M. A. (1964). A revised listing of the insects associated with stored products in Nigeria. A. Rep. Nigerian stored Prod. Res. Inst., Tech. Rep. No. 19, pp. 96-119.

—— ADEYEMI, S. A. O. (1964). A survey of grain storage sites in Western Nigeria. A. Rep. Nigerian stored Prod. Res. Inst., Tech. Rep. No. 15, pp. 78-84.

—— —— (1969). A comparison of phosphine and ethylene dibromide for the fumigation of cowpeas in polythene lined hessian sacks. *Bull. entomol. Soc. Nigeria* **2**, 45-50.

—— —— QURESHI, A. H. (1967). An assessment of the value of phosphine and ethylene dibromide for the control of pests in grain stored in polythene lined sacks. A. Rep. Nigerian stored Prod. Res. Inst., Tech. Rep. No. 13, pp. 113-21.

—— RILEY, J. (1961). Small scale storage of maize in Western Nigeria. A. Rep. W. Afr. stored Prod. Res. Unit, Tech. Rep. No. 15, pp. 85-96.

—— —— (1962). An investigation of drying rates and insect control in a maize crib with improved ventilation. A. Rep. W. Afr. stored Prod. Res. Unit, Tech. Rep. No. 12, pp. 72-8.

COYNE, F. P. (1971). Improving the protection of stored maize from insect attack on small farms in Kenya. *Int. Pest Control* **3**, 8-13.

CRAUFURD, R. Q. (1961). Breakage of rice during milling. *Trop. stored Prod. Inf.* **3**, 64-7.

DAVEY, P. M. (1965). The susceptibility of sorghum to attack by the weevil *Sitophilus oryzae* (L.). *Bull. entomol. Res.* **56**, 287-97.

DAVIES, J. C. (1960). Coleoptera associated with stored products in Uganda. *E. Afr. agric. J.* **25**, 199-201.

DELASSUS, M. (1971). Microorganic activity in foodstuffs and methods of control. *Ford Foundation/IRAT/IITA seminar on grain storage in the humid tropics, Ibadan.*

DEORAS, P. J. (1967). Rat problems in India. *Pesticides* **1**, 67-70.

DYTE, C. E. (1970). Insecticide resistance in stored-product insects with special reference to *Tribolium castaneum. Trop. stored Prod. Inf.* **20**, 13-18.

FOOD AND AGRICULTURE ORGANIZATION (1968). Improved storage and its contribution to world food supplies. Chapter 4 in *State of food and agriculture, 1968*, pp. 115-43. FAO, Rome.

—— (1969). Crop Storage. Tech. Rep. No. 1 of the Food Research and Development Unit, Accra, Ghana. Prepared for the Government of Ghana by FAO acting as executing agency for the UNDP, based on the work of J. Rawnsley. PL: SF/GHA 7. FAO, Rome.

FORSYTH, J. (1962). Major food storage problems. In *Agriculture and land use in Ghana* (ed. J. B. Wills), pp. 394–401. Oxford University Press, London.

—— (1966). *Agricultural insects in Ghana*. Ghana University Press, Accra.

GASGA (1973). Group for Assistance on Storage of Grains in Africa, Seminar on 'The methodology of evaluating grain storage losses', TPI. *Trop. stored Prod. Inf.* **24**, 13–16.

GILES, P. H. (1969). Observations in Kenya on the flight activity of stored products insects, particularly *Sitophilus zeamais* (Motsch.). *J. stored Prod. Res.* **4**, 317–29.

—— ASHMAN, F. (1971). A study of pre-harvest infestation of maize by *Sitophilus zeamais* (Motsch.) in the Kenya Highlands. *J. stored Prod. Res.* **7**, 69–83.

GIRISH, G. K., TRIPATHI, B. P., SRIVASTAVA, P. K., and KRISHNAMURTHY, K. (1971). Studies on the behaviour of ethylene dibromide and carbon tetrachloride mixtures. Part I. *Bull. Grain Technol.* **9**, 242–6.

—— —— —— —— (1972). Studies on the behaviour of ethylene dibromide and carbon tetrachloride mixtures. Part II. *Bull. Grain Technol.* **10**, 30–6.

GRAHAM, W. M. (1970). Warehouse ecology studies of bagged maize in Kenya. I. The distribution of adult *Ephestia* (*Cadra*) *cautella* (Walker) (Lepidoptera, Phycitidae). II. Ecological observations of an infestation by *E. cautella*. *J. stored Prod. Res.* **6**, 147–55 (I); 157–67 (II).

GREIG, D. J. (1971). Drying systems available and conditions necessary for successful application. *Ford Foundation/IRAT/IITA seminar on grain storage in the humid tropics, Ibadan*.

HALL, D. W. (1969). Food storage in the developing countries. *J. R. Soc. Arts* **142**, 562–79.

—— (1970). *Handling and storage of food grains in tropical and sub-tropical areas*. FAO Agric. Dev. Paper No. 90. FAO, Rome.

—— (1971a). The importance of storage in the humid tropics. *Ford Foundation/IRAT/IITA seminar on grain storage in the humid tropics, Ibadan*.

—— (1971b). Environmental problems: pests of harvested tropical crops. *Outl. Agric.* **7**, 68–73.

HALLIDAY, D. and PREVETT, P. F. (1963). Fumigation of pyramid stacks of bagged decorticated groundnuts with methyl bromide. *J. Sci. Fd. Agric.* **14**, 586–92.

HOWE, R. W. (1965). A summary of estimates of optimal and minimal conditions for population increase of some stored products insects. *J. stored Prod. Res.* **1**, 177–84.

HYDE, M. B. (1962). Airtight storage of grain. *Ann. appl. Biol.* **50**, 362–4.

—— (1969). Hazards of storing high-moisture grain in airtight silos in tropical countries. *Trop. stored Prod. Inf.* **18**, 9–12.

KOCKUM, S. (1965). Crib storage of maize. A trial with

pyrethrins and lindane formulations. *E. Afr. agric. J.* **30**, 8–10.

KRISHNAMURTHY, K. (1971). Work on rodents and their control in India to prevent losses in foodgrains. *Ford Foundation/IRAT/IITA seminar on grain storage in the humid tropics, Ibadan*.

—— RAMASIVAN, T., and UNIYAL, V. (1971). Studies on rodents and their control. VI. Studies on fluctuations in population and breeding period of *Rattus rattus* in Hapur Region. *Bull. Grain Technol.* **9**, 79–82.

LE CONTE, J. and BOSSOU, C. (1963). Le problème de la conservation du maïs en épis dans le Sud Dahomey. Etude du développement de l'attaque par *Sitophilus oryzae* en champ et en magasin. *Agron. Trop.*, *Nogent* **18**, 969–84.

MAJUMDER, S. V. (1968). *Manual of rodent control*. CFTRI, Mysore.

—— VENUGOPAL, J. S. (1969). *Grain sanitation*. Academy of Pest Control Sciences, India.

MONRO, H. A. U. (1971). *Manual of fumigation for insect control*. FAO Agric. Studies No. 79 (2nd edn., revised). FAO, Rome.

MUCKLE, T. B. and STIRLING, H. G. (1971). Review of the drying of cereals and legumes in the tropics. *Trop. stored Prod. Inf.* **22**, 11–30.

MUNRO, J. W. (1966). *Pests of stored products*. Hutchinson, London.

NELSON, L. R., CUMMINS, D. G., HARRIS, H. B., and CALVERT, G. V. (1972). Grain preservation for storage of high moisture grain. Res. Rep. Univ. Georgia, Coll. Exp. Stn. No. 129.

O'DOWD, E. T. (1971a). Hermetic storage in Nigeria using weldmesh silos lined with butyl rubber. Samaru Misc. Paper No. 30, Institute for Agricultural Research, Ahmadu Bello University, Zaria.

—— (1971b). Hermetic storage of cowpea (*Vigna unguiculata*) in small granaries, silos and pits in northern Nigeria. Samaru Misc. Paper No. 31, Institute for Agricultural Research, Ahmadu Bello University, Zaria.

OKWELOGU, T. N. (1969). Produce inspection in tropical Africa. *Wld Crops* **21**, 260–3.

OZBURN, G. W., WARD, P., GEORGE, M. R., RILEY, J., and CORRES, M. (1960). Hermetic storage of guinea corn. A. Rep. W. Afr. stored Prod. Res. Unit, Tech. Rep. No. 11, p. 45.

PATTINSON, I. (1971). Existing quality standards of produce in tropical Africa. *Ford Foundation/IRAT/IITA seminar on grain storage in the humid tropics, Ibadan*.

PILLAI, S. P., MUTHU, M., MAJUMDER, S. K., SHARANGAPANI, M. V., and AMLA, B. L. (1970). Development of a gas-proof strip-pack of ethylene dibromide for use on small-scale fumigations. *Int. Pest Control* **12**, 23–7.

POINTEL, J. G. (1967). Contribution à la conservation du niébé. *Agron. Trop.*, *Nogent* **22**, 925–32.

—— (1969). Essai et enquête sur greniers à maïs togolais. *Agron. Trop.*, *Nogent* **24**, 709–18.

POINTEL, J. G. (1971). Control of insects during storage by means of fumigation. *Ford Foundation/IRAT/IITA seminar on grain storage in the humid tropics, Ibadan.*

PREVETT, P. F. (1959). *An investigation into storage problems of rice in Sierra Leone. Colon. Res. Stud.* **28.**

—— (1961*a*). An investigation into the bulk storage of sorghum in five cylindrical aluminium silos. A. Rep. W. Afr. stored Prod. Res. Unit, Tech. Rep. No. 13, pp. 73–82.

—— (1961*b*). Field infestation of cowpea (*Vigna unguiculata*) pods by beetles of the families Bruchidae and Curculionidae in Northern Nigeria. *Bull. entomol. Res.* **52,** 635–45.

—— (1971*a*). Storage of paddy and rice (with particular reference to pest infestation). *Trop. stored Prod. Inf.* **22,** 35–49.

—— (1971*b*). Some laboratory observations on the development of two African strains of *Plodia interpunctella* (Hubn.) (Lepidoptera:Phycitidae), with particular reference to the incidence of diapause. *J. stored Prod. Res.* **7,** 253–60.

PROCTOR, D. L. and ASHMAN, F. (1972). The control of insects in exported Zambian groundnuts using phosphine and polyethylene lined sacks. *J. stored Prod. Res.* **8,** 127–37.

RAMBO, E. K. (1971). Efficiency and economics of locally produced and imported structures. *Ford Foundation/IRAT/IITA seminar on grain storage in the humid tropics, Ibadan.*

RAWNSLEY, J. (1968). Biological studies in Ghana on the control of *Cadra cautella* (Wlk.), the Tropical Warehouse Moth. *Ghana J. agr. Sci.* **1,** 155–9.

RILEY, J. (1963). Distribution of methyl bromide gas in a stack of 1000 ton of bagged cocoa beans during fumigation at Apapa. A. Rep. W. Afr. stored Prod. Res. Unit, Tech. Rep. No. 3, pp. 41–7.

—— MATHESON, K. C. (1960). Protection by insecticides of maize on the cob stored in cribs. A. Rep. W. Afr. stored Prod. Res. Unit, Tech. Rep. No. 12, pp. 48–54.

—— SIMMONS, E. A. (1967*a*). The fumigation of large cocoa stacks in a specially designed cocoa warehouse using phosphine. A. Rep. Nigerian stored Prod. Res. Inst., Tech. Rep. No. 1, pp. 17–27.

—— —— (1967*b*). The fumigation of large cocoa stacks at the Ikeja cocoa stores using methyl bromide in the 1966–67 cocoa season. A. Rep. Nigerian stored Prod. Res. Inst., Tech. Rep. No. 2, pp. 29–38.

ROBERTS, E. H. (1962). Storage of cereal seed in the humid tropics with special reference to rice. Paper presented at the *CCTA/FAO symposium on stored food, Freetown.*

SCHULTEN, G. G. M. (1971). Initiation and progress of infestation in maize. *Ford Foundation/IRAT/IITA seminar on grain storage in the humid tropics, Ibadan.*

—— WESTWOOD, D. (1972). Grain Storage Project, Malawi. December 1969–June 1972. Reports. A Ministry of Agriculture project, financed by OXFAM, the Agricultural Development and Marketing corporation and Chancellor College, University of Malawi. Extension Aids Branch, Ministry of Agriculture, Malawi, Zambia.

SMITH, C. V. (1969). Meteorology and grain storage. Tech. Note UN Wld. Met. Org. No. 101 (WMO No. 243 TP 133). Secretariat of World Meteorological Organization, Geneva.

SMITH, R. B. L. and BOONLONG, S. (1970). Ferro-cement bins for hermetic storage of rice. Asian Inst. of Technol., Bangkok, Res. Rep. No. 12.

STEVENSON, J. H. (1971). *Symposium on food grain marketing in Asia.* Asian Productivity Organization, Tokyo.

SUTHERLAND, J. W., PESCOD, D., AIRAH, M., and GRIFFITHS, H. J. (1970). Refrigeration of bulk stored wheat. *Aust. Refrig. Air Cond. Heat.* Aug. issue, 30–45.

UPTON, M. (1962*a*). WASPRU/UCI Grain Storage Project. Costs of maize storage 1959–60. A. Rep. Afr. stored Prod. Res. Unit, Tech. Rep. No. 15, pp. 83–6.

—— (1962*b*). WASPRU/UCI Grain Storage Project. The cost of guinea corn storage in silos. A. Rep. W. Afr. stored Prod. Res. Unit, Tech. Rep. No. 17, pp. 89–90.

WASPRU (1962). A. Rep. W. Afr. stored Prod. Res. Unit.

WATTERS, F. L. (1971). Control of storage insects by residual chemicals. *Ford Foundation/IRAT/IITA seminar on grain storage in the humid tropics, Ibadan.*

—— (1972). Control of storage insects by physical means. *Trop. stored Prod. Inf.* **23,** 13–28.

WEBB, E. R. (1969). *A crop dryer and grain storage silo for the small farm.* Ministry of Agriculture and Natural Resources, Ibadan, Nigeria.

WHEATLEY, P. E. (1971). Relative susceptibility of maize varieties. *Ford Foundation/IRAT/IITA seminar on grain storage in the humid tropics, Ibadan.*

WHITNEY, W. K. (1971). Control of storage insects by non-chemical methods. *Ford Foundation/IRAT/IITA seminar on grain storage in the humid tropics, Ibadan.*

WILLIAMS, C. N., BEENEY, J., and WEBB, B. H. (1969). A solar heat drier for crops and other products. *Trop. Agric., Trin.* **46,** 47–54.

WILLIAMS, S. K. T. (1971). Grain storage in the Western State of Nigeria; success or failure. *Ford Foundation/IRAT/IITA seminar on grain storage in the humid tropics, Ibadan.*

13. Agricultural mechanization

C. GAURY

13.1. A definition of mechanization of agriculture

The term mechanization is used in this chapter to cover any means of mechanization† from the most simple to the most complicated, for carrying out agricultural operations from the first preparation of the land to the final preparation of the harvested products for sale. The means of mechanization can be

(1) simple hand tools or hand-operated machines;
(2) animal-drawn tools;
(3) motorized equipment, either with tractors or with static or carried engines, including those engines used for crop protection and irrigation, etc.

Corresponding to the three classes of mechanization, three major types of agriculture can be recognized:

(1) agriculture carried out entirely by hand labour;
(2) agriculture carried out using animal traction;
(3) agriculture carried out with the use of internal combustion engines or electric motors.

13.2. The present situation in the mechanization of agriculture in the tropical regions of West Africa

Throughout the region cultural methods employing hand labour continue to predominate. This type of agriculture is practised using mostly family labour and on holdings generally not exceeding 2 ha. Many authors have estimated that this method of agriculture accounts for 80–90 per cent of all cultivated land in West Africa. Before 1945 the only mechanized tools available were very simple instruments made by local artisans to local designs which varied from one village to another. Many of these tools were

† The distinction is very clearly drawn in French between mechanization and motorization. In some English usage the mechanization of agriculture tends to carry the implication that engines are to be used, but this is not so in French.

made from scrap material, particularly from 200-l oil barrels. In a few areas, however, where suitable iron ore occurred, as in north-eastern Nigeria and northern Cameroun, tools were locally forged from this. Locally made tools were usually not strong enough to be used for more than a season or two. After 1945, with better communication and trade, there was a rapid increase in the importation of tools manufactured in industrial countries from good-quality steel, and these progressively replaced those made locally. Later, as extension services were expanded and modern agriculture developed, new types of tools were imported, such as sprayers for crop-protection work in cocoa, coffee, and cotton, coffee pulpers, maize shellers, pedal-operated rice threshers, winnowers, etc.

More recently, small modern factories that produce simple agricultural hand tools have been set up in West Africa: e.g. in Accra, Ghana, hoes, cassava graters, wheelbarrows, etc.; in Abidjan, Ivory Coast, machetes, rice seeders, pedal-operated rice threshers, groundnut decorticators, simple water pumps, etc.; in Dakar, Senegal, sieves, groundnut decorticators, pedal-operated rice threshers, etc.; in Douala, Cameroun, axes, hoes, pickaxes, shovels, machetes, knapsack sprayers, etc. Nearly all these factories also construct equipment for animal traction and simple motorized equipment.

13.3. Country-by-country summary of the degree of the use of draught animals

13.3.1. Senegal

The use of a horse-drawn groundnut seeder for planting single rows has been established since 1930 and also the use of an animal-drawn hoe for weeding. Since 1950 there has been a steady change from horses to oxen, and this has allowed the diversification of equipment beyond those which small horses could pull. Ploughs which could plough to 25 cm deep both with and without wheels, polycultivators for one, two, or three rows, groundnut lifters, and carts have all been developed.

At the Bambey seminar it was reported (Anon. 1971) that in Senegal in 1970 there were 12 000–15 000 lightweight ploughs, 135–150 000 seeders, 120–130 000 hoes, 20–25 000 groundnut lifters, and 50–55 000 animal-drawn carts. The majority of this equipment had been built in local factories and was used largely for groundnut production but, with the exception of the groundnut diggers, was mostly also employed in the cultivation of sorghum and maize. The animal-drawn plough is also used for cotton cultivation on about two-thirds of all the cotton land, comprising 14 000 ha, the remainder of the cotton acreage being still hand-cultivated.

13.3.2. Mali

In Mali animal power has been used for ploughing the heavy rice soils for 50 years in the valley of the Niger. The local zebus are much stronger than those of Senegal. During the last 10 years the Government of Mali has encouraged the use of draught animals for rice, cotton, groundnuts, and sorghum, and has developed credit finance for the purchase of equipment. The programme, which began in 1962, envisaged the progressive equipping of 100 000 family farm units with ploughs, harrows, seeders, hoes, carts, etc. Development effort has been concentrated in certain main areas, where the peasants have been most highly organized in the production of rice, groundnuts, or cotton. The results achieved in the first 10 years were indicated to the Bambey seminar as follows: 82 000 ploughs, 19 000 poly-cultivators, 3300 hoes, 5900 harrows, 4700 seeders, 900 groundnut diggers, and 30 000 carts were now deployed. These have been imported partly from Europe and partly from Senegal. A small modern factory is being built at Bamako which will begin production of some equipment in 1974, but a much older established workshop under the Office du Niger at Markalla has produced practically nothing.

In Mali farms that produce cotton are the best equipped and they also use their equipment for other crops in the rotation. In the 77 000 ha of cotton cultivated in 1972 half was on land which had been ploughed, a quarter of the land had been prepared using ridger bodies mounted on polycultivators, and only a quarter had been cultivated by hand.

13.3.3. Nigeria

In Nigeria draught animals are only used in the savanna zone which occupies the northern part of the country. Groundnuts and cotton are the main crops which are mechanized in this way. Cultivation in this area is mainly on ridges and the most important implements used have been ridgers imported from Britain, but these are also locally manufactured in a small factory in Zaria. The Zaria factory also makes small ox-carts (Food and Agriculture Organization 1966). It was estimated in 1966 that there were 40 000 ridgers in use representing a coverage of approximately one for every 100 farms (CEEMAT internal paper).

Ahmadu Bello University at Zaria has been interested in developing the use of both oxen and donkeys and has developed equipment for donkey traction (harrows, cultivators, and a donkey-cart). It seems however that the Governments of only a few states within Nigeria encourage the development of animal traction. They mostly prefer to move directly from manual cultivation to the use of tractors.

13.3.4. Chad

In Chad the Ministry of Agriculture and several other organizations for rural development, and in particular the National Office of Rural Development (ONDR) have very strongly encouraged the use of animal traction in the areas where groundnuts and cotton are grown. The equipment is mostly ploughs and carts drawn by zebus. The use of multi-purpose toolbars has been increasing since 1968. In 1971, in the cotton zone of Chad, there were already about 30 000 ploughs and 10 000 carts in use, representing a plough for every 10 farms and a cart for every 30 farms. 115 000 ha from a total of 299 000 ha cultivated land sown to cotton during the agricultural season 1971–2 were cultivated with ox-drawn equipment. All the equipment in Chad is imported, some from Europe and some from Cameroun.

13.3.5. Cameroun

For the last 20 years the Government has consistently encouraged the use of animal-drawn equipment, particularly in the savanna zone in the north. Draught animals are mostly zebus, but a few donkeys are also used. Government policy is implemented by two important organizations concerned mostly with the production of groundnuts and cotton respectively, and hence also with sorghum which is always associated in rotations with the two preceding crops. In the cotton zone alone the number of ploughs utilized during the 1971–2 season was about 13 000, and these were used for cultivation of about a third of the total area of cotton, i.e. 32 000 ha from a total of 99 000 ha. All the new animal-drawn equipment used in Cameroun is constructed locally in a factory at Douala which has been operating for 10 years.

13.3.6. *Upper Volta*
In Upper Volta a great effort was made in about 1960 to utilize donkeys, which are very abundant in that country. But these animals are not strong enough to use for pulling ploughs; however, they can pull hoes, which are well suited to the light soils in the centre and north-east of the country, where sorghum, pearl millet, and groundnuts are grown. Donkeys are mostly used for transport; most of the donkey carts are made in a local factory. In the south-west of Upper Volta, ox traction is being used, largely in cotton cultivation.

13.4. The use of tractors in agriculture
The use of tractors in West Africa began only after the Second World War. Previously, the only tractors were to be found in a few agronomy stations and on major commercial agricultural plantations. The motorization of agriculture developed mainly as a result of the initiative of the West African Governments, and the development was facilitated by equipment provided by the USA under the Marshall Aid Plan during 1948–52. However, motorized equipment is still used only in a very small part of the total area available for cultivation throughout all the states of West Africa.

In the beginning, the gifts under the Marshall Aid Plan and material bought from the USA and Europe were distributed mainly among the agronomic stations, among the official extension services, and in major agricultural enterprises. In most of the English-speaking countries of West Africa (such as Gambia, Ghana, Nigeria, and Sierra Leone) Governments, particularly that of Ghana, have always considered that motorization was the key to rapid development of agricultural productivity. And this philosophy lies behind the fact that the motorization of agriculture is much greater in anglophone Africa than in francophone Africa. In francophone Africa, Governments laid much greater stress on the introduction of animal traction as an intermediate phase, and have become interested in motorization only during the last few years. The situation as it

was in 1971 in the five West African states where motorization is now most developed will now be examined briefly.

13.4.1. *Ghana*
The major importation of tractors into Ghana began in 1962, and the scale of operations was greatly increased in 1966. Importations had reached about 5000 tractors by 1964, and from 1968 onwards an average of about 1000 per year were being imported. Judging from two separate figures reported to the Bambey Seminar (Stout 1971) it appears that between 1960 and 1971 Ghana imported of the order of 10 500 tractors, some wheeled and some tracked. Nearly all these tractors, with their equipment, came from western Europe, the remainder being provided from the USA and eastern Europe. This also applies to combine harvesters. About 70 combine harvesters were imported in 1971 and are used for harvesting about 10 per cent of the acreage of rice and maize in Ghana.

Motorization in Ghana was spearheaded by the Government itself and by the various organizations under its control responsible for the cultivation of rice, maize, and of sugar cane, etc. Much of the equipment had been used in the large state farms, e.g. in the Volta River Agricultural Resettlement Unit (VRARU), by the Division of Mechanization and Transport of the Ministry of Agriculture, and by the United Ghana Farmers Cooperative Council (UGFCC). The two latter organizations carried out work for farmers, such as land clearing, ploughing, and harvesting of rice and maize. The VRARU was responsible for carrying out the necessary work for 13 000 farmers in resettlement schemes on new land when their own lands were submerged under the artificial Volta Lake. This scheme alone used about 20 heavy tracked tractors and more than 100 wheeled tractors with complete equipment for all aspects of cultivation. The UGFCC was closed down in 1966.

The Division of Mechanization and Transport resold some of the equipment which had been obtained with financial aid for private use, and much went to the north of Ghana. But this division was still operating in 1971, and, according to Stout (1971), was responsible in 1968 for 29 centres located in 8 regions. For several years large private farming companies have been embarking on motorization. Farmers with 20 ha, for example, may have one or two tractors, and the large Ghanaian farming

N

societies which may farm hundreds of hectares are also well equipped.

The small farmers who have tractors and other equipment usually hire out these for use by their neighbours when they have completed their own work. They thus become a class of private entrepreneurs.

13.4.2. Ivory Coast

Serious studies of the possibilities of mechanization began in the Ivory Coast in 1947 on the initiative of the Chamber of Agriculture, which was set up by the Government, and which provided financial aid for the operations of an experimental committee for the mechanization of production. This committee bought a number of tractors and equipment and used them in experiments in agriculture with the intention of choosing those which were most suitable. However, for lack of sufficient financial credit the committee only functioned for 2 or 3 years. Its activities were very useful. Motorization developed slowly at first, but the pace has increased since 1965. From this time the number of tractors imported each year has been between 120 and 270 (CEEMAT 1970).

At the end of 1971 it was estimated that for agriculture alone there were 1412 wheeled tractors and 212 tracked tractors working in the Ivory Coast. Half the wheeled tractors were owned by private operators, who usually only have one each, and a quarter by the State Development Societies (Societies d'État) which between them own three-quarters of the crawler tractors. Until 1971 most of the tractors used for agriculture were deployed in the forest zone in the south of the country, where they were used for the cultivation of industrial crops such as oil palm, coconuts, rubber, bananas, and pineapples. The last two of these can be considered as food crops as well as industrial crops.

The Government of the Ivory Coast is now developing the use of tractors for the true food crops such as rice and yams in the centre and in the north of the country. Both four-wheeled and single-axle tractors are used. The developments are being carried out mostly in the areas in which the State Societies such as SODERIZ and Bandama Valley Authority (AVB) are operating. In the north-west of the Ivory Coast during the last 10 years small private entrepreneurs have emerged who supply services to rice cultivators, and in the same way that the State Society Motoragri operates in other parts of the country. Motoragri was set up by the Government in 1966 and provides a large amount of equipment for assisting the development of agriculture. It carries out bush clearing, road building, construction of dams, and also, when required, any agricultural work on a repayment basis.

The principal work carried out by tractors in 1969 in addition to mechanized clearing with crawler tractors, lay in transport and ploughing, soil improvement, and the upkeep of plantations. For these uses the following equipment was available: 740 trailers; 290 disc ploughs; 530 disc harrows; 190 rotary mowers.

There has been a tendency to use tractors of greater and greater power. In 1969, 60 per cent of the wheeled tractors had a power of more than 40 horsepower (hp). In 1971, of the purchase of 203 tractors, 93 per cent were of more than 40 hp, comprising 79 of 40–50 hp and 110 of 50–100 hp.

13.4.3. Nigeria

In Nigeria the Federal Government and the State Governments, particularly those of Western Nigeria and Northern Nigeria, appear to consider that the use of tractors in agriculture is the only means of rapidly developing the production of food crops, despite the great rural exodus towards the towns that has occurred in the Western State.

Tractors are nearly always utilized in services provided by the State Ministries of Agriculture or other organizations which depend to a greater or lesser extent on the administration. However, the Government also encourages the creation of private enterprise by providing favourable terms for private operators to purchase tractors.

As well as the very large numbers of small cultivators whose farms are too small to be equipped with motorized implements, there are also a number of farms, many of several hundred hectares, that have been developed by well-to-do Nigerians with sufficient capital to acquire their own machinery. In the Western State in 1971 the operations of large private farmers accounted for the use of more than 400 tractors.

The Government tractor-hire services mostly use 30–50 hp tractors, and these are equipped largely only for ploughing and disc-harrowing.

Single-axle tractors are little used, though some Nigerian agriculturalists are suggesting that they should be used in irrigated rice cultivation for seedbed preparation, using rotary cultivators. The employment of such equipment would appear to be economically worthwhile for people having 2–4 ha of rice land to cultivate, and the equipment could also be shared with their neighbours.

Tractor Hire Units were created in 1971 to assist farmers in a rapid development of production. The Tractor Hire Units of the Western State possessed 200 tractors in 1971, distributed among 25 centres, each serving an area within a radius of about 20–25 km.

13.4.4. Mali

In Mali the motorization of agriculture began on a large scale on the irrigated lands managed under the Office du Niger in about 1946. The Office du Niger is a State enterprise principally concerned with the cultivation of rice, but more recently with maize also. Heavy equipment was first used in laying out the irrigation schemes, and agricultural tractors were then used for opening up the savanna lands and carrying out some cultivation. Since 1949 an area of almost 50 000 ha has been brought into cultivation. Much of this was developed for family cultivation with animal traction, but there was also a special development area of 8000 ha where rice was developed entirely with motorized mechanical cultivation. Of these 58 000 ha about 10 000 ha have been mechanically land-planed with great precision to allow the irrigated cultivation of cotton. Another part of the area has been developed for the cultivation of sugar cane. In 1971 the situation had developed little beyond this, and it was estimated that about 40 000 ha was effectively being cultivated by the Office du Niger, of which 10 000 ha were tractor-cultivated and 30 000 ha employed animal traction. The number of tractors was about 160, and there were 43 stationary rice threshers.

Motorized equipment is utilized for land-clearing, for the destruction of wild self-sown rice, and for work needing to be done quickly, but for most ordinary cultural operations animal traction is still far more important and is being encouraged by the Government.

In 1971 the Division of Mechanization of the Ministry of Agriculture stated that the following equipment was available in Mali:

228 wheeled tractors
378 crawler tractors
  8 single-axle tractors
199 ploughs (for tractors)
188 disc harrows

101 spiked tooth harrows
105 seeders
 26 sprayers
 41 mowers
133 static rice threshers
 44 groundnut decorticators
101 trailers for tractors
507 small motor irrigation pumps.

By far the greatest part of this material is being used by the public services. There are a few agriculturalists who have their own equipment, but there are only a very small number of private entrepreneurs who own tractors for hire to farmers. All the motorized equipment used in Mali in 1971 had been imported.

13.4.5. Senegal

Motorization of agriculture began after the Second World War under the programmes for three major State enterprises:

(1) the Groundnut Block of Kaffrine;
(2) the General Company for Tropical Oilseeds (CGOT);
(3) the Senegal Management Commission (MAS).

The Groundnut Block at Kaffrine is situated in the traditional groundnut-producing zone of Senegal in a region of natural tree-savanna vegetation. The Block was created in 1946 for the study of the motorized production of groundnuts in a scheme of 3000 ha of which 300–400 ha were to have integrated mechanized production, the rest to be cultivated by settlers with some level of motorized production. American equipment was obtained, comprising 9 heavy crawler tractors and associated equipment for bush clearing and for cultural operations, 40 wheeled tractors of 22–44 hp, ploughs, disc harrows, and combined fertilizer spreaders and seeders.

After several years it became apparent that the fully integrated tractor-powered mechanization of groundnuts was not economic, because sufficient yield per hectare could not be obtained and the price for the produce was not sufficient. The programme was therefore slowed down, and in 1955 the motorization began to be progressively replaced by animal traction, motorized elements being retained only for certain difficult operations. The Groundnut Block was renamed the Experimental Sector for the Modernization of Agriculture (SEMA).

CGOT began operations on an area of 30 000 ha of forest land in southern Senegal in Casamance, intended for the mechanical cultivation of ground-nuts following the techniques begun at Kaffrine and stimulated by the shortage of vegetable oils which existed at that time in Europe. This operation was planned at the same time and with the same objectives as the well-known abortive East African Groundnut Scheme in Tanzania. Here also the fully mechanized production of groundnuts was not economic and had to be abandoned after a few years. The equipment which was available in 1951 for this project comprised 51 crawler tractors and 4 heavy-wheeled tractors for clearing purposes and 17 wheeled tractors for cultivation equipped with sets of ploughs, harrows, etc. By 1971 most of this equipment had been dispersed for other use and the scheme possessed only 5 crawler tractors, 11 wheeled tractors and 6 combine harvesters.

MAS, which was set up as early as 1938 by the Government for developing production in the Senegal Valley, and particularly in the delta, was managed from 1946 onwards especially for mechanized rice production until 1971 at which time the Government decided to replace this by the more highly economic enterprise of sugar production. Land under the control of the MAS was leased to other enterprises for development, at first to a private society and later to a State organization called the Society for the Development of Rice Culture in Senegal (SDRS). The latter organization had the following equipment deployed in 1971: 12 crawler tractors; 22 wheeled tractors and various ploughs, disc harrows, fertilizer spreaders, seeders, etc.; 24 combine harvesters.

In 1964 another State organization, the Society for the Management and Exploitation of the Delta Lands (SAED) was created to continue the studies begun by MAS with a view to developing 30 000 ha of cultivable alluvial soils on the downstream side of the perimeter of the land occupied by the SDRS. The Government's purpose was to reduce the dependence on imports of food to Senegal and set up groups of small farmers organized in co-operatives for food production. By 1971 29 co-operatives had been formed, each with about 400 members. Until the present time all the work of developing this land has been carried out with tractors of both types. It is expected that later most of the heavy work will continue to be carried out with tractors but all the other work will be carried out with oxen or by single-axle tractors.

In 1971 SAED possessed 60 crawler tractors, 24 four-wheeled tractors, 53 disc ploughs, 67 disc harrows, 6 stationary rice threshers, and 1 combine harvester.

Two further important new agricultural enter-prises which employ motorized cultivation were set up in Senegal in 1971. These were the Senegal Sugar Company, which occupies the irrigated perimeter areas around the SDRS at the Richard Toll research station, and BUD-Senegal which specializes in the production of vegetable crops for fresh market.

In great contrast to the situation in the Far East the use of single-axle tractors has been very little adopted in West Africa, and the only ones which one finds at present are those that have been imported by Taiwanese agricultural missions and used in the development projects with which they have been associated.

13.5. Special problems of agricultural mechanization in tropical regions

The difficulties which are found in West Africa are common to those in most other parts of the developing world. First and foremost is the financial cost of acquiring the expensive machinery, but other problems are as follows:

(1) the generally small size of individual land holdings;

(2) the lack of a spirit of co-operation among small-scale agriculturalists which would allow the efficient shared use of equipment to develop;

(3) the absence of mechanical understanding;

(4) the more unfavourable edaphic and climatic conditions under which the machines have to operate compared with conditions in temperate countries;

(5) the lack of adequate after-sales service, which is an inevitable consequence of the dispersion of small numbers of machines over very large areas;

(6) the absence of small workshops on a village scale able to cope with routine maintenance operations;

(7) the poverty of most of the farmers, associated with the low economic return which they can obtain from their crops and their generally subsistence approach to food-crop production;

(8) the insufficiency of systems for providing credit which could overcome this poverty;

(9) the lack of post-harvest operations on the direct products of agriculture which tends further to reduce the value of agriculture to the primary producers.

To overcome these various obstacles in a short time in West Africa will require enormous efforts on the part of any States which wish to develop motorized agriculture. Some of the essential associated developments will be as follows:

(1) agrarian reform;
(2) improvements in extension services and methods;
(3) the encouragement by agents and manufacturers of adequate after-sales service and the development of local repair workshops;
(4) better organization of credit and of the marketing of produce.

13.6. The necessity for accelerated mechanization of agriculture

While the improvement of hand tools should obviously be encouraged the effects of this on over-all production cannot be expected to be very great with the exception of the use of crop protection equipment. For making a more rapid increase in production it is generally necessary to move to the use of draught animals or especially, after a careful study, to motorized agriculture. In either case, it is important to develop an appropriate package of recommendations relating equipment to the degree of advancement of agronomic technique. A rapid increase in production is needed both for economic and psychological reasons. Economically it is important to reduce the need for imports of food. This is first because in many states, even in average years, there are shortages of cereals, sugar, and food legumes, and secondly because good supplies must be established so that the rapidly increasing population that is expected over the coming decades can be fed from local sources. From a psychological point of view increased mechanization will make less the drudgery of farming associated with heavy manual work and will tend to improve the social status of those who farm the land. A second strong psychological motive for increased mechanization is that for the Governments of many states this *per se* is a symbol of development.

13.7. The choice between animal draught and motorization

This choice must be studied in detail for every particular situation. Motorization will almost always necessitate initial and continuing imports, not only

of equipment but also of fuel, lubricants, and spare parts. Equipment for the development of animal draught, on the other hand, can often be obtained by building local factories, although some of the basic raw materials and certain specialized parts will usually have to be imported. Motorization will therefore tend to be a drain on the foreign exchange of a state (unless the food crops produced are in turn exported), while the development of animal draught may assist the country's development through the parallel encouragement of local industrialization. The fear that, by reducing the need for hand labour, motorization will lead to unemployment has on the whole proved ill-founded, except in rare cases in highly over-populated areas. On the contrary, experience shows that very often the lack of availability of hand labour at certain critical times in the agricultural calendar limits productivity; this is the case, for example, in the south of the Ivory Coast, in Ghana, and in the Western State of Nigeria; in such cases motorization becomes a necessity. In the other areas, fears of unemployment are unnecessary because motorization increases over-all development, and this rapidly leads to a diversification of employment opportunities.

13.8. The organization of research on agricultural mechanization in anglophone and francophone West Africa

The organization of agricultural mechanization is quite different in anglophone and francophone areas. In anglophone areas research for mechanization is sometimes carried out by Ministries of Agriculture but more often is part of the programme of universities and associated experimental stations. Often these operate with collaboration from foreign organizations such as the Overseas Liaison Department of the National Institute of Agricultural Engineering of UK and also with several American universities. In the francophone countries research is most often carried out in collaboration with French Institutes in multi-disciplinary agronomic research stations specializing in particular crops and coming within the control of the Ministries of Agriculture or of Scientific Research. In two francophone countries, Ivory Coast and Mali, research is entirely confined within official organizations which, as necessity arises, collaborate with the French Centre for Experimental Studies in Tropical Agricultural Mechanization (CEEMAT). The amount of support given to research on mechanization varies between different states, both in anglophone and francophone areas.

In addition to the national organizations, there is also the International Institute for Tropical Agriculture (IITA), recently set up at Ibadan in Nigeria, which possesses a division of agricultural mechanization which works in liaison with other international institutes, notably with the International Rice Research Institute in the Philippines (IRRI).

13.8.1. Organizations in anglophone states

Gambia. Research is carried out within the Ministry of Agriculture.

Ghana. Research is carried out by the Division of Agricultural Engineering at the University of Legon (Accra) and by the Department of Agricultural Engineering in the University of Kumasi.

Legon University has three experiment stations, at Kade, Kpong, and Niengua. Kumasi University carries out mechanization research at the station near Nyankpala. The extension of results of research is carried out both by the extension division of the University and also by the Division of Mechanization and Transport of the Ministry of Agriculture. USAID has also been concerned with agricultural mechanization in Ghana through its farm project on sorghum and maize at Ejura, occupying more than 4 000 ha.

Several private companies in Ghana which use tractors in agriculture also do some development work. These include the Northern Engineering Co. Ltd. at Tamale and Mencilo and Co.

Liberia. Research on agricultural mechanization is carried out only in the College of Agriculture in the University, in association with the rice station at Suakoko.

Nigeria. Several universities are actively involved in research in agricultural mechanization and its application, as well as at IITA. The priorities for research are defined by a Federal Government organization, the Agricultural Research Council of Nigeria.

University organizations for research into agricultural engineering are as follows:

The Agricultural Engineering Unit of the University of Ibadan,
The Agricultural Engineering Section of the University of Ahmadu Bello (Zaria),
The Department of Agricultural Engineering of Ife University,
The Department of Agricultural Engineering of Nsukka University.

Research is also carried out at Moor Plantation of the Department of Agriculture at Ibadan and the Nigerian Stored Products Research Institute, which is under the Ministry of Commerce.

Extension of the results of research is carried out by most of those organizations which conduct research and also the Engineering Section of the Ministry of Agriculture, the Western Nigerian Development Corporation, and USAID.

Sierra Leone. Njala University College is involved in research and also the Agricultural Station of Rokupr; the latter particularly for studies on rice.

13.8.2. Organizations in francophone states

Ivory Coast. General agronomic research is under the Ministry of Scientific Research, but research on the mechanization of agriculture and the application of the results is under the control of the Ministry of Agriculture. Research programmes are reviewed each year by a Committee for the Agricultural Mechanization of the Ivory Coast (COMACI) under the Ministry of Agriculture. COMACI controls a Centre of Studies and Experimentation at Abidjan which carries out a programme in collaboration with relevant agronomic stations. As far as research is concerned with food crops in particular, studies are conducted at the agronomic station of Bouaké where management is provided for the Ivory Coast Government by IRAT. Bouaké Station controls several substations.

In 1971, at the main station, research was on mechanization for upland rice, sorghum, maize, and yams. At the substation at Gagnoa for irrigated rice; the substation at Man for upland rice; the substation of Ferkessedougou for irrigated rice and sugar cane; and the substation of Tombokro for irrigated food crops. Research relating to the production of forage crops is carried out at the Centre for Animal Research at Bouaké, which is also managed by the French National Institute of Animal Production and Health for Tropical Countries (IEMVT).

Research relevant to tropical food production is also carried out at the station of Anguédédou for pineapples, and the station at Azaguié for bananas and tree fruits.

The application and extension of research results is carried out by the following organizations: SODERIZ for rice, SODEFEL for fresh market vegetables and fruits, SODESUCRE for sugar cane, AVB (the Bandama Valley Authority) for rice, maize, etc.

The Ivory Coast Government also calls as necessary for assistance in executing certain programmes on other specialized French national institutions such as BDPA, SATEC, and CFDT.

Dahomey. There is relatively little research on agricultural mechanization, but work is ongoing on animal traction and on drying and storage of maize. Research is conducted partly by the Section for Agricultural Mechanization of the Rural Engineering Service of the Ministry of Agriculture and partly by the Rural Engineering Laboratory of the Agricultural College at Sékou. Extension of the results of research is carried out by the Section for Agricultural Mechanization of the Rural Engineering Service, the Agricultural College at Sékou, by SONADER (the National Society for Rural Development) and by SODEVO (Society for the Development of the Ouémé Valley); and also by foreign organizations such as the Swiss Cooperative Union, which is doing extension work with animal traction, and a Taiwan technical assistance mission which is developing the use of single-axle tractors in irrigated rice culture.

Upper Volta. Research is carried out in a specialized section of the agronomic station of Saria and by stations at Farako Ba, Mogtedo, and Kamboinse, which are each concerned with the production of food crops. The management of these stations is all vested by the Government in IRAT.

In 1971 research was chiefly on the mechanization of rice cultivation and the harvest of forage seeds.

Extension of research information is carried out under the direction of the Minister of Agriculture by the extension service and the regional development offices of the Ministry, and also by the specialized French organizations, BDPA, CFDT, and SATEC.

Mali. Government has encouraged mechanization of agriculture for a number of years. The main efforts have been towards the increasing use of draught animals. The research programme is defined each year by a national consultative committee for agricultural mechanization under the Ministry of Pro-

duction. The execution of the research programme is carried out by a Centre for Trials of Agricultural Machinery (CEMA) in collaboration with the relevant agronomic stations and with the specialized sections of the 'Operations for Development'. This latter organization is similar to the one in the Ivory Coast, COMACI. As far as food crops are concerned, CEMA works mainly in liaison with IRAT which operates the following stations for the Mali Government: Sotuba-Bamako, Ibetemi-Mopti (floating rice), Kogoni (for irrigated rice and other food crops).

The application of the results of research is carried out through the extension service and by the CEMA, but mainly by large, specialized agricultural development organizations, including the Office du Niger for irrigated rice, Opération Riz-Ségou for irrigated rice, Opération Riz-Mopti for flooded rice and irrigated rice, Opération Arachide for groundnuts, Opération of the Upper Valley and the State Farm at Baguineda for fresh market crops and food crops, and CFDT for cotton, sorghum, and groundnuts.

Mauritania. Although this country has no organization specifically concerned with mechanization, appropriate research and information obtained in Senegal is put into practice on the rice station at Kaedi through the activities of IRAT acting on behalf of Government.

Niger. A parastatal body in Niger is concerned with the development of agricultural production. This is the Union Nigerienne de Credit et de Cooperation (UNCC). This organization has operated a small research section on agricultural mechanization since 1971 which works in close collaboration with the Kolo and Tarna stations operated by IRAT. Research is mainly concerned with light equipment suitable for animal traction for the cultivation of rice, sorghum, and groundnuts.

Execution of the results of research is ensured by the services of the Ministry of Agriculture, by the UNCC, and by CFDT.

Senegal. Amongst all the francophone states Senegal is the one in which research on agricultural mechanization has been in progress for the longest.

Under the over-all direction of the Ministry of Rural Development, research is essentially carried out through the Division of Agricultural Mechanization and Rural Engineering of the National Agronomic Research Centre at Bambey (CNRA), of which the management has been undertaken since

1971 for the Senegal Government by IRAT. Originally, in about 1930, the activity of this division was mostly orientated towards the mechanization of sowing and weeding of groundnuts, and in developing the use of horses as draught animals. Research has been re-orientated to include the mechanization of sorghum cultivation and also of pearl millet, rice, grain legumes, and forages. Effort has always been concentrated on animal power, but with increasing attention to oxen and towards greater diversification of equipment. Nowadays research is more or less equally concerned with mechanization with both single-axle tractors and four-wheeled tractors, with post-harvest operations on agricultural products and also with the improvement of the use of water resources.

The Division of Agricultural Mechanization carries out research in collaboration with the other Divisions of CNRA and on the trial stations associated with this Institution:

The Bambey station: dry-land cultivation of sorghum, pearl millet, groundnuts, food legumes, forages.
The Richard Toll station: irrigated rice cultivation.
The Djibelor station: irrigated rice cultivation.
The Sefa station: dry-land cultivation (upland rice, sorghum, maize, groundnuts, and food legumes).

When research reaches the satisfactory development of prototypes, the plans are handed over to a local enterprise. The Senegalese Industrial Society for the Construction of Agricultural Machinery (SISCOMA), which is charged with the manufacture and marketing of such equipment.

The application of the results of research are carried out by CNRA and also by the State Societies concerned with agricultural development in Senegal.

Togo. Research on agricultural mechanization is relatively little developed. Research is carried out by the service for Rural Engineering under the Ministry of Agriculture and by the Mechanical section of the National Agricultural College at Tove. Work has been particularly concerned with drying and conservation of agricultural products.

The application of research results is essentially the responsibility of SORAD (The Regional Societies for Management and Rural Development), which is interested in all crops including rice, maize, sorghum, cassava, etc.

13.9. General considerations on the mechanization of tropical agriculture and of food crops in particular

13.9.1. The importance of the mechanization of agriculture in Senegal

The Minister for Rural Development in Senegal claims (Habib Thiam 1971) that research and development must aim to increase the productivity of human labour rather than increasing production per unit area. To achieve this the use of power greater than manpower is essential. The word mechanization, as used in this chapter, implies the use of such power but does not imply the use of engines. Mechanization makes new cultural methods possible. In this way it allows the social as well as the economic advancement of farmers. If the introduction of machines into agriculture is to be economic, whether by the use of animal or motor power, the search for a new style of cultivation rather than merely a higher-power version of the existing style is required. Good management will be essential, and the unit size of plots will have to be increased. This has implications for land reform in most areas which are already settled.

In Senegal the use of seed planters drawn by animals started in about 1930 and at present there are 150 000 in service over wide areas. The Government of Senegal aids the equipping of the rural community by encouraging the development of supply co-operatives and by organizing a credit system suited to the needs of the farmers.

13.9.2. Place and limits of mechanization as one of the factors of agricultural development

In equatorial Africa, according to Stout (1971), the problem lies in choosing the appropriate level of mechanization for the specific needs of each agro-ecological region and taking account also of all the local factors. Some local factors, such as population density, small plot size, rainfall distribution and reliability, and level of commercial development, as well as the intellectual level and social behaviour patterns, may restrict the application of the technical possibilities for mechanization.

If the essential bases for the profitability of tractor-powered production are not, or cannot be, brought together, it is wiser to use draught animals or even use manual cultivation.

13.9.3. *Present situation and future prospects of mechanization in tropical Africa*

Uzureau (Food and Agricultural Organization, 1971c) enumerates, in comparison with the situation in the industrialized countries, the factors unfavourable to mechanization which predominate in the present agriculture of many tropical countries of Africa:

(1) inappropriate agrarian structure;
(2) weakness of technical knowledge in the farmers;
(3) insufficiency of capital resources;
(4) poorly organized marketing of produce;
(5) lack of supporting infrastructure.

Accordingly, the FAO views it as unnecessary to be resigned to the present situation of under-equipping. As far as cultivation by animal draught is concerned, the FAO report (1971c) given at the Bambey seminar emphasizes the geographical as well as technical and economic limitations. On the other hand, FAO considers that the development of cultivation should not be restricted by the memory of past failures nor by any fear of encouraging rural exodus and underemployment. These changes are implied if one wishes to emerge from the subsistence economy which predominates at present. For the emergence of modern economic farming it is necessary to develop industrial cultivation for export, to allow and even favour a reduction in the percentage of the population engaged in farming, to create jobs for the work force freed from physical labour by mechanization (in mechanized agricultural skills and in the industrial development), and to expand, through training, the salaried cadres of extension workers and machinery repairers and distributors.

Tractor-powered agricultural development may take the following forms in Africa:

(1) state enterprises and state farms;
(2) state enterprises to undertake agricultural work for farmers;
(3) commercial enterprises to undertake agricultural work for farmers;
(4) farmers carrying out work under contract for their neighbours who lack equipment;
(5) co-operatives for using agricultural machines;
(6) development of family cultivation using small tractors which would need to be specially constructed.

Co-ordinated research efforts through the exchanging of information relevant to corresponding ecological regions are important.

13.9.4. *General technical conditions of mechanization in Africa*

As much as 35 per cent of agricultural produce is wasted by poor methods of harvesting, treatment after harvest, and storing, according to Garrard (1971). The equipping of African agriculture is slow and lagging behind agronomic research that depends on this development for its utilization. It would, however, be a mistake to rely on the introduction of machines which have been proved satisfactory only in countries where conditions of use are very different.

Where existing arrangements of land tenure are an obstacle to mechanization it is necessary to look for satisfactory solutions and it is possible that improved co-operation between different land holders can provide an alternative to more radical land reform if the latter would take too long to achieve. Rapid introduction of agricultural mechanization before industry is developed so that labour can be absorbed can lead to unemployment and its social consequences. It is important to create jobs by manufacturing as much minor agricultural material locally as possible and to provide training to those becoming redundant on the land so as to allow them to find other productive work.

With the introduction of mechanization it is necessary to ensure the conservation of the soil in countries with low, as well as with high, rainfall. Funds may be necessary to achieve this and mechanization of production should therefore be preceded by an over-all economic study.

It is most desirable to avoid importing into the same country closely corresponding machines of a variety of makes if after-sales service is to be maintained at reasonable cost.

The British National Institute of Agricultural Engineering at Silsoe has a department specifically devoted to a study of special needs in tropical countries, the development of prototypes and their testing, in collaborating countries by local research personnel, and providing technical assistance for developing simple agricultural equipment with materials which can be produced on the spot.

A transportable rice thresher, a multi-purpose animal-drawn toolbar, and a small 15 hp tractor have all been developed in this way.

13.9.5. *The development of mechanization of food crops in Africa*

Labrousse (1971) has reviewed mechanization in French-speaking tropical Africa. Much of his information has been incorporated in sections 13.1.–13.8.

Mechanical land clearing using powerful crawler tractors in forest and savanna areas was extended over large areas between 1946 and 1958 for developing land for large-scale schemes for groundnuts, rice, etc. This work was discontinued because it lacked justification on the basis of conventional economic analyses. Labrousse suggests, however, that such work ought to be considered as an investment which the State could back despite such analysis, because of the accumulation of real wealth involved in transferring idle land into a productive national resource. The cost of clearing could be reduced by employing alternative means such as a combination of mechanical and chemical treatments. Renewed planning of large-scale land development depends above all on the development of both land and water resources required for irrigated rice and sugarcane cultivation. Such work is currently being carried out with ordinary civil engineering equipment.

There are large private sugar enterprises employing heavy machinery in the Congo, Madagascar, Mali, Upper Volta, and the Ivory Coast. A limited amount of cassava cultivation is also mechanized in Madagascar and Togo, and the use of tractor power for yam-growing is under trial in the Ivory Coast.

Labrousse (1971) considers that it is often difficult to justify the mechanization of food crops in areas where food is not scarce and is still of low value in the market economy.

13.10. Effects of mechanical cultivation on the soil in dry tropical areas

Soil cultivation, in particular ploughing, occupies an important place in the mechanized agriculture of temperate countries. The value of introducing deep ploughing to Senegal, where until recently it has never been practised, is sometimes questioned by agronomists who consider that the structure of the light, friable soil of dry zones should be disturbed as little as possible to avoid the risk of making it infertile. The actual effects of ploughing of soils in the Sudan-Sahelian zone were discussed by Nicou and Poulain (1971) of CNRA.

Research by CNRA has shown beyond doubt the great value of cultivation in general, and ploughing in particular, for all crops in the dry zone, if carried

out under favourable conditions. This is sometimes difficult to achieve because of the short duration of the rainy season which enforces early sowing, leaving little time for preparing the soil after the start of the rains; and because of very rapid hardening of the soil after the end of the rains, which necessitates a higher power requirement for ploughing at this time. A sufficient source of machine power to be able to work dry soils is not normally within the resources of peasant farmers. Only animal draught power is realistically available to peasant farmers under the present economic conditions.

Equipment that can be used with local teams of oxen include light ploughs made of steel, with ploughshares 23–25 cm long, and which, according to soil moisture content, can plough at a depth of 12–20 cm. For pseudo-ploughing, i.e. without soil inversion, hoes or scarifiers with flexible or rigid tines and rigid chisel tines can be used. Because of the amount of draught power needed, oxen must be used. Donkeys and horses are less powerful and are suitable only for sowing and light maintenance work.

Ploughing improved both structure and porosity. This encouraged improved root development of plants and thereby enabled higher agricultural yields to be obtained (Chopart 1971). Deep ploughing, carried out at a suitable time, did not encourage erosion as had been suggested but in fact improved water acceptance and decreased run-off, thus contributing to soil conservation.

Apart from improvement in rooting and yields, ploughing reduced weed development. Organic matter can only be satisfactorily ploughed in by animal draught at the end of the rainy season, when the soil is still wet. However, if organic residues have been removed, ploughing can be performed in dry or wet soil, at the beginning or end of the rainy season. Ploughing has a residual effect, improving yields of the next crop after a cereal crop.

In dry tropical regions with only a short rainy season, sowing *must* be carried out early in the rains and, to gain essential time, ploughing must be done, whenever it is possible to do so at the end of the preceding wet season. However, in regions with a prolonged rains period it is more often advantageous to plough at the beginning of the rainy season and to sow late rather than to sow early without ploughing if these are the only alternatives.

Pseudo-ploughing gives much less valuable yield responses than ploughing with soil inversion.

13.11. The adaptation of systems of cultivation to mechanization

Mechanization of agricultural tasks necessitates a new balance between the size of the work force and the amount of work to be carried out. It alters the norms of size of the farm holding which can be cultivated by traditional family units in each ecological zone and for each crop. The need for modifications to the land/labour ratio become increasingly conspicuous as the degree of mechanization is increased. Conversely if the land/labour ratio is rather inflexible, this will affect the level of mechanization that can be considered appropriate.

It is not sufficient however for mechanization just to reduce human effort and allow a greater area to be cultivated. To be worthwhile it must also be more profitable, procuring a higher net income for a farmer who also values a higher cash income more than he values his traditional methods.

Monnier (1971) describes in detail the hours of work (labourers and animal teams) as well as the ratio between the necessary number of man-hours and number of animal-team-hours, by which he defines the degree of mechanization, for the main crops, in three different situations:

(1) donkey or horse draught with light equipment (unintensive cultivation);
(2) ox draught with medium weight equipment (semi-intensive cultivation);
(3) ox draught with equipment requiring all the power of the animal (intensive cultivation).

Monnier's study shows, incidentally, that manpower is the main factor limiting food-crop cultivation in Senegal.

The most important of several labour bottlenecks which justifies research on mechanization is in the time needed for manual weeding. Sowing and harvest time are the two other peak labour periods.

To overcome labour bottlenecks research activities of the CNRA in recent years have been concerned with the substitution of ox power for horse power and perfecting ox-cultivation equipment. Several new items of equipment have been developed and these are now becoming widespread (see Section 13.13). For a 4-year crop rotation comprising maize, cotton, sorghum, and groundnuts, a team of farmer, labourers, and a pair of draught animals can cultivate an area of land as follows: 5·2 ha with two horses and 3–4 labourers; 8·4 ha with two oxen, light equipment, and 5–6 labourers; 12·0 ha with oxen

with heavier equipment and 6–7 labourers. The use of tractors was not considered by Monnier to be appropriate for peasant farmers, neither was the use of chemical weedicides.

13.12. The orientation of crop and livestock research to meet the needs of mechanization

Because of the current trend in Senegal towards cultivation using oxen there has been increasing attention given to the breeding of draught animals of improved size, and hence pulling power and in establishing the working capacity of oxen. Hamon (1971) described the development of the new ox breed, Metis Bambey.

Breeding started with a cross between the Zebu of north Senegal, weighing 450 kg on average, and the N'dama of the east and south, which are *Trypanosoma* tolerant but do not weigh more than about 300 kg. Metis Bambey, developed from this cross, is larger than the Zebu, reaching weights of over 500 kg, and is superior to its parents as a working animal. It is relatively *Trypanosoma* tolerant. Its useful draught life is about 10 years. Set to work at 4 years old it is sold to the butcher at 14 years old for a price higher than the purchase price. Metis Bambey can achieve 100–150 kg draught without excessive fatigue if the work load is kept more or less constant, but frequent, abrupt variations of effort, such as is caused by ploughing dry ground, soon exhausts the animal.

CNRA have developed suitable feed rations, using local products for working animals and also for animals being fattened. This has led to increased interest in forage production, particularly in the groundnut region which is extensively cultivated by oxen and where the climate does not support permanent natural pastures.

Because of the lack of organic matter in tropical soils and particularly in the dry regions, a special interest attaches to the production of manure. This has spectacular and well-known effects on yield. CNRA has developed an inexpensive cattle shed, designed for the efficient collection of manure for use on crops.

13.13. Experiments and prospects for motorization

Monnier and Tourte (1971) review the causes of failure of the first attempts to introduce tractor power in West African food-crop agriculture and suggest factors that would favour the re-introduction of a partial use of tractor power. This may be particularly appropriate in land clearing and improving agricultural use of irrigable land. These authors, however, also emphasize the value, when possible, of having an intermediate stage of using oxen, which enables the peasant farmer to learn progressively the techniques of using a range of implements before moving on to the use of motor power.

The CNRA has moved from primary interest in mechanization to research on models of improved peasant farming systems making a rational use of the intensive techniques developed by research. This important subject is covered in greater depth by Tourte and Moomaw in Chapter 14 of the present volume.

The constraint of manpower availability at harvest and threshing times is particularly important in large-scale (i.e. 20 or more ha per family) peasant cultivation, which is fairly common in certain areas of Senegal.

Monnier and Tourte (1971) discuss studies on a holding of 72 ha employing 7 pairs of oxen, and having four-fifths of the land used for groundnut and one-fifth for cereals. With the same complement of 27 labourers, manual harvesting and threshing took 109 days, while harvesting with mechanical threshing took only 59 days. The threshing period could be even further reduced when newly developed millet and groundnut threshers of greater capacity presently under study by the CNRA are introduced.

Women on the farm studied by Monnier and Tourte took little part in field work but spent most of their time threshing, de-husking, and manually grinding the millet. The use of a de-husking mill could reduce the time necessary for these tasks and free them for outside jobs, where they would replace the temporary work force at peak periods. Hired labour is increasingly difficult to obtain.

These kinds of study enable research to be directed towards the development of prototypes of relevant equipment or the adaptation of imported goods when necessary. Such equipment includes groundnut planters fitted for application of weedicides, high capacity threshers for millet, sorghum, and rice, and strippers and shellers for groundnuts. The value of the study of harvesting equipment for millet and sorghum depends in turn upon success in the development of dwarf cultivars adapted to mechanical

harvesting, which is also currently under consideration by research workers.

13.14. Equipment used in tropical countries

13.14.1. *Animal draught equipment developed by CNRA*

The Agricultural Machinery division of the CNRA, which has devoted much of its activities for the last 40 years to the mechanization of the cultivation of the main crops of Senegal, groundnut, millet, and sorghum, by the use of power of available animals, has essentially finished that work. Alterations of detail of the designs achieved is now considered to lie within the competence of the extension services.

Work has culminated in the development of multi-purpose high-capacity equipment adapted to the strength of the highly bred modern oxen. The equipment (Monnier 1971) is based on the 'polycultivator', a machine which has been in existence for around 12 years, which comprises a toolbar mounted on two wheels with pneumatic tyres with adjustable track. The new machine can be equipped as a seed planter or for hoeing. The planter has sloping disc distributors with the discs being exchangeable for different types of seed. It enables two rows of millet, sorghum, maize, or cotton to be sown at 90 cm apart, or three rows of rice at 45 cm, or three rows of groundnuts at 60 cm to be planted at a single pass of the machine. The hoe attachment enables the same crops to be hoed and the spacing of the tines can be adjusted for the same inter-row spaces as the seed planters.

With this equipment 1 ha of sorghum can be sown in 5 hours instead of the 7 hours which was required using the previous single-row machine. Hoeing 1 ha can be done in 5 hours instead of 8 hours with the older model. Further refinements are being undertaken which should enable these times to be further reduced. Although this multi-purpose equipment, which is made at Dakar, is relatively expensive, its use has been shown to be profitable.

Le Moigne (1971) draws attention to a type of animal-powered rice drill for four or five rows which has been developed from the cereal seed drills of the temperate countries and is being tested in Senegal, Mali, and Madagascar.

13.14.2. *Studies on motorized equipment in Senegal*

The CNRA is particularly concerned with the power

of wheeled tractors necessary for different types of work, as well as their traction in sandy or muddy soil (Le Moigne 1971). CNRA has developed a motor-powered two-row precision planter, a two-row hoe, and several kinds of lifting equipment, all for ground-nuts. Three prototypes of post-harvest equipment were being constructed in 1971 at a local factory. These were a millet thresher with a capacity of 150 kg per hour, a hulling mill for millet with a capacity of 200–250 kg per hour, and a winnower-cleaner for groundnuts.

13.14.3. Comparative studies on small motorized equipment in the Far East

Khan (1971) provided information to the seminar on the development of mechanization in south-eastern Asia. In this area the majority of agricultural holdings, especially those producing rice, are of about 2–10 ha. This tends to be too large for cultivation using animal draught and too small for motorized cultivation with commercial tractors of 30 hp or more. Japanese machines, developed for similar areas, were largely unsuitable because they are too complicated and too expensive for most peasant farmers.

The International Rice Research Institute in the Philippines has developed equipment enabling farms of 2–10 ha to be mechanized for both production of, and post-harvest operations on, rice. This equipment can be constructed in most countries of the region following simple production processes. Fifteen prototype or operational machines have so far been developed at IRRI:

(1) a tractor which uses the thrust of a rotary hoe to work soft ground;

(2) a tractor–cultivator to prepare wet soils, which has four caged wheels, with those at the back and front turning at different speeds which causes the puddling of wet soil;

(3) an anhydrous ammonia distributor for flooded soil, mounted on a single-axle tractor;

(4) a six- and eight-row hand-pulled seed drill for planting pre-sprouted rice in muddy soil;

(5) a multiple-use seed drill (derived from the previous one) for dry or muddy soil, powered either by hand, by draught animals, or by a motor;

(6) a motorized portable rotary hoe for inter-row cultivation in rice;

(7) a drum thresher for threshing very wet rice, powered by a 4-hp motor and having a capacity of 250 kg per hour;

(8) a portable tray thresher for rice weighing 175 kg, and powered by a 3-hp motor;

(9) a tractor-powered thresher that can be adapted for use with any cereal grain;

(10) a light, multi-purpose cereal thresher, with a conical beater, designed to thresh damp grain;

(11) a light rice harvester-stripper, mounted on a single-axle tractor;

(12) a motorized rotary grain cleaner for wet cereals, with a capacity of 3 t per hour;

(13) a rotary hand-powered grain cleaner for small-scale farmers;

(14) a rapid rice drier, using heated sand;

(15) a continuous warm-sand drier for small farms.

Several of these machines are already fully developed and IRRI has given licences for production and marketing to Philippino and Japanese manufacturers. These include the pre-sprouted rice drill, the motorized rotary weeder, the drum thresher, the tray thresher, and the motorized rotary grain cleaner.

Certain models have been used in trials in 1971 at IITA Ibadan in Nigeria and at the CNRA in Senegal. The IRRI representative is prepared to grant freely the use of the designs in West Africa.

13.14.4. Equipment suitable for use with animal draught

Equipment at present in use. Le Moigne (1971) lists the methods and the types of equipment at present used in traditional family farming for the various tasks from clearing to post-harvest processing.

1. *Bush clearing.* Animal power is inadequate for this work, and farmers have to use hand tools such as axes, machetes, and saws. The work is very slow and this method is unsuitable for the clearance of large areas. In Senegal, on lightly wooded savanna, a minimum of 50 man-days per ha is required. In intermediate forest, in the wet zone, 500–600 man-days per ha is required.

2. *Cultivating non-flooded soil.* Several types of cultivation can be distinguished.

(a) *Ploughing with inversion of the soil*, which is carried out with any of the following.

   (i) Very light steel ploughs, without support with 15–20 cm plough shares. These are rather uncommon.

   (ii) Steel ploughs of less than 30 kg, having longitudinal stability assisted by a single wheel, with or without a coulter. A 15–

20 cm plough share is used with a donkey or horse, or a 20–27 cm plough share with oxen. One hectare can be ploughed in 35–40 hours at 10–15 cm depth. These ploughs are quite common in Senegal, Mali, Cameroun, Madagascar, and other francophone countries.

(iii) Ploughs having both longitudinal and lateral stability achieved either by having two wheels of unequal diameter and used for conventional ploughing or made reversible by having two symmetrical frames and two equal wheels. A model of the former type, weighing approximately 50 kg, is used in Mali, and the latter type, weighing approximately 90 kg and requiring two, three, or four pairs of oxen, is used in Madagascar.

(b) *Primary cultivation without soil inversion* is carried out in some countries with tine cultivators, but these are very little used in West Africa. Simple tines and hoes can be fitted on multi-purpose frames, which can be pulled by animals.

(c) *Secondary cultivation* is carried out with:
  (i) simple hoes or multi-purpose frames fitted as required;
  (ii) harrows.

(d) *Ridging and earthing-up using animal power*, which, up to the present time, has only been carried out in West Africa for cotton. Ploughs or multi-purpose frames with longitudinal stability are provided with special ridging bodies with a double mouldboard.

3. *Cultivating of flooded soil.* This is essentially concerned only with rice growing and aims at preparing a suitable wet-bed for transplanting. In Africa techniques have been particularly studied in Madagascar, where the following are in use:

(a) Japanese-type ploughs, often reversible, which partially invert the soil;

(b) metallic 'trampling' rollers, of the caged-wheel type, which imitate the puddling effect of oxen hooves;

(c) various models of wooden and metal harrows.

4. *Seed sowing and planting equipment.* Two types of animal-powered seed drills are used in Senegal and Madagascar for cereals.

(a) Rice drills, which generally drill four or five rows spaced at a maximum of 30–40 cm, are derived from the cereal seed drills of the temperate countries, which have grooved rollers to control seed distribution.

(b) Millet, sorghum, and maize, as well as rice, can also be sown using planters with rotating discs which are interchangeable according to the crop to be planted and the required density distribution.

Certain animal-powered planters, formerly used in market gardening in Europe, could be adapted for cassava, but no trials of such equipment have been reported in West Africa.

There are still no satisfactory rice-transplanting machines pulled either by man or animals. The Japanese motorized machines are at present too complicated to be considered suitable for use by African peasants.

At present, fertilizer spreaders for cereals are not used in small-scale farming.

13.15. Crop management

13.15.1. Weed control

Tropical climatic conditions encourage rapid development of weeds as well as crop plants. Good management to avoid intense competition is therefore of great importance.

Equipment for weeding is basically a hoe pulled by animals: either a single-purpose machine equipped with blades angled or parallel to one another is used or blades can be mounted on the chassis of a multi-purpose toolbar. It is possible to use more hoe blades per implement for ox draught than with horses or donkeys. There are 30 000 ox-drawn hoes used in Senegal, mainly used for hoeing millet, sorghum, maize, and cassava.

Such equipment cannot be used for flooded rice. For this the only suitable tools are manual rotary hoes, of which a type made in Japan is very common in Madagascar.

13.15.2. Crop protection

Up to now special equipment for crop protection has not been used for food crops in West Africa. It is not possible to drive sprayer pumps using draught animals, and to advance beyond manually operated equipment one must resort directly to motorized equipment. Either motorized backpack mistblowers or tractor-mounted and powered sprayers are the alternative possibilities. However, weedicides may be applied with a minimal power requirement for distribution, the problem being reduced to having the liquid or granules carried to the right place.

13.15.3. *Mechanical harvesting*

In the francophone tropical countries of West Africa there has been no serious attempt to mechanize the harvesting of cereal and forage crops. On a research scale animal-drawn cutting and sheaf-making machines and harvester-binders have been tried, but without success, for rice, millet, and sorghum. The machines require too much traction for this power source. Forage harvesters have been used to a limited extent in Madagascar and Senegal.

13.15.4. *Transport*

The use of imported or locally constructed animal-drawn carts, having either metal or pneumatic wheels, is very widespread, particularly where animal-powered cultivation is being developed. Such carts are commonly used to carry loads of 500–1000 kg.

13.15.5. *Post-harvest operations*

Draught animals in tropical countries are not powerful enough to drive mechanical threshers. Rice threshers operated by turning a handle, or sometimes pedal-operated, are still uncommon, but are quite useful and suitable. Both manual and motorized maize shellers are used in certain countries.

A motor-powered thresher and a mill decorticator for millet and sorghum have been produced in Senegal as prototypes. For mechanical decortication of rice, small European or Japanese motorized decorticators are used. These have a capacity of several hundred kilograms per hour.

13.16. Imported tractors and other motorized equipment

In addition to the IRRI equipment already described, there are already several hundred Japanese or Taiwanese single-axle tractors powered by 7–13 hp diesel engines and equipped with rotary hoes for the preparation of irrigated rice fields. The use of such equipment is being developed where pedal-operated rice threshers, which can alternatively be powered by an auxiliary motor, have also been imported. Small motorized combined processing machines of Japanese make have also been imported.

Statistics for the importation and continuing use of conventional tractors in West Africa are probably very incomplete (see p. 275 et seq.), but even from this data it was apparent that West Africa has very little functioning equipment of this sort, and that very often, and probably generally, tractors are badly

used because of lack of training for the drivers and the lack of any effective after-sales service.

According to Parry (1971) only 8000 tractors in the range 30–100 hp were sold in the whole of tropical Africa in 1970.

The need for a small, simple, inexpensive tractor, with a capacity of 15–20 hp, and adapted for family farming, is also apparent. Garrard (1971) gives details of a prototype three-wheel tractor powered by a 7–15 hp motor which has been constructed by the NIAE and has already been tried out in Africa. It was reported in 1971 that the University of Swaziland had constructed a prototype machine that was simple, strong, and reasonably priced and which had performed satisfactorily in its first trials. The construction of small tractors with simplicity and reliability as their main speciality, is not apparently of interest to the big firms. These companies anticipate markets of perhaps only hundreds of units during the first years, which cannot justify a production line large enough to allow prices much lower than that set for classical 25–30 hp tractors (Parry 1971).† It therefore seems to be necessary to find some alternative solution for the manufacture of such much-needed equipment.

13.17. Manufacture and trials of agricultural machinery

Conventional and single-axle tractors, and the majority of motorized agricultural machines used in tropical Africa are imported from the industrialized countries in Europe and North America and from Japan and Taiwan. In contrast, animal-drawn equipment and hand implements are being increasingly made in the local factories, either under licence or according to local designs. In 1971 such factories were working in Cameroun, the Ivory Coast, Ghana, Upper Volta, Nigeria, and Senegal.

Boulanger (1971) has outlined the particular operating conditions for a manufacturer in a tropical country such as Senegal. The Société Industrielle Senegalaise de Construction de Machines Agricoles (SISCOMA) makes animal-drawn equipment in response to demand in the country.

Favourable characteristics of the market for this type of equipment include the absence of competition from any similar local industry and the lack of interest in this type of equipment shown by firms in the

† The recent changes in the price of fuel may improve the attractiveness of such a project by widening the market for machinery which has low fuel consumption. Ed.

developed countries. Obstacles to mechanization depend upon tradition, the highly seasonal character of agricultural activity (and correspondingly of orders), and the general dependence for orders on national and international financial backing which provides credit for farmers.

In Senegal the development of any new industry is viewed with favour, but raw materials are scarce and dependence on imported supplies often involves long delays that are difficult to avoid and which disrupt production schedules.

SISCOMA tries to manufacture equipment adapted to the power sources available, to the type of cultivation required, and to the technical ability and financial means of the users. It works in close liaison with research bodies and notably with the CNRA. It participates in the work of publicizing the use of its equipment and also has a training centre. Finally it guarantees an after-sales service, for which it is trying to enlist the participation of local blacksmiths and hence stimulate rural agri-businesses.

Le Craz (1971) has outlined the procedures needed for trials of machines in Africa and the organization of testing stations. He pointed out that the need for trials in Africa, in addition to those carried out elsewhere, arises from technical, economic, and human considerations.

Equipment designed for use with the soils, climates, and crops of temperate countries usually needs to be adapted to the new environment of Africa, which is characterized by high temperatures, poorly distributed and often insufficient rainfall, and soils with a low organic matter content. Account must also be taken of any special cultural techniques required by the crops in the area where the equipment is to be used. Sometimes no adaptation is practicable, and new machines need to be made.

A technologically advanced machine that is expensive because of its refinements is not always economically good value under the present conditions in local farming. It is therefore essential to extend tests of machinery beyond merely technical trials, to include economic studies on a farm. These tests are most conveniently carried out in close co-operation with agronomical research, but as management of the equipment under these conditions will be well above average, care must be taken in the interpretation of results. African farmers, like others the world over, distrust new techniques until they have personally satisfied themselves of their value. Thus the improvement brought about by a machine should

be of such magnitude that it is immediately and clearly apparent. The machine should also be simple, strong, easy to maintain, and straightforward to use.

Le Craz considers that trials of machinery at an official station which simply copy those already carried out in the manufacturing country should be reduced to a minimum in Africa. Instead, practical trials and demonstrations under local conditions should be emphasized.

13.18. The extent and economics of mechanization

13.18.1. Methods of economic analysis

Total mechanization of cultivation is most unlikely to provide the optimum economic use of resources, particularly in a country where the opportunity cost of labour is low. Partial mechanization to reduce the labour demand for those tasks that require considerable strength or much time and which at present constitute the 'bottlenecks' should be the primary concern.

Winch (1971) suggests that any development of mechanization should be preceded by two different kinds of economic analysis.

1. First, a micro-analysis at the farm level to determine the comparative profitability of different types of mechanization for the farmer. At this stage it is advisable to consider the range of possible resources and the different combinations in farming systems. The manpower requirements and time-flow in the farm calendar of all the crops must be considered. Possibilities for intensifying production and/or extending the area under cultivation should also be studied.

2. Secondly, an analysis should be made of national and regional macro-economic variables, in which considerations of alternative employment, distribution of revenue, requirements for and earnings of foreign exchange, and the need for foreign expertise all interact. The aim of this analysis should be to determine the social gain at the national or regional level that might be expected from any technological changes proposed. The effect on rural employment is outstandingly important in view of the rapid growth of the African population and the urban problems that may be created by a rural exodus.

13.18.2. Tractor-hire services

Stout (1971) reviews the technical and financial structures of the organizations offering tractor services to farmers in four countries of West Africa.

In Ghana, these services are provided by a special department of the Ministry of Agriculture, which receives and studies requests from farmers and then signs a contract which stipulates the payment to be made in advance for the work to be carried out. Contractual services are mainly for ploughing.

In Nigeria work is carried out for farmers by both Government and private enterprise. Stout pointed out that the latter have greater operational flexibility than the Government service and also have lower tariffs.

In Gambia the Government provides the tractor cultivation services, which mainly consist of ploughing rice fields during the dry season. In the first few years of the scheme the work was carried out without charge as a way of making the service known; later the farmers paid an increasing proportion of the cost. The final aim is for them to pay the actual cost.

In the Ivory Coast, services have been offered since 1966 by the State Society, Motoragri, which operates more than 200 tractors. This society operates at a high standard of efficiency and recognizes the fundamental importance of proper training for drivers and mechanics.

13.18.3. Financing the equipping of Senegal farmers

Fall (1971) has outlined the organization adopted by the Senegal Government of the financing of agricultural 'equipping'. The word 'equipping' here implies the possession of a full range of inputs, including agricultural hardware, fertilizers, plant-protection products, seeds, and seed-treatment products.

The National Office of Co-operation and Aid for Development (ONCAD) is entrusted with carrying out the programme worked out by the Government. ONCAD occupies a bridging role between the National Development Bank of Senegal (BNDS), which provides credit, and the co-operatives of peasant farmers.

The equipment requirements for the next season are decided upon at a meeting of the farmers' co-operative. After the products of the co-operatives have been sold BNDS establishes the credit-worthiness of each of them and informs ONCAD. The value of credit to which each co-operative is entitled is calculated on the basis of one-quarter of the average value of sales in the three preceding years, less the annual repayments outstanding from previous loans and a security margin for the bank. The credit value calculated in this way can be used to enable three categories of loan to be obtained:

(1) A short-term loan for food, varying from 10 per cent to 15 per cent of the credit value, according to the region;

(2) a short-term loan for buying fertilizers, fungicides, seed, and light equipment, varying from 55 per cent to 70 per cent of the credit value;

(3) a medium-term loan for buying other agricultural equipment and livestock, up to approximately 25 per cent of the credit value.

Short-term loans have to be repaid at the end of the agricultural year. The repayment of medium-term loans is staggered over 5 years and is repaid in the form of a levy on the value of harvest at sale. The interest rate is 5 per cent, and an additional 2 per cent commission is charged for the ONCAD operation.

The repayments are usually made satisfactorily in a normal year. New loans are not granted to co-operatives which have repaid less than 80 per cent of any repayments that were due. After a bad harvest, the Government, after official checks, can take measures to allow the postponement of payments. Each co-operative is collectively responsible for any defaulting members, and these owe their debts to the co-operative and not to the bank directly. Successful co-operatives may earn bonus benefits after the sale of the crop. The existence of ONCAD ensures that there is adequate trained manpower for proper financial control of the operations of the co-operatives.

13.19. Major topics of discussion which arose during the Bambey seminar on mechanization (1971)

Among the most extensively discussed points were the following:

1. The fear expressed by many of the English-speaking delegates that increasing mechanization of agriculture would lead to unemployment and thus increase the socially undesirable exodus of the rural population to the towns, where industries to employ people do not exist to a sufficient extent to absorb them. From the exchange of views on this topic it seemed that this fear may be exaggerated, because, while

mechanization suppressed the need for certain labouring jobs, it created the need for other productive employment in agri-business in rural areas.

2. The need for appropriate types of tractors and other mechanical tools adapted to the size of the farms.
3. The need to 'rationalize' equipment (i.e. not to have many different models of the equivalent kinds of machine). This could be facilitated by creating national committees for agricultural machinery in each country.
4. The fundamental importance of training the users of machines at all levels, so that machines are not abused.
5. The past lack of co-ordination of research carried out in the countries of tropical Africa. It was felt that this would greatly be reduced by the formation of inter-state regional research centres in the two main ecological regions of West Africa.

During the course of one session at the 1971 Bambey seminar on mechanization each participant was asked to place in order of importance his view of the priorities for research in mechanization. The answers after analysis showed the following order:

(1) soil preparation—threshing;
(2) choice of tractor size as regards suitable power ratings for technical and economic reasons and the study of small tractors in particular;
(3) sowing—crop maintenance.

Crops which should be mechanized, in order of suitability were, according to this assessment:

(1) rice; (2) sorghum; (3) maize; (4) pearl millet; (5) groundnuts; (6) forage crops; (7) vegetable crops.

13.20. Desirable future evolution of research and its applications

Many specialists in the field of mechanization agree that research carried out so far has not yielded the expected results, for the reasons that have been explained already, i.e. because of inadequate physical liaison and exchange of information, particularly between anglo- and francophone countries, and because of the use of national rather than regional-ecological research programmes. A detailed proposal to rectify this has been made by Stout (1971). Stout's suggested scheme for organization for West Africa is essentially similar to that operating in East Africa

under the East African Agriculture and Forestry Research Organization (EAAFRO). At least in theory, research and development work is carried out under regional control, but actual work is effected under contract by the national research institutes, which receive on this account the necessary credits from the East African Community Secretariat. The Secretariat ensures that there is no unnecessary duplication between the national programmes and also organizes an efficient exchange of information.

Corresponding Regional Units, one in West and one in East Africa, could also share information efficiently and most effectively use the limited available resources. Le Craz reminded delegates at the 1971 Bambey seminar of recommendation 23 in document 30 of the Rome conference on the Sudan zone that had been held in November 1968.

Members of the Conference, recognizing the importance of mechanization for the growth of agricultural production, and the need to strengthen regional co-operation in this field, recommend that at least one major research centre should be expanded with the aim of promoting a co-ordinated programme of research on all aspects of agricultural machinery, that is, light equipment, equipment for animal drawn cultivation, tractors and other motorized equipment, in relation to the methods of cultivation and the particular requirements of the Sudan region.

After the Bambey seminar, a conference to establish co-operative programmes for agronomic research in the Guinea region was organized by FAO in August 1971 at Ibadan, Nigeria. At this conference (Food and Agriculture Organization 1971) all the recommendations of earlier conferences, including the Bambey (1971), Abidjan (1968), and Rome (1968) conferences, were reviewed and the research bodies currently in existence in the countries concerned were noted, including IITA and WARDA, which are both regional centres.

Setting up corresponding programmes to that for the Sahel zone to cover both the Sudan and Guinea ecological zones was adopted as an objective, and FAO was asked to take responsibility for promoting this and for securing finance. This programme is planned to consist of the following.

1. Making a preliminary study of results obtained and research in progress in the countries in each of the two zones.
2. Equipping a regional centre for each ecological zone, with regional substations. The regional centres will be the machinery trial station at

Bambey (CNRA) in Senegal for the Sudan zone and at IITA at Ibadan, Nigeria for the Guinea zone.

3. Linking the proposed scheme with the network of information centres provided by CEEMAT for the Sudan zone and IITA for the Guinea zone.

The preliminary study recommended in this was in fact carried out in 1972 by an FAO mission, which then proposed in its report a 5-year project entitled 'Organization for Agricultural Mechanization—ORMA' in West Africa.

The ORMA project is similar to that suggested by Stout (1971) and includes the idea of having two main centres at Bambey and Ibadan, one for each zone, which would work in collaboration with national centres (existing or projected) as well as with any other institutes and universities concerning themselves with agricultural mechanization at a national level. FAO is now trying to obtain financial support for the project.

Too often the interesting results obtained by research into the mechanization of agriculture are incompletely applied, or applied not at all. Increasing delay occurs as a result of an insufficient extension organization. The extension organization, in turn, should be connected with a credit organization, and should also include a training function rather than restricting itself merely to an advisory role. The mechanization of agriculture appears as but one of the many factors that must be effectively integrated if the intensification of agricultural production is to continue.

References

ANON. (1971). Direction des Services Agricoles du Senegal: politique d'équipement du monde rural au Senegal. *Ford Foundation/IITA/IRAT seminar on agricultural mechanization at Bambey, 1971.*

BOULANGER, C. (1971). Problèmes du construction de matériel en pays tropical pour le matériel attelé. *Ford Foundation/IITA/IRAT seminar on agricultural mechanization at Bambey, 1971.*

CEEMAT (1970). *Machinisme agricole tropical.* No. 32. 1970. Antony, France.

CHOPART, J. L. (1971). Influence de l'enfouissement sur les structures des racines des plantes. *Ford Foundation/ IITA/IRAT seminar on agricultural mechanization at Bambey, 1971.*

CRAZ, J. LE (1971). Essai de machines et stations d'essais. *Ford Foundation/IITA/IRAT seminar on agricultural mechanization at Bambey, 1971.*

FALL, B. N. (1971). A policy for mechanization for the rural environment in Senegal. *Ford Foundation/IITA/IRAT seminar on agricultural mechanization at Bambey, 1971.*

FOOD AND AGRICULTURE ORGANIZATION (1971a). Organization and coordination of a research programme concerning agricultural mechanization. Paper presented at the *FAO conference at Ibadan to establish agronomic research programmes for the Guinea ecological zone. August, 1971.*

—— (1971b). Agricultural mechanization research in West Africa. Paper presented at the *FAO conference at Ibadan to establish agronomic research programmes for the Guinea ecological zone. August 1971.*

—— (1971c). La mécanisation de l'agriculture en Afrique tropicale. *Ford Foundation/IITA/IRAT seminar on agricultural mechanization at Bambey, 1971.*

GARRARD, N. M. (1971). Agricultural mechanization in Africa. *Ford Foundation/IITA/IRAT seminar on agricultural mechanization at Bambey, 1971.*

HABIB THIAM, S. E. (Minister of Rural Development in Senegal) (1971). Introductory speech to the *Ford Foundation/IITA/IRAT seminar on agricultural mechanization at Bambey, 1971.*

HAMON, R. (1971). Quelques résultats obtenus en matière d'intégration élevage-agriculture. *Ford Foundation/ IITA/IRAT seminar on agricultural mechanization at Bambey, 1971.*

KHAN, AMIR U. (1971). Equipment for mechanized tropical agriculture. *Ford Foundation/IITA/IRAT seminar on agricultural mechanization at Bambey, 1971.*

LABROUSSE, G. (1971). Notes sur les aspects historico-économiques du développement de la mécanisation des cultures vivrières enpays tropicaux. *Ford Foundation/ IITA/IRAT seminar on agricultural mechanization at Bambey, 1971.*

MOIGNE, M. LE (1971). Quelques données sur le matériel utilisable en culture attelée. *Ford Foundation/IITA/ IRAT seminar on agricultural mechanization at Bambey, 1971.*

MONNIER, J. (1971). Relation entre mécanisation, dimensions et systèmes d'exploitation. *Ford Foundation/IITA/ IRAT seminar on agricultural mechanization at Bambey, 1971.*

NICOU, R. and POULAIN, J. F. (1971). Les effets agronomiques du travail du sol en zone tropicale sèche. *Ford Foundation/IITA/IRAT seminar on agricultural mechanization at Bambey, 1971.*

PARRY, D. W. (1971). Mechanization: the viewpoint of the manufacturer. *Ford Foundation/IITA/IRAT seminar on agricultural mechanization at Bambey, 1971.*

STOUT, B. A. (1971). Large scale mechanization equipment. *Ford Foundation/IITA/IRAT seminar on agricultural mechanization at Bambey, 1971.*

WINCH, F. (1971). A note on employment generation in African agriculture. *Ford Foundation/IITA/IRAT seminar on agricultural mechanization at Bambey, 1971.*

14. Traditional African systems of agriculture and their improvement

R. TOURTE AND J. C. MOOMAW

14.1. Introduction

Man is at the beginning and end of development
(L. S. SENGHOR)

A Green Revolution depends upon the desire of farmers to achieve it. Social as well as technical factors determine what may be brought about. For many Africans the Green Revolution has scarcely started, while parts of Asia have made a success of the dramatic change in technology applied to two basic food crops—wheat and rice.

A seminar on the traditional systems of agriculture in Africa and their improvement held in IITA, Ibadan, in 1970 afforded an appropriate place for examining the reasons for the delay and the lack of enthusiasm in African agriculture for a change similar to the Asian one. The seminar concluded that the Green Revolution would be favoured by:

(1) the acceleration of scientific and technical research to improve the technology in such major areas as soil science, crop improvement, and management techniques;

(2) intensive efforts to improve the natural environment both in terms of physical and human resources in order to realize the excellent potentials for increased productivity demonstrated by research;

(3) dynamism in development organizations responsible for the application of the available new technologies in order to unlock these potentials.

The seminar concluded that present inadequacies lay partly in the formulation of the research message and partly in its application, especially with regard to the development of intensive systems of production. For intensive agricultural methods to be viable in West Africa the thematical results from agricultural research must be better utilized and combined and greater compatibility with the present production patterns of the rural areas must be achieved; the new systems must also be economically more attractive.

The Ibadan seminar paid particular attention to the problems of realizing agricultural potential in the sub-arid and Sahelian zones and the strategies for extending improved practices in rural areas. This chapter will review both ideas and knowledge in this vast field, stressing the conclusions of the Ibadan seminar without, however, limiting itself to them.

Agricultural potentials in West Africa depend so profoundly on the physical and biological environment that these are briefly stated below.

14.1.1. Climatic factors of the environment

The Sudan–Sahelian zone has a very high insolation all the year round and is characterized by the following.

1. Two clear-cut seasons, wet and dry.
2. An exceptionally high photosynthetic potential, hence the possibility of very high productivity when water is supplied. Daily insolation exceeds 600 gm cal/cm^2 for several months at most latitudes.
3. Acute moisture stress, more or less absolute during the dry season and irregular during the rainy season. Rainfall may reach 1200 mm during the season of the rains in the more favoured southern parts, but be as low as 300 mm in the north where cropping is progressively replaced by nomadic grazing.

The humid (or Guinean) zone, in contrast, has only a short dry season and, in general, more than enough water supply. Problems here are posed by an excessive rather than a deficient water supply; such problems involve run-off control, erosion control, and drainage. However, the abundance of rain in conjunction with high temperature endows this zone, where the climax vegetation is tropical forest, with high production potential: nearly 4 times that of the temperate regions (Bradfield 1970). Other things being equal, agriculture in the Sahelian and Guinea zones is capable of producing 4 times as much dry matter per hectare per year as agriculture in temperate zones if water is supplied in sufficient, but not excessive, quantity.

The Sudan–Guinean zone is intermediate in characteristics between the two preceding zones. Two seasons of rainfall occur, often just enough for significant plant production, interrupted or greatly disturbed by two intervening dry seasons.

14.1.2. *Edaphic factors of the environment*

There are many types of soil, starting from the poor soils in the north (e.g. ferruginous tropical) to the black soils with allophane in Cameroun, passing through ferrallitic and hydromorphic soils of the 'intermediate' zones. The fertility of tropical soils in Africa is on the average relatively low, either for natural reasons or as a result of exploitation by man.

The soils of Africa have been classified using several systems. These have been compared by Aubert and Tavernier (1972) and mapped (D'Hoore 1965). The potentials of the soil resource and its management has been studied extensively (Nye and Greenland 1960; Moss 1968; Charreau 1969; Jones and Wild 1975).

14.1.3. *Biotic factors of the environment*

'Man is clearly the determining factor in unlocking the potential of climate and soil in agricultural production. It is up to him to put in his labour, his ingenuity, his techniques and the necessary materials so that all soil is efficiently utilized for maximum production' (Kanwar 1968).

14.1.4. *Land-use capability of ecological zones*

Below are set out the key ideas which delimit the zones to be developed in terms of their agricultural capabilities and potentials and their place relative to the production system of the countries and regions in question.

1. The potentials of tropical agriculture seem to be high, even in the relatively dry regions of Africa south of the Sahara. Thus, in upland farming, we have already seen (Thiam, Tourte, and Pocthier 1970) yields of maize of 4–7 times the present yield per hectare, and in irrigated cultivation (Chaminade 1970) of more than 10 t/ha of paddy per harvest.

But under similar climatic conditions in India, yields of more than 20 t/ha per annum have been obtained in multiple cropping (Kanwar 1970).

2. The disparity between the present levels of production and technology and the potentials foreseen shows clearly the substantial natural and intellectual investment which is required to raise traditional agriculture in the tropics to a level of modernity compatible with its potential. This is especially true if we attempt to narrow the technological gap between the economically advanced countries and the countries considered by economists to be 'at the stage of economic take-off' (i.e. 200 dollars of annual GP *per caput*).

3. As compared to modern agriculture in the temperate regions, intensive tropical agriculture requires a higher degree of technical competence. This is because we have to control several phenomena whose dynamics are highly accelerated in comparison with the same phenomena under temperate conditions: the biological and biochemical cycles of soils are fast, the thermal variations are abrupt, the rains are precipitous and aggressive and the risks of soil degradation are consequently high if the techniques recommended for improvement are not well integrated.

14.1.5. *The ecological basis of agricultural development planning*

The delimitation of agricultural potential into ecological zones is a useful preliminary to rational planning of economic development. This has been done in India (Kanwar 1970). All India has been divided according to soil types and eight climatic belts, representing a gradation of the effective soil moisture index. This index is

$$\frac{100\,(P - PET)}{PET}$$

where P is the cumulative rainfall and (PET) is the cumulative potential evapotranspiration. The climatic belts of India (see Table 14.1) are superimposed on the soil groups. The soil group and climate complex form the unit for planning. All the climatic regions can be further subdivided into five isothermic sub-belts, i.e. 0–10 °C, 10–20 °C, 20–25 °C, 25–28 °C, and 28 °C.

The climatic belts in Senegal can be subdivided into isothermic zones. Each zone can then be given an index of efficiency of the improvements recommended for any particular crop by comparing the actual yield being achieved with the potential yields indicated

TABLE 14.1

The climatic belts in India

Climatic belts	Moisture deficit index $\dfrac{100(P-PET)}{PET}$ (per cent)	Important soil groups
1	-80 to -100	Alluvial
2	-60 to -80	Alluvial
3	-40 to -60	Black
4	-20 to -40	Black
5	0 to -20	Red and mixed red and black
6	$+0$ to 50	Coastal lateritic alluvial
7	50 to 100	Laterite, coastal alluvial
8	100	Alluvial, coastal alluvial, and hill soils

by experiments under optimal husbandry. Also, one can assign an inter-zonal index, based on the present yields per unit area in the zone, as a proportion of the mean yields per unit area for all regions where the crop is grown. The product of the efficiency and inter-zonal indices gives a comparative measure of the incentive to investment in the different zones. This incentive index can then be related by costs of inputs to the over-all investment required and can help to establish a rational development policy (Tourte 1966).

The zoning by moisture deficit clearly dominates the agricultural production map of India. The moisture deficit index is inversely related to potential evapotranspiration. In wet districts where the monsoon is regular and rainfall is high, agricultural production during the monsoon season may be limited by the reduced solar radiation due to cloud cover and not by soil fertility level. One reason for the remarkable increase in wheat yields in India may be that much of the crop is grown under irrigation in the dry season when gains from efficient use of water, fertilizer, and other inputs can be greatest.

14.1.6. *Indices of crop efficiency and spread*

To exemplify the indices of efficiency and inter-zonal relative productivity, Kanwar (1970) discusses the Indian strategy for rice and wheat development. Of the 158 rice-producing districts in India, 73 produced 70 per cent of the total rice crop. Most of the districts having high indices are in climatic belts characterized by a positive soil moisture index and occupy laterite, alluvial, or coastal alluvial soils. Efficient wheat-producing zones, on the other hand, fall largely in climatic belts with strong negative

moisture deficit indices and occur on alluvial soils which are able to be irrigated during most of the growing season. The concepts of relative yield and spread are applicable mainly as a planning tool.

Kanwar indicates that, although Indian wheat production has more than doubled since 1964 (to 20 million t per year), the most efficient cropping zone alone, if supplied with necessary inputs, could produce 37 million t of wheat, and hence this is where the resources should be applied.

Although efficient rice production occurs mainly in high-rainfall zones most of the production is during the dry season. All of the inputs such as fertilizer, water, and pest control can be more efficiently used in this season. Rice, jute, and sugar cane tend all to be efficiently produced in the same wet zones, whereas wheat, sorghum, groundnuts, and some pulses are most efficiently produced in zones with only moderate moisture supply, on black, red, and lateritic soils.

On a regional level in West Africa, we can already broadly map out extensive agricultural zones (Greenland 1970) for the major crops, with the following broad use potentials.

1. In the *forest zone*, where it is difficult to introduce the techniques of intensive farming on seasonal plants, development effort should be assigned to tree-crop production (oil palm, coffee, coconut, *Hevea* rubber, etc.).

2. In the *intermediate Sudan-Guinea zone*, the greatest potential is for root crops and those cereals requiring a lot of water (maize, rain-fed rice).

3. The *savanna zone* is best suited to dry-land cereals able to tolerate some water stress, grain-legume crops, and meat production.

14.2. The need for change

14.2.1. *Phases of agricultural development in tropical Africa*

The agricultural development of the tropical regions of Africa has evolved broadly through three phases.

The first phase was subsistence agriculture, which is oriented towards satisfying the basic needs of the population, with each community producing its own requirements. This phase was, and still is, characterized in certain regions by a generally satisfactory balance between:

(1) population;
(2) natural fertility of the soil;
(3) food-crop production.

This is the 'old balance' (Hanson 1970). In this system each member of the agricultural community produced what he, or his close relatives and neighbours, needed and not what he could most efficiently produce (Kanwar 1970).

The old equilibrium in traditional agriculture was suddenly changed in some parts of Asia by introducing new crop cultivars with high yield potential together with the associated technology of improved fertilizer, cultural practices, and water control. The availability of the same technical inputs in African agriculture has not yet resulted in the dramatic increases in food-crop yields which occurred in parts of Asia.

In the second phase of West African agricultural development access to the world market economy, through the introduction of export crops, upset the traditional systems. They were often replaced by 'mining' systems of farming, which generate a more-or-less rapid decline of the soil fertility. This soil deterioration is accelerated by the introduction of light mechanization, especially in extensive farming with its excellent product–labour requirement relationship.

The Green Revolution in Africa, which began 25 or 30 years ago, has, as yet, been confined to cash crops—mainly those which are exported. The changes in production levels of oil palm, rubber, cocoa, coffee, groundnuts, and cotton have been substantial and have made major differences to farmers' incomes. But the production of maize, rice, wheat, cassava, yam, cowpeas, and vegetable crops has not yet responded in the same way. This second phase of agricultural development corresponds to the growth of semi-shifting cultivation systems (Charreau 1970).

In this second phase, land becomes a limiting factor through the reduction and disappearance of waste land and fallows, while natural fertility diminishes. Maximization of profit is aimed for, but the capital resource of the land is likely to be degraded by extensive cultivation. Apparent costs of production are reduced if this hidden destruction of capital is ignored or not costed (Cepede 1973). Capital is injected in this phase in the form of light mechanical equipment, which can overcome labour bottlenecks. Simple mechanical equipment can be used with animal power in some ecological zones.

In the third phase of development, permanent or fixed cultivation emerges, with a more elaborate farm structure and organization and a more intensive system of production.

While the three phases should logically succeed each other in time, in reality they often exist side by side in the same area, sometimes even on the same farm. This partially explains the complexity of present patterns of production and the problem of popularizing a new improved method throughout a given area.

14.2.2. *Contrasts between Africa and Asia*

Hanson (1970) has offered several explanations for the differences between the Asian and African tropics. The African consumer is not yet sufficiently hungry for hunger to be a major incentive for change. Agricultural production at the present low levels supports the current population at an adequate level of supply of starchy food. And although a real protein deficit exists in the humid forest zone, it does not result in the urgent need for increased production that derives from the famine and chronic food shortage which is widespread in Asia.

14.2.3. *Some characteristics of African farming systems*

African farming systems are extremely complex, resulting from the combinations of root crops, tree crops, cereals, vegetables, and other crops that are grown in combination on a normal subsistence farm. In addition, the land–clearing system and bush-fallow rotation normally result in a high residual population of trees and forest plants on cultivated plots. This facilitates the return of the land to bush or forest, as is required for the maintenance of fertility and prevention of erosion. Mixed cropping also insures the farmer against crop failure. The wide variety of crops permits the harvest of some food in spite of many different types of natural disaster.

African agriculture is normally conducted without either irrigation or supplemental power being available to the farmer. In the humid regions, an adequate rainfall level normally ensures at least one crop a year, and traditional practice in the arid areas normally involves livestock raising with a low level of crop production during the single short rainy season. While animals constitute a large reservoir of wealth and high-protein food, they are little used as draught animals. Animal pests and diseases and the complexity of livestock production make for serious difficulty in introducing the use of animal power, while problems of technical capability and capital investment limit the introduction of machines.

Marketing and infrastructure problems, including lack of credit, poor storage facilities, and few processing industries, all contribute to the uncertainty of reasonable prices for any excess food that might be produced by the subsistence farmer. Some of these facilities have been provided for certain cash crops, but very rarely for food crops.

14.2.4. *The need for revolutionary change in food-crop production*

The provision of suitable new technology for food production for African farmers has not yet been clearly achieved by crop research institutions. The African people, by their food habits and preferences, have chosen a group of food crops on which attention of research institutions must now be focused:

(1) tuber and roots (cassava, yam, sweet potato);
(2) coarse grains (maize, sorghum, pearl millet (*Pennisetum*));
(3) rice;
(4) grain legumes;
(5) tropical vegetables;
(6) tropical grasses and legumes suitable for animal production.

The intensification of production in these six crop groups will, in turn, require intensified research activity for each of them. Substantial social and economic research will be needed to determine the acceptability of change in agricultural production technology in the rural communities in which the new methods of production are to be used.

If African agriculture is to feed the additional population, which is projected to rise from the present 235 million people to 600 million by A.D. 2000, the attitudes not only of farmers and family groups but also of Governments (and especially the agricultural planners of most of the independent countries in tropical Africa) have to be changed dramatically. There has rarely been an experimental approach to the setting up of strategies for development or to determining adapted socio-economic systems of production. Experimental science has been largely confined to defining the technical possibilities for development, and the approach has been more of analysis in agricultural stations rather than of synthesis in the areas of application.

In developing a technology that will enable farmers to break away from subsistence-farming patterns, Albrecht (1970) has advised that the development of combinations of improved farming

practices is essential. Not only the farmer but the community must maximize its return from adopted changes.

The Green Revolution in Asia has been criticized for providing insufficient employment and providing too low a share of benefit to the poor farmer; it is said that some farmers are forced to leave the land and migrate to already-overcrowded cities. There has been some criticism of the quality of new food crops, and production technologists have been blamed for creating a whole host of 'second-generation' problems, but it is equally true that the sociologists, economists, and political scientists must take their share of the blame in not foreseeing the consequences of technological change nor understanding the potential for rapid development which it holds.

Much remains to be done both in the development of new and better-adapted agricultural technology for tropical Africa and in the study of the limitations, problems, and advantages of changing agricultural practices. New food-crop production technology and the nutritional improvement to be derived from changing food supplies must be studied through existing national research agencies, including those in the social sciences.

14.3. The agricultural potential of the shifting cultivation systems

The comparison between shifting cultivation and continuous-production systems made by Greenland (1970) clearly shows the long-range possibilities for higher levels of yield both per unit area and per unit time using continuous systems with adequate inputs. No agricultural system should be considered stable or desirable if it does not maintain or improve the fertility of the soil.

The shifting cultivation system successfully maintains biological equilibrium at a relatively high fertility level by restoring vegetation and nutrients over most of the productive land area. Nutrient reserves are made available to crops by clearing and burning before cropping, and soil erosion and destruction of the biological populations occurs on only a relatively small part of the total area. The light tillage and incomplete removal of natural vegetation allows rapid regeneration of forest. Thus nutrient and organic matter levels are restored to well above the minimum for crop production before the land is next cleared for agricultural use.

However, using the shifting cultivation system

O

it is not possible for crop yields substantially larger than those obtained at present to be produced from a given area of land since much land at any one time will be agriculturally unproductive. While fertilizers offer no great advantage if the fallow period has been adequate, reduction in the length of the fallow period with no fertilizer use can lead to a continuous decline in productivity.

Continuous-management systems will require fertilizer practices which continuously restore nutrients removed by cropping, leaching, and burning. These systems will also require erosion-control practices and practices to maintain or restore organic matter to provide favourable soil structure for crop growth.

Organic-matter decline not only reduces total nutrients in the rooting zone of the soil but also reduces the base (cation) exchange capacity. Short-term legume fallows appear to offer little advantage over natural fallows since they have low cash value and require substantial investment in establishment and management for maintenance. Nitrification inhibitors may offer some advantage in controlling losses of organic and reduced nitrogen from tropical soils where nitrate nitrogen is easily leached away under high rainfall.

The present state of knowledge does not yet appear to offer any safe, economic, and stable system for continuous agricultural production for many soils in the forest regions. Forest areas thus may be best utilized for perennial tree crops such as cocoa, oil palm, and rubber.

Establishment of a continuous and stable management system in the savanna area is thought to be less difficult. Here there is much less risk of permanent damage in attempting continuous cultivation, and some substantial apparent successes have already been recorded. On the gentler slopes and with less frequent rains, the risks of erosion are reduced. Lowering the frequency of burning reduces initial soil fertility levels obtained on opening land for cropping, and grass-legume rotations may offer substantial advantages, as may the judicious use of fertilizers.

The role of iron or aluminium oxides and hydroxides in tropical soils—and especially hydroxides, which dry irreversibly to form laterite in some soils—is not yet well understood. Recent evidence, however, shows that formation of laterite is reduced or prevented by organic materials (and the associated micro-organisms) and additional

studies of clay minerals may assist in the understanding of problems of soil management that can occur under more intensive cropping practices.

Charreau (1970) describes how the development of continuous cropping and production systems in the savanna areas can be achieved. Extensive work in the Casamance area of Senegal has shown that in this tropical zone, continuous cultivation which has substantially increased and maintained levels of production has been accomplished on large areas without substantial unfavourable effects on basic soil properties. The success of these intensive cultivation systems has been achieved with:

(1) forest clearing and consolidation of cultivated areas into holdings under controlled management;
(2) careful tillage—mainly ploughing;
(3) early sowing and the careful use of fertilizers;
(4) appropriate crop rotations.

Some vegetation types have been reported to be nearly as productive of organic matter in the savanna as in the forest zone, in spite of reduced rainfall. Open savanna forest produces 7–8 t/ha per annum of organic matter, compared with levels of 10–70 t/ha per annum in the humid tropical forests, but many grassland fallows produce levels of 12–20 t/ha per annum. In addition, erosion protection can be adequately achieved with appropriate type of vegetation, provided that bare soil is not subjected to erosive storms.

TABLE 14.2

Type of vegetation cover	Erosion (t/ha)
Forest	0·2
Fallow	4·9
Cultivated	7·3
Bare soil	21·3

Losses of nutrients are generally also slower in the savanna than in the forest zone for nitrogen and sulphur, especially if moderate management is applied.

In many arid zone soils physical problems arise from the often sandy texture of the soil. In such soils deep, but not too frequent, ploughing with the incorporation of organic matter often results in the improvement of soil structure and porosity, which is favourable to the root development of crop plants.

Level tillage is considered advantageous compared with the classic ridge ploughing or the traditional heap preparation, since level tillage tends to improve water infiltration and facilitates better weed control.

Early sowing, by giving rapid early vegetation cover on the soil at the start of a rainy period, reduces the erosion risk. This alone, however, does not account for all the proven superiority of this practice. Possibly also, early sowing permits the plants to take advantage of a flush of mineralized nitrogen from the burst of nitrifying activity at the break of the rains, and soil moisture conditions may be more favourable early in the rains. The differences between the success of continuous-management systems in forest and savanna zones are still inadequately understood. The pessimism with which continuous food-crop production is often regarded may, however, be overcome in a relatively short time with appropriate research effort.

14.4. Technical possibilities for improved systems

14.4.1. *The importance of a systems concept and approach*

Certain terms must be defined for the present context. As our definitions are unlikely to be universally accepted, however, we point out that it is not our intention to try to impose a universal recognition of these but simply to make our meaning plain and unequivocal.

A *systems approach* is characterized by the interest taken in the interactions between components that may be severally altered (Spedding 1972). In agriculture, an *ecological approach* is one relevant to problems of the ecosystems and to the relationships between several living components. The importance of any single component cannot be fully realized unless it is related to a larger system of which it is an integral part.

The search for an optimal combination of different components, and hence of their best interactions, implies the optimizing of the system as a whole (Heady 1970). The construction of a mathematical model to describe the different constraints and relationships of the system (Brockington 1970; Kanwar 1970) is now being more frequently used as an elegant and powerful means of:

(1) identifying the most coherent and most profitable combinations of inputs;
(2) simulating new situations;
(3) exposing areas of ignorance.

This final point is especially important as it provides

a profitable feedback, promoting re-orientation of research and re-examination of its priorities.

Basically, systems analysis requires a statement of the objectives of the total system, a quantitative analysis of the environment of the system, a listing of the resources of the system, a schedule of the components of the system (especially activities to be carried out in determining measures of performance), and finally an analysis of the management of the system (Churchman 1968).

Two types of model can be distinguished (Henin and Desfontaines 1971; Monnier and Ramond 1970):

(1) models which combine the specific factors of production and the crops already familiar or popularized in the rural areas;
(2) models that involve technical innovation and new speculations.

The elaboration of systems, be it by intuition or by formal model construction, produces rough drafts that need to be 'tested on a large scale so as to justify deep thinking and a careful organization' (Spedding 1972).

Systems under consideration or being modelled may be simple or complex. A *farming system* has as its objective only the practical aspects of the production of crops (i.e. it combines all single factors and their interactions such as rotation × production tool × labour, etc.).

A *system of production* in agriculture may involve the association of different stock or crop enterprise (livestock, forestry, etc.) systems. A system of production can, to begin with, be worked out in the absence of the constraints of any particular farming situation, e.g. on an agricultural station, and can there be evaluated simply in terms of technical and economic criteria; this leads to the formulation of a prototype production system (package).

When the technical system is modified to take account of the constraints of any particular area of application (human in particular), it is a *socio-economic system of production* (Malassis 1972).

A socio-economic system introduces another important concept: that of the pattern of units of production and farming techniques which are characterized by particular physical limitations such as land, buildings, or tools, and also by human and economic factors such as labour and capital.

At the technical level the pattern of a particular production system is formulated around an existing type of production. At the socio-economic level, where the model is tested in an area of application, the technical model has to be modified in the light of constraints operating in the existing farming patterns, and conversely the latter should evolve so as to allow innovative systems of production to be incorporated. It is therefore useful to distinguish clearly between systems of production and practices on actual farms which, by being repeatedly tested against one another in the area of practical application, eventually make it possible to arrive at the best practical and productive farming system.

In the simplest situation, represented by opening a new farm on virgin land, we can directly apply a chosen system of production. The construction of the model for the system then enables us to determine the major characteristics and, in particular, to quantify the parameters of the model in an actual farm situation.

Finally, the existence of different or complementary systems of agriculture on similar land in the same area will inevitably bring about an interaction of ideas and methods. Such an area will appear as a mosaic of different types of farming, representing varying degrees of innovation which are, to a greater or lesser extent, coherent, systematized, and dynamic. Such an assemblage will be called an *agrarian structure*, which characterizes the particular agro-social environment.

14.4.2. *Mixed cropping systems in traditional and intensified agriculture*

Most traditional agricultural systems in Africa are characterized by mixed cropping, in which several cultivated crops are intermixed among each other. They usually differ greatly in productivity, habit, and phenological and agro-ecological characteristics. For example, cereals, grain legumes, and root and tuber crops may all be grown together.

Many authors have pointed out the possible advantages of mixed cropping in a subsistence type of agriculture (Norman 1970), but until recently research on mixed cropping in relation to business farming has been rare. This is partly due to the necessary complexity of the experimentation and partly due to the commonly held view that, since such systems are associated with subsistence farmers and have for many years been considered primitive, they should not be a serious topic for research. There have been some notable exceptions in Africa and elsewhere, however, some of which have been reviewed by Leakey (1970), including studies on cotton-groundnut cropping. Evans (1960) showed that a benefit was derived from intercropping groundnuts with maize and sorghum, and under arid conditions Monro (1960) obtained a 29 per cent increase from a crop of cotton and maize grown together.

Some of the potential advantages suggested for mixed cropping include maximal use of light, water, and nutritive elements through the best combinations in space and time of leaf and root development of the various crops, continuity of soil cover, and the continuous permeation of soil by roots, by which erosion is considerably reduced. To these advantages we must add the value of synergic relationships such as the rhizobial symbiosis of a legume which may provide nitrogen indirectly to an associated crop. On the other hand, there will also be competition between associated crops for nutrients. Another general advantage of mixed crops over single-crop stands lies in the possibilities for avoiding creating conditions likely to damage the ecosystem.

In mixed cropping attacks by pests and diseases tend to be reduced in severity, although they are more varied, and invasion by weeds may be checked by the aggressiveness of the mixed-crop population. The multiplicity of cultivated plants spreads the risk since all component crops of the association are not equally susceptible to the climatic and phytosanitary hazards of any particular season. This adds a measure of security for the farmers.

However, the value of mixed cropping must be judged on the basis of productivity of the combination of crops in comparison either with that of the same amounts of each crop grown on separate plots totalling the same area or with the yields produced by the same labour inputs in the pure-stand arrangement. Whether we choose the return-to-land or return-to-labour criterion as a basis for judging the advantages or disadvantages should depend upon whether land or labour is limiting in a particular agro-social situation. It is clear that a more qualitative appraisal, followed up with a more precise quantitative evaluation, is required under both land- and labour-limiting conditions.

Mixed cropping under agronomically unimproved conditions. Studies have been conducted in three villages in Northern Nigeria (Norman 1970) through a systematic and statistical comparison of mixed croppings and pure cultures practised by farmers in the region where the following crops were grown: pearl millet, sorghum, maize, groundnut, cowpea, cassava, and sweet potatoes. Only dry-land conditions were included in the study. The following were the major findings.

1. The pure crop stands occupied only 17 per cent of the cultivated area.
2. There were many combinations of mixed cropping, but six common combinations together occupied 60 per cent of the cultivated land area. Millet-sorghum alone represented 31 per cent of the area used for mixed cropping.
3. The spatial arrangements of the mixed crops was usually systematic in contrast to the random intra-planting patterns sometimes said to obtain in the other regions of Africa (Geerte 1963).
4. The major production input was labour, with all agricultural operations being carried out manually; 80 per cent of the labour was by members of the family.
5. Of the annual average of working hours, 26 per cent were during the June–July peak period. During this peak period mixed cropping requires 29 per cent more labour per unit area than pure-crop stands. This is equivalent to 62 per cent more labour over the whole year. Yields of each individual crop were all lower per unit area under mixed cropping than under pure culture, as might be expected.
6. Profit per working hour is slightly less for mixed cropping than for pure culture, but the gross return from mixed cropping was 60 per cent higher than if the same crops were grown alone. This increase was attributed to higher effective population densities (of mixed crops) compared to those which are used when each crop is grown in pure stand.

The reasons for intercropping given by farmers, while not specific or detailed, indicated that their main concern was to obtain a higher output from a given amount of land. The need for security of food production and the beneficial effects of legumes were also considered reasons. These studies verify that farmers consider that growing crops in mixtures rather than in pure stands is more profitable.

Norman suggests that farmers should be given improved technology that will increase their returns under intercropping practices before they will be ready to accept the more radical change of mono-cropping. Mechanization may be a key factor that will persuade farmers to adopt mono-cropping practices, since no present mechanized system is known that will handle mixed crops successfully.

Tardieu (1970) and others in the wet highlands of western Cameroun demonstrated that intercropping maize, taro (*Colocasia*), and macabo (*Xanthosoma*) under both traditional and improved methods of culture resulted in improvements over single-crop cultures, except under hydromorphic soil conditions. On allophanous black soils the yields were high and the income from mixed cropping surpassed that from the same sole-cropped species at the same level of fertilization.

Andrews (1970) conducted research in the Guinean zone of Nigeria under favourable agronomic conditions on multiple cropping (or catch-crop growing) and mixed cropping. In multiple cropping with two fertilized plantings in succession during a single rainy season he produced higher yields as well as greater income per hectare with maize followed by cowpeas than from sorghums ratooned or from guinea sorghum. A dwarf sorghum grown as a sole crop yielded twice as much as a sorghum grown in a mixture with either pearl millet or maize, but the three-way combination of millet and maize with the sorghum produced not only higher yields but nearly doubled the gross return.

In intercropping experiments with different ratios of pearl millet and sorghum, mixing within these in the ratio of two millet plants for every sorghum plant produced substantially higher gross yields than corresponding numbers of plants grown as pure-stand rows. The millet yield was reduced by only about 4 per cent by the interplanted sorghum in comparison with millet grown as a pure stand.

Andrews considered that for successful intercropping:

(1) intercrop competition must be less than intra-crop competition;
(2) the arrangement and relative numbers of the contributing crop plants must be carefully chosen to take account of the difference in competitive ability of each species;
(3) the effect of competition between crops is greatly reduced when their maximum demands on the environment occur at different times; minimizing competition can thus be achieved either by selecting crops with differing growth cycles or by planting them at different times;
(4) the season available for plant growth should be long enough to allow each crop to have

suitable conditions at appropriate times in its ontogeny;

(5) legumes are probably a necessary component of efficient multiple cropping systems under most soil conditions.

Under unimproved conditions, therefore, mixed farming has much to justify it. In spite of extensive efforts, mixed cropping has given way to pure cultures in only a few regions in Africa, and it is relevant to consider whether technological progress could be achieved more widely through the improvement of mixed cropping than by persisting in attempts to replace this practice by pure cultures.

Mixed cropping under improved conditions (with higher levels of inputs). Some of the constraints preventing efficient crop production under intensive management may be removed or reduced by combining improved cultivars of crops and by the use of adequate levels of fertilizer, etc. in multiple-cropping and crop-rotation systems. The possibility of substantial improvements in mixed- and multiple-cropping farming has been investigated in research work conducted in the north of the Guinean zone of Nigeria (Andrews 1970, 1972). Multiple cropping, or successive cropping, means growing two crops during a single rainy season and with a possible overlapping in time. Mixed cropping, or intercropping, under improved conditions normally means planting different crops, either at the same time or overlapping in time, in alternate lines or strips. The best combination of maize and cowpeas gave an improved monetary return of over 60 per cent per unit area in comparison with dwarf late hybrid sorghum planted alone, as an economic control. Osiru and Willey (1972) have developed a useful technique for the study of crop mixtures.

We may conclude that mixed cropping can improve peasant agriculture within the limits of present or slightly increased farm sizes and especially where there is intense pressure on cultivable land.

The adoption of systems of mixed and multiple cropping can undoubtedly lead to a high degree of intensification in regions where the ecological conditions, especially the duration of the rainy season, allow.

14.4.3. *Systems of pure culture in savanna agriculture*

Historically the savanna zone has been oriented mainly towards pure culture agricultural systems.

The major reasons for stands of single crops in the savanna appear to be:

(1) the low rainfall and a strong alternation between the two dry and rainy seasons;
(2) a low number of adapted crops;
(3) large cultivable area per inhabitant;
(4) necessity for increased mechanization of cultural practices, particularly ploughing, to prepare the soils in readiness for planting in the shortest possible time for early establishment at the break of the rains.

Intensification of agriculture in these zones has been reached by passing through shifting or semi-shifting cultivation to a permanent, fixed agriculture which ensures maintenance and improvement of the soil capital. According to Greenland (1970) no community can expect to survive as a prosperous and developing community unless it develops a stable and productive system of agriculture. High productivity cannot be stable unless soil fertility is maintained or improved.

This intensification of agriculture in the savanna area requires the use of similar methodology to that of most other conventional types of agriculture, i.e.

(1) selection and crop protection of high-yielding plant materials;
(2) developing techniques and equipment adapted to different conditions and methods of production;
(3) enhancing the productivity of labour through mechanization and organization.

There is sufficient land in the savanna area in extent, since population density is low, but its quality in these areas with respect to fertility, water, mineral, organic, and biological properties is greatly in need of improvement. Much research has been directed towards this end. The region that is mainly concerned is that characterized by one rainy season of shorter than 5 months duration. Most of the soils found belong to the ferruginous tropical soils, ferrisols and ferralitic soil groups comprising immature soils over alluvium or colluvium (D'Hoore 1965). These soils have common agronomic characteristics. Texture is usually sandy or sandy clay in the surface horizons with a well-marked predominance of kaolinite in the clay fraction of the soil. The absence of montmorillonitic clays makes the creation of a true structure dependent entirely on tillage. In addition, these soils have the following properties.

1. Low exchange capacity (generally between 2 m eq and 5 m eq per 100 g).
2. Rates of base saturation of between 40 per cent and 100 per cent.
3. Moderate to very acidic pH.
4. Frequent mineral deficiency; in particular, phosphorus deficiency is general, with levels below 200 mg/kg (in P_2O_5 equivalent units) and effective potassium deficiency is easily induced by nitrogen additions when the soils are cultivated. A thorough study is especially needed on the cyclic changes in nitrogen availability, which need to be further understood and taken into account.
5. Extensive leaching resulting from the freely draining nature of the soils.
6. Great variation in microbial populations, which decrease during the dry season, but expand rapidly at the beginning of the rainy season and are associated with intensive mineralization (mineralization peak) of nitrogen at the start of the rains.
7. A high degree of erosivity, due more to the intensity of the rains than the particular susceptibility of the soils. This requires the maintenance of a crop canopy, to cover the soil during the period of storms, and other anti-erosion measures.

14.4.4. *The transition from traditional to semi-intensive systems*

The passage from traditional agricultural systems to semi-intensive systems† is being achieved successfully in the savanna region in Senegal by putting into effect the recommendations arising from research into relatively large-scale operation. The recommended practices (Charreau 1970) in Senegal are probably valid for the whole of the Sudan-Sahelian zone.

The major characteristics of the traditional farming systems, from which modifications need to be applied by the majority of farmers in the region, are as follows.

1. Progressive disforestation with inadequate uprooting of stumps.
2. Semi-shifting cultivation with cycles of cropping

---

† The term 'semi-intensive' has been retained to denote already modernized agricultural systems using elaborate techniques such as deep tillage and varied mechanical tools, but using only oxen to provide draught.

interrupted by fallow periods which become shorter and shorter.
3. A lack of animal husbandry among cultivating communities.

These, and other, characteristics together lead to the following consequences:

1. Deep tillage is impossible.
2. Organic matter is not maintained in the soil through ploughing-in of compost.
3. Weed control is inadequate due to lack of sufficient tools.
4. Mineral fertility becomes insufficient.
5. The range of crops grown in the farming system is limited.
6. There is very little organized planning of crop sequences on any given piece of land and crops are normally mixed.
7. Selected crop cultivars are little used.
8. Phytosanitary control is unknown or ignored.

Not surprisingly, low yields are normal in such systems (400–600 kg/ha for cereals; 800–1000 kg/ha for groundnuts) even for the crops that are apparently able to thrive under the poor conditions offered by the environment. The soil capital is not preserved, with the result that after a few years the fields are abandoned and new lands, if available, have to be cleared.

The first efforts by agronomists to improve the existing system without radically altering it adopted only techniques within the poor financial resources of the peasants. Efforts were made on combinations of the following innovations:

1. The use of selected cultivars with higher yield and improved disease resistance.
2. Seed-treatment against pests and diseases, and occasionally also crop spraying or dusting.
3. Improvement of the cultural techniques, such as raising plant populations, row planting, weeding, sowing, and harvesting.
4. The use of rather low rates of mineral fertilizer on the most high-yielding crops consistent with the highest financial return to the cost of this input.
5. Defining cropping successions and ensuring periodical short fallows.

The adoption of these techniques, often combined with the use of draught animals, resulted in improved production and higher yields. But this was mainly for the less exacting crops such as groundnut (and to a lesser extent cotton); results were less encouraging for cereals.

The modifications to the system have proved inadequate for the long-term maintenance or improvement of soil fertility. For this reason we do not consider them as a satisfactory agronomic solution. The situation after a short period of such 'improvements' is scarcely different from the original traditional systems. Crop yields drop. Groundnut yields, for example, stabilize at yields of only about 1500 kg/ha, whereas cereal yields fall back to a level of about 500–800 kg/ha.

The second stage in attempting to establish stable semi-intensive agriculture depends upon heavy animal traction, which makes deep tillage, and especially ploughing, possible (Charreau 1970; Charreau and Nicou 1970; Tourte, Pocthier, Raymond, Monnier, Nicou, Poulain, Hamon and Charreau 1971). Ploughing creates a better soil structure and porosity conducive to the rooting of planted crops, and ensures a better water and mineral supply to the plants.

Much experimental data from Senegal and other francophone West African countries has been reviewed by Monnier and Ramond (1970) and Jones and Wild (1975). All cultivated crops respond favourably to ploughing, but the degree of response varies according to the type of plant. Cereals, particularly maize, sorghum, and rain-fed rice, profit most from structural improvement. Groundnuts, however, appear to respond least, especially when organic matter is ploughed in. Other plants, such as cotton and *Vigna*, often have intermediate responses.

The average yield increases are between 10 per cent (groundnuts) and more than 100 per cent (rain-fed rice), depending on the type of crop and relevant level of fertilization. Instances of lack of response or of negative response are few and could usually be attributed to deficiencies in other cultural operations.

Apart from what are considered to be the direct consequences of improved rooting, ploughing is also important in weed control. Finally, ploughing makes possible the incorporation of organic matter through returning harvest residues to the soil.

Despite all these advantages, deep tillage, and especially ploughing, still has a bad reputation in the tropical regions. Notes of warning have been sounded in tropical agriculture against the plough as a destroyer of the soil. Such fears are not without foundation, because a tillage that is badly executed, or carried out at an unsuitable time, can have disastrous consequences for the soil, especially when combined with violent rain and intense insolation.

Experimental data in support of ploughing can be cited from southern Senegal, where it has been shown that ploughing can help to prevent erosion on bare soil (Charreau 1970). Run-off was reduced and the specific turbidity in the gullying water was also low when the soil had been ploughed. Erosion was reduced by a factor of 3 times in ploughed soil compared with that in an unploughed control (6·5 t/ha against 18·1 t/ha). It is certain that ploughing does not *ipso facto* aggravate erosion, and often the contrary is the case. The bad reputation earned by ploughing seems to be due mainly to the confusion which has for a long time existed in the tropics between *depth* of tillage and *intensity* of tillage. The two factors are independent and should not be confused. Too great an intensity of tillage destroys soil structure and thus seriously aggravates erosion.

The effect of deep ploughing on soil and crop yields does not cease after the year in which the ploughing is carried out. Residual effects which persist for several years play an immense role in continuing productivity. From studies conducted in Senegal, Monnier and Ramond (1970) attributed a beneficial residual action to two major factors:

(1) ploughing, when combined with the burning of organic matter, has a more lasting effect on the structure of the soil and on yields than ploughing without previous burning;
(2) cereals preserve the structure created by the ploughing while other crops, particularly the groundnut, do not.

'Back-end' ploughing. Research has indicated that there may be some advantage in carrying out ploughing immediately after cropping rather than at the beginning of the following season. At the break of the rains there is so much to be done that inevitably ploughing at this time delays much of the sowing, and that in turn should be avoided as much as possible since it reduces yield.

Ploughing at the end of the season, however, is only possible if the vegetative cycle of the plant is shorter than the rainy season so that the soil is still moist at ploughing time. This is possible for some crops such as groundnut and short-cycle cereals but not for the traditional cereals, pearl millet, and late sorghum, nor for cotton.

Fertilizer use. Only when a reasonable depth of topsoil (agricultural profile) has been created can chemical fertilization, which is perhaps the most efficient and most powerful means of intensification, play its full part.

Methods of establishing fertilizer requirements are well known. Hence they will only briefly be mentioned here:

(1) the detection and classification of the mineral deficiencies for each soil type;

(2) determination of the most appropriate type of fertilizer and rates of use;

(3) determination of maintenance fertilizer levels needed to replace uptake by plants removed in harvest and losses due to drainage and as gas, e.g. by denitrifying bacteria.

A combination of improved water-holding capacity and nutrient status usually leads to an obvious improvement of the organic and biological properties of the soil, to such an extent that the use of fallowing no longer appears indispensable and fixed farming appears feasible.

Improvement of soil has three important results for the adoption of more intensive farming:

(1) It makes possible a diversification of the crops grown in the savanna regions (which, traditionally, were limited to millet, cowpeas, groundnuts, and sorghum) to include also maize, cotton, and rain-fed rice;

(2) economic animal production becomes a further possibility;

(3) it ensures a buffering, to some extent, against climatic hazards.

At the agricultural station in Sefa (Casamance, Senegal), where the foregoing techniques have been systematically utilized since 1960, it was not long before the effects were felt. Normal yields to be expected regularly are now of the order of 5 t/ha for cob maize, 4 t/ha for paddy rice, 3 t/ha for groundnut in shell, 3 t/ha for threshed bulrush millet. Such levels of production were not dreamed of during the 1950s.

A future for mechanization? The stage that should follow the successful adoption of the semi-intensive stage of farming is the introduction of tractors. This was prematurely started a quarter of a century ago, but in fact the advance to full mechanization is only realistic in economic terms when levels of production have advanced sufficiently. This stage, which we call intensive agriculture, is basically no different from semi-intensive farming, except in that the methods are more elaborate. Farming with tractors

is at present being developed on a large scale by research stations (Monnier and Tourte 1971; see also Chapter 13 of this book).

14.4.5. Intensification through irrigation

Irrigation constitutes the most costly and complex technological factor for intensifying tropical crop production. Water resources in West Africa are enormous, and some irrigation facilities have been installed, but with a highly variable record of success. (See Chapter 9 of this book.)

In general, a large initial planning and engineering input is required, and high costs arise from the requirement for land levelling, ditches, canals, pumping stations, drainage systems, and various kinds of distribution and application techniques. Average installation costs are of the order of US \$1000–5000 per ha, with annual maintenance costs of US \$100–250 per ha in addition. These costs must be recovered in the long run by the increased levels of production and value of the crop, and therefore require greatly genetically improved plant materials adapted to this kind of agriculture, sophisticated cultural practices, reliable input supplies, and an enormous investment in education to develop the management skills and technical capabilities of the farmer.

Chabrolin (1970) has pointed out that the cost and complexity of the operation normally requires a choice between two approaches:

(1) an industrial–plantation agriculture with a few highly qualified technicians, maximum technical and engineering input, and high capital investment; or

(2) a labour-intensive, but still technically sophisticated, system with large investments in training and education.

The choice between the two depends on social objectives in the community and is primarily political.

In most economies, the social and economic objectives of improving the living conditions and meeting the aspirations of the people require that the second alternative be attempted. In addition, the changes must take place in a 'quantum jump' fashion, which requires truly revolutionary changes in the professional activity of the farmer and his ideas and attitudes, as well as his way of life. This has inevitably led to irrigation systems being attempted without sufficient planning and pre-implementation education. In such circumstances, yields and financial returns are initially low, wasteful use is made of scarce resources, and large losses in production result. The difficulties of training traditional farmers in the maintenance and management of irrigation facilities for agriculture and the use of mechanized and technically advanced cultural methods has led to attempts at intermediate systems. These systems have the drawbacks of both the traditional and the modern methods and few of the advantages of either.

Bradfield (1970) has presented a proposal for a labour-intensive irrigated cropping system which has been demonstrated to be enormously productive on an experimental basis. At the International Rice Research Institute in the Philippines in the warm tropics at latitude 14° N, a multiple-cropping experiment has been based on the continuous production of irrigated rice. Using continuous production over 20 t/ha of paddy was produced in a year. With such large potential production of the basic crop additional crops as a partial substitute for rice must be sought in order to maximize income; this is because the price of rice will probably decline if there is a surplus in production. On a clay soil with flood irrigation, a multiple-cropping system required close attention to several aspects of tillage and management as follows.

1. Bedding the soil to accelerate the drying of the top layer where crops other than rice are to be planted and cultivated.
2. Keeping the volume of soil tilled and the number of tillage operations to a minimum.
3. Using early-maturing cultivars of crops which produce high average yields per day over the duration of the crop.
4. Growing ratoon crops where feasible, thus eliminating one or two planting operations.
5. Starting slow-starting vegetable crops in compact propagation beds before transplanting to the field when they reach the period of more rapid growth.
6. Growing some crops each season which can be harvested and utilized in an immature stage, such as sweet corn and vegetable cultivars of soya beans.
7. Intercropping whenever possible. Some sacrifice in yield of either crop can be justified if over-all production is increased substantially.

The principal crops around which this system has been established are shown below and a gross product of over US $3000 per ha was realized.

Average yield (t/ha)	
Rice	4–6
Sweet potato	20–30
Soya bean	2–3·6
Sweet corn	4 (in the ear)
Sorghum 1	5–7
Sorghum 2	6–7
Sorghum 3	5–6

A number of rotations with these crops have been employed, all of which utilize a basic wet-season rice crop in the season of highest rainfall.

Because heavy-textured soils and frequent rains limit the time that the soil is dry enough to cultivate, mechanization must be used very efficiently and intensively over a short period. A single-axle 6–8 hp tractor can perform all the operations envisaged, and land preparation at times is done in only a few hours between the harvest of one crop and the planting of its successor. During the rainy season, an enforced delay of a tillage operation, if insufficient tractors are available, may result in a lag of 2–4 weeks in getting the next crop established if the weather becomes unfavourable. This may be equivalent to 25–30 per cent of the life-cycle of the succeeding crop, and under these conditions an investment in mechanization becomes enormously profitable.

Bradfield (1970) suggests that this system can support about 180 people per square mile, in contrast to shifting cultivation under the same climatic conditions, which is estimated to be capable of supporting only a stable population of about 18 people per square mile. If, in contrast, we merely modify the shifting-cultivation system in order to attempt to support more people merely by reducing the forest fallow, crop yields decline after a few cycles.

A multiple-cropping system designed to suit the local environment, concentrating limited resources into a small area of high potential, appears to be an attractive possibility. Inevitably this will require some imported inputs, but these can be supplemented by making efficient use of land, water, crop residues, labour, and machines.

14.5. Extension and development

14.5.1. Extension of new farming systems

Whatever systems of intensification may be chosen in any particular circumstances, i.e. mixed cropping, multiple cropping, pure cultures, etc., we consider that it is now clear that agriculture in West Africa, including both the forest and the savanna regions, can be considerably intensified both with safety and to the economic advantage of the farming community.

Thus the prerequisite for a Green Revolution in this area seems to exist. Moreover, a few large-scale agricultural development projects have already been successfully carried out in areas of tropical Africa that at present mainly support subsistence peasant farming. The development message from research in Africa is at present badly communicated to the farmers. The reasons for this must be discovered and corrected as quickly as possible if we are to justify present research activities in economic terms. We reject *a priori* the reason sometimes given for this poor communication: that it is due solely to lack of competence in the extension service. There is, of course, no doubt that the service in some areas is better than in others, but it is unthinkable that the extension services are incapable almost everywhere of inducing change. There is no doubt that in a number of places there exists both the competence and the means and the willingness to succeed, but still no radical change is forthcoming. The reasons for failure must be found elsewhere. The problem, in our view, lies mainly in the type of recommendations made so far by research workers.

Efforts to improve agricultural production based on agronomic research in tropical Africa have largely concentrated on the separate improvements in different crops, e.g. by breeding improved cultivars and developing plant protection schedules and by suggestions concerned with different single factors of production such as land, capital, and labour. Emphasis has been on soil fertilization, the development of a certain mechanical tool, or the advocacy of a single weed-control practice. There has been little emphasis so far on complete 'packages' of combined practices. It is only recently that agronomists in tropical agriculture have realized clearly the need for working out systems which integrate the different elementary recommendations from specialists.

The achievement of a practical package of practices makes three important demands on research:

(1) verification of technical feasibility of the combination of the several schemes recommended;
(2) a search for an optimal combination, both technically and economically;
(3) broad testing of the resulting schemes in actual farming situations.

Because of the relative success of agricultural innovation in Senegal in comparison with many

other West African countries, we describe the methodology there; it comprises six clearly defined phases and involves an integration of research and extension activity.

The analytical phase. Each crop or factor of production is separately studied and all research is aimed at maximum productivity. Areas of study include soil and plant improvement, plant protection, mechanization, etc. But in each, in contrast to much of the more traditional research, there is particular attention to costs and returns and labour requirements (Monnier and Ramond 1970). The labour requirements of all the major activities are determined by the actual timing of all operations. An estimate of the number of available working days (taking account of climatic conditions and the physical ability and general health of the work-force) enables the feasibility of the resulting work schedule to be assessed for each combination of technical proposals. The capabilities of the work-force are the major constraint to change, since labour and not land is limiting in most dry tropical regions. Calculations based on the amount of work to be done must also make due allowance for the defects in organization which must realistically be expected.

The synthetic phase of evaluation of technical systems of production. It is not at all certain that the best combined model can be constructed simply by adding together each specific technique for maximizing yield. For example, the best time for sowing, which is theoretically as early as possible in the rainy season, is usually incompatible with the need for thorough preparation of the seedbed. During the second phase of research into a cultural system the compatibility of individual suggested improvements is tested. The following are studied in particular (Monnier and Ramond 1970).

1. The productivity possible from using a limited number of working hours for hand labour, animal power, or mechanized operations as alternatives. This shows the degree of mechanization appropriate for each activity, and hence the research effort needed to devise an appropriate use of man hours for the activity under consideration.
2. The different work activities which can conveniently be considered in 'blocks' of work which have to be accomplished during a full crop cycle. In the savanna region there are three such blocks.

1st block: superficial dry-land tillage on bare land.

2nd block: seeding, establishment and maintenance of the growing crop.

3rd block: harvesting, and ploughing at the end of the cycle, requiring deeper tillage and greater traction.

The precise timing of the 1st block is relatively unimportant; timing of the second is crucial to success; timing of the third is variable over a considerable period.

Research should aim to allow as much latitude as possible in operations so as to try and minimize labour bottlenecks.

Classification of existing patterns of production. This phase can overlap in time or even be carried out simultaneously with the previous phase. This study will embrace existing crops and their husbandry, extent and yields, social organization, method of land tenure, internal and external communications, and the pressure or otherwise of any professional and commercial organizations. The study will provide a base for constant review of progress and provide the all-important feedback by which the methodology of development can be appraised.

Phase of model construction. This may involve various methods for constructing these models. The methods range from the simplest, such as intuition, to the most complex computer methods now available, using the techniques of linear programming, polyvalent matrices, etc. (Heady 1970; Brockington 1970; Heady and Williams 1958; Attonaty and Hautcolas 1970; Brower and de Wit 1969; Boussard 1970; Cordonnier, Carles, and Marsal 1970).

The phase of dissemination by extension. In Senegal models already exist which integrate the innovating data from research and the constraints of the area of application.

This phase of development activity, as currently carried out in Senegal, is derived from some simple considerations (Tourte 1965, 1967), which can be easily summarized as follows.

1. The development must involve systems of production which have been validated by testing in the environment for which they are recommended.
2. The gap between fully modern technology, as now often recommended by research in West Africa, and the present level of traditional agriculture is so great that the wholesale and in-

adequately considered introduction of this new technology could only result in a serious upset of systems of production. Great care is needed to guarantee both security and rapid penetration of the technologically feasible developments.
3. The existing cross-links between agrarian, economic, social, and political development and the behaviour of different human groups are so intimate that agronomists cannot afford to ignore the social and structural repercussions engendered by the technology he recommends.
4. The application of research cannot be totally successful unless there is genuine continuing dialogue in a common language between the planners and farmers.

It was on the basis of these reflections that the Government of Senegal in 1968 undertook to establish trial experimental units called 'Regional Action for Integrated Development' on two co-operatives situated in the major agricultural region of Sine Saloun.

On these co-operatives the following have been tackled with enormous success and sometimes in an original way.

1. Diversification into new enterprises such as stock raising, market gardening, and fruit production, and their incorporation into the annual work calendar with improved working efficiency.
2. Improvement of the infrastructures of the farms and villages, and especially the improvement of water supplies by providing wells and pumping water by means of animal traction.
3. Reorganization and enlargement of the functions of the co-operatives.
4. Establishment of new commercial networks.
5. Land and rural reform involving regrouping, plot exchanges, and land consolidation.
6. Improvement of health and other social services.

14.15.2. Experimental-unit appraisal

These co-operatives in Senegal represent the application of the 'experimental unit' which can, in certain respects, be compared with the 'Puebla' project carried out by CIMMYT (although small in size and perhaps more detailed) and the Government agencies in Mexico (CIMMYT 1969, 1970). A similar approach has also been formulated as the 'Masagana 99 Project' by IRRI and the Government of the Phillippines (IRRI 1972); this project is adapted to local social and political structures and grew out

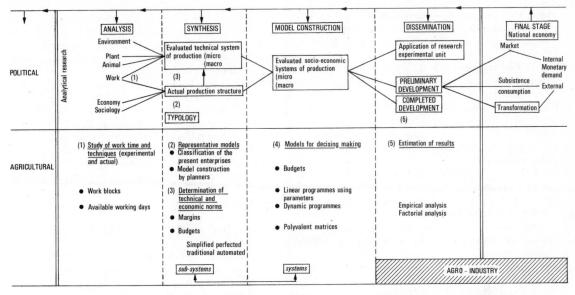

FIG. 14.1. An integrated system of agronomic research. (Drawn at the seminar on agricultural economics, GERDAT (1972).)

of earlier attempts to convey advanced technology to the primary producer.

The approach being used in Senegal is summarized in Fig. 14.1, which also relates the research to a larger system of planning and the policy of the agricultural development of a country.

References

ALBRECHT, H. R. (1970). The evolution of systems of subsistence cropping in the tropics. *Ford Foundation/ IRAT/IITA seminar on traditional systems of African agriculture, Ibadan.*

ANDREWS, D. J. (1970). Successive cultivation and inter-cropping with sorghum at Samaru. *Ford Foundation/ IRAT/IITA seminar on traditional systems of African agriculture, Ibadan.*

—— (1972). Intercropping with sorghum in Nigeria. *Expl. Agric.* 8, 139-50.

ATTONATY, J. M. and HAUCOLAS, J. C. (1970). Du pro-gramme linéaire au calcul automatisé d'exploitation agricole de budgets d'exploitation agricole. *Bull. tech. Inf. Ingrs Servs agric. fasc.* 255, 711-40.

AUBERT, G. and TAVERNIER, R. (1972). Soil survey. In *Soils of the humid tropics*, pp. 17-44. National Academy of Sciences, Publ. No. 1948, Washington D.C., USA.

BOUSSARD, J. M. (1970). *Programmation mathematique et théorie de la production agricole.* Cujas, Paris.

BRADFIELD, R. (1970). Systems of multiple cropping in the tropics. *Ford Foundation/IRAT/IITA seminar on tradi-tional systems of African agriculture, Ibadan.*

BROCKINGTON, N. R. (1970). An ecological approach to agricultural research. *Ford Foundation/IRAT/IITA seminar on traditional systems of African agriculture, Ibadan.*

BROWER, R. and DE WIT, C. T. (1969). A simulation model of plant growth with special attention to root growth and its consequences. In *Root growth* (ed. W. J. Whitting-ton). Butterworths, London.

CEPEDE, M. M. (1973). (Ed.) *Colloque FAO-OCDE sur les échanges mondiaux de produits agricoles et alimentaires.* Centre National des Expositions et Concours Agricoles, Paris.

CHABROLIN, R. (1970). Intensification of agriculture through irrigation in the dry tropical zone. *Ford Founda-tion/IRAT/IITA seminar on traditional systems of African agriculture, Ibadan.*

CHAMINADE, R. (1970). The need to intensify cultivation systems in West Africa. *Ford Foundation/IRAT/IITA seminar on traditional systems of African agriculture, Ibadan.*

CHARREAU, C. (1969). Influence des techniques culturales sur le developpement du ruissellement et de l'erosion en Casamance. *Agron. Trop., Nogent* 24, 836-42.

—— NICOU, R. (1970). L'amélioration du profil cultural dans les sols sableux et sablo-argileux de la zone tropi-cale sèche Ouest-Africaine et ses incidences agrono-miques (d'après les travaux des chercheurs de l'IRAT en Afrique de l'Ouest. *Agron. Trop., Nogent* 26, 209-55, 565-631, 903-78, 1183-1237.

CHURCHMAN, C. W. (1968). *The systems approach.* Dell Publishing Co., New York.

CIMMYT (1969). *The Puebla Project 1967-69.* Progress

report on a programme to rapidly increase corn yields on smallholdings. Centro internacional de mejoramiento de maiz y trigo (CIMMYT) September 1969, Londres, Mexico.

CIMMYT (1970). *International Conference on strategies for increasing agricultural production on small holdings, Puebla, Mexico.* CIMMYT, Londres, Mexico.

CORDONNIER, P., CARLES, R., and MARSAL, P. (1970). *Economie de l'entreprise agricole.* Cujas, Paris.

D'HOORE, J. L. (1965). *La carte des sols d'Afrique au 1 : 5.000.000.* Commission de Cooperation Technique en Afrique, Lagos.

GEERTE, R. (1963). *Agricultural innovation.* University of California Press.

GREENLAND, D. J. (1970). The maintenance of shifting cultivation versus the development of continuous management systems. *Ford Foundation/IRAT/IITA seminar on traditional systems of African agriculture, Ibadan.*

HANSON, H. (1970). A Green Revolution for tropical Africa in the 70s and the role of the agronomy researcher. *Ford Foundation/IRAT/IITA seminar on traditional systems of African agriculture, Ibadan.*

HEADY, E. O. (1970). Systems approach in agricultural research. *Ford Foundation/IRAT/IITA seminar on traditional systems of African agriculture, Ibadan.*

—— WILLIAMS, T. E. (1958). *Linear programming methods.* Iowa State University Press.

HENIN, S. and DESFONTAINES, J. P. (1971) Principe et utilité de l'étude des potentialités agricoles régionales. *C. r. Séanc Acad. Agric. Fr.* No. 10.

IRRI (1962). *Masagana 99 Project Report.* IRRI, Las Baños, Philippines.

JONES, M. J. and WILD, A. (1975). *Soils of the West African Savanna.* (Tech. Comm. No. 55, Commonwealth Bureau of Soils Harpenden.) C.A.B. Farnham Royal, Bucks.

KANWAR, J. S. (1968). Cropping patterns in India—scope and prospects. *Symposium on cropping patterns.* ICAR.

—— (1970). Relationship of agricultural systems to ecological zones in India. *Ford Foundation/IRAT/IITA seminar on traditional systems of African agriculture, Ibadan.*

LEAKEY, C. L. A. (1970). Heterogeneous agricultural populations. *Agric. Prog.* 45, 34-42.

MALASSIS, L. (1972). *Seminaire GERDAT d'Economie rurale.* ENSAM Montpellier Mai, 1972 et publications de la Station d'Economie rurale INRA, Montpellier.

MONNIER, J. and RAMOND, C. (1970). Intensive agricultural systems of production in Senegal and their effect on the evolution of farming structures. *Ford Foundation/IRAT/IITA seminar on traditional systems of African agriculture, Ibadan.*

MONNIER, J. and TOURTE, R. (1971). Expériences et perspectives de motorization; la motorization en milieu paysan, pourquoi pas? *Ford Foundation/IRAT/IITA seminar on agricultural mechanization, Bambey.*

MOSS, R. P. (1968) (Ed.). *The soil resources of tropical Africa.* Cambridge University Press.

NORMAN, D. W. (1970). Mixed farming. *Ford Foundation/IRAT/IITA seminar on traditional systems of African agriculture, Ibadan.*

NYE, P. H. and GREENLAND, D. J. (1960). *The soil under shifting cultivation.* Tech. Comm. No. 51, Commonwealth Bureau of Soils. CAB, Farnham Royal, England.

OSIRU, D. S. O. and WILLEY, R. W. (1972). Studies of mixtures of dwarf sorghum and beans (*Phaseolus vulgaris*) with particular reference to plant populations. *J. agric. Sci., Camb.* 79, 531-40.

SPEDDING, C. R. W. (1972). Synthesis of systems in agriculture. *Fourrages* 51, 3-18.

TARDIEU, M. (1970). Fertilization trials on mixed crops in the highlands of West Cameroon. Paper presented at the *Ford Foundation/IRAT/IITA seminar on traditional systems of African agriculture, Ibadan.*

THIAM, N., TOURTE, R., and POCTHIER, G. (1970). The Senegalese experience in the modernization of agriculture. Paper presented at the *Ford Foundation/IRAT/IITA seminar on traditional systems of African agriculture, Ibadan.*

TOURTE, R. (1965). Suggestions pour une politique d'application de la Recherche agronomique dans les pays en voie de developpement. *Agron. Trop., Nogent* 20, 1163-76.

—— (1966). *Premières réflexions sur les liaisons Recherche agronomique—Aménagement du territoire.* CNRA, Bambey.

—— (1967). *Eléments pour une planification de la Recherche rurale au Sénégal.* Doc. mimeo. IRAT, Sénégal.

—— POCTHIER, G., RAMOND, C., MONNIER, J., NICOU, R., POULAIN, J-F., HAMON, R., and CHARREAU, C. (1971). Thèmes légers, thèmes lourds, systèmes intensifs. Voies différentes ouvertes au développement agricole due Sénégal. *Agron. Trop., Nogent* 26, 632-71.

15. Land tenure

R. O. ADEGBOYE

15.1. Introduction

15.1.1. Definitions of land tenure

Land tenure, as discussed in this chapter, embraces man's control over rural and agricultural land-use. Timmons (1964) defined land tenure as 'the body of rights and relationships between men that have been developed to govern their behaviour in the use and control of land and its resources'. Dorner (1972, p. 17) explained that

> the land tenure system embodies those legal and contractual or customary arrangements whereby people in farming gain access to productive opportunities on the land. It constitutes the rules and procedures governing the rights, duties, liberties and exposures of individuals and groups in the use and control over the basic resources of land and water. In short, land tenure institutions help to shape the pattern of income in the farm sector. However, land tenure institutions do not exist in isolation. The dimension and the future security of farming opportunities are critically affected by labour, capital and product markets.

From these descriptions of land tenure it is implied that land in itself possesses economic value by virtue of individuals or groups attempting to gain access to its use. There is a necessary element of competition for its use as the population of potential users becomes larger. Regulatory measures may then be needed. These measures are different from society to society depending on how much, in what way, and how soon, those currently in control of land resources would like to yield to or share with others the access, economic or otherwise, to the productive use of the land.

15.1.2. The principal components and importance of land tenure

The mere sharing, it must be noted, of satisfaction in land-ownership extended by the land-owner could change the life, outlook, and performance of a tenant. The three principal aspects of tenure in order of importance are ownership, transfer, and use; because it is ownership that creates access to using, occupying, leasing, and redeeming a piece of land. Land-ownership plays an important role in determining the status of each individual in a rural agricultural community. Additionally, regularity of food supply is linked with land tenure in so far as a landlord can dictate to a tenant conditions of land-use. Land-use and tenure may also be closely governed by moral and religious principles. The major importance of land tenure in the development of agriculture was stressed by Anyane (1962) who wrote: 'Perhaps the most important problem of agricultural production in Ghana is that of land tenure arrangement. It underlies many of the other problems; and it would seem imperative that before any headway can be made in developing Ghana's agriculture, the land tenure question has to be solved first.'

15.1.3. Traditional land utilization

In most of West Africa land provides for the entire food needs of those who live on it, i.e. that there is a subsistence food economy, but this does not imply that all the crops produced on it are consumed locally. Much of the world's supply of cocoa, palm oil, and groundnut is produced on land held within an essentially subsistence farming system. Any local excess of the food crops produced, such as rice, yams, maize, cassava, beans, millets, and guinea corn, are mainly consumed in West Africa but sometimes quite a distance from the area of production. Rotational bush- or grass-fallowing is still normal in areas where population pressure has not built up to levels preventing this, and technology has not advanced. In this practice the farmer moves periodically to virgin or 'rested' land in order to give the plots currently under cultivation time to regenerate. The period allowed for regeneration varies from 3 years to 10 years, depending on population pressure and on climatic and soil conditions. In a system of shifting cultivation, the farmer with much land traditionally establishes his continued ownership through use of the land by being able to show the heaps, huts, fire places, planted trees, and recent bush regeneration. Sometimes the farmer uses his fallow land for hunting and wood-gathering and maintains a continuing physical presence to ward off possible trespassing.

There is scarcely any area in West Africa which is not agriculturally useful. And the big rivers such as the Senegal, Gambia, Niger, Volta, and Benue provide irrigated farming as well as fishing opportunities to those living near them.

15.1.4. Factors changing the pattern of land-use

The pattern of land-use within the agro-ecological zones in West Africa has been greatly affected by the changes in ownership resulting from wars, migration, and resettlements. The French and the British introduced cocoa, coffee, rubber, oil palm, and groundnuts to West Africa, and it is generally believed that crops such as cotton, pearl (bulrush) millet, sorghum, and possibly rice must have moved southwards with migrating populations. The spread of Islam from North to West Africa has changed the ways of life, including land-use, among the peoples of West Africa. The land inheritance laws in particular have been greatly influenced by the Islamic religion, as is pointed out below in relation to Ghana, Nigeria, Senegal, and the Ivory Coast.

The creation of national reserves, including forest reserves, under colonial administrations was a great departure from the customary 'bush fallow' system whereby the peasant may always return to the land after a period of forest regeneration. To the peasants the national reserves and forests are viewed as uneconomic tracts of land. Farmers are forbidden entry into such areas, even though they, or their families, were the original owners. The establishment of large-scale cash cropping, either by foreign companies or by local enterprising farmers, has also altered the concept of customary tenure in the sense that the introduction of individual ownership of large tracts of land displaced smallholders.

Consequent landlessness of some of those displaced is a condition quite unknown under customary tenure. Infrastructural improvements in the name of economic development also bring a number of disruptions of or alterations to customary tenure. Provision of feeder roads, for example, even though intended to facilitate and encourage the production and marketing of export crops sometimes disrupts customary tenure by cutting across family lands, the integrity of which used to be a source of family cohesion.

15.2. Concept of group ownership of land

15.2.1. General observations

In customary land tenure, particularly in West Africa, group ownership of land has been almost universal, as pointed out by Elias (1951), Meek (1957), Ollennu (1962), Köbben (1963), and Thomas (1963). Detailed arrangements in ownership structure may differ from community to community but, essentially, individuals derive rights of ownership from the group to which they belong. One belongs to an ownership group either by being born into or absorbed by the group. Elias (1962) pointed out that

> ownership is that of the group and the individual member has mere possession. But this possession is really more than sheer physical control by the allottee or his allocated portion of the land; he can exclude from it not only strangers to the group but also others of the same group, provided that in the latter event he can show that he himself has committed no breach of customary rules relative to holdings by group members generally.

Instead of describing the customary system of land holding as 'communal' Elias suggested the term 'corporate' since, he argued 'the relation between the group and the land is invariably complex in that the rights of the individual members often coexist with those of the group in the same parcel of land. But the individual members hold definitely ascertained and well recognised rights within the comprehensive holding of the group.'

15.2.2. Group ownership in Ghana

Ownership by chiefs. In most parts of Ghana the chief is the over-all owner of the land. Each village derives its rights of use from whatever amount of land is apportioned to the chief. In the Ashanti land around Kumasi, the Asantihene (king or paramount chief of the Ashantis) demands and receives annual homage from every subchief within his domain. It is generally believed that any family can farm anywhere as long as the permission of the chief of the particular place is sought and obtained. It is also believed that any stranger who wants land simply has to consult the chief, and upon being granted a use right, the stranger pays the village chief a token annual rent.

It may often be a problem to find the appropriate chief for the area where the desired land is located or to obtain permission for the use of a tract of land of 'economic' size. While chiefs are generally willing to grant permission for the use of about a hectare for the production of seasonal food crops they are usually reluctant to grant similar permission for the use of larger areas of land for this purpose. The reason for such reluctance is due to the scattered disposition of plots already belonging to the members of the village and to the chief's task of persuading all the present users to vacate those plots on the land wanted for the larger holding and finding attractive vacant plots to give them in exchange.

Family ownership. Among the Frafra of the north and Ewe of the east of Ghana the paramount chief does not own any land beyond that of his family, of which he is the head. Land of these tribes is vested in the family. Land is said to 'belong' to a common ancestor, and when vacant the land can be used by any member of the family, even when that member comes from a different village from that in which the land is located. Family permission is sought when any member wants to use a vacant plot but, in many cases, the family council, with the chief as chairman, can grant the request on behalf of the family. The chief as a family member is allotted a piece of land but if and when he desires to have an additional piece of land he is expected to ask for it through the family council. The consent of the council to the chief's request is usually expected to be given, but it is never taken as being automatically granted.

Community ownership under the stool.† Community ownership as a concept implies that the community retains the absolute (or allodial) title and grants only cultivation rights to individuals. When a man from one community wishes to acquire cultivation rights on land belonging to another community he must negotiate through his chief. The consent of the community where the land is located is thus passed through its own chief and the chief of the would-be user. This is desirable, even if both communities are under a common higher-ranking chief, in order to avoid future disputes. Any subordinate chief who grants rights of use to individuals or communities must do so with the consent of the stool. The fact that a piece of land has been given for use by another community does not remove it from the jurisdiction of the stool under which the land has been traditionally located. The stool continues to exercise jurisdiction regardless of the state of alienation of the land. While use-rights may pass from person to

† Stool is the 'office' of the chief rather than the individual currently holding the post.

person, the controlling authority of the stool is regarded as continuing unchanged.

The indefiniteness of boundary lines creates problems. The traditional authorities do not have accurate knowledge of the boundaries of the land already authorized for use, since there are no detailed maps. New pieces of land being allocated may often encroach onto other land previously allocated, especially when a chief just walks into the bush to show some new person his boundary. This makes it possible for many people to hold properly registered and conveyed titles for the use of the same piece of land.

When chiefs have authority over land change, the new chief usually does not know all the boundaries of all the land already allocated. It is therefore likely that the new chief will allocate land that has already been allocated several years earlier. A new chief is installed whenever the old one dies or if a chief has been destooled for any offence which is not customarily excusable. A chief may sometimes be forced to resign to save face after his stool has lost a lawsuit over land. Generally there exists, between two villages, a tract of land which is not claimed by anyone, partly because no definite boundary exists and partly to avoid disputes. If, however, one village attempts to clear such a piece of land the other village is almost certain to go to court. People take pride in pursuing land lawsuits vigorously, and the loser does not usually accept the judgement of the first court. Any chief who does not want to continue the case by appeal to a higher court may be asked to resign his chiefship.

Communal land. Land may be set aside for family or national shrines, for burial ground, or for communal grazing and hunting purposes. Sometimes an uncleared forest may be reserved for posterity or village expansion. Crops, fences, and standing huts are not permitted to exist on such communal land. And, in order not to introduce strange elements into a community, the land set aside for communal purposes may not be sold to settle family or individual debts.

15.2.3. *Group ownership in Nigeria*

In Nigeria also, ownership of land is by groups and not individuals. The group may be a family, a village, or a community of villages. Individuals within this group have varying rights to the land as dictated by the respective local institutions. Certain members may have unlimited rights with respect to what types of crops they can grow on the land while other members, such as tenants and strangers, are restricted to the production of food or seasonal crops only.

Except in the Lagos area, where the presence of British rule introduced the idea of individual ownership early in the twentieth century, practically all Nigerians did, and still do, recognize group ownership, especially of farm lands. The importance attached to group survival was, and is, reflected in rural areas in the allocation of plots to group members to raise food crops sufficient for the upkeep of each member's immediate family. Also where a member, through illness or crop failure, appeared not to have enough food for his family, the group, or certain members in the group, would always stand ready to help.

The British did not usually interfere with the customary administration of land except for compulsory acquisition of land intended for public interest. Nor did the Fulanis, who entered Nigeria earlier than the British; they were pre-occupied with the spread of Islam and little interested in settled farm life.

Thus nearly all land in Nigeria still belongs to groups. Individuals still manage rather than own land, sometimes individually, sometimes collectively. The individual cultivator has control over the crops he produces, while the group has power to decide to what use a particular area of land is put. Long-term perennials, especially oil palms in the south and shea-butter nut trees in the north, belong to the community, but arable crops, e.g. maize, beans, cassava, and yams, are regarded as the property of the individuals who planted them. It is therefore not uncommon to find both the group and the individuals claiming varying rights over the crops on the same piece of land.

The group's control of sanction over its land and its members extends to matters affecting what types and number of tenants and strangers to accept, absorb, or accommodate on the group's land. The group is usually consulted when any tenant or stranger is to be introduced into a village, and the group takes it upon itself to watch the tenant or stranger for some time before giving consent. But in many cases the new person introduces himself as a hired labourer. Since the sale of land is forbidden or regarded with disapproval in many parts of Nigeria, the stranger's easiest access to agricultural land is through tenancy.

The right of anyone who plants a crop to harvest it is recognized by custom, and this may be part of the reason why a tenant farmer is generally allowed to plant only food crops. Tree crops (i.e. cocoa, rubber, shea-butter nut), would tie the land up for long periods, and the landlord would probably be unable in practice to recover land he had allowed to be planted to such crops. The landlord's behaviour should thus not be interpreted as being against investment in tree crops as such but as being conditioned by his desire to preserve effective ownership.

In Southern Nigeria tenant's labour is often used to establish cocoa trees, but on the clear understanding that these are the property of the landlord. When an area of land is given to a tenant to grow food crops the landlord requests the tenant to inter-plant his first food crops with cocoa trees. These trees are husbanded together with the tenant's food crops and, after 2 or 3 years, the cocoa trees will have grown sufficiently to discourage any further interplanting. The landlord then takes over the young cocoa farm and resettles the tenant on another area of virgin or fallow land where the process is repeated. It is not uncommon to find a tenant helping his landlord to establish as many as five plots, following the same process. No other tribute or rent is collected from a tenant who helps the landlord to establish cocoa plots in this way.

Among some pastoral groups all land is, or could be, grazed by all people, be they members of the same family or not. Common occupation of land among nomads is a stronger bond than that of family and is similar to blood brotherhood.

15.2.4. *Group ownership in Senegal*

In Senegal, land is used mainly for the cultivation of arable crops, a situation which allows annual or seasonal review of ownership. Additionally, the population of Senegal is predominantly Moslem, and this affects land tenure, particularly through the pattern of land inheritance that is part of this religion.

The general practice in Senegal is that land is controlled and cultivated by the person who first cleared the land, or his descendants. This is the concept of lineage land. How the early lineage heads acquired their land is not always accurately known but Boutillier (1963), discussing the land-tenure systems among the Toucouleurs of the lower valley of the river Senegal, writes

A variety of rights are recognised. The right of cultivation, acquired by whoever clears the ground and by his descendants, may be lent or rented out by its holder in return for a yearly payment. The right of the master of the land is limited by the right of cultivation and by the fact that the master of the land, as a general rule, is only the representative of a lineage with inalienability as a corollary.

The various kinds of tenure were (Boutillier 1963):

fields held as personal property	32 per cent
fields held as joint family posses-sions	11 per cent
fields occupied by virtue of rights of cultivation	21·5 per cent
fields occupied by renting	37 per cent

From Boutillier's writing it is not shown whether the Moslem community rented out much of their land to non-Muslims. Disputes arising from the system are as follows (Boutillier 1963, p. 135): 'personal versus joint family ownership; a right of cultivation versus a right to rent the land; full personal property versus a right only as a master of the land; conflicting claims to a right of master of the land or to a right of cultivation; disputes regarding land boundaries'. These comprise the routine problems created by customary tenure in modern society; i.e. the absence of registration of valid titles to land, inalienability and litigation, and conflicts between individual and group rights in land.

In discussing the land-tenure relationships among the Diola of the Lower Casamance in Senegal Thomas (1963) explained that the Diola people believe that God, the possessor, gave land to them to produce rice but the spirits were to preside over the land distribution performed by the chiefs who were said to be the earthly representatives of the spirits. Individuals derived their rights of use of the land from their respective lineages, but such rights were granted only to adult males who were married. The lineages were both patrilineal and patrilocal.

There are a number of problems associated with the Diola system, similar to those described for the Toucouleur. With the introduction of groundnuts as the main cash crop it became possible to farm land which was not suitable for the production of rice. Such groundnut land suddenly became of value, and more people wished to cultivate it. Whereas with rice production it was possible to use the same swamp lands year after year with consequent permanency of occupancy, this was not so with groundnuts because this annual crop normally exhausts the land and allows severe build up of pests and diseases if it is cultivated year after year on the same land. It

became necessary therefore to re-allocate rights of use periodically as some plots had to be fallowed and other plots were re-opened to cultivation after a fallow break. Families no longer needed to stay together as they did on the swamp-rice lands. Also, with the spread of Islam and the gradual introduction of Western formal education, the ideas of sacredness of family land were gradually diminished and the concept of commercial (economic) use of land developed.

15.2.5. *Group ownership in the Ivory Coast*

In the Ivory Coast the uncultivated forests belong to the entire family or village. In discussing the land-tenure systems with respect to boundaries among the Bete and the Dida of the Ivory Coast, Köbben (1963, p. 248) observes

> Each village has its own territory, the boundaries of which are fairly precisely defined: a certain tree referred to by name, or a small stream, serving as a land mark. One is continually amazed by the precision with which everyone knows the boundaries in the midst of the forest; not only those between a man's own village and the neighbouring ones, but also those of many other villages with which he has no direct dealings.

Alienation of family land can be authorized only by the eldest man but with the consent of the entire family, sometimes represented only by the heads of the nuclear families. Similarly a recovery of family land hitherto on loan is also done by the eldest man and in consultation with the other members of the family, as the necessity arises.

However, customary tenure is, in the Ivory Coast, as responsive to change as it can be anywhere else, and the introduction of cocoa and coffee has created the need for a concept of land-use for permanent tree crops. This requires a modification to the power of periodic allocation exercised by the eldest man. Similarly, population growth will no doubt shorten fallow periods and thereby cause farmers to use the same land for longer periods and create practices for seasonal crops demanding permanent rather than shifting occupation of land. At present in the Ivory Coast the concept of land as a commodity for sale and purchase is not widely accepted.

15.3. Land transfer and tenancy

It is well accepted that one cannot validly transfer that which one does not own. Ownership creates rights to transfer in the same way as transfer becomes a source of ownership. Transfer through monetary sale is not well accepted in customary land tenure, and it is generally regarded with disapproval.

15.3.1. *Transfer by conquest*

In the past much transfer of land was through conquest. As a result of invasion the defeated occupiers of a piece of land would either flee or become slaves to their conquerors. There were many such invasions, migrations, and consequent transfers of land in the history of West Africa. According to Mabogunje (1971)

> some of the most densely settled areas in West Africa today are to be found in this [forest] belt. Many of the more important groups, however, have traditions of having moved in, generally from the north, and of having displaced or absorbed less organised and more primitive aboriginal people. We are thus led to conclude that in the early periods of West African history the role of the forest areas has been that of a refugee zone, providing, albeit temporarily, some degree of security from the aggressiveness of stronger, better organised groups in the grassland region to the north.

15.3.2. *Transfer by sale*

Land for agricultural use is acquired mainly through inheritance within the extended family, and tenure is often inalienable. To a great extent, in most parts of Nigeria the people disapprove of sale of agricultural land, but there are some areas where sale of land is being gradually accepted, especially where city expansion continues to encroach on agricultural land. Also, there are some places where land has always been sold and purchased by individuals, yet the group as a whole would still object to sale were they given an opportunity to do so (Adegboye 1966b).

The strongest and most often mentioned criticism made of customary tenure is its inalienability through sale and that, as such, it cannot serve as security for a loan by which to raise capital.

15.3.3. *Transfer by pledging*

Leasing, borrowing, and exchanging are some more acceptable methods of land transfer, and each has been used in varying degrees and in many parts of West Africa. The most commonly used, and by far the most traditional of them, is pledging.

Customary tenure has always permitted the pawning or pledging of articles, including the crops planted on the land, for the purposes of providing security for loans. Cocoa trees in Western Nigeria were found by Adegboye (1969) to be the principal pledge objects for security against loans raised for children's education, marriage, funerals, litigation, old age and sickness, tax payment, and occasionally for agricultural improvement. The land is not so pledged, but only the trees on it that are owned individually.

An important difference between the customary tenure concept of pledge and the concept of pledge in the Anglo-Saxon sense is that the customary pledge is permanently redeemable. There may thus be situations in which grandchildren or other relations are called upon to repay the loan extended to a deceased person in order to recover the cocoa trees pledged. Land itself, even if not inalienable, would not command much value in terms of security since generally those who have the money to lend also have plenty of land. The money lender is not interested in working the land to produce certain products, but in harvesting and selling produce for a cash return on his loan. Borrowing of land takes place when there is a temporary shortage of land for the production of food crops. This temporary shortage may be brought about (1) by a situation in which the land in fallow has not regenerated enough by the time the owner or user needs land for the production of a seasonal crop; (2) when the farmer may want to take advantage of cheap migrant labour and does not have enough land of his own to keep them busy; or (3) when a labourer decides to stay and become a tenant and he borrows land from season to season until he is absorbed into the land-owning group. Exchange occurs when farmers decide to create contiguous plots of land.

15.3.4. *Farm tenancy*

When a tenant in cultivating the land starts by clearing virgin forest and bears all the expenses of planting or seeding, cultivating, harvesting, and marketing, then the proceeds are usually shared in the proportion of one-third to the landlord and two-thirds to the tenant. Oil palm, kola, and other tree crops planted by the tenant become the property of the landlord. The tenant is expected to give foodstuffs as tribute until cocoa or whatever tree crop he plants reaches maturity.

Where, however, a tenant is responsible for maintaining a cocoa plot that is already established, the landlord and the tenant usually share the net yields equally. Where the cultivation of food is involved, the tenant usually takes two-thirds while the landlord takes one-third of the produce. Before 1962 there were laws preventing foreigners from owning land, but customarily there were some situations under which the farm itself, rather than the produce, was divided equally between the landlord and the tenant. Now, however, whatever arrangements may be made must comply with the local (customary) conditions.

Voluntary migration has always led to occupancy of land through farm tenancy. Migrations occur when inhabitants of a particular place discover that their land is insufficiently productive to meet the demand of the expanding population, or when the inhabitants of an area cannot secure employment to earn money to pay taxes or participate in an exchange market. Sometimes also, the imposition of strange and foreign methods of administration may cause an indigenous population to leave temporarily in search of a less restrictive bureaucracy or to emigrate permanently. A record of such forced migration was mentioned by Crowther (1968):

> The French administration, beginning with the Villages of Liberty, and continuing through taxation, compulsory cultivation, recruitment and requisitioning of labour, forced hundreds of thousands of the ablest-bodied men of Niger, Sudan, Upper Volta and Guinea to migrate over distances up to 2,000 kilometres.

Such distances placed many of the migrants in areas totally unknown to them where, for practical purposes, they had to learn new languages and adjust to other ways of life. After such migrants had settled, they started their new lives as labourers and thence became full-time tenants, while gradually integrating into the community.

15.3.5. Tenancy on timber and mineral rights
Landlords have no rights to lease concessions to timber contractors. Only Government, through the Ministry of Lands, can authorize the leases and collect royalties. Part of the royalties are paid to the stool (as they are meant to be used to develop the stool area). For concessions leased from reserved forest, royalties are paid to the Forestry Department. Government takes all the payments from tenants granted mineral rights, passing on nothing to the landlord.

15.4. Transfer of land by inheritance
15.4.1. General
The most common form of land transfer at the present time is through inheritance, i.e. the intergeneration transfer of property rights to the use of land. In the customary tenure system the group, as the ultimate owner of land, is presumed never to die. Inheritance is predominantly patrilineal, i.e. the rights in land-holding are passed on death from the father to the son (the word 'son' being used very flexibly to include also younger brother, nephew, and cousin). A land-owner in customary tenure is invariably regarded as dying intestate, i.e. the wishes of the deceased before his death are discounted, so that the customary rules of descent and distribution may be allowed to operate without this constraint. By these rules the elders are permitted to administer as they see fit the supervision, protection, and the final partition-ing, where necessary, of the landed property left behind.

15.4.2. Land inheritance in Ghana
Descendants and successions. By far the highest proportion of land transfer in Ghana today is through inheritance. Ghana offers a unique example in this type of land transfer in that its population is divided almost equally between patrilineal and matrilineal societies. If land has been purchased previously, the stool can exercise its reversionary rights if a purchaser dies without a successor. In the matrilineal societies land is owned by women but negotiations for new additions or litigations to retain or recover old ones are made through men. Children inherit from their mothers and when any child dies his real successor is his mother. Bentsi-Enchill (1964) listed the persons in the line of succession in a matrilineal group. The matrilineal group includes all the Akan-speaking group (i.e. Akwapim, Akim, Fanti, Ashanti, Brong, and Nzima) plus a few other non-Akan tribes, e.g. Birifor, Tampolense, Vagala, and Ca-Hashi. In these societies women do some farming, though generally not as a main occupation but for family upkeep and partly for pocket money, since the husband in many cases will have an ample income from the cultivation of cocoa.

The patrilineal group include such tribes as Ewe, Kusasi, Frafra, Grushi, Wala, Dagarti, Gonja, and Dagombe. Among the Ewes the land is divided among all children equally but, upon marriage, the female child loses her share to her brothers, or, if there are no brothers, to her uncle. Among the other patrilineal tribes the land is divided only among the male members and by size according to seniority. Where there is no grown-up son, a successor is appointed by the family council. Bentsi-Enchill (1964) noted that

> A successor, according to customary law, is essentially a person appointed to the office—an elective office from which he is removable—of standing in the shoes of the deceased member of the family and carrying out his responsibilities. These responsibilities include the management of the property of the said deceased for the benefit of those upon whom the right to possess and enjoy the property of the deceased devolves.

Widow inheritance. The widow in Ghana, as in most African countries, is expected by custom to remarry the deceased's brother or nephew and by such remarriage continues to enjoy the care and protection offered her by the deceased. However, should the widow be too old to remarry she will be given the freedom to harvest from her late husband's farm from time to time. A widow's children are generally looked after by those who are capable of inheriting from the deceased. Widows who are unwilling to submit to customary remarriage may lose their right to maintenance, as pointed out by Bentsi-Enchill (1964). This practice of widow inheritance is widespread and not limited to the patrilineal society in Ghana alone.

Land inheritance for communal purposes. Land on which a deceased person had rights of use can become an addition to the land set aside for communal purposes if the recognized owner dies without issue or if, in the opinion of the family council, no family law of descent and distribution appears to be adequate and acceptable to the would-be inheritors. In general, any land, once set aside for communal purposes is not allowed to be individually owned or inherited. Also, if a piece of land is recovered from a stranger or reclaimed from swamps by communal effort (financial or otherwise), such land is not allowed to be inherited by individuals.

15.4.3. Land inheritance in Nigeria

General. Most inheritance in Nigeria is patrilineal, i.e. one may inherit land from a father, an uncle, or a brother, but not from a mother, an aunt, or a niece. In addition one may succeed to family land through the permission of the family head. One may inherit land only from one's family or clan. It is not customarily permissible, whatever the relationship, for an older man to inherit land by the decease of one younger than himself.

Only males can inherit land. Women who are unmarried are regarded as being under the care of their father and under the control and care of their husband after marriage. Women are permitted, however, to hold land in trust for those sons who are not old enough to inherit at the time of their husband's death, but such holding in trust is practised only when there are no adult sons or male relatives as alternatives. It is commonly accepted throughout Nigeria that women who become widows are inherited along with the property left by the deceased. This serves to provide security of land to farm as well as care for the widow. Where a widow is too old to remarry or there is no relation of the late husband who wishes to marry her, such a widow, if she chooses to stay, is given a life interest in her late husband's estate.

Where the deceased leaves behind two or more widows the practice is to divide his agricultural land into as many portions as there are wives, without taking into consideration possible differences in the number of children each widow may have. But commonly, only wives who have borne male children by the deceased are taken into account.

Inheritance practices peculiar to particular tribes or religions in Nigeria. Some exceptions to the general account given above are important enough to require special mention. Some of these exceptions arise out of religion, others out of the different concepts of tenure and experiences peculiar to certain tribes.

1. *Muslim inheritance.* Meek (1957) observes

> Mohammadan law, with its insistence on the equal division of property among the deceased's children (females included) is opposed to the principle of primogeniture, and for this reason primogeniture is much less prominent among the Muslim communities of Northern Nigeria than among the non Mohammadan, whether of the north or of the south.

However, in many parts of Nigeria the first-born male takes on the responsibilities of the late father and holds the farm land in one unit. This has many implications.

2. *Hausa inheritance practices.* Among the Hausas certain offices have an associated entitlement to land. The children or relations of the holder of such an office thus cannot inherit this land because it is inherited by his successor in office. But, whatever land the office-holder has, other than by virtue of his office, passes to his own immediate family in the normal way.

When a property-holder dies, customary tenure

permits the sale of part of the property (generally the movables) to defray burial and funeral expenses and to pay any debts. If one of his sons can bear the expenses without the need to sell property, that son is usually proclaimed the principal inheritor. Also, any son who has been given land by his father while still alive is not permitted a share of more land from the holding remaining which belongs to the extended family.

3. *Yoruba inheritance practices.* It is a general practice among the Yorubas of south-western Nigeria to leave the administration of the landed property of a deceased person in the hands of the family council. The family council may comprise all the male members of the extended family where the number is small. But the council may include only male unit representatives, a unit being a family within the extended one. It is the council that applies the relevant customary rules of descent and distribution. In general the brothers of the deceased are given first consideration in matters concerning family land, with the hope or expectation that these brothers will take care of the immediate family of the deceased from the proceeds of the land in question. Where the eldest son is old enough the brothers consult with him before taking any major decisions. But if the eldest son is far away, or not interested, the next eldest is invited.

Similarly the brothers are asked to inherit the wives left behind, especially the older ones, while the deceased's eldest son, if he is old enough, takes the youngest or the younger wife. But a child cannot inherit his mother as a wife. So if the eldest son happens to be born to the youngest wife, or the only wife, then all wives (now widows) go to the brothers or male relatives of the deceased. It is not a general rule that all who share widows must share the land, but it may happen that this will occur if the need is felt to raise more food, and in practice this is fairly common.

The right to harvest cash crops such as cocoa pass to the children of the deceased person who planted the trees. This follows customary rights of the individuals to reap all the crops he has planted. The land, however, as already explained is not owned by the planter of the cocoa trees or his successors. Wild trees such as oil palms continue to be community property, generation after generation. They are considered in effect as part of the land rather than as crop plants.

4. *Ibo inheritance practices.* In Ibo custom when a man dies an administrator is appointed to assume all his responsibilities. The choice of this administrator follows the customary practice of primogeniture which, however, as Obi (1963) observed may imply any of three different things:

(1) succession to a man's estate by his eldest son, if all his sons were born to one and the same wife;
(2) succession jointly by a man's eldest sons by each of his several wives irrespective of their ages;
(3) in the absence of an eldest son, succession by the eldest man among any of those who could for any reason be described as the deceased's nearest patrilineal blood relation.

This administrator's responsibilities include the care of children and widows. The eldest sons of each wife would inherit the use of land jointly but if the land in question is too small for more than one then the eldest sons may farm the land in rotation, either season by season or year by year, depending on the type of crops raised. Each eldest son taking part in this rotation has as his first responsibility the care, maintenance, and upkeep of his younger brothers and sisters by his own mother and the care of the homestead for the duration of his period of tenure.

Economic trees are inherited separately from the land on which they grow. In many societies among the Ibos these belong to the community rather than to individuals and are therefore not subject to inheritance. In some communities trees are divided equally among the several wives' families so long as each has male children. In others they are shared only among those who agree to accept certain family obligations such as the annual worship of certain idols and the making of payments towards the bride price for the younger male members of the family.

5. *Idoma inheritance practices.* According to Armstrong (1965) 'the family system is strongly patrilineal, and the families are grouped into corporate, patrilineal lineages and sub-lineages. The lineage, which may contain several thousand people, is the basic land-holding unit.' The Idoma people recognize only intestate succession. Since the land is held by the lineage, anybody from the lineage can farm anywhere he can find available space. This space may have been caused by other people who put the land to fallow or, perhaps, by the death of the original occupier.

Where the deceased left an adult son or sons his right of occupancy automatically passes to the son or sons. A man can inherit from his brother's either older or younger, but never from his son. This is different from the Yoruba custom by which a man cannot ever inherit from younger relatives. A full brother of the deceased, as administrator, is expected to hold the land for the young sons against the claims of the older sons or other brothers of the deceased. Where there is no full brother a half-brother by the same father as the deceased is appointed trustee in preference to a maternal half-brother.

Since land is neither bought nor sold, everyone depends upon his lineage to acquire any land. Since most of the crops raised are either annual or seasonal arable crops, the attachment a man has to the land relates only to the period the land bears crops or it is in continuous use for one purpose or the other. Economic trees, which here are mainly bananas, are inherited jointly by all the children of the deceased; the produce is used for the upkeep of the deceased's family, or sold to repay any debts unpaid at the time of death. Whatever a married woman has acquired during the lifetime of the husband can also pass as the man's property.

15.4.4. Land inheritance in the Ivory Coast
As in most countries where customary tenure operates, the main method of land acquisition is through inheritance. The Ivory Coast is exceptional because one age group succeeds another rather than an individual succeeding his or her father, uncle, mother, aunt, or any other relations. Land continues to remain the property (in custody) of the surviving members of an age group until that age group dies off completely. The family land is the responsibility of the eldest man who, jointly with some other elders, exercises the power of periodic re-allocation of land among the members of the family. The eldest man also sees to the upbringing of the young people, including making appropriate arrangements concerning the marriage of each boy or girl within the group. It is even believed by some people that the eldest man possessed the power to determine and point to the best fields for rice cultivation. He also guides his people as to the best time of the year to open up new forests for fresh food-crop cultivation.

Even though gift *inter vivos* has always been recognized, in practice a father concedes one or two economic (usually kola) trees to his son while the son is still working under him. The whole farm, with the

exception of these few trees, passes to the age group upon the death of the current holder-user of the land.

15.5. Policy measures to adjust customary tenure to current needs

15.5.1. Plantation agriculture—general
Many West African countries, including Ghana, Nigeria, Senegal, Liberia, and the Ivory Coast, have embarked upon plantation agriculture for various reasons. Many of the plantation schemes grew out of suggestions from former colonial powers or technical advisers. Some of the advantages claimed by advocates of the plantation system include:

(1) commercial orientation of production and increased marketable surplus;
(2) availability of processing and marketing facilities and consequent improvement in the quality of the agricultural products, especially for export markets;
(3) economy in the use of highly indivisible inputs such as tractors and large-scale machines for processing;
(4) employment of farm management experts and the introduction of scientific farming;
(5) employment of hired labour and encouragement of labour mobility from the labour-surplus to the land-surplus areas.

Some of the disadvantages of the plantation system are said to include:

(1) capital-intensive investment in a capital-scarce, but labour-surplus, economy;
(2) promotion of structural dualism in agriculture which makes plantations resemble mere enclaves in a vast sea of subsistence agriculture, i.e. the historical experiences in the Congo (now Zaire), Liberia, Malaya, and Indonesia;
(3) fostering of a dangerously mono-crop economy in environments susceptible to extremes of weather conditions and to severe and potentially devastating epidemics of pests and diseases;
(4) creation of an agrarian proletariat with attendant economic and social problems;
(5) dependence on foreign management.

The use of plantation methods has not usually led to any significant improvement in farming technology, particularly in respect of the tools used. Since both plantation crops and food crops compete for the same land, the more land that is put to use as plantations the less there is available for food crops. Land-pressure problems have therefore been created with no significant increase in the efficiency of technology. Output of produce has not increased over recent years for most of West Africa since the increases in yield obtained, for instance, by the use of chemical sprays has only just been sufficient to offset the yield lost through the old age and death of other trees.

15.5.2. *Nigeria: policy measures for plantation agriculture in relation to land tenure*

In Nigeria early policy aimed to increase total agricultural output through plantation farming both by Government and foreign companies. Some development corporations, financed by Government, adopted large-scale farming methods on plantations in the south of the country, where tree crops such as cocoa, rubber, and oil palm attracted the attention of different corporations at different times. For example, the United Africa Company established oil-palm plantations in order to produce sufficient palm oil to feed the soap and other factories owned by the same company. The Government, through its own extension and its incentive-creating efforts, planted cocoa and rubber on demonstration plots in many forest areas. But these plantation farming activities cannot be said to have added significantly to total agricultural output. Neither can it be said that they have, or had, any customary land-tenure implications. All land used for plantation farming has generally been acquired compulsorily: In spite of customary tenure rather than by its modification. It is difficult to justify alienation for plantations as lying within the concept of 'public purpose' of customary tenure. The creation of new employment has in practice been limited because labour needs become less and less as the trees mature. Since it is neither possible nor desirable to set up plantations every year, the number of employees declines with time, or at best remains about the same.

15.5.3. *Other policy efforts in Nigeria*

Both Federal and State Governments of Nigeria are now making conscious efforts to 'improve' agriculture in the sense of increasing output by providing better access to capital and improving the social conditions of the farm people, among which are the following:

(1) the provision of soft loans (supervised credit);
(2) the supply at little or no cost of the seeds and seedlings from Government multiplication plots and the subsidizing of inputs such as fertilizers and spray chemicals;

(3) the provision for peasant farmers of a tractor-hire service on a credit basis at low cost;
(4) the setting-up and encouragement of the growth of co-operative societies;
(5) the establishment and running of farm institutes;
(6) the setting-up of farm settlements.

While many, if not all, of these efforts are geared towards increased production, and probably increased income, they all appear to have land-tenure implications in varying degrees. For example, soft loans, subsidized inputs, and tractor-hire services can singly and collectively raise the total amount of land a man can cultivate, thereby causing him to buy, borrow, lease, or seek to inherit land from a relation. Although farm institutes do not provide land the candidates who receive their training there automatically become active participants in land transactions since they need land to utilize their training. With the setting-up of new co-operatives in Nigeria, a different approach to collective marketing of produce, i.e. finding and sustaining a lively market for the produce, has been introduced. Members of the co-operatives sell their produce through their associations, and each member, knowing or being in a position to know what quantities other members have sold, will try each year to produce more and earn more as a means of competing with fellow co-operators. This affects land-use and land-acquisition.

The policies that have made, or have come nearest to making, the greatest effect on land tenure have been the setting up of farm settlements. The principal objectives of farm settlements in Nigeria, from the Niger (Mokwa) Project (1949-54) to the present farm settlements which were inaugurated in western and eastern Nigeria in 1960, have always been to expand, modernize, and generally increase productivity in agriculture. The settlements are also expected to provide jobs for young school-leavers. But they have not served well as demonstration plots for neighbouring farmers, especially in the Western State (Adegboye, Basu, and Olantunbosun 1969).

Farm settlements are usually sited on productive but sparsely populated land. Government acquisition is effected through the Ministry of Lands and Housing and, when the land is not given free, some compensation may be paid through the village heads. In western Nigeria the young settlers are chosen from those who have undergone some training at farm institutes. Government assists the settler by providing housing, farm tools and equipment, extension services, and a marketing scheme. Settlers are also drawn from among the children of the neighbouring villages. The displacement and dispossession caused through the act of acquisition by Government, may, in some cases be offset by the agreeable fact that the settlers are the children of the area where the settlement is located. But there have been several places where the villagers did not feel very happy about being displaced.

In many cases some of those whose lands were acquired for farm settlements were forced to look for alternative non-agricultural employment since they found it difficult to obtain farms of economic size nearby, or could not face the problem of having to start all over again. Also, those who remained in the area by moving to the neighbouring villages, were made to change their status, namely from owner to tenant. It is also known that a change in society took place and new problems emerged from the new relationship between the settlers and the former owners of the land.

15.5.4. *Ghana: methods of adjusting land-tenure systems*

The Government of Ghana has for some time been conscious of the need to adjust the land-tenure systems to meet the demands for increased agricultural output and a better standard of living for the people. To this effect a number of approaches have been followed, e.g. compulsory acquisition, projects inducing voluntary grants, formation of land-reform committees, studying problems of registration of land, and getting farming land of economic size.

Compulsory acquisition by Government. The Ghana Government has always made use of its powers to compulsorily acquire land for public use. The lands used for state farms, settlements, and workers' brigade farms were taken compulsorily from the peasants. This type of acquisition is generally based on the Government's declared intention of ensuring that all people who can benefit from farming have ample land to farm. It cannot, however, be established that the acquisitions made by Government always carried with them appropriate compensation. Besides, current research has not suggested that state farms and workers' brigade farms have in fact added enough to total agricultural output in Ghana to encourage their continuance or expansion.

Projects inducing voluntary grants. People are always ready and willing to give land free to Government for the purposes of building schools, hospitals, etc., since it is believed that such land-acquisition is of direct advantage to the people.

Land-reform committee. In Ghana the goals of land reform are not very clear. The Government attempted in 1966 to define the goals by setting up a Lands Committee but this committee did not enjoy the necessary support and encouragement from the people in general nor even from the Government which set it up. Hence there were few if any meetings of the committee, and no report emerged. Such a committee might, if given the chance, have produced a concept of land reform based on a wide perspective that would probably have been useful at least as a basis for future modifications to customary tenure. Of course, the argument may be that Government is exercising caution in order to avoid creating landlessness, as this is generally considered politically explosive and economically undesirable. Even though caution may be regarded as expedient, it sometimes results in the postponement of decision making and, if postponement later becomes politically impossible, it may then be difficult to prove there was wisdom in permitting the earlier delay.

Is land reform necessary? Any society seeking land reform must make a choice between economic efficiency and the retention of the traditional ties and institutions. That society must ascertain what type or types of ownership it desires, i.e. individual or collective, state or private, or a mixture of any or all of these. It must find out if the existing customary tenure provides scope for enterprise, fixity of tenure, fair rent, and fair compensation, and if it is flexible enough to yield to the pressures of commercial agriculture, taking into consideration the varying plot-size requirements of the different types of enterprise. It must explore whether the peasant small-scale farmers can work more efficiently if the extension service were directed to making them more efficient. It is necessary to see if the title to land is easy to prove in case of court challenge and if the boundaries can be constantly reviewed in order to avoid or minimize litigations.

Problems of registration of titles to land. Why do we register titles to land? And what type of interest has always been registered? It is believed that titles are registered to avoid overlapping claims. But registration is expensive. And registration is not immune from dispute of ownership, especially when it can be proved that the grantor is not the owner.

Very often a chief who poses as an owner may be merely a trustee of the land. In Ghana it is usually difficult to find out who are the legal authorities who can allocate land without any possible future successful legal challenge. The chief and a few favourites may sign in a land transfer and be said to have signed, even if not explicitly stated, 'for and on behalf of the family or community'. Court cases would usually follow when those who claim to be those with real power to sign but who were not consulted, or when any who were not satisfied with the distribution of the proceeds from the land so transferred, raise their protests. The Registrar of Lands once asked all the communities in Ghana to submit lists of people who had power to sign in community land transactions, but this request met with no success. If registration proves not to be possible nationally or regionally, efforts could at least be made within the smallest administrative areas to register the true traditional authorities over land and send their signatures to the Ministry of Lands in order to make future acquisition easier. Clan title or stool title may be registered if it is difficult to register individual titles. Individual ownership *per se* should not, after all, be the all-important goal of land reform. Interest rather than ownership could also be made registrable.

A result of registration is that a traditional ruler cannot start giving out land until he can accurately locate plots that do not partly or entirely belong to someone else. Also, a leaser or buyer of registered land can ensure that all litigation accompanying registration has been settled among the land-owning groups before the transaction is completed. When registered land has been acquired ownership is established, or if it is to be challenged, then the burden of proof of the contrary would lie with traditional authorities.

Land consolidation to obtain farming plots of economic size. It is generally accepted that not enough food is being produced in Ghana and that there seems to be a large market awaiting any foodstuff produced. But it is difficult to acquire agricultural land of sufficient size to produce such needs on a modern economic scale. The Government is appealing to agriculture graduates and others with some training in agriculture to go back to the land, but it seems that the Govern-

ment does not appreciate the problems involved in securing large enough parcels of land. It is not easy even to lease and develop an area of land of adequate size for a continuous period since the traditional concept of a lease is limited to seasonal or annual periods or, at most, until the expiry of the time needed for certain cash crops planted by the leasee to mature.

It could be suggested that all unused or idle land should be vested in Government and that the Government should lease out such land to individuals who have the knowledge and ability to use it economically. The problem here is to find an acceptable formula for declaring a piece of land to be idle or unused. This suggestion, even if it were accepted, could lead to over-centralization and might result in only 'favourites' of the Government in power obtaining land. It is therefore perhaps more advisable that chiefs should continue to hold land in trust for the people, but that by means of provision for proper registration a constant review of economic interest in land-use can be achieved. The Government's power of compulsory acquisition implies the payment of compensation, and this necessitates the identification of the appropriate authorities to receive the compensation on behalf of others.

It might also be suggested that the Government could acquire unused land of agricultural potential through lease or purchase, and then lease this out in plots of 8, 20, 40, or 80 ha to interested or needy local people or organizations. It would be easier for people to approach the Government for land than it would be to approach chiefs. And security may thereby be introduced. To achieve this, the Government may need to train and increase the number of qualified surveyors. Land already acquired for state farms, which is now being given back to the people, could now be properly surveyed and registered before such re-transfer is completed and, by this, some measure of reform would have been achieved.

15.6. Requirements for future development

15.6.1. General

It is useless to assume in modern West Africa that a piece of land is ownerless because it is vacant or unoccupied. Any acquisition of land through conquest or capture in war would not now be considered as final or appropriate. Disputes over land are supposed to be settled only in law courts. Apart from accession through reclamation and settlement, the main method of acquisition is through inheritance, especially since the sale and gift of land rights are not widely accepted or practised in the customary tradition.

It is therefore necessary in the future to emphasize inquiries into various forms of inheritance of land, its nature, methods, problems, and solutions. Change of ownership, whenever it occurs, through inheritance, provides a special opportunity for improvements in land-use. To achieve more economic use the new user–successor must gain assured rights of use, duly conferred by a group recognized by government whether at village, family, or clan level.

New research efforts could be directed towards looking into the composition of the authorities generally involved in re-allocation rights and by the education of such people towards better land-use. Several points which in particular need clarification are discussed in the following sections.

15.6.2. The concept of alienability

Is the concept of inalienability of land rights undergoing change? If so, what are the trends? What are the motives behind and consequences of inalienability? What factors are aiding or preventing inalienability? What role should the Government play? Should the Government help the owners to preserve and keep to themselves what they have or should the Government, in the name of welfare, look at the interests of the community in general?

15.6.3. Ownership and accessibility

How much land does a family, village, or clan really own? How much access does the family, village, or clan have to land outside that which it owns? Does the area of land an individual member of the family, village, or clan owns or has access to, grow, diminish, or fluctuate from year to year? What control does any individual have in keeping or changing the size allotted to him from year to year? How much validity or flexibility is there in the concepts of group ownership with accessibility being all that an individual can hope to obtain? Since accessibility is economic while ownership is political, what role should Government play?

15.6.4. Inequality of holdings of agricultural land

Effort should be focused on the desirability, or lack of it, of inequality of holdings, with particular regard to national or individual gains in total agricultural output. The result of such research would help

Government to consider land redistribution as a political move towards minimizing inequality of opportunity. Alternatively, the Government could concentrate on providing education and technical know-how, thereby enabling people either to improve their farming methods or to leave the land and practise other trades. As it is agreed that Governments generally are as aware of the possible economic gains deriving from land reform as they are of the politically unfavourable consequences, what role should the economist play in such a situation?

15.6.5. Population growth and agricultural land

As cities grow more and more in size, agriculturally useful land is increasingly used for non-agricultural purposes and an increasing proportion of the population becomes landless. This means that, with time, there are relatively fewer land-owning families, villages, and clans. What becomes of the rights to land that city dwellers would have enjoyed? What is the trend of adjustment? What are current attitudes towards the succeeding generations who will eventually inherit the land that will be overtaken by urban growth? What provisions are made for those affected, both now and in the future?

As more land is acquired for city expansion there is bound to be speculation concerning further expansion, causing more farm land to be lost into temporary non-use, i.e. a period between the acquisition and the effective non-agricultural use of the land. What should be the rights of the previous owners as against those of the new owners during this period? How should the Government help those who have to continue to give up their land to make way for city expansion?

On the farm the fallow period becomes shortened and perhaps eliminated as the real population pressure increases. Since an individual family does not necessarily increase at the same rate as the over-all population, the squeezing out of fallow periods does not affect all land at the same time. What role can and should the Government play in population redistribution, possibly through land reform?

15.6.6. Provision of infrastructure

The provision of good roads opens up new opportunities to land users and reduces their and their children's attachment to land with already existing services. New roads also change the environment by exposing the occupiers to outside influences such as new crops, new marketing systems, and even new ways of life. Such changes are all likely to necessitate a review of existing tenure arrangements. For example, where a tenant discovers that the conditions of tenancy in other areas are better than in his own home area, he may prefer to move to more attractive areas.

Similarly, direct investment in agriculture by Government may change the farmer's attitude to land-use and may encourage him to move out into areas where more land can be acquired in order to practise larger-scale agriculture. What implications do these have for land tenure? What should Government do first to preserve existing traditions in customary tenure on the grounds of their merits or, secondly, to change such tenure because of its faults?

15.7. Summary

An attempt has been made to give a general picture of customary tenure, followed by a review of literature on a few selected and specific tenure arrangements of tribes in Ghana, Nigeria, Senegal, and the Ivory Coast.

The review has revealed a widespread concept of group ownership and indicated that transfers by sale or lease are not widely accepted. The main method of acquisition of land rights is through inheritance, and it is therefore on inheritance of land and its use that future research effort needs to be focused. Particularly important are the concepts of alienability, ownership, and accessibility and the social and political consequences of inequality of holdings and the effect of population growth on agricultural land. The effect of the provision of infrastructure on land-tenure arrangements is also suggested for possible attention by research workers. Even though customary land-tenure arrangements discussed here related strictly to West Africa, it is well known that they are closely similar in other countries in tropical Africa and that the group, rather than the individual, is the key factor in customary tenure.

References

ADEGBOYE, R. O. (1966). An analysis of land tenure structure in some selected areas in Nigeria. *Nigerian J. econ. soc. Stud.* 8, 259–68.

—— (1969). Procuring loans through pledging of cocoa trees. *J. geogr. Ass. Nigeria* 12, 63–76.

—— BASU, A. C., and OLATUNBOSUN, D. (1969). Impact of Western Nigerian farm settlements on surrounding farmers. *Nigerian J. econ. soc. Stud.* 11, 229–39.

ANYANE, S. LA (1962). Agriculture in the general economy. In *Agriculture and land use in Ghana* (ed. J. B. Wills). Oxford University Press, London.

ARMSTRONG, R. G. (1965). Intestate succession among the Idoma. In *Studies in the laws of succession in Nigeria* (ed. J. D. M. Derrett). Oxford University Press, London.

BENTSI-ENCHILL, K. (1964). *Ghana land law*. Sweet and Maxwell, London.

BOUTILLIER, J. L. (1963). Les rapports du système foncier Toucouleur et de l'organisation sociale et économique traditionelle. Leur évolution actuelle. In *African agrarian systems* (ed. D. Biebuyck). Oxford University Press, London.

CROWTHER, M. (1968). *West Africa under colonial rule*. Hutchinson, London.

DORNER, P. (1972). *Land reform and economic development*. Kingsport Press, Tennessee.

ELIAS, T. O. (1951). *Nigerian land law and custom*. Routledge and Kegan Paul, London.

—— (1962). *The nature of African customary law*. Manchester University Press.

KÖBBEN, A. J. (1963). Land as an object of gain in a non-literate society. Land tenure among the Bete and Dida (Ivory Coast, West Africa). In *African agrarian systems* (ed. D. Biebuyck), pp. 245–66. Oxford University Press, London.

MABOGUNJE, A. L. (1971). The land and peoples of West Africa. In *History of West Africa* (eds J. F. A. Ajayi and M. Crowther), Vol. 1. Longmans, London.

MEEK, C. K. (1957). *Land tenure and land administration in Nigeria and the Cameroons*. Her Majesty's Stationery Office, London.

OBI, S. N. C. (1963). *The Ibo law of property*. Butterworths, London.

OLLENNU, N. A. (1962). *Customary land law in Ghana*. Sweet and Maxwell, London.

THOMAS, L. V. (1963). Essai sur quelques problèmes relatifs au régime foncier des Diola de Basse-Casamance (Sénégal). In *African agrarian systems* (ed. D. Biebuyck), pp. 314–30. Oxford University Press, London.

TIMMONS, J. F. (1964). Agricultural development through modifications in land tenure structures. Paper presented at the *Conference on Economic Development of Agriculture*, Ames, Iowa.

Appendix: Maize production in the lowland tropics

K. J. TREHARNE AND D. J. GREENLAND

Origin and introduction

Maize originated in the highland tropics of the Americas (Sprague 1955). It has, however, been widely adapted and grown in the lowland tropics as a major staple, although the yields obtainable in these areas are considerably lower than its potential in the highland regions.

In West Africa, maize was introduced in the early sixteenth century by Portuguese traders and it has been an important crop in the forest region for some 300 years. More recently it has penetrated progressively farther north into the savannas, becoming increasingly important in the traditional sorghum and millet areas. Although the drought tolerance characteristics of maize are not as good as those of sorghum, drought avoidance through shorter maturity, with higher yield potential, makes this possible and attractive.

Production statistics

The production of maize in different tropical regions in recent years, compiled from FAO statistics, is given in Table A.1. Its relative importance in

TABLE A.1
Maize production in three tropical regions

	Production (10^6)				Yield (kg/ha)			
	1961–5	1972	1973	1974	1961–5	1972	1973	1974
Africa	9·09	12·82	11·06	13·16	934	1065	942	1108
Latin America	27·00	35·18	37·51	38·86	1228	1384	1450	1392
Far East	11·29	13·72	15·16	14·74	1000	1071	1040	1046

Source: FAO statistics 1974.
The increase in production during the last decade is attributable to greater land area rather than to increase in unit yield.

Nigeria, the most populous African country, can be judged from production statistics for 1974–5 and the proposed production for 1979–80 (Table A.2) as given in the Third National Development Plan, published by the Central Planning Office, Federal Ministry of Economic Development, Lagos.

TABLE A.2
Major crop production in Nigeria, 1974–5, and proposed 1979–80

	1974–5		1979–80	
	(10^6 tonnes)	Protein (per cap. per day)	(10^6 tonnes)	Protein (per cap. per day)
Maize	1·62	4·22	2·35	5·71
Millet	3·05	11·2	3·57	11·6
Sorghum	4·02	12·3	4·7	13·0
Rice	0·4	0·8	0·8	1·1
Wheat	0·02	1·12	0·2	1·6
Yams	14·4	5·75	16·8	5·93
Cassava	5·5	1·36	6·45	1·41
Pulses	1·08	6·4	1·5	7·82

From Third National Development Plan, p. 68 (Central Planning Office, Lagos).

Maize as a crop

Because of its large-scale production in subtropical and temperate zones, details of the genetics, physiology, and agronomic requirements of maize have been very fully described in several books. Although this large volume of published material provides a wide background to the species, the same cannot be said of maize production in tropical areas, and particularly the lowland tropics. Moreover, the information available on tropical maize has not been published in a single volume, but such a volume is in preparation.

The following comments outline some of the differences between maize production in the tropics and in subtropical and temperate areas.

Physiology and agronomy

The superior yields of high-altitude tropical and temperate maize compared with those of the lowland tropics may depend on many factors but temperature is probably the most important. The rate of development from planting to anthesis is a function of temperature, and while photosynthetic activity is influenced essentially by day temperature, growth rate is determined also by night temperature. Thus plants experiencing similar day temperatures but lower night temperatures have slower development and, therefore, more photosynthate is available for growth during any given developmental stage. In the highlands, falling temperatures after the onset of rains can limit yield. Lowland varieties are usually harvested within 4 months of planting, whereas in the highlands of Kenya, for example, the crop may take 12 months to reach maturity. It is therefore not unexpected that high-altitude crops produce greater growth and yield than those in the warm night environment of the lowlands. The higher solar radiation in the highlands may also contribute to superior production through increased photosynthesis. Grain yield is determined by the number of kernels pollinated and the supply of assimilate, either directly from crop photosynthesis or by mobilization during grain filling of previously stored stem reserves, or both. The physiological events determining potential ear and kernel numbers at anthesis are not fully understood. Moreover, the environmental effects on the processes which subsequently occur and further reduce this potential also form an important area of research need (Duncan 1975).

Canopy characteristics are important in terms of achieving good penetration and distribution of light so that the efficient C_4 photosynthetic pathway may operate. A plant type having an erect upper leaf canopy without excessive mutual shading and a lower spreading canopy for maximum interception can be considered desirable, particularly in terms of agronomic plant spacing. Short plants with small tassels are also advantageous for proper light interception. Other desirable agronomic features to be selected for are lodging resistance, ear placement, ability to respond to improved soil fertility and fertilizers, heat and drought tolerance, and ability to produce an ear at high plant densities. All of these characters need to be linked to satisfactory yield and a maturity suitable to the environment. Because much maize in the lowland tropics is grown in mixed stand with other species, suitability to such conditions is also required.

Maize is generally recognized to be a crop that requires rather high levels of nutrient supply to yield well, and is particularly demanding with respect to nitrogen. The improved varieties of maize developed for use in highland and temperate areas have also been selected for production under mechanized farming conditions, free from weeds or other plant competition. In the lowland humid tropics it is probable that most maize production for many years to come will be by peasant farmers. This implies that it will normally be grown under rather poor fertility conditions and as a component of a mixed crop combination. Further, tillage will often be by hand methods.

Mixed cropping and minimal tillage techniques do in fact have considerable advantages in terms of control of soil erosion, and will remain important considerations in the lowland humid tropics where liability to erosion is great.

Pests and diseases

Many of the potentially desirable plant attributes mentioned above may also influence vulnerability to pests and diseases, and the degree of effective disease resistance to these determines the extent of realization of the genetic potential of the crop (Hooker 1973). The introduction of temperate high-yielding types to the tropics has been largely unsuccessful owing to their high susceptibility to diseases and pests.

The genetic diversity inherent in tropical open-pollinated composites and inter-varietal crosses is considered an important factor in helping to combat the development and spread of disease. The major pathogens of the lowland humid tropics include species of *Pythium*, causing seedling blights and ear rots, and several leaf parasites such as *Helminthosporium maydis* and *Puccinia polysora*. In the highland tropics *H. turcicum* and *Puccinia sorghi* assume greater importance. Stalk rots such as that caused by *Fusarium moniliforme* are most commonly associated with senescence, particularly in plants subjected to stresses, and susceptibility seems to be related to loss of cell vigour and reduction in sugar content. Thus the physiologically desirable transfer of stem reserves to grain formation may be undesirable from the point of view of resistance to stalk rot. A comparison of diseases of tropical and temperate maize has been given by Renfro and Ullstrup (1973). The nature of the inheritance of resistance is discussed in depth by Hooker (1975), who indicates that a more complete interpretation of the genetic basis of resistance is needed, particularly the importance of major genes, modifiers, and polygenic systems.

The abundance of insects, especially leaf hoppers of the genus *Cicadulina*, in the humid tropics, which act as vectors may well account for the prevalence in particular of 'Storey's streak virus' in Africa. Maize mosaic, dwarf mosaic, and stripe virus diseases are not regarded as serious economically.

The most severe direct insect damage to maize is caused by stem borers (*Sesamia calamistis* and *Busseola fusca*), while the ground beetle *Buphonella africana* and others are severe limiting factors in some locations. Nematodes are also known to cause severe damage.

As maize culture extends to the sub-Sahelian zones of West Africa, it is clear that continual monitoring and selection for resistance to disease and insect attack and relationships of these to changing environments and cultural practices will be needed. Multilocation testing of germplasm and careful selection of resistant improved varieties will help to overcome these problems.

Quality aspects

Since the report by Mertz *et al.* (1964) on the improved nutritive value of opaque-2 maize, successful efforts have been made to incorporate this gene for high-quality protein into varieties with other characteristics suitable to many environments. It is to be expected that the protein quality in varieties adapted for high yield in the humid tropics will gradually improve as a result of these efforts. Much additional work is necessary in relation to consumer acceptance and preference, for maize is used in different ways by the many diverse communities of the humid tropics.

Prospects

As shown in Table A.1, maize production in the tropics increased only slowly between 1965 and 1974, and much of the increase was due to the planting of greater areas. Current yields are well below the 6 to 7 t/ha achieved on the better plots of trials at research stations in the lowland humid tropics, and far below the 12 to 14 t/ha achieved on plots in the highland tropics. There is obvious scope for

substantial improvement in the yields and production of maize in the humid tropics. This will be achieved only by a combination of efforts by the agronomist, physiologist, and plant breeder, as well as the entomologist and pathologist, by identifying the factors that limit production in the areas where the crop is to be grown and their breeding improved varieties to increase yield in those conditions.

References cited

DUNCAN, W. G. (1975). Maize. In *Crop physiology, some case histories* (ed. L. T. Evans), pp. 23–50. Cambridge University Press.

HOOKER, A. L. (1973). Maize. In *Breeding plants for disease resistance* (ed. R. R. Nelson), pp. 132–54. Pennsylvania State University Press.

MERTZ, E. T., BATES, L. S., and NELSON, O. E. (1964). Mutant gene that changes protein composition and increases lysine content of maize endosperm. *Science*, **145**, 279–80.

RENFRO, B. L. and ULLSTRUP, A. J. (1973). A comparison of maize diseases in temperate and in tropical environments. In *Second International Congress of Phytopathology*. Minnesota.

SPRAGUE, G. F. (ed.) (1955). *Corn and corn improvement.* Academic Press, New York.

Additional reading

BERGER, J. (1962). *Maize production and the manuring of maize.* Centre D'Étude de L'Azote, Geneva.

FOOD AND AGRICULTURE ORGANIZATION (1972). *Improvement and production of maize, sorghum and millets.* FAO, Rome.

INGLETT, G. E. (1970). *Corn: culture, processing, product.* Avi Publishing Co., Westpart, Connecticut.

MANGELSDORF, P. C. (1974). *Corn: its origin, evolution, and improvement.* Belknap Press of Harvard University Press, Cambridge, Massachusetts.

MATUCHEN, N. J. and SCARECROW, N. (1971). *Bibliography of corn.* (3 vols.) CIMMYT, El Batan, Mexico.

PIERRE, W. J., ALDRICH, S. R., and MARTIN, W. P. (1966). *Advances in corn production; principles and practices.* Iowa State University Press, Ames, Iowa.

SCHRIMPF, K. (1960). *Maize, cultivation and fertilization.* Aktiengessellschaft, Ruhrstickstoff, Germany.

WALLACE, H. A. and BROWN, W. L. (1956). *Corn and its early fathers.* Michigan State Press, Michigan.

List of organizations

ADRAO (= WARDA)	Association for the Development of Rice Cultivation in West Africa
AICSIP	All India Cooperative Sorghum Improvement Programme
APO	Asian Productivity Organization
ASECNA	Agence pour la Sécurité de la Navigation Aerienne en Afrique et à Madagascar
AVB (Ivory Coast)	Bandama Valley Authority
BDPA (France)	Bureau for Development Bank of Senegal
BNDS (Senegal)	National Development Bank of Senegal
CEEMAT (France)	Centre for Experimental Studies in Tropical Agricultural Mechanization
CEDT (France)	French Company for the Development of Textile Fibres
CEMA (Mali)	Centre for Trials of Agricultural Machinery
CGOT (Senegal)	General Company for Tropical Oilseeds
CIMMYT (Mexico)	Centro Internacional de Mejoramieto de Maiz y Trigo
CNRA (France) and (Senegal—IRAT)	National Centre for Agronomic Research
COMACI (Ivory Coast)	Committee for the Agricultural Mechanization of the Ivory Coast
CSIRO (Australia)	Commonwealth Scientific and Industrial Research Organization
EAAFRO (Kenya)	East African Agricultural and Forestry Research Organization
FAO	Food and Agricultural Organization of the United Nations
GASGA	Group for Assistance on the Storage of Grains in Africa
IAEA (Vienna)	International Atomic Energy Authority of the United Nations
IARI (New Delhi, India)	Indian Agricultural Research Institute
ICRISAT (Hyderabad, India)	International Crops Research Institute for the Semi-Arid Tropics
IEMVT (France)	National Institute of Animal Production and Health for Tropical Countries
IITA (Ibadan, Nigeria)	International Institute of Tropical Agriculture
IRAT	Institute de la Recherches Agronomiques Tropicales et des Cultures Vivrières
IRRI (Los Banôs, Philippines)	International Rice Research Institute
MAS (Senegal)	Senegal Management Commission
NIAE (U.K.)	National Institute of Agricultural Engineering
OAU	Organization for African Unity
ODM (U.K.)	Ministry of Overseas Development
ONCAD (France)	National Office for Cooperation and Aid for Development
ONDR (Chad)	National Office of Rural Development
OPR (Madagascar)	Operation for Rice Productivity
ORD (Upper Volta)	Regional Development Offices
ORMA (FAO)	Organization for Research in Agricultural Mechanization
ORSTOM (France)	Office for Scientific Research and Technology Overseas
SAED (Senegal)	Society for the Management of the Delta Lands
SATC (France)	Society for Technical Aid and Cooperation
SDRS (Senegal)	Society for the Development of Rice Cultivation in Senegal
SEMA (Senegal)	Experimental Sector for the Modernization of Agriculture
SEMRY (Cameroun)	Society for the Expansion and Modernization of Rice Cultivation in Yagoua
SODAICA (Senegal)	Society for Agricultural and Industrial Development in Casamance
SODEFEL (Ivory Coast)	Society for the Development of Roots and Vegetables
SODERIZ (Ivory Coast)	Society for the Development of Rice Production
SODESUCRE (Ivory Coast)	Society for the Development of Sugar Production
SODEVA (Senegal)	Society for the Development and Agricultural Extension
SODEVO (Dahomey)	Society for the Development of the Oueme Valley
UGFCC (Ghana)	United Ghana Farmer's Cooperative Council

UNCC (Niger)	Niger Union for Cooperatives and Credit
USAID	United States Agency for International Development
VRARU (Ghana)	Volta River Agricultural Resettlement Unit
WARDA (= ADRAO) (Monrovia, Liberia)	West African Rice Development Association
WHO	World Health Organization of the United Nations

Index